WITHDRAWN
UTSA LIBRARIES

From The Library Of -4574

Harold G. Longbotham, PhD

PROPERTY OF
DR. HAROLD LONGBOTHAM
111 VILLA ANN
SAN ANTONIO, TX 78213
(210) 341-6556

Digital Video Processing

PRENTICE HALL SIGNAL PROCESSING SERIES

Alan V. Oppenheim, Series Editor

ANDREWS & HUNT *Digital Image Restoration*
BRACEWELL *Two Dimensional Imaging*
BRIGHAM *The Fast Fourier Transform and Its Applications*
BURDIC *Underwater Acoustic System Analysis 2/E*
CASTLEMAN *Digital Image Processing*
CROCHIERE & RABINER *Multirate Digital Signal Processing*
DUDGEON & MERSEREAU *Multidimensional Digital Signal Processing*
HAYKIN *Advances in Spectrum Analysis and Array Processing. Vols. I, II & III*
HAYKIN, ED. *Array Signal Processing*
JOHNSON & DUDGEON *Array Signal Processing*
KAY *Fundamentals of Statistical Signal Processing*
KAY *Modern Spectral Estimation*
KINO *Acoustic Waves: Devices, Imaging, and Analog Signal Processing*
LIM *Two-Dimensional Signal and Image Processing*
LIM, ED. *Speech Enhancement*
LIM & OPPENHEIM, EDS. *Advanced Topics in Signal Processing*
MARPLE *Digital Spectral Analysis with Applications*
MCCLELLAN & RADER *Number Theory in Digital Signal Processing*
MENDEL *Lessons in Estimation Theory for Signal Processing Communications and Control 2/E*
NIKIAS *Higher Order Spectra Analysis*
OPPENHEIM & NAWAB *Symbolic and Knowledge-Based Signal Processing*
OPPENHEIM, WILLSKY, WITH YOUNG *Signals and Systems*
OPPENHEIM & SCHAFER *Digital Signal Processing*
OPPENHEIM & SCHAFER *Discrete-Time Signal Processing*
PHILLIPS & NAGLE *Digital Control Systems Analysis and Design, 3/E*
PICINBONO *Random Signals and Systems*
RABINER & GOLD *Theory and Applications of Digital Signal Processing*
RABINER & SCHAFER *Digital Processing of Speech Signals*
RABINER & JUANG *Fundamentals of Speech Recognition*
ROBINSON & TREITEL *Geophysical Signal Analysis*
STEARNS & DAVID *Signal Processing Algorithms in Fortran and C*
TEKALP *Digital Video Processing*
THERRIEN *Discrete Random Signals and Statistical Signal Processing*
TRIBOLET *Seismic Applications of Homomorphic Signal Processing*
VIADYANATHAN *Multirate Systems and Filter Banks*
WIDROW & STEARNS *Adaptive Signal Processing*

Digital Video Processing

A. Murat Tekalp
University of Rochester

Prentice Hall PTR
Upper Saddle River, NJ 07458

Tekalp, A. Murat.
Digital video processing / A. Murat Tekalp.
p. cm. -- (Prentice-Hall signal processing series)
ISBN 0-13-190075-7 (alk. paper)
1. Digital video. I. Title. II. Series.
TK6680.5.T45 1995
621.388'33--dc20 95-16650
CIP

Editorial/production supervision: **Ann Sullivan**
Cover design: **Design Source**
Manufacturing manager: **Alexis R. Heydt**
Acquisitions editor: **Karen Gettman**
Editorial assistant: **Barbara Alfieri**

Printed on Recycled Paper

©1995 by Prentice Hall PTR
Prentice-Hall, Inc.
A Simon and Schuster Company
Upper Saddle River, NJ 07458

The publisher offers discounts on this book when ordered in bulk quantities.

For more information, contact:

Corporate Sales Department
Prentice Hall PTR
One Lake Street
Upper Saddle River, NJ 07458

Phone: 800-382-3419
Fax: 201-236-7141

email: corpsales@prenhall.com

All rights reserved. No part of this book may be reproduced, in any form or by any means, without permission in writing from the publisher.

Printed in the United States of America

10 9 8 7 6 5 4 3 2 1

ISBN: 0-13-190075-7

Prentice-Hall International (UK) Limited, *London*
Prentice-Hall of Australia Pty. Limited, *Sydney*
Prentice-Hall Canada Inc., *Toronto*
Prentice-Hall Hispanoamericana, S.A., *Mexico*
Prentice-Hall of India Private Limited, *New Delhi*
Prentice-Hall of Japan, Inc., *Tokyo*
Simon & Schuster Asia Pte. Ltd., *Singapore*
Editora Prentice-Hall do Brasil, Ltda., *Rio de Janeiro*

Library
University of Texas
at San Antonio

To Sevim and Kaya Tekalp, my mom and dad
and to Özge, my beloved wife

Contents

IV VIDEO FILTERING

VI VIDEO COMPRESSION

APPENDICES

Preface

At present, development of products and services offering full-motion digital video is undergoing remarkable progress, and it is almost certain that digital video will have a significant economic impact on the computer, telecommunications, and imaging industries in the next decade. Recent advances in digital video hardware and the emergence of international standards for digital video compression have already led to various desktop digital video products, which is a sign that the field is starting to mature. However, much more is yet to come in the form of digital TV, multimedia communication, and entertainment platforms in the next couple of years. There is no doubt that digital video processing, which began as a specialized research area in the 70s, has played a key role in these developments. Indeed, the advances in digital video hardware and processing algorithms are intimately related, in that it is the limitations of the hardware that set the possible level of processing in real time, and it is the advances in the compression algorithms that have made full-motion digital video a reality.

The goal of this book is to provide a comprehensive coverage of the principles of digital video processing, including leading algorithms for various applications, in a tutorial style. This book is an outcome of an advanced graduate level course in Digital Video Processing, which I offered for the first time at Bilkent University, Ankara, Turkey, in Fall 1992 during my sabbatical leave. I am now offering it at the University of Rochester. Because the subject is still an active research area, the underlying mathematical framework for the leading algorithms, as well as the new research directions as the field continues to evolve, are presented together as much as possible. The advanced results are presented in such a way that the application-oriented reader can skip them without affecting the continuity of the text.

The book is organized into six parts: i) *Representation of Digital Video*, including modeling of video image formation, spatio-temporal sampling, and sampling lattice conversion without using motion information; ii) *Two-Dimensional (2-D) Motion Estimation*; iii) *Three-Dimensional (3-D) Motion Estimation and Segmentation*; iv) *Video Filtering*; v) *Still Image Compression*; and vi) *Video Compression*, each of which is divided into four or five chapters. Detailed treatment of the mathematical principles behind representation of digital video as a form of computer data, and processing of this data for 2-D and 3-D motion estimation, digital video standards conversion, frame-rate conversion, de-interlacing, noise filtering, resolution enhancement, and motion-based segmentation are developed. The book also covers the fundamentals of image and video compression, and the emerging world standards for various image and video communication applications, including high-definition TV, multimedia workstations, videoconferencing, videophone, and mobile image communications. A more detailed description of the organization and the contents of each chapter is presented in Section 1.3.

As a textbook, it is well-suited to be used in a one-semester advanced graduate level course, where most of the chapters can be covered in one 75-minute lecture. A complete set of visual aids in the form of transparency masters is available from the author upon request. The instructor may skip Chapters 18-21 on still-image compression, if they have already been covered in another course. However, it is recommended that other chapters are followed in a sequential order, as most of them are closely linked to each other. For example, Section 8.1 provides background on various optimization methods which are later referred to in Chapter 11. Chapter 17 provides a unified framework to address all filtering problems discussed in Chapters 13-16. Chapter 24, "Model-Based Coding," relies on the discussion of 3-D motion estimation and segmentation techniques in Chapters 9-12. The book can also be used as a technical reference by research and development engineers and scientists, or for self-study after completing a standard textbook in image processing such as *Two-Dimensional Signal and Image Processing* by J. S. Lim. The reader is expected to have some background in linear system analysis, digital signal processing, and elementary probability theory. Prior exposure to still-frame image-processing concepts should be helpful but is not required. Upon completion, the reader should be equipped with an in-depth understanding of the fundamental concepts, able to follow the growing literature describing new research results in a timely fashion, and well-prepared to tackle many open problems in the field.

My interactions with several exceptional colleagues had significant impact on the development of this book. First, my long time collaboration with Dr. Ibrahim Sezan, Eastman Kodak Company, has shaped my understanding of the field. My collaboration with Prof. Levent Onural and Dr. Gozde Bozdagi, a Ph.D. student at the time, during my sabbatical stay at Bilkent University helped me catch up with very-low-bitrate and object-based coding. The research of several excellent graduate students with whom I have worked Dr. Gordana Pavlovic, Dr. Mehmet Ozkan, Michael Chang, Andrew Patti, and Yucel Altunbasak has made major contributions to this book. I am thankful to Dr. Tanju Erdem, Eastman Kodak Company, for many helpful discussions on video compression standards, and to Prof. Joel Trussell for his careful review of the manuscript. Finally, reading of the entire manuscript by Dr. Gozde Bozdagi, a visiting Research Associate at Rochester, and her help with the preparation of the pictures in this book are gratefully acknowledged. I would also like to extend my thanks to Dr. Michael Kriss, Carl Schauffele, and Gary Bottger from Eastman Kodak Company, and to several program directors at the National Science Foundation and the New York State Science and Technology Foundation for their continuing support of our research; Prof. Kevin Parker from the University of Rochester and Prof. Abdullah Atalar from Bilkent University for giving me the opportunity to offer this course; and Chip Blouin and John Youngquist from the George Washington University Continuing Education Center for their encouragement to offer the short-course version.

A. Murat Tekalp "tekalp@ee.rochester.edu"
Rochester, NY February 1995

About the Author

A. Murat Tekalp received B.S. degrees in electrical engineering and mathematics from Boğaziçi University, Istanbul, Turkey, in 1980, with the highest honors, and the M.S. and Ph.D. degrees in electrical, computer, and systems engineering from Rensselaer Polytechnic Institute (RPI), Troy, New York, in 1982 and 1984, respectively.

From December 1984 to August 1987, he was a research scientist and then a senior research scientist at Eastman Kodak Company, Rochester, New York. He joined the Electrical Engineering Department at the University of Rochester, Rochester, New York, as an assistant professor in September 1987, where he is currently a professor. His current research interests are in the area of digital image and video processing, including image restoration, motion and structure estimation, segmentation, object-based coding, content-based image retrieval, and magnetic resonance imaging.

Dr. Tekalp is a Senior Member of the IEEE and a member of Sigma Xi. He was a scholar of the Scientific and Technical Research Council of Turkey from 1978 to 1980. He received the NSF Research Initiation Award in 1988, and IEEE Rochester Section Awards in 1989 and 1992. He has served as an Associate Editor for IEEE Transactions on Signal Processing (1990-1992), and as the Chair of the Technical Program Committee for the 1991 MDSP Workshop sponsored by the IEEE Signal Processing Society. He was the organizer and first Chairman of the Rochester Chapter of the IEEE Signal Processing Society. At present he is the Vice Chair of the IEEE Signal Processing Society Technical Committee on Multidimensional Signal Processing, and an Associate Editor for IEEE Transactions on Image Processing, and Kluwer Journal on Multidimensional Systems and Signal Processing. He is also the Chair of the Rochester Section of IEEE.

About the Notation

Before we start, a few words about the notation used in this book are in order. Matrices are always denoted by capital bold letters, e.g., $\mathbf{R}$. Vectors, defined as column vectors, are represented by small or capital bold letters, e.g., $\mathbf{x}$ or $\mathbf{X}$. Small letters refer to vectors in the image plane or their lexicographic ordering, whereas capitals are used to represent vectors in the 3-D space. The distinction between matrices and 3-D vectors will be clear from the context. The symbols F and f are reserved to denote frequency. F (cycles/mm or cycles/sec) indicates the frequency variable associated with continuous signals, whereas f denotes the unitless normalized frequency variable.

In order to unify the presentation of the theory for both progressive and interlaced video, time-varying images are assumed to be sampled on 3-D lattices. Time-varying images of continuous variables are denoted by $s_c(x_1, x_2, t)$. Those sampled on a lattice can be considered as either functions of continuous variables with actual units, analogous to multiplication by an impulse train, or functions of discrete variables that are unitless. They are denoted by $s_p(x_1, x_2, t)$ and $s(n_1, n_2, k)$, respectively. Continuous or sampled still images will be represented by $s_k(x_1, x_2)$, where (x_1, x_2) denotes real numbers or all sites of the lattice that are associated with a given frame/field index k, respectively. The subscripts "c" and "p" may be added to distinguish between the former and the latter as need arises. Sampled still images will be represented by $s_k(n_1, n_2)$. We will drop the subscript k in $s_k(x_1, x_2)$ and $s_k(n_1, n_2)$ when possible to simplify the notation. Detailed definitions of these functions are provided in Chapter 3.

We let $\mathbf{d}(x_1, x_2, t; \ell\Delta t)$ denote the spatio-temporal displacement vector field between the frames/fields $s_p(x_1, x_2, t)$ and $s_p(x_1, x_2, t + \ell\Delta t)$, where ℓ is an integer, and Δt is the frame/field interval. The variables (x_1, x_2, t) are assumed to be either continuous-valued or evaluated on a 3-D lattice, which should be clear from the context. Similarly, we let $\mathbf{v}(x_1, x_2, t)$ denote either a continuous or sampled spatio-temporal velocity vector field, which should be apparent from the context. The displacement field between any two particular frames/fields k and $k + \ell$ will be represented by the vector $\mathbf{d}_{k,k+\ell}(x_1, x_2)$. Likewise, the velocity field at a given frame k will be denoted by the vector $\mathbf{v}_k(x_1, x_2)$. Once again, the subscripts on $\mathbf{d}_{k,k+\ell}(x_1, x_2)$ and $\mathbf{v}_k(x_1, x_2)$ will be dropped when possible to simplify the notation.

A quick summary of the notation is provided below, where $\mathbf{R}$ and $\mathbf{Z}$ denote the real numbers and integer numbers, respectively.

Time-varying images

Continuous spatio-temporal image:
$s_c(x_1, x_2, t) = s_c(\mathbf{x}, t), \quad (\mathbf{x}, t) \in \mathbf{R}^3 = \mathbf{R} \times \mathbf{R} \times \mathbf{R}$

Image sampled on a lattice - continuous coordinates:
$s_p(x_1, x_2, t) = s_p(\mathbf{x}, t), \quad \begin{bmatrix} \mathbf{x} \\ t \end{bmatrix} = \mathbf{V} \begin{bmatrix} \mathbf{n} \\ k \end{bmatrix} \in \Lambda^3$

Discrete spatio-temporal image:
$s(n_1, n_2, k) = s(\mathbf{n}, k), \quad (\mathbf{n}, k) \in \mathbf{Z}^3 = \mathbf{Z} \times \mathbf{Z} \times \mathbf{Z}.$

Still images

Continuous still image:
$s_k(x_1, x_2) = s_c(\mathbf{x}, t)|_{t=k\Delta t}, \quad \mathbf{x} \in R^2, \quad k$ fixed integer

Still image sampled on a lattice:
$s_k(x_1, x_2) = s_p(\mathbf{x}, t)|_{t=k\Delta t}, \quad k$ fixed integer
$\mathbf{x} = [v_{11}n_1 + v_{12}n_2 + v_{13}k, v_{21}n_1 + v_{22}n_2 + v_{23}k]^T,$
where v_{ij} denotes elements of the matrix $\mathbf{V}$,

Discrete still image:
$s_k(n_1, n_2) = s(\mathbf{n}, k), \quad \mathbf{n} \in Z^2, \quad k$ fixed integer

The subscript k may be dropped, and/or subscripts "c" and "p" may be added to $s(x_1, x_2)$ depending on the context.
$\mathbf{s}_k$ denotes lexicographic ordering of all pixels in $s_k(x_1, x_2)$.

Displacement field from time t to $t + \ell\Delta t$:

$\mathbf{d}(x_1, x_2, t; \ell\Delta t) = [d_1(x_1, x_2, t; \ell\Delta t), d_2(x_1, x_2, t; \ell\Delta t)]^T,$
$\ell \in \mathbf{Z}, \quad \Delta t \in R, \quad (x_1, x_2, t) \in \mathbf{R}^3$ or $(x_1, x_2, t) \in \Lambda^3$

$\mathbf{d}_{k,k+\ell}(x_1, x_2) = \mathbf{d}(x_1, x_2, t; \ell\Delta t)|_{t=k\Delta t}, \quad k, \ell$ fixed integers

$\mathbf{d}_1$ and $\mathbf{d}_2$ denote lexicographic ordering of the components of the motion vector field for a particular $(k, k + \ell)$ pair.

Instantaneous velocity field

$\mathbf{v}(x_1, x_2, t) = [v_1(x_1, x_2, t), v_2(x_1, x_2, t)]^T,$
$(x_1, x_2, t) \in \mathbf{R}^3$ or $(x_1, x_2, t) \in \Lambda^3$

$\mathbf{v}_k(x_1, x_2) = \mathbf{v}(x_1, x_2, t)|_{t=k\Delta t} \quad k$ fixed integer

$\mathbf{v}_1$ and $\mathbf{v}_2$ denote lexicographic ordering of the components of the motion vector field for a given k.

Chapter 1

BASICS OF VIDEO

Video refers to pictorial (visual) information, including still images and time-varying images. A still image is a spatial distribution of intensity that is constant with respect to time. A time-varying image is such that the spatial intensity pattern changes with time. Hence, a time-varying image is a spatio-temporal intensity pattern, denoted by $s_c(x_1, x_2, t)$, where x_1 and x_2 are the spatial variables and t is the temporal variable. In this book video refers to time-varying images unless otherwise stated. Another commonly used term for video is "image sequence," since a time-varying image is represented by a time sequence of still-frame images (pictures). The "video signal" usually refers to a one-dimensional analog or digital signal of time, where the spatio-temporal information is ordered as a function of time according to a predefined scanning convention.

Video has traditionally been recorded, stored, and transmitted in analog form. Thus, we start with a brief description of analog video signals and standards in Section 1.1. We then introduce digital representation of video and digital video standards, with an emphasis on the applications that drive digital video technology, in Section 1.2. The advent of digital video opens up a number of opportunities for interactive video communications and services, which require various amounts of digital video processing. The chapter concludes with an overview of the digital video processing problems that will be addressed in this book.

1.1 Analog Video

Today most video recording, storage, and transmission is still in analog form. For example, images that we see on TV are recorded in the form of analog electrical signals, transmitted on the air by means of analog amplitude modulation, and stored on magnetic tape using videocasette recorders as analog signals. Motion pictures are recorded on photographic film, which is a high-resolution analog medium, or on laser discs as analog signals using optical technology. We describe the nature of the analog video signal and the specifications of popular analog video standards in the following. An understanding of the limitations of certain analog video formats is important, because video signals digitized from analog sources are usually limited by the resolution and the artifacts of the respective analog standard.

1.1.1 Analog Video Signal

The analog video signal refers to a one-dimensional (1-D) electrical signal $f(t)$ of time that is obtained by sampling $s_c(x_1, x_2, t)$ in the vertical x_2 and temporal coordinates. This periodic sampling process is called scanning. The signal $f(t)$, then, captures the time-varying image intensity $s_c(x_1, x_2, t)$ only along the scan lines, such as those shown in Figure 1.1. It also contains the timing information and the blanking signals needed to align the pictures correctly.

The most commonly used scanning methods are progressive scanning and interlaced scanning. A progressive scan traces a complete picture, called a frame, at every Δt sec. The computer industry uses progressive scanning with $\Delta t = 1/72$ sec for high-resolution monitors. On the other hand, the TV industry uses 2:1 interlace where the odd-numbered and even-numbered lines, called the odd field and the even field, respectively, are traced in turn. A 2:1 interlaced scanning raster is shown in Figure 1.1, where the solid line and the dotted line represent the odd and the even fields, respectively. The spot snaps back from point B to C, called the horizontal retrace, and from D to E, and from F to A, called the vertical retrace.

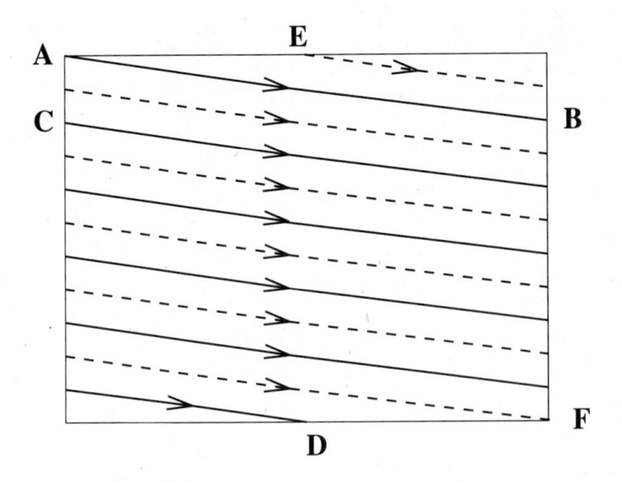

Figure 1.1: Scanning raster.

An analog video signal $f(t)$ is shown in Figure 1.2. Blanking pulses (black) are inserted during the retrace intervals to blank out retrace lines on the receiving CRT. Sync pulses are added on top of the blanking pulses to synchronize the receiver's horizontal and vertical sweep circuits. The sync pulses ensure that the picture starts at the top left corner of the receiving CRT. The timing of the sync pulses are, of course, different for progressive and interlaced video.

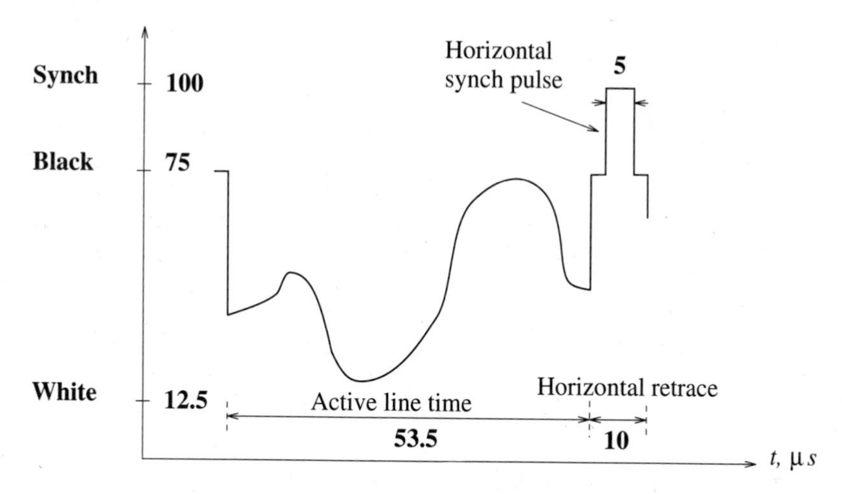

Figure 1.2: Video signal for one full line.

Some important parameters of the video signal are the vertical resolution, aspect ratio, and frame/field rate. The vertical resolution is related to the number of scan lines per frame. The aspect ratio is the ratio of the width to the height of a frame. Psychovisual studies indicate that the human eye does not perceive flicker if the refresh rate of the display is more than 50 times per second. However, for TV systems, such a high frame rate, while preserving the vertical resolution, requires a large transmission bandwidth. Thus, TV systems utilize interlaced scanning, which trades vertical resolution to reduced flickering within a fixed bandwidth.

An understanding of the spectrum of the video signal is necessary to discuss the composition of the broadcast TV signal. Let's start with the simple case of a still image, $s_c(x_1, x_2)$, where $(x_1, x_2) \in R^2$. We construct a doubly periodic array of images, $\tilde{s}_c(x_1, x_2)$, which is shown in Figure 1.3. The array $\tilde{s}_c(x_1, x_2)$ can be expressed in terms of a 2-D Fourier series,

$$\tilde{s}_c(x_1, x_2) = \sum_{k_1=-\infty}^{\infty} \sum_{k_2=-\infty}^{\infty} S_{k_1 k_2} \exp\left\{ j2\pi\left(\frac{k_1 x_1}{L} + \frac{k_2 x_2}{H}\right)\right\} \tag{1.1}$$

where $S_{k_1 k_2}$ are the 2-D Fourier series coefficients, and L and H denote the horizontal and vertical extents of a frame (including the blanking intervals), respectively.

The analog video signal $f(t)$ is then composed of intensities along the solid line across the doubly periodic field (which corresponds to the scan line) in Figure 1.3.

Assuming that the scanning spot moves with the velocities v_1 and v_2 in the horizontal and vertical directions, respectively, the video signal can be expressed as

$$f(t) = \sum_{k_1=-\infty}^{\infty} \sum_{k_2=-\infty}^{\infty} S_{k_1 k_2} \exp\left\{ j2\pi\left(\frac{k_1 v_1 t}{L} + \frac{k_2 v_2 t}{H}\right)\right\} \tag{1.2}$$

where L/v_1 is the time required to scan one image line and H/v_2 is the time required to scan a complete frame. The still-video signal is periodic with the fundamentals $F_h = v_1/L$, called the horizontal sweep frequency, and $F_v = v_2/H$. The spectrum of a still-video signal is depicted in Figure 1.4. The horizontal harmonics are spaced at F_h Hz intervals, and around each harmonic is a collection of vertical harmonics F_v Hz apart.

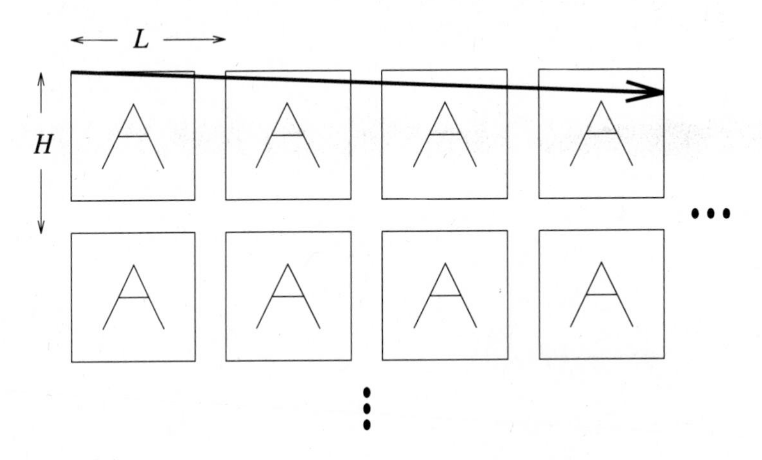

Figure 1.3: Model for scanning process.

In practice, for a video signal with temporal changes in the intensity pattern, every frame in the field shown in Figure 1.3 has a distinct intensity pattern, and the field is not doubly periodic. As a result, we do not have a line spectrum. Instead, the spectrum shown in Figure 1.4 will be smeared. However, empty spaces still exist between the horizontal harmonics at multiples of F_h Hz. For further details, the reader is referred to [Pro 94, Mil 92].

1.1.2 Analog Video Standards

In the previous section, we considered a monochromatic video signal. However, most video signals of interest are in color, which can be approximated by a superposition of three primary intensity distributions. The tri-stimulus theory of color states that almost any color can be reproduced by appropriately mixing the three additive primaries, red (R), green (G) and blue (B). Since display devices can only generate nonnegative primaries, and an adequate amount of luminance is required, there is, in practice, a constraint on the gamut of colors that can be reproduced. An in-depth discussion of color science is beyond the scope of this book. Interested readers are referred to [Net 89, Tru 93].

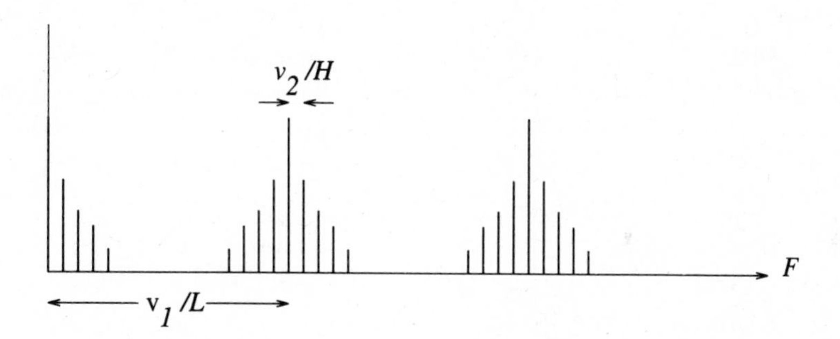

Figure 1.4: Spectrum of the scanned video signal for still images.

There exist several analog video signal standards, which have different image parameters (e.g., spatial and temporal resolution) and differ in the way they handle color. These can be grouped as:

- Component analog video
- Composite video
- S-video (Y/C video)

In component analog video (CAV), each primary is considered as a separate monochromatic video signal. The primaries can be either simply the R, G, and B signals or a luminance-chrominance transformation of them. The luminance component (Y) corresponds to the gray level representation of the video, given by

$$Y = 0.30R + 0.59G + 0.11B \tag{1.3}$$

The chrominance components contain the color information. Different standards may use different chrominance representations, such as

$$\begin{aligned} I &= 0.60R + 0.28G - 0.32B \\ Q &= 0.21R - 0.52G + 0.31B \end{aligned} \tag{1.4}$$

or

$$\begin{aligned} Cr &= R - Y \\ Cb &= B - Y \end{aligned} \tag{1.5}$$

In practice, these components are subject to normalization and gamma correction. The CAV representation yields the best color reproduction. However, transmission of CAV requires perfect synchronization of the three components and three times more bandwidth.

Composite video signal formats encode the chrominance components on top of the luminance signal for distribution as a single signal which has the same bandwidth as the luminance signal. There are different composite video formats, such as NTSC (National Television Systems Committee), PAL (Phase Alternation Line), and SECAM (Systeme Electronique Color Avec Memoire), being used in different countries around the world.

NTSC

The NTSC composite video standard, defined in 1952, is currently in use mainly in North America and Japan. NTSC signal is a 2:1 interlaced video signal with 262.5 lines per field (525 lines per frame), 60 fields per second, and 4:3 aspect ratio. As a result, the horizontal sweep frequency, F_h, is $525 \times 30 = 15.75$ kHz, which means it takes $1/15,750$ sec $= 63.5$ μs to sweep each horizontal line. Then, from (1.2), the NTSC video signal can be approximately represented as

$$f(t) \approx \sum_{k_1=-\infty}^{\infty} \sum_{k_2=-\infty}^{\infty} S_{k_1 k_2} \exp\{j2\pi(15,750k_1 + 30k_2)t\} \tag{1.6}$$

Horizontal retrace takes 10 μs, that leaves 53.5 μs for the active video signal per line. The horizontal sync pulse is placed on top of the horizontal blanking pulse, and its duration is 5 μs. These parameters were shown in Figure 1.2. Only 485 lines out of the 525 are active lines, since 20 lines per field are blanked for vertical backtrace [Mil 92]. Although there are 485 active lines per frame, the vertical resolution, defined as the number of resolvable horizontal lines, is known to be

$$485 \times 0.7 = 339.5 \text{ (340) lines/frame}, \tag{1.7}$$

where 0.7 is known as the Kell factor, defined as

$$\text{Kell factor} = \frac{\text{number of perceived vertical lines}}{\text{number of total active scan lines}} \approx 0.7$$

Using the aspect ratio, the horizontal resolution, defined as the number of resolvable vertical lines, should be

$$339 \times \frac{4}{3} = 452 \text{ elements/line}. \tag{1.8}$$

Then, the bandwidth of the luminance signal can be calculated as

$$\frac{452}{2 \times 53.5 \times 10^{-6}} = 4.2 \text{ MHz}. \tag{1.9}$$

The luminance signal is vestigial sideband modulated (VSB) with a sideband, that extends to 1.25 MHz below the picture carrier, as depicted in Figure 1.5.

The chrominance signals, I and Q, should also have the same bandwidth. However, subjective tests indicate that the I and Q channels can be low-pass filtered to 1.6 and 0.6 MHz, respectively, without affecting the quality of the picture due to the inability of the human eye to perceive changes in chrominance over small areas (high frequencies). The I channel is separated into two bands, 0-0.6 MHz and 0.6-1.6 MHz. The entire Q channel and the 0-0.6 MHz portion of the I channel are quadrature amplitude modulated (QAM) with a color subcarrier frequency 3.58 MHz above the picture carrier, and the 0.6-1.6 MHz portion of the I channel is lower side band (SSB-L) modulated with the same color subcarrier. This color subcarrier frequency falls in midway between $227F_h$ and $228F_h$; thus, the chrominance spectra shift into the gaps midway between the harmonics of F_h. The audio signal is frequency modulated (FM) with an audio subcarrier frequency that is 4.5 MHz above the picture carrier. The spectral composition of the NTSC video signal, which has a total bandwidth of 6 MHz, is depicted in Figure 1.5. The reader is referred to a communications textbook, e.g., Lathi [Lat 89] or Proakis and Salehi [Pro 94], for a discussion of various modulation techniques including VSB, QAM, SSB-L, and FM.

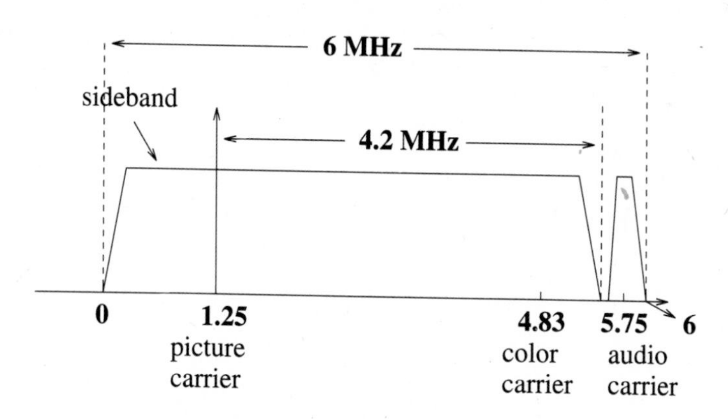

Figure 1.5: Spectrum of the NTSC video signal.

PAL and SECAM

PAL and SECAM, developed in the 1960s, are mostly used in Europe today. They are also 2:1 interlaced, but in comparison to NTSC, they have different vertical and temporal resolution, slightly higher bandwidth (8 MHz), and treat color information differently. Both PAL and SECAM have 625 lines per frame and 50 fields per second; thus, they have higher vertical resolution in exchange for lesser temporal resolution as compared with NTSC. One of the differences between PAL and SECAM is how they represent color information. They both utilize Cr and Cb components for the chrominance information. However, the integration of the color components with the luminance signal in PAL and SECAM are different. Both PAL and SECAM are said to have better color reproduction than NTSC.

In PAL, the two chrominance signals are QAM modulated with a color subcarrier at 4.43 MHz above the picture carrier. Then the composite signal is filtered to limit its spectrum to the allocated bandwidth. In order to avoid loss of high-frequency color information due to this bandlimiting, PAL alternates between +Cr and -Cr in successive scan lines; hence, the name phase alternation line. The high-frequency luminance information can then be recovered, under the assumption that the chrominance components do not change significantly from line to line, by averaging successive demodulated scan lines with the appropriate signs [Net 89]. In SECAM, based on the same assumption, the chrominance signals Cr and Cb are transmitted alternatively on successive scan lines. They are FM modulated on the color subcarriers 4.25 MHz and 4.41 MHz for Cb and Cr, respectively. Since only one chrominance signal is transmitted per line, there is no interference between the chrominance components.

The composite signal formats usually result in errors in color rendition, known as hue and saturation errors, because of inaccuracies in the separation of the color signals. Thus, S-video is a compromise between the composite video and the component analog video, where we represent the video with two component signals, a luminance and a composite chrominance signal. The chrominance signal can be based upon the (I,Q) or (Cr,Cb) representation for NTSC, PAL, or SECAM systems. S-video is currently being used in consumer-quality videocasette recorders and camcorders to obtain image quality better than that of the composite video.

1.1.3 Analog Video Equipment

Analog video equipment can be classified as broadcast-quality, professional-quality, and consumer-quality. Broadcast-quality equipment has the best performance, but is the most expensive. For consumer-quality equipment, cost and ease of use are the highest priorities.

Video images may be acquired by electronic live pickup cameras and recorded on videotape, or by motion picture cameras and recorded on motion picture film (24 frames/sec), or formed by sequential ordering of a set of still-frame images such as in computer animation. In electronic pickup cameras, the image is optically focused on a two-dimensional surface of photosensitive material that is able to collect light from all points of the image all the time. There are two major types of electronic cameras, which differ in the way they scan out the integrated and stored charge image. In vacuum-tube cameras (e.g., vidicon), an electron beam scans out the image. In solid-state imagers (e.g., CCD cameras), the image is scanned out by a solid-state array. Color cameras can be three-sensor type or single-sensor type. Three-sensor cameras suffer from synchronicity problems and high cost, while single-sensor cameras often have to compromise spatial resolution. Solid-state sensors are particularly suited for single-sensor cameras since the resolution capabilities of CCD cameras are continuously improving. Cameras specifically designed for television pickup from motion picture film are called telecine cameras. These cameras usually employ frame rate conversion from 24 frames/sec to 60 fields/sec.

Analog video recording is mostly based on magnetic technology, except for the laser disc which uses optical technology. In magnetic recording, the video signal is modulated on top of an FM carrier before recording in order to deal with the nonlinearity of magnetic media. There exist a variety of devices and standards for recording the analog video signal on magnetic tapes. The Betacam is a component analog video recording standard that uses 1/2" tape. It is employed in broadcast- and professional-quality applications. VHS is probably the most commonly used consumer-quality composite video recording standard around the world. U-matic is another composite video recording standard that uses 3/4" tape, and is claimed to result in a better image quality than VHS. U-matic recorders are mostly used in professional-quality applications. Other consumer-quality composite video recording standards are the Beta and 8 mm formats. S-VHS recorders, which are based on S-video, recently became widely available, and are relatively inexpensive for reasonably good performance.

1.2 Digital Video

We have been experiencing a digital revolution in the last couple of decades. Digital data and voice communications have long been around. Recently, hi-fi digital audio with CD-quality sound has become readily available in almost any personal computer and workstation. Now, technology is ready for landing full-motion digital video on the desktop [Spe 92]. Apart from the more robust form of the digital signal, the main advantage of digital representation and transmission is that they make it easier to provide a diverse range of services over the same network [Sut 92]. Digital video on the desktop brings computers and communications together in a truly revolutionary manner. A single workstation may serve as a personal computer, a high-definition TV, a videophone, and a fax machine. With the addition of a relatively inexpensive board, we can capture live video, apply digital processing, and/or print still frames at a local printer [Byt 92]. This section introduces digital video as a form of computer data.

1.2.1 Digital Video Signal

Almost all digital video systems use component representation of the color signal. Most color video cameras provide RGB outputs which are individually digitized. Component representation avoids the artifacts that result from composite encoding, provided that the input RGB signal has not been composite-encoded before. In digital video, there is no need for blanking or sync pulses, since a computer knows exactly where a new line starts as long as it knows the number of pixels per line [Lut 88]. Thus, all blanking and sync pulses are removed in the A/D conversion.

Even if the input video is a composite analog signal, e.g., from a videotape, it is usually first converted to component analog video, and the component signals are then individually digitized. It is also possible to digitize the composite signal directly using one A/D converter with a clock high enough to leave the color subcarrier components free from aliasing, and then perform digital decoding to obtain the desired RGB or YIQ component signals. This requires sampling at a rate three or four times the color subcarrier frequency, which can be accomplished by special-purpose chip sets. Such chips do exist in some advanced TV sets for digital processing of the received signal for enhanced image quality.

The horizontal and vertical resolution of digital video is related to the number of pixels per line and the number of lines per frame. The artifacts in digital video due to lack of resolution are quite different than those in analog video. In analog video the lack of spatial resolution results in blurring of the image in the respective direction. In digital video, we have pixellation (aliasing) artifacts due to lack of sufficient spatial resolution. It manifests itself as jagged edges resulting from individual pixels becoming visible. The visibility of the pixellation artifacts depends on the size of the display and the viewing distance [Lut 88].

The arrangement of pixels and lines in a contiguous region of the memory is called a bitmap. There are five key parameters of a bitmap: the starting address in memory, the number of pixels per line, the pitch value, the number of lines, and number of bits per pixel. The pitch value specifies the distance in memory from the start of one line to the next. The most common use of pitch different from the number of pixels per line is to set pitch to the next highest power of 2, which may help certain applications run faster. Also, when dealing with interlaced inputs, setting the pitch to double the number of pixels per line facilitates writing lines from each field alternately in memory. This will form a "composite frame" in a contiguous region of the memory after two vertical scans. Each component signal is usually represented with 8 bits per pixel to avoid "contouring artifacts." Contouring results in slowly varying regions of image intensity due to insufficient bit resolution. Color mapping techniques exist to map 2^{24} distinct colors to 256 colors for display on 8-bit color monitors without noticeable loss of color resolution. Note that display devices are driven by analog inputs; therefore, D/A converters are used to generate component analog video signals from the bitmap for display purposes.

The major bottleneck preventing the widespread use of digital video today has been the huge storage and transmission bandwidth requirements. For example, digital video requires much higher data rates and transmission bandwidths as compared to digital audio. CD-quality digital audio is represented with 16 bits/sample, and the required sampling rate is 44kHz. Thus, the resulting data rate is approximately 700 kbits/sec (kbps). In comparison, a high-definition TV signal (e.g., the AD-HDTV proposal) requires 1440 pixels/line and 1050 lines for each luminance frame, and 720 pixels/line and 525 lines for each chrominance frame. Since we have 30 frames/s and 8 bits/pixel per channel, the resulting data rate is approximately 545 Mbps, which testifies that a picture is indeed worth 1000 words! Thus, the viability of digital video hinges upon image compression technology [Ang 91]. Some digital video format and compression standards will be introduced in the next subsection.

1.2.2 Digital Video Standards

Exchange of digital video between different applications and products requires digital video format standards. Video data needs to be exchanged in compressed form, which leads to compression standards. In the computer industry, standard display resolutions; in the TV industry, digital studio standards; and in the communications industry, standard network protocols have already been established. Because the advent of digital video is bringing these three industries ever closer, recently standardization across the industries has also started. This section briefly introduces some of these standards and standardization efforts.

Table 1.1: Digital video studio standards

Parameter	CCIR601 525/60 NTSC	CCIR601 625/50 PAL/SECAM	CIF
Number of active pels/line			
Lum (Y)	720	720	360
Chroma (U,V)	360	360	180
Number of active lines/pic			
Lum (Y)	480	576	288
Chroma (U,V)	480	576	144
Interlacing	2:1	2:1	1:1
Temporal rate	60	50	30
Aspect ratio	4:3	4:3	4:3

Digital video is not new in the broadcast TV studios, where editing and special effects are performed on digital video because it is easier to manipulate digital images. Working with digital video also avoids artifacts that would be otherwise caused by repeated analog recording of video on tapes during various production stages. Another application for digitization of analog video is conversion between different analog standards, such as from PAL to NTSC. CCIR (International Consultative Committee for Radio) Recommendation 601 defines a digital video format for TV studios for 525-line and 625-line TV systems. This standard is intended to permit international exchange of production-quality programs. It is based on component video with one luminance (Y) and two color difference (Cr and Cb) signals. The sampling frequency is selected to be an integer multiple of the horizontal sweep frequencies in both the 525- and 625-line systems. Thus, for the luminance component,

$$f_{s,lum} = 858 f_{h,525} = 864 f_{h,625} = 13.5\text{MHz}, \tag{1.10}$$

and for the chrominance,

$$f_{s,chr} = f_{s,lum}/2 = 6.75\text{MHz}. \tag{1.11}$$

The parameters of the CCIR 601 standards are tabulated in Table 1.1. Note that the raw data rate for the CCIR 601 formats is 165 Mbps. Because this rate is too high for most applications, the CCITT (International Consultative Committee for Telephone and Telegraph) Specialist Group (SGXV) has proposed a new digital video format, called the Common Intermediate Format (CIF). The parameters of the CIF format are also shown in Table 1.1. Note that the CIF format is progressive (noninterlaced), and requires approximately 37 Mbps. In some cases, the number of pixels in a line is reduced to 352 and 176 for the luminance and chrominance channels, respectively, to provide an integer number of 16 × 16 blocks.

In the computer industry, standards for video display resolutions are set by the Video Electronics Standards Association (VESA). The older personal computer (PC) standards are the VGA with 640 pixels/line × 480 lines, and TARGA with 512 pixels/line × 480 lines. Many high-resolution workstations conform with the S-VGA standard, which supports two main modes, 1280 pixels/line × 1024 lines or 1024 pixels/line × 768 lines. The refresh rate for these modes is 72 frames/sec. Recognizing that the present resolution of TV images is well behind today's technology, several proposals have been submitted to the Federal Communications Commission (FCC) for a high-definition TV standard. Although no such standard has been formally approved yet, all proposals involve doubling the resolution of the CCIR 601 standards in both directions.

Table 1.2: Some network protocols and their bitrate regimes

Network	Bitrate
Conventional Telephone	0.3-56 kbps
Fundamental BW Unit of Telephone (DS-0)	56 kbps
ISDN (Integrated Services Digital Network)	64-144 kbps (px64)
Personal Computer LAN (Local Area Network)	30 kbps
T-1	1.5 Mbps
Ethernet (Packet-Based LAN)	10 Mbps
Broadband ISDN	100-200 Mbps

Various digital video applications, e.g., all-digital HDTV, multimedia services, videoconferencing, and videophone, have different spatio-temporal resolution requirements, which translate into different bitrate requirements. These applications will most probably reach potential users over a communications network [Sut 92]. Some of the available network options and their bitrate regimes are listed in Table 1.2. The main feature of the ISDN is to support a wide range of applications over the same network. Two interfaces are defined: basic access at 144 kbps, and primary rate access at 1.544 Mbps and 2.048 Mbps. As audio-visual telecommunication services expand in the future, the broadband integrated services digital network (B-ISDN), which will provide higher bitrates, is envisioned to be the universal information "highway" [Spe 91]. Asynchronous transfer mode (ATM) is the target transfer mode for the B-ISDN [Onv 94].

Investigation of the available bitrates on these networks and the bitrate requirements of the applications indicates that the feasibility of digital video depends on how well we can compress video images. Fortunately, it has been observed that the quality of reconstructed CCIR 601 images after compression by a factor of 100 is comparable to analog videotape (VHS) quality. Since video compression is an important enabling technology for development of various digital video products, three video compression standards have been developed for various target bitrates, and efforts for a new standard for very-low-bitrate applications are underway. Standardization of video compression methods ensures compatibility of digital video equipment by different vendors, and facilitates market growth. Recall that the boom in the fax market came after binary image compression standards. Major world standards for image and video compression are listed in Table 1.3.

Table 1.3: World standards for image compression.

Standard	Application
CCITT G3/G4	Binary images (nonadaptive)
JBIG	Binary images
JPEG	Still-frame gray-scale and color images
H.261	p × 64 kbps
MPEG-1	1.5 Mbps
MPEG-2	10-20 Mbps
MPEG-4	4.8-32 kbps (underway)

CCITT Group 3 and 4 codes are developed for fax image transmission, and are presently being used in all fax machines. JBIG has been developed to fix some of the problems with the CCITT Group 3 and 4 codes, mainly in the transmission of halftone images. JPEG is a still-image (monochrome and color) compression standard, but it also finds use in frame-by-frame video compression, mostly because of its wide availability in VLSI hardware. CCITT Recommendation H.261 is concerned with the compression of video for videoconferencing applications over ISDN lines. The target bitrates are p × 64 kbps, which are the ISDN rates. Typically, videoconferencing using the CIF format requires 384 kbps, which corresponds to $p = 6$. MPEG-1 targets 1.5 Mbps for storage of CIF format digital video on CD-ROM and hard disk. MPEG-2 is developed for the compression of higher-definition video at 10-20 Mbps with HDTV as one of the intended applications. We will discuss digital video compression standards in detail in Chapter 23.

Interoperability of various digital video products requires not only standardization of the compression method but also the representation (format) of the data. There is an abundance of digital video formats/standards, besides the CCITT 601 and CIF standards. Some proprietary format standards are shown in Table 1.4. A committee under the Society of Motion Picture and Television Engineers (SMPTE) is working to develop a universal header/descriptor that would make any digital video stream recognizable by any device. Of course, each device should have the right hardware/software combination to decode/process this video stream once it is identified. There also exist digital recording standards such as D1 for recording component video and D2 for composite video.

Table 1.4: Examples of proprietary video format standards

Video Format	Company
DVI (Digital Video Interactive), Indeo	Intel Corporation
QuickTime	Apple Computer
CD-I (Compact Disc Interactive)	Philips Consumer Electronics
Photo CD	Eastman Kodak Company
CDTV	Commodore Electronics

Rapid advances have taken place in digital video hardware over the last couple of years. Presently, several vendors provide full-motion video boards for personal computers and workstations using frame-by-frame JPEG compression. The main limitations of the state-of-the-art hardware originate from the speed of data transfer to and from storage media, and available CPU cycles for sophisticated real-time processing. Today most storage devices are able to transfer approximately 1.5 Mbps, although 4 Mbps devices are being introduced most recently. These numbers are much too slow to access uncompressed digital video. In terms of CPU capability, most advanced single processors are in the range of 70 MIPS today. A review of the state-of-the-art digital video equipment is not attempted here, since newer equipment is being introduced at a pace faster than this book can be completed.

1.2.3 Why Digital Video?

In the world of analog video, we deal with TV sets, videocassette recorders (VCR) and camcorders. For video distribution we rely on TV broadcasts and cable TV companies, which transmit predetermined programming at a fixed rate. Analog video, due to its nature, provides a very limited amount of interactivity, e.g., only channel selection in the TV, and fast-forward search and slow-motion replay in the VCR. Besides, we have to live with the NTSC signal format. All video captured on a laser disc or tape has to be NTSC with its well-known artifacts and very low still-frame image quality. In order to display NTSC signals on computer monitors or European TV sets, we need expensive transcoders. In order to display a smaller version of the NTSC picture in a corner of the monitor, we first need to reconstruct the whole picture and then digitally reduce its size. Searching a video database for particular footage may require tedious visual scanning of a whole bunch of videotapes. Manipulation of analog video is not an easy task. It usually requires digitization of the analog signal using expensive frame grabbers and expertise for custom processing of the data.

New developments in digital imaging technology and hardware are bringing together the TV, computer, and communications industries at an ever-increasing rate. The days when the local telephone company and the local cable TV company, as well as TV manufactures and computer manufacturers, will become fierce competitors are near [Sut 92]. The emergence of better image compression algorithms, optical fiber networks, faster computers, dedicated video boards, and digital recording promise a variety of digital video and image communication products. Driving the research and development in the field are consumer and commercial applications such as:

- All-digital HDTV [Lip 90, Spe 95]
  @ 20 Mbps over 6 MHz taboo channels

- Multimedia, desktop video [Spe 93]
  @ 1.5 Mbps CD-ROM or hard disk storage

- Videoconferencing
  @ 384 kbps using p $\times$ 64 kbps ISDN channels

- Videophone and mobile image communications [Hsi 93]
  @ 10 kbps using the copper network (POTS)

Other applications include surveillance imaging for military or law enforcement, intelligent vehicle highway systems, harbor traffic control, cine medical imaging, aviation and flight control simulation, and motion picture production. We will overview some of these applications in Chapter 25.

Digital representation of video offers many benefits, including:
i) Open architecture video systems, meaning the existence of video at multiple spatial, temporal, and SNR resolutions within a single scalable bitstream.
ii) Interactivity, allowing interruption to take alternative paths through a video database, and retrieval of video.
iii) Variable-rate transmission on demand.
iv) Easy software conversion from one standard to another.
v) Integration of various video applications, such as TV, videophone, and so on, on a common multimedia platform.
vi) Editing capabilities, such as cutting and pasting, zooming, removal of noise and blur.
vii) Robustness to channel noise and ease of encryption.
All of these capabilities require digital processing at various levels of complexity, which is the topic of this book.

1.3 Digital Video Processing

Digital video processing refers to manipulation of the digital video bitstream. All known applications of digital video today require digital processing for data compression. In addition, some applications may benefit from additional processing for motion analysis, standards conversion, enhancement, and restoration in order to obtain better-quality images or extract some specific information.

Digital processing of still images has found use in military, commercial, and consumer applications since the early 1960s. Space missions, surveillance imaging, night vision, computed tomography, magnetic resonance imaging, and fax machines are just some examples. What makes digital video processing different from still image processing is that video imagery contains a significant amount of temporal correlation (redundancy) between the frames. One may attempt to process video imagery as a sequence of still images, where each frame is processed independently. However, utilization of existing temporal redundancy by means of multiframe processing techniques enables us to develop more effective algorithms, such as motion-compensated filtering and motion-compensated prediction. In addition, some tasks, such as motion estimation or the analysis of a time-varying scene, obviously cannot be performed on the basis of a single image. It is the goal of this book to provide the reader with the mathematical basis of multiframe and motion-compensated video processing. Leading algorithms for important applications are also included.

Part 1 is devoted to the representation of full-motion digital video as a form of computer data. In Chapter 2, we model the formation of time-varying images as perspective or orthographic projection of 3-D scenes with moving objects. We are mostly concerned with 3-D rigid motion; however, models can be readily extended to include 3-D deformable motion. Photometric effects of motion are also discussed. Chapter 3 addresses spatio-temporal sampling on 3-D lattices, which covers several practical sampling structures including progressive, interlaced, and quincunx sampling. Conversion between sampling structures without making use of motion information is the subject of Chapter 4.

Part 2 covers nonparametric 2-D motion estimation methods. Since motion compensation is one of the most effective ways to utilize temporal redundancy, 2-D motion estimation is at the heart of digital video processing. 2-D motion estimation, which refers to optical flow estimation or the correspondence problem, aims to estimate motion projected onto the image plane in terms of instantaneous pixel velocities or frame-to-frame pixel correspondences. We can classify nonparametric 2-D motion estimation techniques as methods based on the optical flow equation, block-based methods, pel-recursive methods, and Bayesian methods, which are presented in Chapters 5-8, respectively.

Part 3 deals with 3-D motion/structure estimation, segmentation, and tracking. 3-D motion estimation methods are based on parametric modeling of the 2-D optical flow field in terms of rigid motion and structure parameters. These parametric models can be used for either 3-D image analysis, such as in object-based image compression and passive navigation, or improved 2-D motion estimation. Methods that use discrete point correspondences are treated in Chapter 9, whereas optical-flow-based or direct estimation methods are introduced in Chapter 10. Chapter 11 discusses segmentation of the motion field in the presence of multiple motion, using direct methods, optical flow methods, and simultaneous motion estimation and segmentation. Two-view motion estimation techniques, discussed in Chapters 9-11, have been found to be highly sensitive to small inaccuracies in the estimates of point correspondences or optical flow. To this effect, motion and structure from stereo pairs and motion tracking over long monocular or stereo sequences are addressed in Chapter 12 for more robust estimation.

Filtering of digital video for such applications as standards conversion, noise reduction, and enhancement and restoration is addressed in Part 4. Video filtering differs from still-image filtering in that it generally employs motion information. To this effect, the basics of motion-compensated filtering are introduced in Chapter 13. Video images often suffer from graininess, especially when viewed in freeze-frame mode. Intraframe, motion-adaptive, and motion-compensated filtering for noise suppression are discussed in Chapter 14. Restoration of blurred video frames is the subject of Chapter 15. Here, motion information can be used in the estimation of the spatial extent of the blurring function. Different digital video applications have different spatio-temporal resolution requirements. Appropriate standards conversion is required to ensure interoperability of various applications by decoupling the spatio-temporal resolution requirements of the source from that of the display. Standards conversion problems, including frame rate conversion and de-interlacing (interlaced to progressive conversion), are covered in Chapter 16. One of the limitations of CCIR 601, CIF, or smaller-format video is the lack of sufficient spatial resolution. In Chapter 17, a comprehensive model for low-resolution video acquisition is presented as well as a novel framework for superresolution which unifies most video filtering problems.

Compression is fundamental for all digital video applications. Parts 5 and 6 are devoted to image and video compression methods, respectively. It is the emergence of video compression standards, such as JPEG, H.261, and MPEG and their VLSI implementations, that makes applications such as all-digital TV, multimedia, and videophone a reality. Chapters 18-21 cover still-image compression methods, which form the basis for the discussion of video compression in Chapters 22-24. In particular, we discuss lossless compression in Chapter 18, DPCM and transform coding in Chapter 19, still-frame compression standards, including binary and gray-scale/color image compression standards, in Chapter 20, and vector quantization and subband coding in Chapter 21. Chapter 22 provides a brief overview of interframe compression methods. International video compression standards such as H.261, MPEG-1, and MPEG-2 are explained in Chapter 23. Chapter 24 addresses very-low-bitrate coding using object-based methods. Finally, several applications of digital video are introduced in Chapter 25.

Bibliography

[Ang 91] P. H. Ang, P. A. Ruetz, and D. Auld, "Video compression makes big gains," *IEEE Spectrum*, pp. 16–19, Oct. 1991.

[Byt 92] "Practical desktop video," *Byte*, Apr. 1992.

[Hsi 93] T. R. Hsing, C.-T. Chen, and J. A. Bellisio, "Video communications and services in the copper loop," *IEEE Comm. Mag.*, pp. 63–68, Jan. 1993.

[Lat 89] B. P. Lathi, *Modern Digital and Analog Communication Systems*, Second Edition, HRW Saunders, 1989.

[Lip 90] A. Lippman, "HDTV sparks a digital revolution," *Byte*, Dec. 1990.

[Lut 88] A. C. Luther, *Digital Video in the PC Environment*, New York, NY: McGraw-Hill, 1988.

[Mil 92] G. M. Miller, *Modern Electronic Communication*, Fourth Edition, Regents, Prentice Hall, 1992.

[Net 89] A. N. Netravali and B. G. Haskell, *Digital Pictures - Representation and Compression*, New York, NY: Plenum Press, 1989.

[Onv 94] R. O. Onvural, *Asynchronous Transfer Mode Networks: Performance Issues*, Norwood, MA: Artech House, 1994.

[Pro 94] J. G. Proakis and M. Salehi, *Communication Systems Engineering*, Englewood Cliffs, NJ: Prentice Hall, 1994.

[Spe 91] "B-ISDN and how it works," *IEEE Spectrum*, pp. 39–44, Aug. 1991.

[Spe 92] "Digital video," *IEEE Spectrum*, pp. 24–30, Mar. 1992.

[Spe 93] "Special report: Interactive multimedia," *IEEE Spectrum*, pp. 22–39, Mar. 1993.

[Spe 95] "Digital Television," *IEEE Spectrum*, pp. 34–80, Apr. 1995.

[Sut 92] J. Sutherland and L. Litteral, "Residential video services," *IEEE Comm. Mag.*, pp. 37–41, July 1992.

[Tru 93] H. J. Trussell, "DSP solutions run the gamut for color systems," *IEEE Signal Processing Mag.*, pp. 8–23, Apr. 1993.

Chapter 2

TIME-VARYING IMAGE FORMATION MODELS

In this chapter, we present models (in most cases simplistic ones) for temporal variations of the spatial intensity pattern in the image plane. We represent a time-varying image by a function of three continuous variables, $s_c(x_1, x_2, t)$, which is formed by projecting a time-varying three-dimensional (3-D) spatial scene into the two-dimensional (2-D) image plane. The temporal variations in the 3-D scene are usually due to movements of objects in the scene. Thus, time-varying images reflect a projection of 3-D moving objects into the 2-D image plane as a function of time. Digital video corresponds to a spatio-temporally sampled version of this time-varying image. A block diagram representation of the time-varying image formation model is depicted in Figure 2.1.

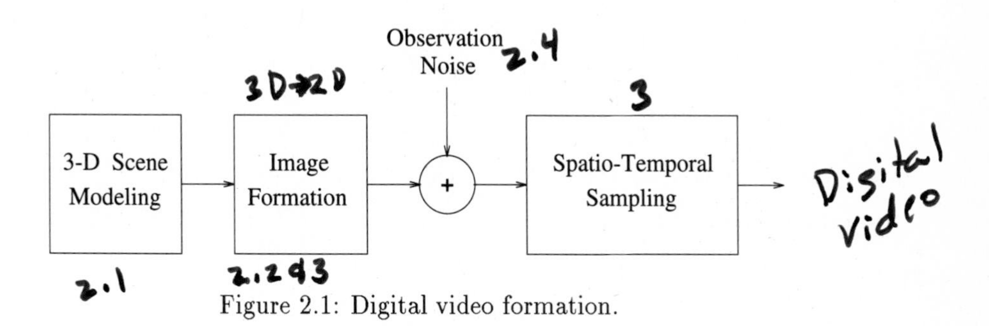

Figure 2.1: Digital video formation.

In Figure 2.1, "3-D scene modeling" refers to modeling the motion and structure of objects in 3-D, which is addressed in Section 2.1. "Image formation," which includes geometric and photometric image formation, refers to mapping the 3-D scene into an image plane intensity distribution. Geometric image formation, discussed in Section 2.2, considers the projection of the 3-D scene into the 2-D image plane. Photometric image formation, which is the subject of Section 2.3, models variations in the image plane intensity distribution due to changes in the scene illumination in time as well as the photometric effects of the 3-D motion. Modeling of the observation noise is briefly discussed in Section 2.4. The spatio-temporal sampling of the time-varying image will be addressed in Chapter 3. The image formation model described in this chapter excludes sudden changes in the scene content.

2.1 Three-Dimensional Motion Models

In this section, we address modeling of the relative 3-D motion between the camera and the objects in the scene. This includes 3-D motion of the objects in the scene, such as translation and rotation, as well as the 3-D motion of the camera, such as zooming and panning. In the following, models are presented to describe the relative motion of a set of 3-D object points and the camera, in the Cartesian coordinate system (X_1, X_2, X_3) and in the homogeneous coordinate system (kX_1, kX_2, kX_3, k), respectively. The depth X_3 of each point appears as a free parameter in the resulting expressions. In practice, a surface model is employed to relate the depth of each object point to reduce the number of free variables (see Chapter 9).

According to classical kinematics, 3-D motion can be classified as rigid motion and nonrigid motion. In the case of rigid motion, the relative distances between the set of 3-D points remain fixed as the object evolves in time. That is, the 3-D structure (shape) of the moving object can be modeled by a nondeformable surface, e.g., a planar, piecewise planar, or polynomial surface. If the entire field of view consists of a single 3-D rigid object, then a single set of motion and structure parameters will be sufficient to model the relative 3-D motion. In the case of independently moving multiple rigid objects, a different parameter set is required to describe the motion of each rigid object (see Chapter 11). In nonrigid motion, a deformable surface model (also known as a deformable template) is utilized in modeling the 3-D structure. A brief discussion about modeling deformable motion is provided at the end of this section.

2.1.1 Rigid Motion in the Cartesian Coordinates

It is well known that 3-D displacement of a rigid object in the Cartesian coordinates can be modeled by an affine transformation of the form [Rog 76, Bal 82, Fol 83]

$$\mathbf{X}' = \mathbf{R}\mathbf{X} + \mathbf{T} \tag{2.1}$$

where $\mathbf{R}$ is a 3×3 rotation matrix,

$$\mathbf{T} = \begin{bmatrix} T_1 \\ T_2 \\ T_3 \end{bmatrix}$$

is a 3-D translation vector, and

$$\mathbf{X} = \begin{bmatrix} X_1 \\ X_2 \\ X_3 \end{bmatrix} \quad \text{and} \quad \mathbf{X}' = \begin{bmatrix} X_1' \\ X_2' \\ X_3' \end{bmatrix}$$

denote the coordinates of an object point at times t and t' with respect to the center of rotation, respectively. That is, the 3-D displacement can be expressed as the sum of a 3-D rotation and a 3-D translation. The rotation matrix $\mathbf{R}$ can be specified in various forms [Hor 86]. Three of them are discussed next.

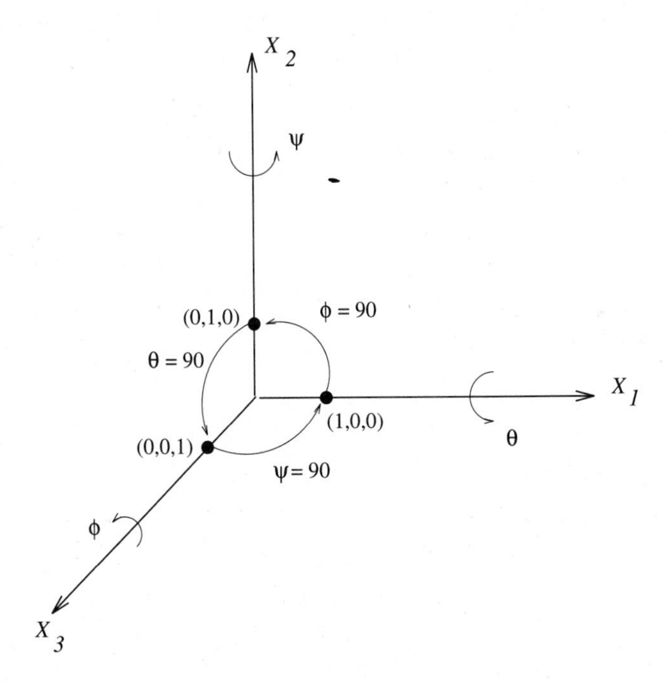

Figure 2.2: Eulerian angles of rotation.

The Rotation Matrix

Three-dimensional rotation in the Cartesian coordinates can be characterized either by the Eulerian angles of rotation about the three coordinate axes, or by an axis of rotation and an angle about this axis. The two descriptions can be shown to be equivalent under the assumption of infinitesimal rotation.

- *Eulerian angles in the Cartesian coordinates:* An arbitrary rotation in the 3-D space can be represented by the Eulerian angles, θ, ψ, and ϕ, of rotation about the X_1, X_2, and X_3 axes, respectively. They are shown in Figure 2.2.

The matrices that describe clockwise rotation about the individual axes are given by

$$\mathbf{R}_\theta = \begin{bmatrix} 1 & 0 & 0 \\ 0 & \cos\theta & -\sin\theta \\ 0 & \sin\theta & \cos\theta \end{bmatrix} \tag{2.2}$$

$$\mathbf{R}_\psi = \begin{bmatrix} \cos\psi & 0 & \sin\psi \\ 0 & 1 & 0 \\ -\sin\psi & 0 & \cos\psi \end{bmatrix} \tag{2.3}$$

and

$$\mathbf{R}_\phi = \begin{bmatrix} \cos\phi & -\sin\phi & 0 \\ \sin\phi & \cos\phi & 0 \\ 0 & 0 & 1 \end{bmatrix} \tag{2.4}$$

Assuming that rotation from frame to frame is infinitesimal, i.e., $\phi = \Delta\phi$, etc., and thus approximating $\cos\Delta\phi \approx 1$ and $\sin\Delta\phi \approx \Delta\phi$, and so on, these matrices simplify as

$$\mathbf{R}_\theta = \begin{bmatrix} 1 & 0 & 0 \\ 0 & 1 & -\Delta\theta \\ 0 & \Delta\theta & 1 \end{bmatrix}$$

and

$$\mathbf{R}_\psi = \begin{bmatrix} 1 & 0 & \Delta\psi \\ 0 & 1 & 0 \\ -\Delta\psi & 0 & 1 \end{bmatrix}$$

$$\mathbf{R}_\phi = \begin{bmatrix} 1 & -\Delta\phi & 0 \\ \Delta\phi & 1 & 0 \\ 0 & 0 & 1 \end{bmatrix}$$

Then the composite rotation matrix $\mathbf{R}$ can be found as

$$\begin{aligned} \mathbf{R} &= \mathbf{R}_\phi \mathbf{R}_\theta \mathbf{R}_\psi \\ &= \begin{bmatrix} 1 & -\Delta\phi & \Delta\psi \\ \Delta\phi & 1 & -\Delta\theta \\ -\Delta\psi & \Delta\theta & 1 \end{bmatrix} \end{aligned} \tag{2.5}$$

Note that the rotation matrices in general do not commute. However, under the infinitesimal rotation assumption, and neglecting the second and higher-order cross-terms in the multiplication, the order of multiplication makes no difference.

• *Rotation about an arbitrary axis in the Cartesian coordinates:* An alternative characterization of the rotation matrix results if the 3-D rotation is described by an angle α about an arbitrary axis through the origin, specified by the directional cosines n_1, n_2, and n_3, as depicted in Figure 2.3.

Then it was shown, in [Rog 76], that the rotation matrix is given by

$$\mathbf{R} = \begin{bmatrix} n_1^2 + (1 - n_1^2)cos\alpha & n_1 n_2(1 - cos\alpha) - n_3 sin\alpha & n_1 n_3(1 - cos\alpha) + n_2 sin\alpha \\ n_1 n_2(1 - cos\alpha) + n_3 sin\alpha & n_2^2 + (1 - n_2^2)cos\alpha & n_2 n_3(1 - cos\alpha) - n_1 sin\alpha \\ n_1 n_3(1 - cos\alpha) - n_2 sin\alpha & n_2 n_3(1 - cos\alpha) + n_1 sin\alpha & n_3^2 + (1 - n_3^2)cos\alpha \end{bmatrix} \tag{2.6}$$

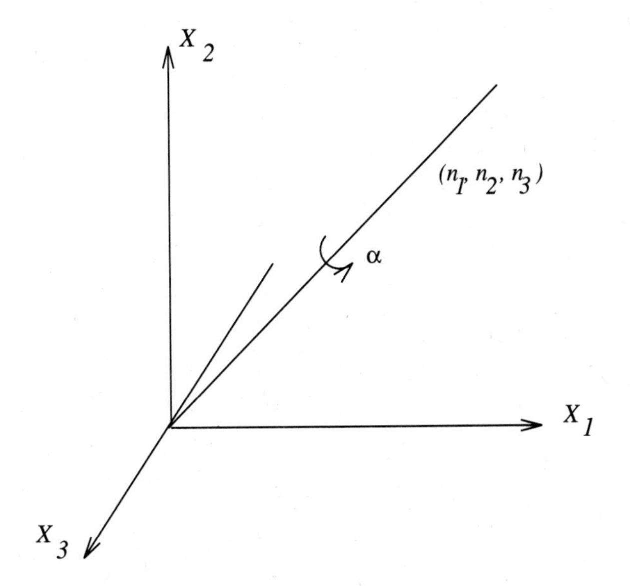

Figure 2.3: Rotation about an arbitrary axis.

For an infinitesimal rotation by the angle $\Delta\alpha$, $\mathbf{R}$ reduces to

$$\mathbf{R} = \begin{bmatrix} 1 & -n_3\Delta\alpha & n_2\Delta\alpha \\ n_3\Delta\alpha & 1 & -n_1\Delta\alpha \\ -n_2\Delta\alpha & n_1\Delta\alpha & 1 \end{bmatrix} \tag{2.7}$$

Thus, the two representations are equivalent with

$$\Delta\theta = n_1\Delta\alpha$$
$$\Delta\psi = n_2\Delta\alpha$$
$$\Delta\phi = n_3\Delta\alpha$$

In video imagery, the assumption of infinitesimal rotation usually holds, since the time difference between the frames are in the order of 1/30 seconds.

• *Representation in terms of quaternions:* A quaternion is an extension of a complex number such that it has four components [Hog 92],

$$\mathbf{q} = q_0 + q_1\mathbf{i} + q_2\mathbf{j} + q_3\mathbf{k} \tag{2.8}$$

where q_0, q_1, q_2, and q_3 are real numbers, and

$$\mathbf{i}^2 = \mathbf{j}^2 = \mathbf{k}^2 = \mathbf{ijk} = -1$$

A unit quaternion, where $q_0^2 + q_1^2 + q_2^2 + q_3^2 = 1$, can be used to describe the change in the orientation of a rigid body due to rotation. It has been shown that the unit quaternion is related to the directional cosines n_1, n_2, n_3 and the solid angle of rotation α in (2.6) as [Hor 86]

$$\mathbf{q} = \begin{pmatrix} n_1 \sin(\alpha/2) \\ n_2 \sin(\alpha/2) \\ n_3 \sin(\alpha/2) \\ \cos(\alpha/2) \end{pmatrix} \tag{2.9}$$

The rotation matrix $\mathbf{R}$ can then be expressed as

$$\mathbf{R} = \begin{bmatrix} q_0^2 - q_1^2 - q_2^2 - q_3^2 & 2(q_0q_1 + q_2q_3) & 2(q_0q_2 - q_1q_3) \\ 2(q_0q_1 - q_2q_3) & -q_0^2 + q_1^2 - q_2^2 + q_3^2 & 2(q_1q_2 + q_0q_3) \\ 2(q_0q_2 + q_1q_3) & 2(q_1q_2 - q_0q_3) & -q_0^2 - q_1^2 + q_2^2 + q_3^2 \end{bmatrix} \tag{2.10}$$

The representation (2.10) of the rotation matrix in terms of the unit quaternion has been found most helpful for temporal tracking of the orientation of a rotating object (see Chapter 12).

Two observations about the model (2.1) are in order:
i) If we consider the motion of each object point $\mathbf{X}$ independently, then the 3-D displacement vector field resulting from the rigid motion can be characterized by a different translation vector for each object point. The expression (2.1), however, describes the 3-D displacement field by a single rotation matrix and a translation vector. Hence, the assumption of a rigid configuration of a set of 3-D object points is implicit in this model.
ii) The effects of camera motion, as opposed to object motion, can easily be expressed using the model (2.1). The camera pan constitutes a special case of this model, in that it is a rotation around an axis parallel to the image plane. Zooming is in fact related to the imaging process, and can be modeled by a change of the focal length of the camera. However, it is possible to incorporate the effect of zooming into the 3-D motion model if we assume that the camera has fixed parameters but the object is artificially scaled up or down. Then, (2.1) becomes

$$\mathbf{X}' = \mathbf{SRX} + \mathbf{T} \tag{2.11}$$

where

$$\mathbf{S} = \begin{bmatrix} S_1 & 0 & 0 \\ 0 & S_2 & 0 \\ 0 & 0 & S_3 \end{bmatrix}$$

is a scaling matrix.

Modeling 3-D Instantaneous Velocity

The model (2.1) provides an expression for the 3-D displacement between two time instants. It is also possible to obtain an expression for the 3-D instantaneous velocity by taking the limit of the 3-D displacement model (2.1) as the interval between the two time instants Δt goes to zero. Expressing the rotation matrix $\mathbf{R}$ in terms of infinitesimal Eulerian angles, we have

$$\begin{bmatrix} X_1' \\ X_2' \\ X_3' \end{bmatrix} = \begin{bmatrix} 1 & -\Delta\phi & \Delta\psi \\ \Delta\phi & 1 & -\Delta\theta \\ -\Delta\psi & \Delta\theta & 1 \end{bmatrix} \begin{bmatrix} X_1 \\ X_2 \\ X_3 \end{bmatrix} + \begin{bmatrix} T_1 \\ T_2 \\ T_3 \end{bmatrix} \tag{2.12}$$

Decomposing the rotation matrix as

$$\begin{bmatrix} 1 & -\Delta\phi & \Delta\psi \\ \Delta\phi & 1 & -\Delta\theta \\ -\Delta\psi & \Delta\theta & 1 \end{bmatrix} = \begin{bmatrix} 0 & -\Delta\phi & \Delta\psi \\ \Delta\phi & 0 & -\Delta\theta \\ -\Delta\psi & \Delta\theta & 0 \end{bmatrix} + \begin{bmatrix} 1 & 0 & 0 \\ 0 & 1 & 0 \\ 0 & 0 & 1 \end{bmatrix} \tag{2.13}$$

substituting (2.13) into (2.12), and rearranging the terms, we obtain

$$\begin{bmatrix} X_1' - X_1 \\ X_2' - X_2 \\ X_3' - X_3 \end{bmatrix} = \begin{bmatrix} 0 & -\Delta\phi & \Delta\psi \\ \Delta\phi & 0 & -\Delta\theta \\ -\Delta\psi & \Delta\theta & 0 \end{bmatrix} \begin{bmatrix} X_1 \\ X_2 \\ X_3 \end{bmatrix} + \begin{bmatrix} T_1 \\ T_2 \\ T_3 \end{bmatrix} \tag{2.14}$$

Dividing both sides of (2.14) by Δt and taking the limit as Δt goes to zero, we arrive at the 3-D velocity model to represent the instantaneous velocity of a point (X_1, X_2, X_3) in the 3-D space as

$$\begin{bmatrix} \dot{X}_1 \\ \dot{X}_2 \\ \dot{X}_3 \end{bmatrix} = \begin{bmatrix} 0 & -\Omega_3 & \Omega_2 \\ \Omega_3 & 0 & -\Omega_1 \\ -\Omega_2 & \Omega_1 & 0 \end{bmatrix} \begin{bmatrix} X_1 \\ X_2 \\ X_3 \end{bmatrix} + \begin{bmatrix} V_1 \\ V_2 \\ V_3 \end{bmatrix} \tag{2.15}$$

where Ω_i and V_i denote the angular and linear velocities in the respective directions, $i = 1, 2, 3$. The model (2.15) can be expressed in compact form as

$$\dot{\mathbf{X}} = \mathbf{\Omega} \times \mathbf{X} + \mathbf{V} \tag{2.16}$$

where $\dot{\mathbf{X}} = [\dot{X}_1\ \dot{X}_2\ \dot{X}_3]^T$, $\mathbf{\Omega} = [\Omega_1\ \Omega_2\ \Omega_3]^T$, $\mathbf{V} = [V_1\ V_2\ V_3]^T$, and $\times$ denotes the cross-product. Note that the instantaneous velocity model assumes that we have a continuous temporal coordinate since it is defined in terms of temporal derivatives.

2.1.2 Rigid Motion in the Homogeneous Coordinates

We define the homogeneous coordinate representation of a Cartesian point $\mathbf{X} = [X_1 X_2 X_3]^T$ as

$$\mathbf{X}_h = \begin{bmatrix} kX_1 \\ kX_2 \\ kX_3 \\ k \end{bmatrix} \tag{2.17}$$

Then the affine transformation (2.11) in the Cartesian coordinates can be expressed as a linear transformation in the homogeneous coordinates

$$\mathbf{X}_h^{'} = \tilde{\mathbf{A}}\mathbf{X}_h \tag{2.18}$$

where

$$\tilde{\mathbf{A}} = \begin{bmatrix} a_{11} & a_{12} & a_{13} & T_1 \\ a_{21} & a_{22} & a_{23} & T_2 \\ a_{31} & a_{32} & a_{33} & T_3 \\ 0 & 0 & 0 & 1 \end{bmatrix}$$

and the matrix $\mathbf{A}$

$$\mathbf{A} = \begin{bmatrix} a_{11} & a_{12} & a_{13} \\ a_{21} & a_{22} & a_{23} \\ a_{31} & a_{32} & a_{33} \end{bmatrix} = \mathbf{SR}$$

Translation in the Homogeneous Coordinates

Translation can be represented as a matrix multiplication in the homogeneous coordinates given by

$$\mathbf{X}_h^{'} = \tilde{\mathbf{T}}\mathbf{X}_h \tag{2.19}$$

where

$$\tilde{\mathbf{T}} = \begin{bmatrix} 1 & 0 & 0 & T_1 \\ 0 & 1 & 0 & T_2 \\ 0 & 0 & 1 & T_3 \\ 0 & 0 & 0 & 1 \end{bmatrix}$$

is the translation matrix.

Rotation in the Homogeneous Coordinates

Rotation in the homogeneous coordinates is represented by a 4×4 matrix multiplication in the form

$$\mathbf{X}_h^{'} = \tilde{\mathbf{R}}\mathbf{X}_h \tag{2.20}$$

where

$$\tilde{\mathbf{R}} = \begin{bmatrix} r_{11} & r_{12} & r_{13} & 0 \\ r_{21} & r_{22} & r_{23} & 0 \\ r_{31} & r_{32} & r_{33} & 0 \\ 0 & 0 & 0 & 1 \end{bmatrix}$$

and r_{ij} denotes the elements of the rotation matrix $\mathbf{R}$ in the Cartesian coordinates.

Zooming in the Homogeneous Coordinates

The effect of zooming can be incorporated into the 3-D motion model as

$$\mathbf{X}_h^{'} = \tilde{\mathbf{S}}\mathbf{X}_h \tag{2.21}$$

where

$$\tilde{\mathbf{S}} = \begin{bmatrix} S_1 & 0 & 0 & 0 \\ 0 & S_2 & 0 & 0 \\ 0 & 0 & S_3 & 0 \\ 0 & 0 & 0 & 1 \end{bmatrix}$$

2.1.3 Deformable Motion

Modeling the 3-D structure and motion of nonrigid objects is a complex task. Analysis and synthesis of nonrigid motion using deformable models is an active research area today. In theory, according to the mechanics of deformable bodies [Som 50], the model (2.1) can be extended to include 3-D nonrigid motion as

$$\mathbf{X}' = (\mathbf{D} + \mathbf{R})\mathbf{X} + \mathbf{T} \tag{2.22}$$

where $\mathbf{D}$ is an arbitrary deformation matrix. Note that the elements of the rotation matrix are constrained to be related to the sines and cosines of the respective angles, whereas the deformation matrix is not constrained in any way. The problem with this seemingly simple model arises in defining the $\mathbf{D}$ matrix to represent the desired deformations.

Some examples of the proposed 3-D nonrigid motion models include those based on free vibration or deformation modes [Pen 91a] and those based on constraints induced by intrinsic and extrinsic forces [Ter 88a]. Pentland *et al.* [Pen 91a] parameterize nonrigid motion in terms of the eigenvalues of a finite-element model of the deformed object. The recovery of 3-D nonrigid motion using this model requires the knowledge of the geometry of the undeformed object. Terzopoulos *et al.* [Ter 88a] have exploited two intrinsic constraints to design deformable models: surface coherence and symmetry seeking. The former is inherent in the elastic forces prescribed by the physics of deformable continua, and the latter is an attribute of many natural and synthetic objects. Terzopoulos and Fleischer [Ter 88b] also proposed a physically based modeling scheme using mechanical laws of continuous bodies whose shapes vary in time. They included physical features, such as mass and damping, in their models in order to simulate the dynamics of deformable objects in response to applied forces. Other models include the deformable superquadrics [Ter 91] and extensions of the physically based framework [Met 93]. The main applications of these models have been in image synthesis and animation.

A simple case of 3-D nonrigid models is that of flexibly connected rigid patches, such as a wireframe model where the deformation of the nodes (the so-called local motion) is allowed. In this book, we will consider only 2-D deformable models that is, the effect of various deformations in the image plane; except for the simple case of 3-D flexible wireframe models, that are discussed in Chapter 24.

2.2 Geometric Image Formation

Imaging systems capture 2-D projections of a time-varying 3-D scene. This projection can be represented by a mapping from a 4-D space to a 3-D space,

$$\begin{aligned} f: \quad & R^4 \rightarrow R^3 \\ & (X_1, X_2, X_3, t) \rightarrow (x_1, x_2, t) \end{aligned} \tag{2.23}$$

where (X_1, X_2, X_3), the 3-D world coordinates, (x_1, x_2), the 2-D image plane coordinates, and t, time, are continuous variables. Here, we consider two types of projection, perspective (central) and orthographic (parallel), which are described in the following.

2.2.1 Perspective Projection

Perspective projection reflects image formation using an ideal pinhole camera according to the principles of geometrical optics. Thus, all the rays from the object pass through the center of projection, which corresponds to the center of the lens. For this reason, it is also known as "central projection." Perspective projection is illustrated in Figure 2.4 when the center of projection is between the object and the image plane, and the image plane coincides with the (X_1, X_2) plane of the world coordinate system.

The algebraic relations that describe the perspective transformation for the configuration shown in Figure 2.4 can be obtained based on similar triangles formed by drawing perpendicular lines from the object point (X_1, X_2, X_3) and the image point $(x_1, x_2, 0)$ to the X_3 axis, respectively. This leads to

$$\frac{x_1}{f} = -\frac{X_1}{X_3 - f} \quad \text{and} \quad \frac{x_2}{f} = -\frac{X_2}{X_3 - f}$$

or

$$x_1 = \frac{fX_1}{f - X_3} \quad \text{and} \quad x_2 = \frac{fX_2}{f - X_3} \tag{2.24}$$

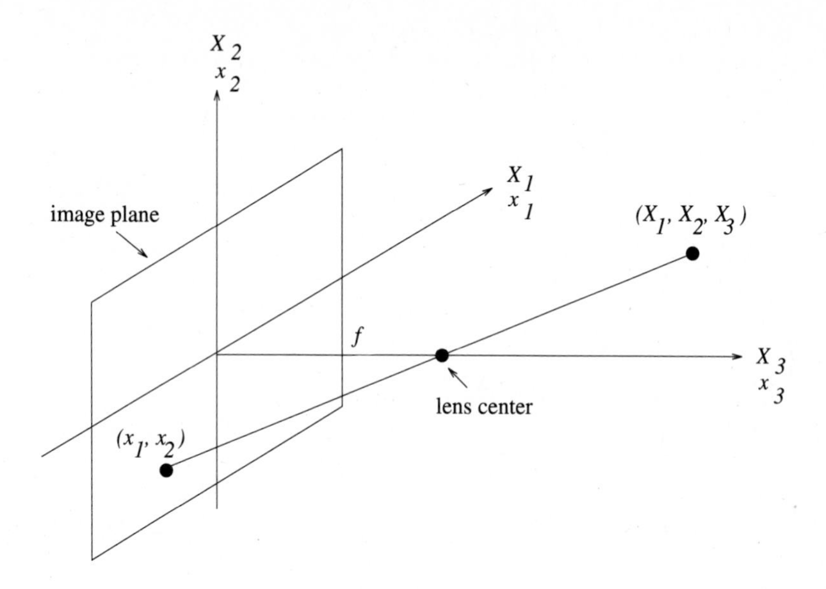

Figure 2.4: Perspective projection model.

where f denotes the distance from the center of projection to the image plane.

If we move the center of projection to coincide with the origin of the world coordinates, a simple change of variables yields the following equivalent expressions:

$$x_1 = \frac{fX_1}{X_3} \quad \text{and} \quad x_2 = \frac{fX_2}{X_3} \tag{2.25}$$

The configuration and the similar triangles used to obtain these expressions are shown in Figure 2.5, where the image plane is parallel to the (X_1, X_2) plane of the world coordinate system. Observe that the latter expressions can also be employed as an approximate model for the configuration in Figure 2.4 when $X_3 >> f$ with the reversal of the sign due to the orientation of the image being the same as the object, as opposed to being a mirror image as in the actual image formation. The general form of the perspective projection, when the image plane is not parallel to the (X_1, X_2) plane of the world coordinate system, is given in [Ver 89].

We note that the perspective projection is nonlinear in the Cartesian coordinates since it requires division by the X_3 coordinate. However, it can be expressed as a linear mapping in the homogeneous coordinates, as

$$\begin{bmatrix} \ell x_1 \\ \ell x_2 \\ \ell \end{bmatrix} = \begin{bmatrix} f & 0 & 0 & 0 \\ 0 & f & 0 & 0 \\ 0 & 0 & 1 & 0 \end{bmatrix} \begin{bmatrix} kX_1 \\ kX_2 \\ kX_3 \\ k \end{bmatrix} \tag{2.26}$$

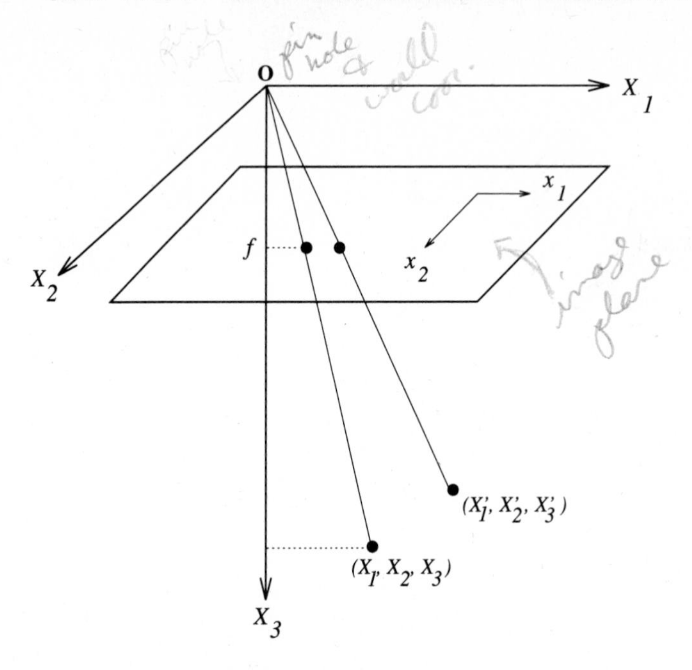

Figure 2.5: Simplified perspective projection model.

where

$$\mathbf{X}_h = \begin{bmatrix} kX_1 \\ kX_2 \\ kX_3 \\ k \end{bmatrix}$$

and

$$\mathbf{x}_h = \begin{bmatrix} \ell x_1 \\ \ell x_2 \\ \ell \end{bmatrix}$$

denote the world and image plane points, respectively, in the homogeneous coordinates.

2.2.2 Orthographic Projection

Orthographic projection is an approximation of the actual imaging process where it is assumed that all the rays from the 3-D object (scene) to the image plane travel parallel to each other. For this reason it is sometimes called the "parallel projection." Orthographic projection is depicted in Figure 2.6 when the image plane is parallel to the $X_1 - X_2$ plane of the world coordinate system.

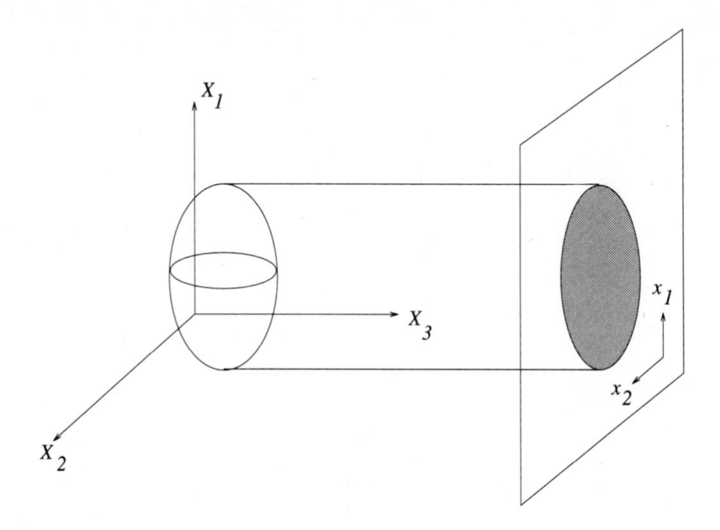

Figure 2.6: Orthographic projection model.

Provided that the image plane is parallel to the $X_1 - X_2$ plane of the world coordinate system, the orthographic projection can be described in Cartesian coordinates as

$$x_1 = X_1 \quad \text{and} \quad x_2 = X_2 \tag{2.27}$$

or in vector-matrix notation as

$$\begin{bmatrix} x_1 \\ x_2 \end{bmatrix} = \begin{bmatrix} 1 & 0 & 0 \\ 0 & 1 & 0 \end{bmatrix} \begin{bmatrix} X_1 \\ X_2 \\ X_3 \end{bmatrix} \tag{2.28}$$

where x_1 and x_2 denote the image plane coordinates.

The distance of the object from the camera does not affect the image plane intensity distribution in orthographic projection. That is, the object always yields the same image no matter how far away it is from the camera. However, orthographic projection provides good approximation to the actual image formation process when the distance of the object from the camera is much larger than the relative depth of points on the object with respect to a coordinate system on the object itself. In such cases, orthographic projection is usually preferred over more complicated but realistic models because it is a linear mapping and thus leads to algebraically and computationally more tractable algorithms.

2.3 Photometric Image Formation

Image intensities can be modeled as proportional to the amount of light reflected by the objects in the scene. The scene reflectance function is generally assumed to contain a Lambertian and a specular component. In this section, we concentrate on surfaces where the specular component can be neglected. Such surfaces are called Lambertian surfaces. Modeling of specular reflection is discussed in [Dri 92]. More sophisticated reflectance models can be found in [Hor 86, Lee 90].

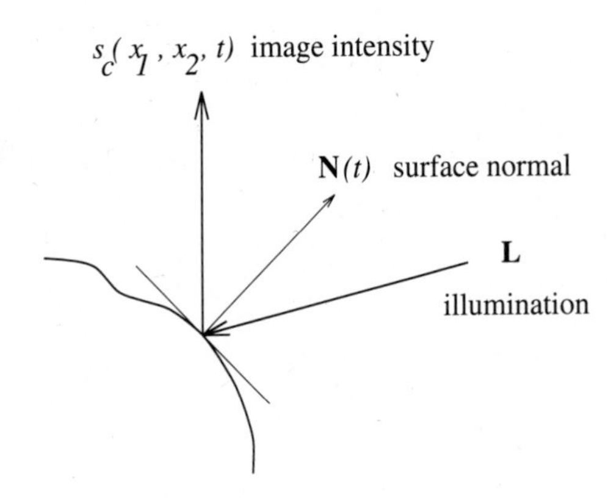

Figure 2.7: Photometric image formation model.

2.3.1 Lambertian Reflectance Model

If a Lambertian surface is illuminated by a single point-source with uniform intensity (in time), the resulting image intensity is given by [Hor 86]

$$s_c(x_1, x_2, t) = \rho \mathbf{N}(t) \cdot \mathbf{L} \tag{2.29}$$

where ρ denotes the surface albedo, i.e., the fraction of the light reflected by the surface, $\mathbf{L} = (L_1, L_2, L_3)$ is the unit vector in the mean illuminant direction, and $\mathbf{N}(t)$ is the unit surface normal of the scene, at spatial location $(X_1, X_2, X_3(X_1, X_2))$ and time t, given by

$$\mathbf{N}(t) = (-p, -q, 1)/(p^2 + q^2 + 1)^{1/2} \tag{2.30}$$

in which $p = \partial X_3/\partial x_1$ and $q = \partial X_3/\partial x_2$ are the partial derivatives of depth $X_3(x_1, x_2)$ with respect to the image coordinates x_1 and x_2, respectively, under the orthographic projection. Photometric image formation for a static surface is illustrated in Figure 2.7.

The illuminant direction can also be expressed in terms of tilt and slant angles as [Pen 91b]

$$\mathbf{L} = (L_1, L_2, L_3) = (\cos\tau \sin\sigma, \sin\tau \sin\sigma, \cos\sigma) \tag{2.31}$$

how?

how?

where τ, the tilt angle of the illuminant, is the angle between $\mathbf{L}$ and the $X_1 - X_3$ plane, and σ, the slant angle, is the angle between $\mathbf{L}$ and the positive X_3 axis.

2.3.2 Photometric Effects of 3-D Motion

As an object moves in 3-D, the surface normal changes as a function of time; so do the photometric properties of the surface. Assuming that the mean illuminant direction $\mathbf{L}$ remains constant, we can express the change in the intensity due to photometric effects of the motion as

$$\frac{ds_c(x_1, x_2, t)}{dt} = \rho \mathbf{L} \cdot \frac{d\mathbf{N}(t)}{dt} \tag{2.32}$$

The rate of change of the normal vector $\mathbf{N}$ at the point (X_1, X_2, X_3) can be approximated by

$$\frac{d\mathbf{N}}{dt} \approx \frac{\Delta \mathbf{N}}{\Delta t}$$

where $\Delta\mathbf{N}$ denotes the change in the direction of the normal vector due to the 3-D motion from the point (X_1, X_2, X_3) to (X_1', X_2', X_3') within the period Δt. This change can be expressed as

$$\begin{aligned} \Delta \mathbf{N} &= \mathbf{N}(X_1', X_2', X_3') - \mathbf{N}(X_1, X_2, X_3) \\ &= \frac{(-p', -q', 1)}{(p'^2 + q'^2 + 1)^{1/2}} - \frac{(-p, -q, 1)}{(p^2 + q^2 + 1)^{1/2}} \end{aligned} \tag{2.33}$$

where p' and q' denote the components of $\mathbf{N}(X_1', X_2', X_3')$ given by

$$\begin{aligned} p' &= \frac{\partial X_3'}{\partial x_1'} = \frac{\partial X_3'}{\partial x_1} \frac{\partial x_1}{\partial x_1'} \\ &= \frac{-\Delta\psi + p}{1 + \Delta\psi p} \\ q' &= \frac{\partial X_3'}{\partial x_2'} = \frac{\Delta\theta + q}{1 - \Delta\theta q} \end{aligned} \tag{2.34}$$

Pentland [Pen 91b] shows that the photometric effects of motion can dominate the geometric effects in some cases.

2.4 Observation Noise

Image capture mechanisms are never perfect. As a result, images generally suffer from graininess due to electronic noise, photon noise, film-grain noise, and quantization noise. In video scanned from motion picture film, streaks due to possible scratches on film can be modeled as impulsive noise. Speckle noise is common in radar image sequences and biomedical cine-ultrasound sequences. The available signal-to-noise ratio (SNR) varies with the imaging devices and image recording media. Even if the noise may not be perceived at full-speed video due to the temporal masking effect of the eye, it often leads to poor-quality "freeze-frames."

The observation noise in video can be modeled as additive or multiplicative noise, signal-dependent or signal-independent noise, and white or colored noise. For example, photon and film-grain noise are signal-dependent, whereas CCD sensor and quantization noise are usually modeled as white, Gaussian distributed, and signal-independent. Ghosts in TV images can also be modeled as signal-dependent noise. In this book, we will assume a simple additive noise model given by

$$g_c(x_1, x_2, t) = s_c(x_1, x_2, t) + v_c(x_1, x_2, t) \tag{2.35}$$

where $s_c(x_1, x_2, t)$ and $v_c(x_1, x_2, t)$ denote the ideal video and noise at time t, respectively.

The SNR is an important parameter for most digital video processing applications, because noise hinders our ability to effectively process the data. For example, in 2-D and 3-D motion estimation, it is very important to distinguish the variation of the intensity pattern due to motion from that of the noise. In image resolution enhancement, noise is the fundamental limitation on our ability to recover high-frequency information. Furthermore, in video compression, random noise increases the entropy hindering effective compression. The SNR of video imagery can be enhanced by spatio-temporal filtering, also called noise filtering, which is the subject of Chapter 14.

2.5 Exercises

1. Suppose a rotation by 45 degrees about the X_2 axis is followed by another rotation by 60 degrees about the X_1 axis. Find the directional cosines n_1, n_2, n_3 and the solid angle α to represent the composite rotation.

2. Given a triangle defined by the points (1,1,1), (1,-1,0) and (0,1,-1). Find the vertices of the triangle after a rotation by 30 degrees about an axis passing through (0,1,0) which is parallel to the X_1 axis.

3. Show that a rotation matrix is orthonormal.

4. Derive Equation (2.6).

5. Show that the two representations of the rotation matrix $\mathbf{R}$ given by (2.6) and (2.10) are equivalent.

6. Discuss the conditions under which the orthographic projection provides a good approximation to imaging through an ideal pinhole camera.

7. Show that the expressions for p' and q' in (2.34) are valid under the orthographic projection.

Bibliography

[Bal 82] D. H. Ballard and C. M. Brown, *Computer Vision*, Englewood Cliffs, NJ: Prentice-Hall, 1982.

[Dri 92] J. N. Driessen, *Motion Estimation for Digital Video*, Ph.D. Thesis, Delft University of Technology, 1992.

[Fol 83] J. D. Foley and A. Van Dam, *Fundamentals of Interactive Computer Graphics*, Reading, MA: Addison-Wesley, 1983.

[Hog 92] S. G. Hoggar, *Mathematics for Computer Graphics*, Cambridge University Press, 1992.

[Hor 86] B. K. P. Horn, *Robot Vision*, Cambridge, MA: MIT Press, 1986.

[Lee 90] H. C. Lee, E. J. Breneman, and C. P. Schutte, "Modeling light reflection for computer color vision," *IEEE Trans. Patt. Anal. Mach. Intel.*, vol. PAMI-12, pp. 402–409, Apr. 1990.

[Met 93] D. Metaxas and D. Terzopoulos, "Shape and nonrigid motion estimation through physics-based synthesis," *IEEE Trans. Patt. Anal. Mach. Intel.*, vol. 15, pp. 580–591, June 1993.

[Pen 91a] A. Pentland and B. Horowitz, "Recovery of nonrigid motion and structure," *IEEE Trans. Patt. Anal. Mach. Intel.*, vol. 13, pp. 730–742, 1991.

[Pen 91b] A. Pentland, "Photometric motion," *IEEE Trans. Patt. Anal. Mach. Intel.*, vol. PAMI-13, pp. 879–890, Sep. 1991.

[Rog 76] D. F. Rogers and J. A. Adams, *Mathematical Elements for Computer Graphics*, New York, NY: McGraw Hill, 1976.

[Som 50] A. Sommerfeld, *Mechanics of Deformable Bodies*, 1950.

[Ter 88a] D. Terzopoulos, A. Witkin, and M. Kass, "Constraints on deformable models: Recovering 3-D shape and nonrigid motion," *Artif. Intel.*, vol. 36, pp. 91–123, 1988.

[Ter 88b] D. Terzopoulos and K. Fleischer, "Deformable models," *Visual Comput.*, vol. 4, pp. 306–331, 1988.

[Ter 91] D. Terzopoulos and D. Metaxas, "Dynamic 3-D models with local and global deformations: Deformable superquadrics," *IEEE Trans. Patt. Anal. Mach. Intel.*, vol. PAMI-13, pp. 703–714, July 1991.

[Ver 89] A. Verri and T. Poggio, "Motion field and optical flow: Qualitative properties," *IEEE Trans. Patt. Anal. Mach. Intel.*, vol. PAMI-11, pp. 490–498, May 1989.

Chapter 3

SPATIO-TEMPORAL SAMPLING

In order to obtain an *analog* or *digital* video signal representation, the continuous time-varying image $s_c(x_1, x_2, t)$ needs to be sampled in both the spatial and temporal variables. An analog video signal representation requires sampling $s_c(x_1, x_2, t)$ in the vertical and temporal dimensions. Recall that an analog video signal is a 1-D continuous function, where one of the spatial dimensions is mapped onto time by means of the scanning process. For a digital video representation, $s_c(x_1, x_2, t)$ is sampled in all three dimensions. The spatio-temporal sampling process is depicted in Figure 3.1, where (n_1, n_2, k) denotes the discrete spatial and temporal coordinates, respectively.

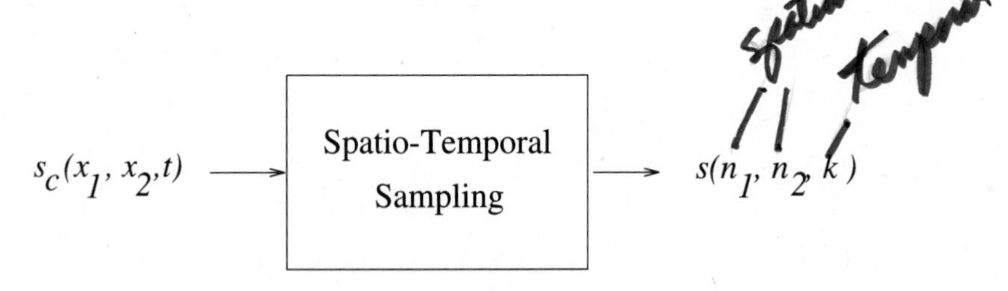

Figure 3.1: Block diagram.

Commonly used 2-D and 3-D sampling structures for the representation of analog and digital video are shown in Section 3.1. Next, we turn our attention to the frequency domain characterization of sampled video. In order to motivate the main principles, we start with the sampling of still images. Section 3.2 covers the case of sampling on a 2-D rectangular grid, whereas Section 3.3 treats sampling on arbitrary 2-D periodic grids. In Section 3.4, we address extension of this theory to sampling of multidimensional signals on lattices and other periodic sampling structures. However, the discussion is limited to sampling of spatio-temporal signals $s_c(x_1, x_2, t)$ on 3-D lattices, considering the scope of the book. Finally, Section 3.5 addresses the reconstruction of continuous time-varying images from spatio-temporally sampled representations. The reader is advised to review the remarks about the notation used in this book on page xxi, before proceeding with this chapter.

3.1 Sampling for Analog and Digital Video

Some of the more popular sampling structures utilized in the representation of analog and digital video are introduced in this section.

3.1.1 Sampling Structures for Analog Video

An analog video signal is obtained by sampling the time-varying image intensity distribution in the vertical x_2, and temporal t directions by a 2-D sampling process known as scanning. Continuous intensity information along each horizontal line is concatenated to form the 1-D analog video signal as a function of time. The two most commonly used vertical-temporal sampling structures are the orthogonal sampling structure, shown in Figure 3.2, and the hexagonal sampling structure, depicted in Figure 3.3.

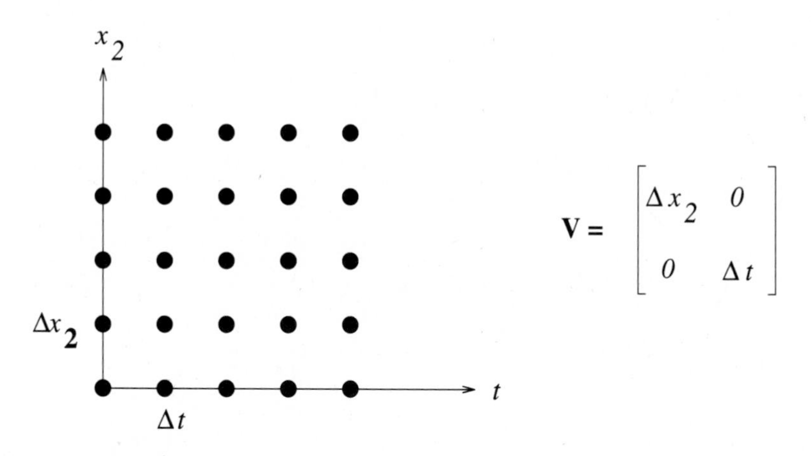

Figure 3.2: Orthogonal sampling structure for progressive analog video.

In these figures, each dot indicates a continuous line of video perpendicular to the plane of the page. The matrices $\mathbf{V}$ shown in these figures are called the sampling matrices, and will be defined in Section 3.3. The orthogonal structure is used in the representation of progressive analog video, such as that shown on workstation monitors, and the hexagonal structure is used in the representation of 2:1 interlaced analog video, such as that shown on TV monitors. The spatio-temporal frequency content of these signals will be analyzed in Sections 3.2 and 3.3, respectively.

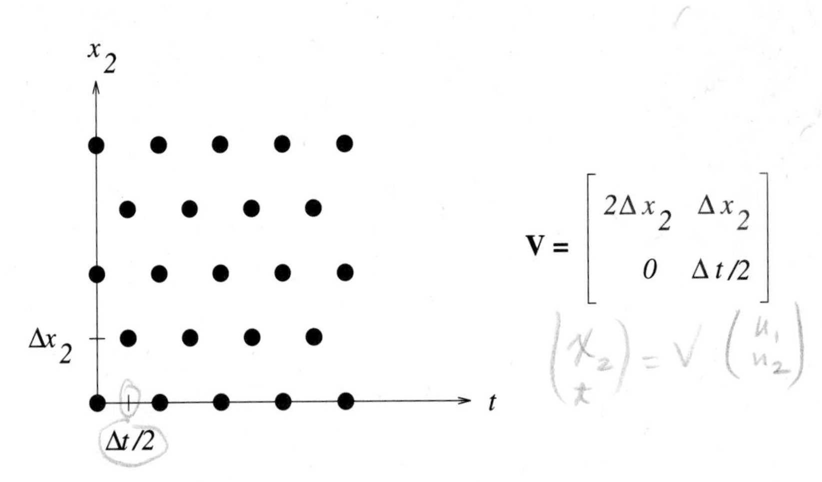

Figure 3.3: Hexagonal sampling structure for 2:1 interlaced analog video.

3.1.2 Sampling Structures for Digital Video

Digital video can be obtained by sampling analog video in the horizontal direction along the scan lines, or by applying an inherently 3-D sampling structure to sample the time-varying image, as in the case of some solid-state sensors. Examples of the most popular 3-D sampling structures are shown in Figures 3.4, 3.5, 3.6, and 3.7. In these figures, each circle indicates a pixel location, and the number inside the circle indicates the time of sampling. The first three sampling structures are lattices, whereas the sampling structure in Figure 3.7 is not a lattice, but a union of two cosets of a lattice. The vector $\mathbf{c}$ in Figure 3.7 shows the displacement of one coset with respect to the other. Other 3-D sampling structures can be found in [Dub 85].

The theory of sampling on lattices and other special M-D structures is presented in Section 3.4. It will be seen that the most suitable sampling structure for a time-varying image depends on its spatio-temporal frequency content. The sampling structures shown here are field- or frame-instantaneous; that is, a complete field or frame is acquired at one time instant. An alternative strategy is time-sequential sampling, where individual samples are taken one at a time according to a prescribed ordering which is repeated after one complete frame. A theoretical analysis of time-sequential sampling can be found in [Rah 92].

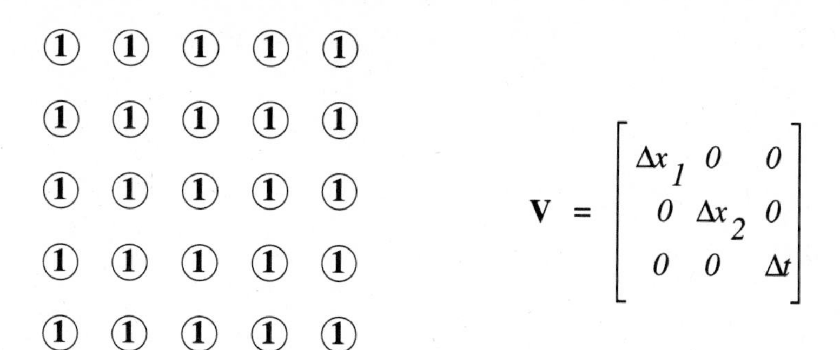

Figure 3.4: Orthogonal sampling lattice [Dub 85] (©1985 IEEE).

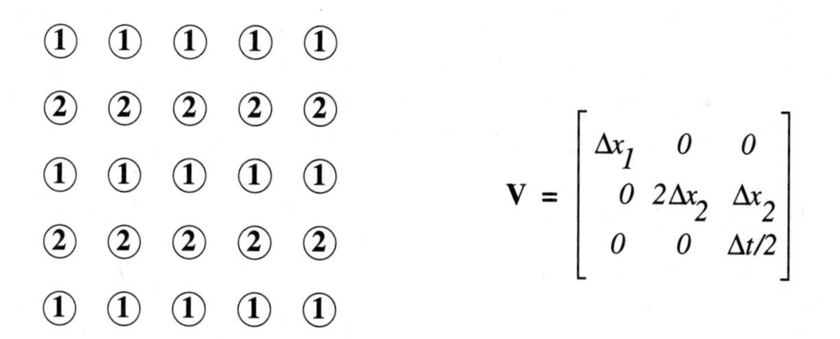

Figure 3.5: Vertically aligned 2:1 line-interlaced lattice [Dub 85] (©1985 IEEE).

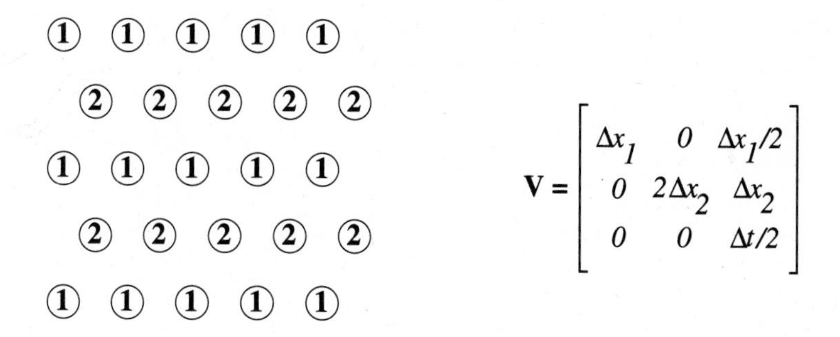

Figure 3.6: Field-quincunx sampling lattice [Dub 85] (©1985 IEEE).

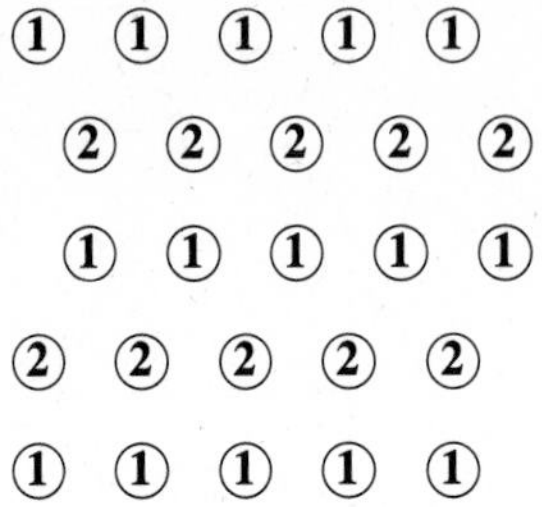

① ① ① ① ①

② ② ② ② ②

① ① ① ① ①

② ② ② ② ②

① ① ① ① ①

$$\mathbf{V} = \begin{bmatrix} \Delta x_1 & \Delta x_1/2 & 0 \\ 0 & 2\Delta x_2 & 0 \\ 0 & 0 & \Delta t \end{bmatrix} \qquad \mathbf{c} = \begin{bmatrix} 0 \\ \Delta x_2 \\ \Delta t/2 \end{bmatrix}$$

Figure 3.7: Line-quincunx sampling lattice [Dub 85] (©1985 IEEE).

3.2 Two-Dimensional Rectangular Sampling

In this section, we discuss 2-D rectangular sampling of a still image, $s_c(x_1, x_2)$ in the two spatial coordinates. However, the same analysis also applies to vertical-temporal sampling (as in the representation of progressive analog video). In spatial rectangular sampling, we sample at the locations

$$\begin{aligned} x_1 &= n_1 \Delta x_1 \\ x_2 &= n_2 \Delta x_2 \end{aligned} \tag{3.1}$$

where Δx_1 and Δx_2 are the sampling distances in the x_1 and x_2 directions, respectively. The 2-D rectangular sampling grid is depicted in Figure 3.8. The sampled signal can be expressed, in terms of the unitless coordinate variables (n_1, n_2), as

$$s(n_1, n_2) = s_c(n_1 \Delta x_1, n_2 \Delta x_2), \quad (n_1, n_2) \in \mathbf{Z}^2. \tag{3.2}$$

In some cases, it is convenient to define an intermediate sampled signal in terms of the continuous coordinate variables, given by

$$\begin{aligned} s_p(x_1, x_2) &= s_c(x_1, x_2) \sum_{n_1} \sum_{n_2} \delta(x_1 - n_1 \Delta x_1, x_2 - n_2 \Delta x_2) \\ &= \sum_{n_1} \sum_{n_2} s_c(n_1 \Delta x_1, n_2 \Delta x_2) \delta(x_1 - n_1 \Delta x_1, x_2 - n_2 \Delta x_2) \\ &= \sum_{n_1} \sum_{n_2} s(n_1, n_2) \delta(x_1 - n_1 \Delta x_1, x_2 - n_2 \Delta x_2) \end{aligned} \tag{3.3}$$

Note that $s_p(x_1, x_2)$ is indeed a sampled signal because of the presence of the 2-D Dirac delta function $\delta(\cdot, \cdot)$.

3.2.1 2-D Fourier Transform Relations

We start by reviewing the 2-D continuous-space Fourier transform (FT) relations. The 2-D FT $S_c(F_1, F_2)$ of a signal with continuous variables $s_c(x_1, x_2)$ is given by

$$S_c(F_1, F_2) = \int_{-\infty}^{\infty}\int_{-\infty}^{\infty} s_c(x_1, x_2) \exp\{-j2\pi(F_1x_1 + F_2x_2)\}dx_1dx_2 \tag{3.4}$$

where $(F_1, F_2) \in \mathbf{R}^2$, and the inverse 2-D Fourier transform is given by

$$s_c(x_1, x_2) = \int_{-\infty}^{\infty}\int_{-\infty}^{\infty} S_c(F_1, F_2) \exp\{j2\pi(F_1x_1 + F_2x_2)\}dF_1dF_2 \tag{3.5}$$

Here, the spatial frequency variables F_1 and F_2 have the units in cycles/mm and are related to the radian frequencies by a scale factor of 2π.

In order to evaluate the 2-D FT $S_p(F_1, F_2)$ of $s_p(x_1, x_2)$, we substitute (3.3) into (3.4), and exchange the order of integration and summation, to obtain

$$\begin{aligned} S_p(F_1, F_2) &= \sum_{n_1}\sum_{n_2} s_c(n_1\Delta x_1, n_2\Delta x_2) \\ &\quad \int\int \delta(x_1 - n_1\Delta x_1, x_2 - n_2\Delta x_2) \exp\{-j2\pi(F_1x_1 + F_2x_2)\}dx_1dx_2 \end{aligned}$$

which simplifies as

$$S_p(F_1, F_2) = \sum_{n_1}\sum_{n_2} s_c(n_1\Delta x_1, n_2\Delta x_2) \exp\{-j2\pi(F_1n_1\Delta x_1 + F_2n_2\Delta x_2)\} \tag{3.6}$$

Notice that $S_p(F_1, F_2)$ is periodic with the fundamental period given by the region $F_1 < |1/(2\Delta x_1)|$ and $F_2 < |1/(2\Delta x_2)|$.

Letting $f_i = F_i\Delta x_i$, $i = 1, 2$, and using (3.2), we obtain the discrete-space Fourier transform relation, in terms of the unitless frequency variables f_1 and f_2, as

$$S(f_1, f_2) = S_p(\frac{f_1}{\Delta x_1}, \frac{f_2}{\Delta x_2}) = \sum_{n_1=-\infty}^{\infty}\sum_{n_2=-\infty}^{\infty} s(n_1, n_2) \exp\{-j2\pi(f_1n_1 + f_2n_2)\} \tag{3.7}$$

The 2-D discrete-space inverse Fourier transform is given by

$$s(n_1, n_2) = \int_{-\frac{1}{2}}^{\frac{1}{2}}\int_{-\frac{1}{2}}^{\frac{1}{2}} S(f_1, f_2) \exp\{j2\pi(f_1n_1 + f_2n_2)\}df_1df_2 \tag{3.8}$$

Recall that the discrete-space Fourier transform $S(f_1, f_2)$ is periodic with the fundamental period $f_1 < |1/2|$ and $f_2 < |1/2|$.

3.2.2 Spectrum of the Sampled Signal

We now relate the Fourier transform, $S_p(F_1, F_2)$ or $S(f_1, f_2)$, of the sampled signal to that of the continuous signal. The standard approach is to start with (3.3), and express $S_p(F_1, F_2)$ as the 2-D convolution of the Fourier transforms of the continuous signal and the impulse train [Opp 89, Jai 90], using the modulation property of the FT,

$$S_p(F_1, F_2) = S_c(F_1, F_2) ** \mathcal{F}\{\sum_{n_1}\sum_{n_2} \delta(x_1 - n_1\Delta x_1, x_2 - n_2\Delta x_2)\} \tag{3.9}$$

where $\mathcal{F}$ denotes 2-D Fourier transformation, which simplifies to yield (3.11).

Here, we follow a different derivation which can be easily extended to other periodic sampling structures [Dud 84, Dub 85]. First, substitute (3.5) into (3.2), and evaluate x_1 and x_2 at the sampling locations given by (3.1) to obtain

$$s(n_1, n_2) = \int_{-\infty}^{\infty}\int_{-\infty}^{\infty} S_c(F_1, F_2) \exp\{j2\pi(F_1 n_1 \Delta x_1 + F_2 n_2 \Delta x_2)\} dF_1 dF_2$$

After the change of variables $f_1 = F_1 \Delta x_1$ and $f_2 = F_2 \Delta x_2$, we have

$$s(n_1, n_2) = \frac{1}{\Delta x_1 \Delta x_2}\int_{-\infty}^{\infty}\int_{-\infty}^{\infty} S_c(\frac{f_1}{\Delta x_1}, \frac{f_2}{\Delta x_2}) \exp\{j2\pi(f_1 n_1 + f_2 n_2)\}\ df_1 df_2$$

Next, break the integration over the (f_1, f_2) plane into a sum of integrals each over a square denoted by $SQ(k_1, k_2)$,

$$s(n_1, n_2) = \sum_{k_1}\sum_{k_2} \frac{1}{\Delta x_1 \Delta x_2}\int\int_{SQ} S_c(\frac{f_1}{\Delta x_1}, \frac{f_2}{\Delta x_2}) \exp\{j2\pi(f_1 n_1 + f_2 n_2)\} df_1 df_2$$

where $SQ(k_1, k_2)$ is defined as

$$-\frac{1}{2} + k_1 \le f_1 \le \frac{1}{2} + k_1 \quad \text{and} \quad -\frac{1}{2} + k_2 \le f_2 \le \frac{1}{2} + k_2$$

Another change of variables, $f_1 = f_1 - k_1$ and $f_2 = f_2 - k_2$, shifts all the squares $SQ(k_1, k_2)$ down to the fundamental period $(-\frac{1}{2}, \frac{1}{2}) \times (-\frac{1}{2}, \frac{1}{2})$, to yield

$$\begin{aligned} s(n_1, n_2) &= \int_{-\frac{1}{2}}^{\frac{1}{2}}\int_{-\frac{1}{2}}^{\frac{1}{2}} \{\frac{1}{\Delta x_1 \Delta x_2}\sum_{k_1}\sum_{k_2} S_c(\frac{f_1 - k_1}{\Delta x_1}, \frac{f_2 - k_2}{\Delta x_2})\} \\ &\quad \exp\{j2\pi(f_1 n_1 + f_2 n_2)\} \exp\{-j2\pi(k_1 n_1 + k_2 n_2)\} df_1 df_2 \end{aligned} \tag{3.10}$$

But $\exp\{-j2\pi(k_1 n_1 + k_2 n_2)\} = 1$ for k_1, k_2, n_1, n_2 integers. Thus, the frequencies $(f_1 - k_1, f_2 - k_2)$ map onto (f_1, f_2). Comparing the last expression with (3.8), we therefore conclude that

$$S(f_1, f_2) = \frac{1}{\Delta x_1 \Delta x_2}\sum_{k_1}\sum_{k_2} S_c(\frac{f_1 - k_1}{\Delta x_1}, \frac{f_2 - k_2}{\Delta x_2}) \tag{3.11}$$

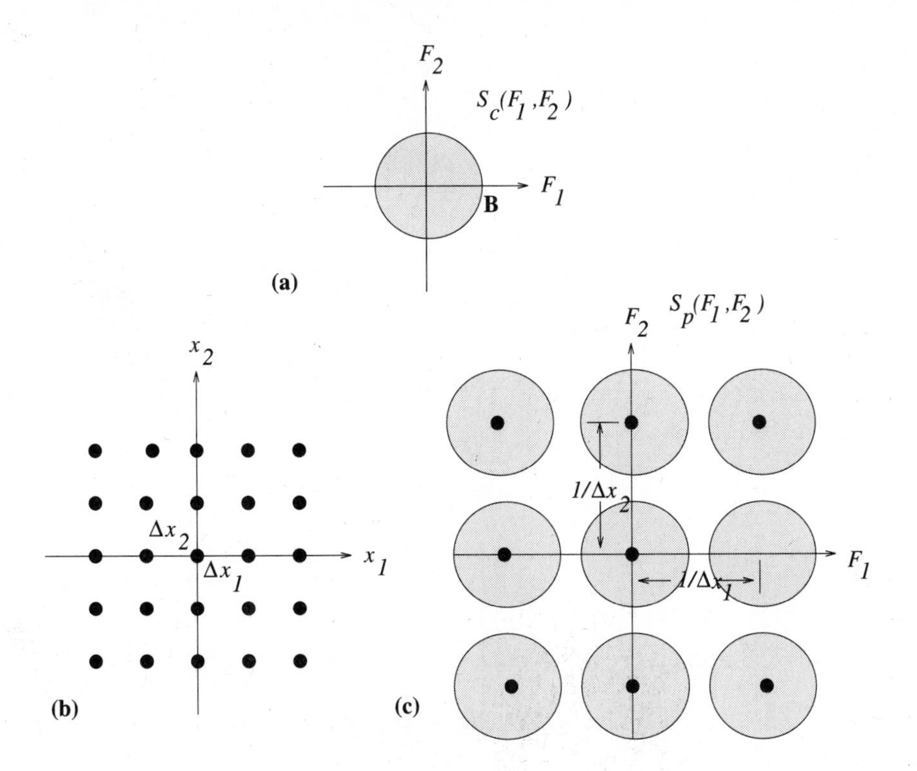

Figure 3.8: Sampling on a 2-D rectangular grid: a) support of the Fourier spectrum of the continuous image; b) the sampling grid; c) spectral support of the sampled image.

or, equivalently,

$$S_p(F_1, F_2) = \frac{1}{\Delta x_1 \Delta x_2} \sum_{k_1} \sum_{k_2} S_c(F_1 - \frac{k_1}{\Delta x_1}, F_2 - \frac{k_2}{\Delta x_2}) \qquad (3.12)$$

We see that, as a result of sampling, the spectrum of the continuous signal replicates in the 2-D frequency plane according to (3.11). The case when the continuous signal is bandlimited with a circular spectral support of radius $B < \max\{1/(2\Delta x_1), 1/(2\Delta x_2)\}$ is illustrated in Figure 3.8.

3.3 Two-Dimensional Periodic Sampling

In this section we extend the results of the previous section to arbitrary 2-D periodic sampling grids.

3.3.1 Sampling Geometry

An arbitrary 2-D periodic sampling geometry can be defined by two basis vectors $\mathbf{v}_1 = (v_{11}\ v_{21})^T$ and $\mathbf{v}_2 = (v_{12}\ v_{22})^T$, such that every sampling location can be expressed as a linear combination of them, given by

$$\begin{aligned} x_1 &= v_{11}n_1 + v_{12}n_2 \\ x_2 &= v_{21}n_1 + v_{22}n_2 \end{aligned} \tag{3.13}$$

In vector-matrix form, we have

$$\mathbf{x} = \mathbf{V}\mathbf{n} \tag{3.14}$$

where

$$\mathbf{x} = (x_1\ x_2)^T, \quad \mathbf{n} = (n_1\ n_2)^T$$

and

$$\mathbf{V} = [\mathbf{v}_1 | \mathbf{v}_2]$$

is the sampling matrix. The sampling locations for an arbitrary periodic grid are depicted in Figure 3.9.

Then, analogous to (3.2) and (3.3), the sampled signal can be expressed as

$$s(\mathbf{n}) = s_c(\mathbf{V}\mathbf{n}), \quad \mathbf{n} \in \mathbf{Z}^2 \tag{3.15}$$

or as

$$\begin{aligned} s_p(\mathbf{x}) &= s_c(\mathbf{x}) \sum_{\mathbf{n} \in \mathbf{Z}^2} \delta(\mathbf{x} - \mathbf{V}\mathbf{n}) \\ &= \sum_{\mathbf{n}} s_c(\mathbf{V}\mathbf{n})\delta(\mathbf{x} - \mathbf{V}\mathbf{n}) = \sum_{\mathbf{n}} s(\mathbf{n})\delta(\mathbf{x} - \mathbf{V}\mathbf{n}) \end{aligned} \tag{3.16}$$

3.3.2 2-D Fourier Transform Relations in Vector Form

Here, we restate the 2-D Fourier transform relations given in Section 3.2.1 in a more compact vector-matrix form as follows:

$$S_c(\mathbf{F}) = \int_{-\infty}^{\infty} s_c(\mathbf{x}) \exp\{-j2\pi\mathbf{F}^T\mathbf{x}\}\, d\mathbf{x} \tag{3.17}$$

$$s_c(\mathbf{x}) = \int_{-\infty}^{\infty} S_c(\mathbf{F}) \exp\{j2\pi\mathbf{F}^T\mathbf{x}\}\, d\mathbf{F} \tag{3.18}$$

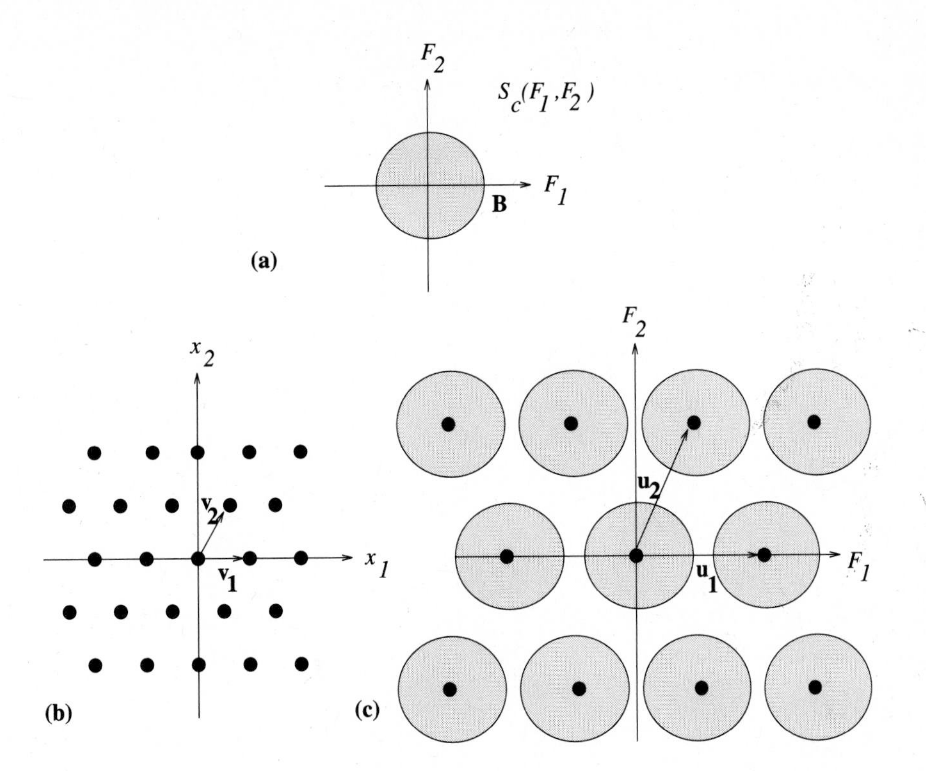

Figure 3.9: Sampling on an arbitrary 2-D periodic grid: a) support of the Fourier spectrum of the continuous image; b) the sampling grid; c) spectral support of the sampled image.

where $\mathbf{x} = (x_1 \ x_2)^T$ and $\mathbf{F} = (F_1 \ F_2)^T$. We also have

$$S_p(\mathbf{F}) = \sum_{\mathbf{n}=-\infty}^{\infty} s_c(\mathbf{V}\mathbf{n}) \ \exp\{-j2\pi\mathbf{F}^T\mathbf{V}\mathbf{n}\} \tag{3.19}$$

or

$$S(\mathbf{f}) = \sum_{\mathbf{n}=-\infty}^{\infty} s(\mathbf{n}) \ \exp\{-j2\pi\mathbf{f}^T\mathbf{n}\} \tag{3.20}$$

$$s(\mathbf{n}) = \int_{-\frac{1}{2}}^{\frac{1}{2}} S(\mathbf{f}) \ \exp\{j2\pi\mathbf{f}^T\mathbf{n}\} \ d\mathbf{f} \tag{3.21}$$

where $\mathbf{f} = (f_1 \ f_2)^T$. Note that the integrations and summations in these relations are in fact double integrations and summations.

3.3.3 Spectrum of the Sampled Signal

To derive the relationship between $S(\mathbf{f})$ and $S_c(\mathbf{F})$, we follow the same steps as in Section 3.2. Thus, we start by substituting (3.18) into (3.15) as

$$s(\mathbf{n}) = s_c(\mathbf{Vn}) = \int_{-\infty}^{\infty} S_c(\mathbf{F}) \exp\{j2\pi \mathbf{F}^T \mathbf{Vn}\} d\mathbf{F}$$

Making the change of variables $\mathbf{f} = \mathbf{V}^T \mathbf{F}$, we have

$$s(\mathbf{n}) = \int_{-\infty}^{\infty} \frac{1}{|det\mathbf{V}|} S_c(\mathbf{Uf}) \exp\{j2\pi \mathbf{f}^T \mathbf{n}\} d\mathbf{f}$$

where $\mathbf{U} = \mathbf{V}^{T^{-1}}$ and $d\mathbf{f} = |det\mathbf{V}| d\mathbf{F}$.

Expressing the integration over the $\mathbf{f}$ plane as a sum of integrations over the squares $(-1/2, 1/2) \times (-1/2, 1/2)$, we obtain

$$s(\mathbf{n}) = \int_{-\frac{1}{2}}^{\frac{1}{2}} \sum_{\mathbf{k}} \frac{1}{|det\mathbf{V}|} S_c(\mathbf{U}(\mathbf{f}-\mathbf{k})) \exp\{j2\pi \mathbf{f}^T \mathbf{n}\} \exp\{-j2\pi \mathbf{k}^T \mathbf{n}\} d\mathbf{f} \quad (3.22)$$

where $\exp\{-j2\pi \mathbf{k}^T \mathbf{n}\} = 1$ for $\mathbf{k}$ an integer valued vector.

Thus, comparing this expression with (3.21), we conclude that

$$S(\mathbf{f}) = \frac{1}{|det\mathbf{V}|} \sum_{\mathbf{k}} S_c(\mathbf{U}(\mathbf{f}-\mathbf{k})) \quad (3.23)$$

or, equivalently,

$$S_p(\mathbf{F}) = \frac{1}{|det\mathbf{V}|} \sum_{\mathbf{k}} S_c(\mathbf{F} - \mathbf{Uk}) \quad (3.24)$$

where the periodicity matrix in the frequency domain $\mathbf{U}$ satisfies

$$\mathbf{U}^T \mathbf{V} = I \quad (3.25)$$

and $\mathbf{I}$ is the identity matrix. The periodicity matrix can be expressed as $\mathbf{U} = [\mathbf{u}_1 | \mathbf{u}_2]$, where $\mathbf{u}_1$ and $\mathbf{u}_2$ are the basis vectors in the 2-D frequency plane.

Note that this formulation includes rectangular sampling as a special case with the matrices $\mathbf{V}$ and $\mathbf{U}$ being diagonal. The replications in the 2-D frequency plane according to (3.24) are depicted in Figure 3.9.

3.4 Sampling on 3-D Structures

The concepts related to 2-D sampling with an arbitrary periodic geometry can be readily extended to sampling time-varying images $s_c(\mathbf{x}, t) = s_c(x_1, x_2, t)$ on 3-D sampling structures. In this section, we first elaborate on 3-D lattices, and the spectrum of 3-D signals sampled on lattices. Some specific non-lattice structures are also introduced. The theory of sampling on lattices and other structures has been generalized to M dimensions elsewhere [Dub 85].

3.4.1 Sampling on a Lattice

We start with the definition of a 3-D lattice. Let $\mathbf{v}_1, \mathbf{v}_2, \mathbf{v}_3$ be linearly independent vectors in the 3-D Euclidean space $\mathbf{R}^3$. A lattice Λ^3 in $\mathbf{R}^3$ is the set of all linear combinations of $\mathbf{v}_1$, $\mathbf{v}_2$, and $\mathbf{v}_3$ with integer coefficients

$$\Lambda^3 = \{n_1\mathbf{v}_1 + n_2\mathbf{v}_2 + k\mathbf{v}_3 \mid n_1,\ n_2,\ k\ \in\ \mathbf{Z}\} \tag{3.26}$$

The set of vectors $\mathbf{v}_1$, $\mathbf{v}_2$, and $\mathbf{v}_3$ is called a basis for Λ^3.

In vector-matrix notation, a lattice can be defined as

$$\Lambda^3 = \{\mathbf{V}\begin{bmatrix}\mathbf{n}\\ k\end{bmatrix} \mid \mathbf{n} \in \mathbf{Z}^2,\ k \in \mathbf{Z}\} \tag{3.27}$$

where $\mathbf{V}$ is called a 3×3 sampling matrix defined by

$$\mathbf{V} = [\mathbf{v}_1 \mid \mathbf{v}_2 \mid \mathbf{v}_3] \tag{3.28}$$

The basis, and thus the sampling matrix, for a given lattice is not unique. In particular, for every sampling matrix $\mathbf{V}$, $\hat{\mathbf{V}} = \mathbf{E}\mathbf{V}$, where $\mathbf{E}$ is an integer matrix with $\det\mathbf{E} = \pm 1$, forms another sampling matrix for Λ^3. However, the quantity $d(\Lambda^3) = |\det\mathbf{V}|$ is unique and denotes the reciprocal of the sampling density.

Then, similar to (3.15) and (3.16), the sampled spatio-temporal signal can be expressed as

$$s(\mathbf{n}, k) = s_c(\mathbf{V}\begin{bmatrix}\mathbf{n}\\ k\end{bmatrix}), \quad (\mathbf{n}, k) = (n_1, n_2, k) \in \mathbf{Z}^3 \tag{3.29}$$

or as

$$\begin{aligned} s_p(\mathbf{x}, t) &= s_c(\mathbf{x}, t) \sum_{(\mathbf{n},k)\in\mathbf{Z}^3} \delta(\begin{bmatrix}\mathbf{x}\\ t\end{bmatrix} - \mathbf{V}\begin{bmatrix}\mathbf{n}\\ k\end{bmatrix}) \\ &= \sum_{(\mathbf{n},k)\in Z^3} s(\mathbf{n}, k)\delta(\begin{bmatrix}\mathbf{x}\\ t\end{bmatrix} - \mathbf{V}\begin{bmatrix}\mathbf{n}\\ k\end{bmatrix}) \end{aligned} \tag{3.30}$$

The observant reader may already have noticed that the 2-D sampling structures discussed in Sections 3.2 and 3.3 are also lattices. Hence, Sections 3.2 and 3.3 constitute special cases of the theory presented in this section.

3.4.2 Fourier Transform on a Lattice

Based on (3.19), we can define the spatio-temporal Fourier transform of $s_p(\mathbf{x}, t)$, sampled on Λ^3, as follows:

$$\begin{aligned} S_p(\mathbf{F}) &= \sum_{(\mathbf{x},t)\in\Lambda^3} s_c(\mathbf{x}, t)\ \exp\left\{-j2\pi\mathbf{F}^T\begin{bmatrix}\mathbf{x}\\ t\end{bmatrix}\right\}, \quad \mathbf{F} = [F_1\ F_2\ F_t]^T \in \mathbf{R}^3 \\ &= \sum_{(\mathbf{n},k)\in\mathbf{Z}^3} s_c(\mathbf{V}\begin{bmatrix}\mathbf{n}\\ k\end{bmatrix})\ \exp\left\{-j2\pi\mathbf{F}^T\mathbf{V}\begin{bmatrix}\mathbf{n}\\ k\end{bmatrix}\right\} \end{aligned} \tag{3.31}$$

In order to quantify some properties of the Fourier transform defined on a lattice, we next define the reciprocal lattice and the unit cell of a lattice. Given a lattice Λ^3, the set of all vectors $\mathbf{r}$ such that $\mathbf{r}^T \begin{bmatrix} \mathbf{x} \\ t \end{bmatrix}$ is an integer for all $(\mathbf{x}, t) \in \Lambda^3$ is called the reciprocal lattice Λ^{3*} of Λ^3. A basis for Λ^{3*} is the set of vectors $\mathbf{u}_1$, $\mathbf{u}_2$, and $\mathbf{u}_3$ determined by

$$\mathbf{u}_i^T \mathbf{v}_j = \delta_{ij}, \quad i, j = 1, 2, 3$$

or, equivalently,

$$\mathbf{U}^T \mathbf{V} = \mathbf{I}_3$$

where $\mathbf{I}_3$ is a 3×3 identity matrix.

The definition of the unit cell of a lattice is not unique. Here we define the Voronoi cell of a lattice as a unit cell. The Voronoi cell, depicted in Figure 3.10, is the set of all points that are closer to the origin than to any other sample point.

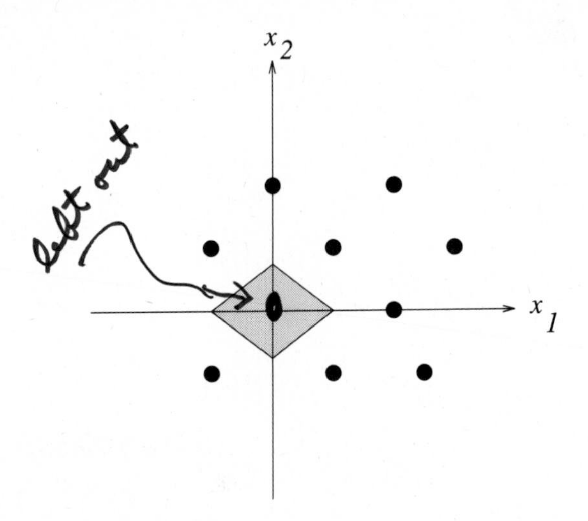

Figure 3.10: The Voronoi cell of a 2-D lattice.

The Fourier transform of a signal sampled on a lattice is a periodic function over $\mathbf{R}^3$ with periodicity lattice Λ^{3*},

$$S_p(\mathbf{F}) = S_p(\mathbf{F} + \mathbf{r}), \qquad \mathbf{r} \in \Lambda^{3*}$$

This follows from $\mathbf{r}^T \begin{bmatrix} \mathbf{x} \\ t \end{bmatrix}$ is an integer for $(\mathbf{x}, t) \in \Lambda^3$ and $\mathbf{r} \in \Lambda^{3*}$, by definition of the reciprocal lattice. Because of this periodicity, the Fourier transform need only be specified over one unit cell $\mathcal{P}$ of the reciprocal lattice Λ^{3*}. Thus, the inverse Fourier transform of $S_p(\mathbf{F})$ is given by

$$s_p(\mathbf{x}, t) = d(\Lambda^3) \int_{\mathcal{P}} S_p(\mathbf{F}) \ \exp\left\{ j2\pi \mathbf{F}^T \begin{bmatrix} \mathbf{x} \\ t \end{bmatrix} \right\} d\mathbf{F}, \quad (\mathbf{x}, t) \in \Lambda^3 \tag{3.32}$$

Note that in terms of the normalized frequency variables, $\mathbf{f} = \mathbf{V}^T\mathbf{F}$, we have

$$S(\mathbf{f}) = S_p(\mathbf{U}\mathbf{f}) = \sum_{(\mathbf{n},k)\in\mathbf{Z}^3} s(\mathbf{n}, k) \ \exp\left\{ -j2\pi \mathbf{f}^T \begin{bmatrix} \mathbf{n} \\ k \end{bmatrix} \right\} \tag{3.33}$$

where $\mathbf{U} = \mathbf{V}^{T^{-1}}$, with the fundamental period given by the unit cell $f_1 < |1/2|$, $f_2 < |1/2|$, and $f_t < |1/2|$.

3.4.3 Spectrum of Signals Sampled on a Lattice

In this section, we relate the Fourier transform $S_p(\mathbf{F})$ of the sampled signal to that of the continuous signal. Suppose that $s_c(\mathbf{x}, t) \in L^1(\mathbf{R}^3)$ has the Fourier transform

$$S_c(\mathbf{F}) = \int_{\mathbf{R}^3} s_c(\mathbf{x}, t) \exp\left\{ -j2\pi \mathbf{F}^T \begin{bmatrix} \mathbf{x} \\ t \end{bmatrix} \right\} d\mathbf{x} \, dt, \quad \mathbf{F} \in \mathbf{R}^3 \tag{3.34}$$

with the inverse transform

$$s_c(\mathbf{x}, t) = \int_{\mathbf{R}^3} S_c(\mathbf{F}) \exp\left\{ j2\pi \mathbf{F}^T \begin{bmatrix} \mathbf{x} \\ t \end{bmatrix} \right\} d\mathbf{F}, \quad (\mathbf{x}, t) \in \mathbf{R}^3 \tag{3.35}$$

We substitute (3.35) into (3.30), and express this integral as a sum of integrals over displaced versions of a unit cell $\mathcal{P}$ of Λ^{3*} to obtain

$$\begin{aligned} s_p(\mathbf{x}, t) &= \int_{\mathbf{R}^3} S_c(\mathbf{F}) \exp\left\{ j2\pi \mathbf{F}^T \begin{bmatrix} \mathbf{x} \\ t \end{bmatrix} \right\} d\mathbf{F}, \quad \mathbf{x} \in \Lambda^3 \\ &= \sum_{\mathbf{r}\in\Lambda^{3*}} \int_{\mathcal{P}} S_c(\mathbf{F} + \mathbf{r}) \exp\left\{ j2\pi (\mathbf{F} + \mathbf{r})^T \begin{bmatrix} \mathbf{x} \\ t \end{bmatrix} \right\} d\mathbf{F} \end{aligned}$$

Since $\exp(j2\pi \mathbf{r}^T \begin{bmatrix} \mathbf{x} \\ t \end{bmatrix}) = 1$ by the property of the reciprocal lattice, exchanging the order of summation and integration yields

$$s_p(\mathbf{x}, t) = \int_{\mathcal{P}} [\sum_{\mathbf{r}\in\Lambda^{3*}} S_c(\mathbf{F} + \mathbf{r})] \exp\left\{ j2\pi \mathbf{F}^T \begin{bmatrix} \mathbf{x} \\ t \end{bmatrix} \right\} d\mathbf{F}, \quad (\mathbf{x}, t) \in \Lambda^3 \tag{3.36}$$

Thus, we have

$$S_p(\mathbf{F}) = \frac{1}{d(\Lambda^3)} \sum_{\mathbf{r}\in\Lambda^{3*}} S_c(\mathbf{F} + \mathbf{r}) = \frac{1}{d(\Lambda^3)} \sum_{\mathbf{k}\in\mathbf{Z}^3} S_c(\mathbf{F} + \mathbf{U}\mathbf{k}) \tag{3.37}$$

or, alternatively,

$$S(\mathbf{f}) = S_p(\mathbf{F})|_{\mathbf{F}=\mathbf{U}\mathbf{f}} = \frac{1}{d(\Lambda^3)} \sum_{\mathbf{k} \in \mathbf{Z}^3} S_c(\mathbf{U}(\mathbf{f}+\mathbf{k})) \tag{3.38}$$

where $\mathbf{U}$ is the sampling matrix of the reciprocal lattice Λ^{3*}. As expected, the Fourier transform of the sampled signal is the sum of an infinite number of replicas of the Fourier transform of the continuous signal, shifted according to the reciprocal lattice Λ^{3*}.

Example

This example illustrates sampling of a continuous time-varying image, $s_c(\mathbf{x}, t)$, $(\mathbf{x}, t) \in \mathbf{R}^3$, using the progressive and the 2:1 line interlaced sampling lattices, shown in Figures 3.4 and 3.5 along with their sampling matrices $\mathbf{V}$, respectively.

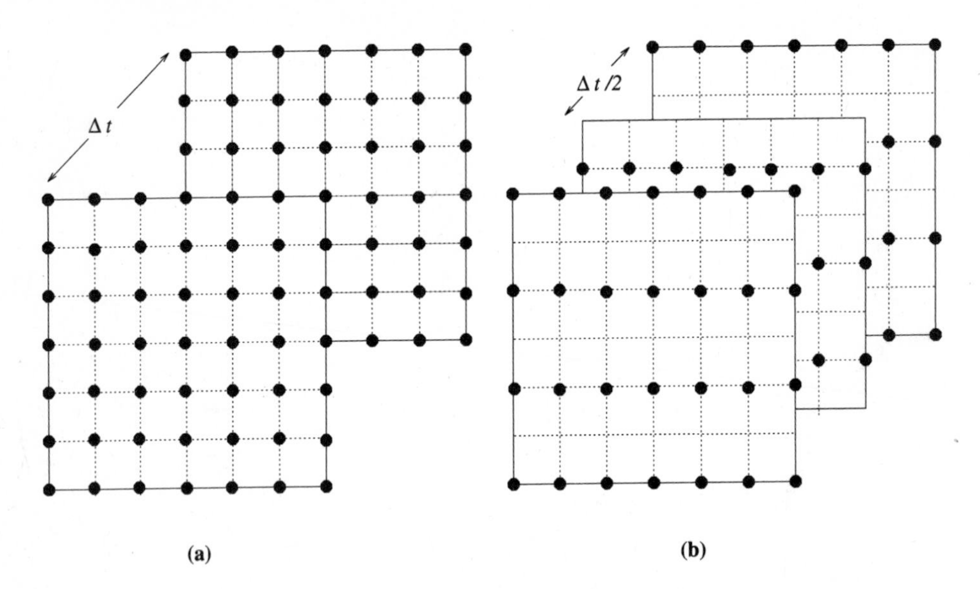

Figure 3.11: Sampling lattices for a) progressive and b) interlaced video.

The locations of the samples of $s_p(\mathbf{x}, t)$, $\begin{bmatrix} \mathbf{x} \\ t \end{bmatrix} = \mathbf{V} \begin{bmatrix} \mathbf{n} \\ k \end{bmatrix} \in \Lambda^3$, or $s(\mathbf{n}, k)$, $(\mathbf{n}, k) \in \mathbf{Z}^3$, for the cases of progressive and interlaced sampling are depicted in Figure 3.11 (a) and (b), respectively. Observe that the reciprocal of the sampling density $d(\Lambda^3) = \Delta x_1 \Delta x_2 \Delta t$ is identical for both lattices. However, the periodicity matrices of the spatio-temporal Fourier transform of the sampled video, indicating the locations of the replications, are different, and are given by

$$\mathbf{U} = \mathbf{V}^{-1^T} = \begin{bmatrix} \frac{1}{\Delta x_1} & 0 & 0 \\ 0 & \frac{1}{\Delta x_2} & 0 \\ 0 & 0 & \frac{1}{\Delta t} \end{bmatrix}$$

and

$$\mathbf{U} = \mathbf{V}^{-1^T} = \begin{bmatrix} \frac{1}{\Delta x_1} & 0 & 0 \\ 0 & \frac{1}{2\Delta x_2} & 0 \\ 0 & -\frac{1}{\Delta t} & \frac{2}{\Delta t} \end{bmatrix}$$

for the progressive and interlaced cases, respectively.

3.4.4 Other Sampling Structures

In general, the Fourier transform cannot be defined for sampling structures other than lattices; thus, the theory of sampling does not extend to arbitrary sampling structures [Dub 85]. However, an extension is possible for those sampling structures which can be expressed as unions of cosets of a lattice.

Unions of Cosets of a Lattice

Let Λ^3 and Γ^3 be two 3-D lattices. Λ^3 is a sublattice of Γ^3 if every site in Λ^3 is also a site of Γ^3. Then, $d(\Lambda^3)$ is an integer multiple of $d(\Gamma^3)$. The quotient $d(\Lambda^3)/d(\Gamma^3)$ is called the index of Λ^3 in Γ^3, and is denoted by $(\Lambda^3 : \Gamma^3)$. We note that if Λ^3 is a sublattice of Γ^3, then Γ^{3*} is a sublattice of Λ^{3*}.

The set

$$\mathbf{c} + \Lambda^3 = \left\{ \mathbf{c} + \begin{bmatrix} \mathbf{x} \\ t \end{bmatrix} \mid \begin{bmatrix} \mathbf{x} \\ t \end{bmatrix} \in \Lambda^3 \text{ and } \mathbf{c} \in \Gamma^3 \right\} \tag{3.39}$$

is called a coset of Λ^3 in Γ^3. Thus, a coset is a shifted version of the lattice Λ^3.

The most general form of a sampling structure Ψ^3 that we can analyze for spatio-temporal sampling is the union of P cosets of a sublattice Λ^3 in a lattice Γ^3, defined by

$$\Psi^3 = \bigcup_{i=1}^{P} (\mathbf{c}_i + \Lambda^3) \tag{3.40}$$

where $\mathbf{c}_1, \ldots, \mathbf{c}_P$ is a set of vectors in Γ^3 such that

$$\mathbf{c}_i - \mathbf{c}_j \notin \Lambda^3 \text{ for } i \neq j$$

An example of such a sampling structure is depicted in Figure 3.12. Note that Ψ^3 becomes a lattice if we take $\Lambda^3 = \Gamma^3$ and $P = 1$.

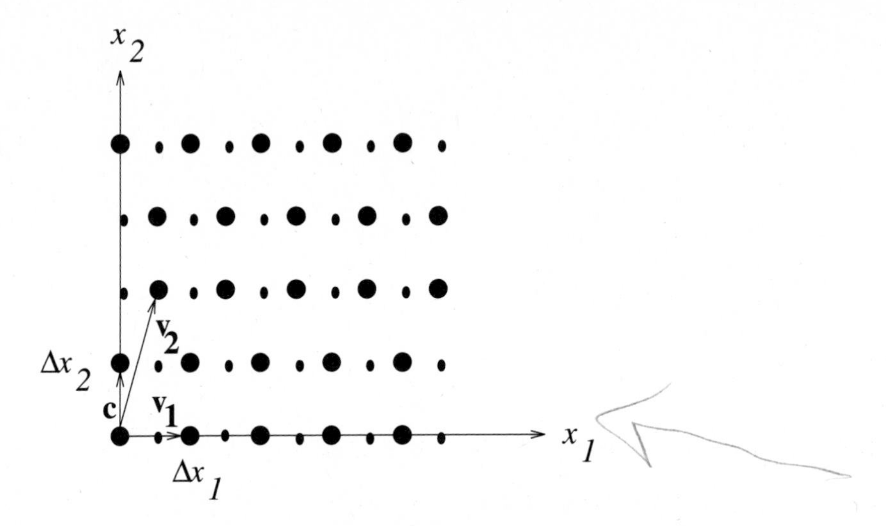

Figure 3.12: Union of cosets of a lattice [Dub 85] (©1985 IEEE).

Spectrum of Signals on Unions of Cosets of a Lattice

As stated, the Fourier transform is, in general, not defined for a sampling structure other than a lattice. However, for the special case of the union of cosets of a sublattice Λ^3 in Γ^3, we can assume that the signal is defined over the parent lattice Γ^3 with certain sample values set to zero. Then,

$$\begin{aligned} S_p(\mathbf{F}) &= \sum_{i=1}^{P} \sum_{\mathbf{x} \in \Lambda^3} s_p(\mathbf{c}_i + \begin{bmatrix} \mathbf{x} \\ t \end{bmatrix}) \exp\left\{-j2\pi\mathbf{F}^T(\mathbf{c}_i + \begin{bmatrix} \mathbf{x} \\ t \end{bmatrix})\right\} \\ &= \sum_{i=1}^{P} \exp\left\{-j2\pi\mathbf{F}^T\mathbf{c}_i\right\} \sum_{(\mathbf{x},t) \in \Lambda^3} s_p(\mathbf{c}_i + \begin{bmatrix} \mathbf{x} \\ t \end{bmatrix}) \exp\left\{-j2\pi\mathbf{F}^T \begin{bmatrix} \mathbf{x} \\ t \end{bmatrix}\right\} \end{aligned} \quad (3.41)$$

The periodicity of this Fourier transform is determined by the reciprocal lattice Γ^{3*}.

It can be shown that the spectrum of a signal sampled on a structure which is in the form of the union of cosets of a lattice is given by [Dub 85]

$$S_p(\mathbf{F}) = \frac{1}{d(\Lambda^3)} \sum_{\mathbf{r} \in \Lambda^{3*}} g(\mathbf{r}) S_c(\mathbf{F} + \mathbf{r}) \quad (3.42)$$

where $\mathbf{V}$ is the sampling matrix of Λ^3, and the function

$$g(\mathbf{r}) = \sum_{i=1}^{P} \exp(j2\pi\mathbf{r}^T\mathbf{c}_i) \quad (3.43)$$

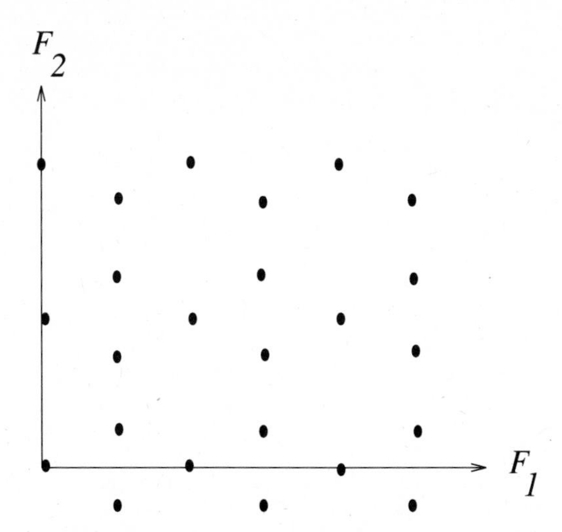

Figure 3.13: Reciprocal structure Ψ^{3*} [Dub 85] (©1985 IEEE).

is constant over cosets of Γ^{3*} in Λ^{3*}, and may be equal to zero for some of these cosets, so the corresponding shifted versions of the basic spectrum are not present.

> *Example* The line-quincunx structure shown in Figure 3.7, which occasionally finds use in practical systems, is in the form of a union of cosets of a lattice. A 2-D version of this lattice was illustrated in Figure 3.12 where $P = 2$. The reciprocal structure of the sampling structure Ψ^3 shown in Figure 3.12 is depicted in Fig. 3.13.

3.5 Reconstruction from Samples

Digital video is usually converted back to analog video for display purposes. Furthermore, various digital video systems have different spatio-temporal resolution requirements which necessitate sampling structure conversion. The sampling structure conversion problem, which is treated in the next chapter, can alternatively be posed as the reconstruction of the underlying continuous spatio-temporal video, followed by its resampling on the desired spatio-temporal lattice. Thus, in the remainder of this chapter, we address the theory of reconstruction of a continuous video signal from its samples.

3.5.1 Reconstruction from Rectangular Samples

Reconstruction of a continuous signal from its samples is an interpolation problem. In ideal bandlimited interpolation, the highest frequency that can be represented in the analog signal without aliasing, according to the Nyquist sampling theorem, is equal to one-half of the sampling frequency. Then, a continuous image, $s_r(x_1, x_2)$, can be reconstructed from its samples taken on a 2-D rectangular grid by ideal low-pass filtering as follows:

$$S_r(F_1, F_2) = \begin{cases} \Delta x_1 \Delta x_2 S(F_1 \Delta x_1, F_2 \Delta x_2) & \text{for } |F_1| < \frac{1}{2\Delta x_1} \text{ and } |F_2| < \frac{1}{2\Delta x_2} \\ 0 & \text{otherwise.} \end{cases} \quad (3.44)$$

The support of the ideal reconstruction filter in the frequency domain is illustrated in Figure 3.14.

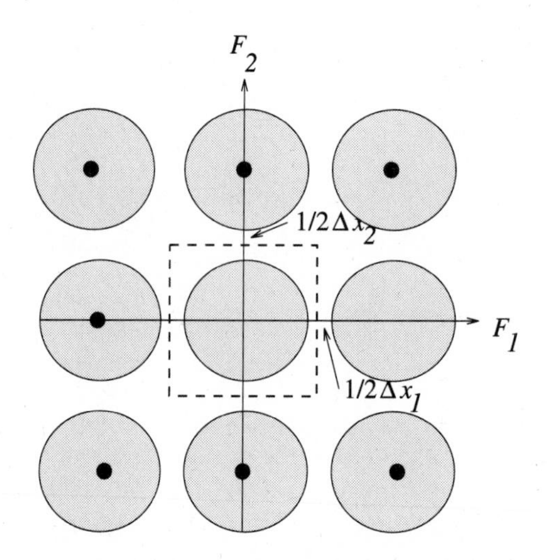

Figure 3.14: Reconstruction filter.

The reconstruction filter given by (3.44) is sometimes referred to as the "ideal bandlimited interpolation filter," since the reconstructed image would be identical to the original continuous image, that is

$$s_r(x_1, x_2) = s_c(x_1, x_2) \quad (3.45)$$

provided that the original continuous image was bandlimited, and Δx_1 and Δx_2 were chosen according to the Nyquist criterion.

The ideal bandlimited interpolation filtering can be expressed in the spatial domain by taking the inverse Fourier transform of both sides of (3.44)

$$s_r(x_1, x_2) = \int_{\frac{-1}{2\Delta x_1}}^{\frac{1}{2\Delta x_1}} \int_{\frac{-1}{2\Delta x_2}}^{\frac{1}{2\Delta x_2}} \Delta x_1 \Delta x_2 \; S(F_1 \Delta x_1, F_2 \Delta x_2) \; \exp\{j2\pi(F_1 x_1 + F_2 x_2)\} \; dF_1 dF_2$$

Substituting the definition of $S(F_1\Delta x_1, F_2\Delta x_2)$ into this expression, and rearranging the terms, we obtain

$$s_r(x_1,x_2)=\Delta x_1\Delta x_2\sum_{n_1}\sum_{n_2} s(n_1,n_2)\int_{\frac{-1}{2\Delta x_1}}^{\frac{1}{2\Delta x_1}}\int_{\frac{-1}{2\Delta x_2}}^{\frac{1}{2\Delta x_2}}\exp\{-j2\pi(F_1\Delta x_1 n_1 + F_2\Delta x_2 n_2)\}$$
$$\exp\{j2\pi(F_1x_1+F_2x_2)\}dF_1dF_2$$
$$= \Delta x_1\Delta x_2\sum_{n_1}\sum_{n_2} s(n_1,n_2)h(x_1-n_1\Delta x_1, x_2-n_2\Delta x_2) \qquad (3.46)$$

where $h(x_1, x_2)$ denotes the impulse response of the ideal interpolation filter for the case of rectangular sampling, given by

$$h(x_1,x_2) = \frac{\sin\left(\frac{\pi}{\Delta x_1}x_1\right)}{\frac{\pi}{\Delta x_1}x_1}\,\frac{\sin\left(\frac{\pi}{\Delta x_2}x_2\right)}{\frac{\pi}{\Delta x_2}x_2} \qquad (3.47)$$

3.5.2 Reconstruction from Samples on a Lattice

Similar to the case of rectangular sampling, the reconstructed time-varying image $s_r(\mathbf{x}, t)$ can be obtained through the ideal low-pass filtering operation

$$S_r(\mathbf{F}) = \begin{cases} |det\mathbf{V}|S(\mathbf{V}^T\mathbf{F}) & \text{for } \mathbf{F}\in\mathcal{P} \\ 0 & \text{otherwise.}\end{cases}$$

Here, the passband of the ideal low-pass filter is determined by the unit cell $\mathcal{P}$ of the reciprocal sampling lattice.

Taking the inverse Fourier transform, we have the reconstructed time-varying image

$$s_r(\mathbf{x},t) = \sum_{(\mathbf{n},k)\in\mathbf{Z}^3} s(\mathbf{n},k)\; h(\begin{bmatrix}\mathbf{x}\\ t\end{bmatrix} - \mathbf{V}\begin{bmatrix}\mathbf{n}\\ k\end{bmatrix})$$
$$= \sum_{(\mathbf{z},\tau)\in\Lambda^3} s_p(\mathbf{z},\tau)\; h(\begin{bmatrix}\mathbf{x}\\ t\end{bmatrix} - \begin{bmatrix}\mathbf{z}\\ \tau\end{bmatrix}) \qquad (3.48)$$

where

$$h(\mathbf{x},t) = |det\mathbf{V}|\int_{\mathcal{P}}\exp\left\{j2\pi\mathbf{F}^T\begin{bmatrix}\mathbf{x}\\ t\end{bmatrix}\right\}d\mathbf{F} \qquad (3.49)$$

is the impulse response of the ideal bandlimited spatio-temporal interpolation filter for the sampling structure used. Unlike the case of rectangular sampling, this integral, in general, cannot be reduced to a simple closed-form expression. As expected, exact reconstruction of a continuous signal from its samples on a lattice Λ^3 is possible if the signal spectrum is confined to a unit cell $\mathcal{P}$ of the reciprocal lattice.

3.6 Exercises

1. Derive (3.11) using (3.9).

2. The expression (3.11) assumes impulse sampling; that is, the sampling aperture has no physical size. A practical camera has a finite aperture modeled by the impulse response $h_a(x_1, x_2)$. How would you incorporate the effect of the finite aperture size into (3.11)?

3. Suppose a camera samples with 20-micron intervals in both the horizontal and vertical directions. What is the highest spatial frequency in the sampled image that can be represented with less than 3 dB attenuation if a) $h_a(x_1, x_2)$ is a 2-D Dirac delta function, and b) $h_a(x_1, x_2)$ is a uniform circle with diameter 20 microns?

4. Strictly speaking, images with sharp spatial edges are not bandlimited. Discuss how you would digitize an image that is not bandlimited.

5. Find the locations of the spectral replications for each of the 3-D sampling lattices depicted in Figures 3.4 through 3.7.

6. Evaluate the impulse response (3.49) if $\mathcal{P}$ is the unit sphere.

Bibliography

[Dub 85] E. Dubois, "The sampling and reconstruction of time-varying imagery with application in video systems," *Proc. IEEE*, vol. 73, no. 4, pp. 502–522, Apr. 1985.

[Dud 84] D. E. Dudgeon and R. M. Mersereau, *Multidimensional Digital Signal Processing*, Englewood Cliffs, NJ: Prentice Hall, 1984.

[Jai 89] A. K. Jain, *Fundamentals of Digital Image Processing*, Englewood Cliffs, NJ: Prentice Hall, 1989.

[Lim 90] J. S. Lim, *Two-Dimensional Signal and Image Processing*, Englewood Cliffs, NJ: Prentice Hall, 1990.

[Opp 89] A. V. Oppenheim and R. W. Schafer, *Discrete-Time Signal Processing*, Englewood Cliffs, NJ: Prentice Hall, 1989.

[Rah 92] M. A. Rahgozar and J. P. Allebach, "A general theory of time-sequential sampling," *Signal Proc.*, vol. 28, no. 3, pp. 253–270, Sep. 1992.

Chapter 4

SAMPLING STRUCTURE CONVERSION

Various digital video systems, ranging from all-digital high-definition TV to videophone, have different spatio-temporal resolution requirements leading to the emergence of different format standards to store, transmit, and display digital video. The task of converting digital video from one format to another is referred to as the standards conversion problem. Effective standards conversion methods enable exchange of information among various digital video systems, employing different format standards, to ensure their interoperability.

Standards conversion is a 3-D sampling structure conversion problem, that is, a spatio-temporal interpolation/decimation problem. This chapter treats sampling structure conversion as a multidimensional digital signal processing problem, including the characterization of sampling structure conversion in the 3-D frequency domain and filter design for sampling structure conversion, without attempting to take advantage of the temporal redundancy present in the video signals. The theory of motion-compensated filtering and practical standards conversion algorithms specifically designed for digital video, which implicitly or explicitly use interframe motion information, will be presented in Chapters 13 and 16, respectively.

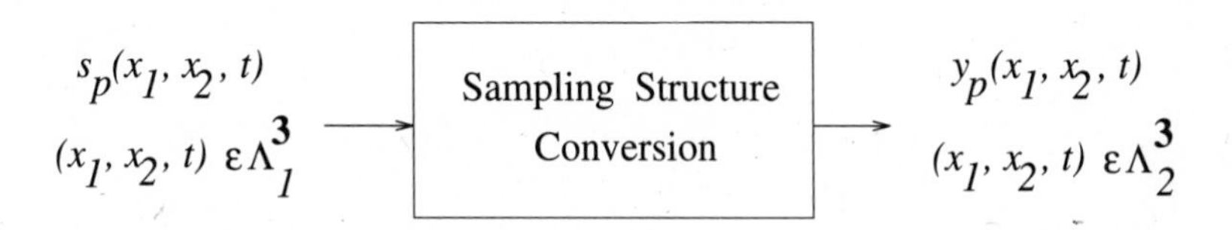

Figure 4.1: Block diagram for sampling structure conversion.

Sampling structure conversion can be treated in two steps: reconstruction of the underlying continuous spatio-temporal signal, followed by its resampling on the desired 3-D sampling structure. An alternative all-digital formulation of the problem, which is equivalent to the two-step approach, is provided in this chapter. The block diagram of the all-digital sampling structure conversion system, which takes a 3-D input signal $s(\mathbf{x}, t)$ sampled on Λ_1^3 and outputs $y(\mathbf{x}, t)$ sampled on Λ_2^3, where Λ_1^3 and Λ_2^3 denote the 3-D input and output sampling structures, respectively, is shown in Figure 4.1. Section 4.1 reviews the 1-D counterpart of the sampling structure conversion problem, the sampling rate change problem. The general sampling structure conversion from a 3-D input lattice to a 3-D output lattice is treated in Section 4.2.

4.1 Sampling Rate Change for 1-D Signals

In 1-D signal processing, interpolation and decimation operations are effectively used to match the sampling rate of the signal with the bandwidth requirements of the processing. This is known as multirate digital signal processing [Cro 83]. In sampling rate change problems, interpolation refers to the process of upsampling followed by appropriate filtering, while decimation refers to downsampling. In the following, we first discuss interpolation and then decimation by an integer factor. Next, sampling rate change by a rational factor is presented.

4.1.1 Interpolation of 1-D Signals

The interpolation process can be analyzed in two steps: (i) upsampling by zero filling (usually referred to as "filling-in"), and (ii) low-pass filtering of the "filled-in" signal. Here, we first discuss upsampling by zero filling and characterize the frequency spectrum of the "filled-in" signal.

Given a signal $s(n)$, we define a signal $u(n)$ that is upsampled by L as

$$u(n) = \begin{cases} s(\frac{n}{L}) & \text{for} \quad n = 0, \pm L, \pm 2L, \ldots \\ 0 & \text{otherwise.} \end{cases} \tag{4.1}$$

The process of upsampling by zero filling is demonstrated in Figure 4.2 for the case $L = 3$.

To study the frequency spectrum of the "filled-in" signal, we take the Fourier transform of (4.1). Using the definition of the Fourier transform for discrete-time signals [Opp 89]

$$\begin{aligned} U(f) &= \sum_{n=-\infty}^{\infty} u(n) \exp\{-j2\pi f n\} \\ &= \sum_{n=-\infty}^{\infty} s(n) \exp\{-j2\pi f L n\} = S(fL) \end{aligned} \tag{4.2}$$

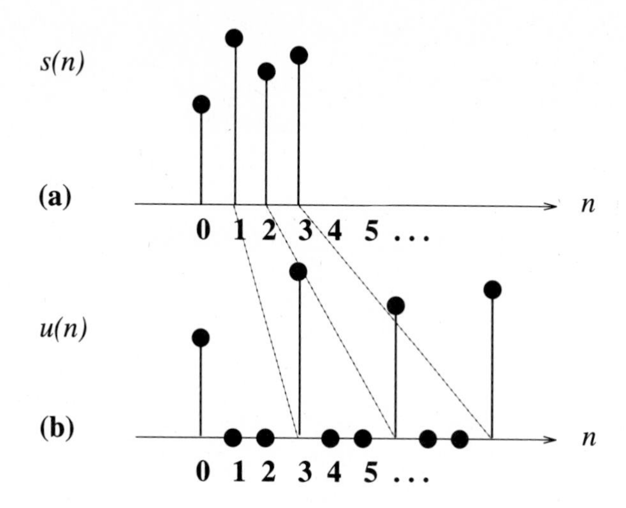

Figure 4.2: Upsampling by $L = 3$: a) input signal; b) zero-filled signal.

It can be seen from (4.2) that the spectrum of the filled in signal is related to the spectrum of the input signal by a compression of the frequency axis. This is illustrated in Figure 4.3, again for the case of $L = 3$. Note that the spectrum of the input $S(f)$ is assumed to occupy the full bandwidth, which is the interval $(-1/2, 1/2)$ since the Fourier transform of discrete signals are periodic with period equal to 1.

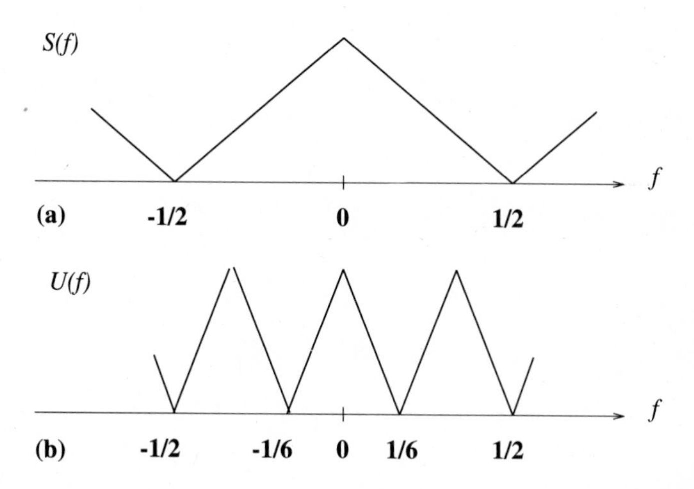

Figure 4.3: Spectra of the a) input and b) upsampled signals for $L = 3$.

Interpolation filtering aims at eliminating the replications caused by the zero-filling process. This requires low-pass filtering of the filled-in signal. In the time domain, the filtering operation can be viewed as replacing the zero samples with nonzero values by means of a smoothing operation.

Ideal Interpolation

Ideal interpolation can be achieved by ideal low-pass filtering. The ideal low-pass filter would have a DC gain factor of L, and its cutoff frequency would be $f_c = 1/2L$. The frequency response of the ideal interpolation filter is shown by the dotted lines in Figure 4.4.

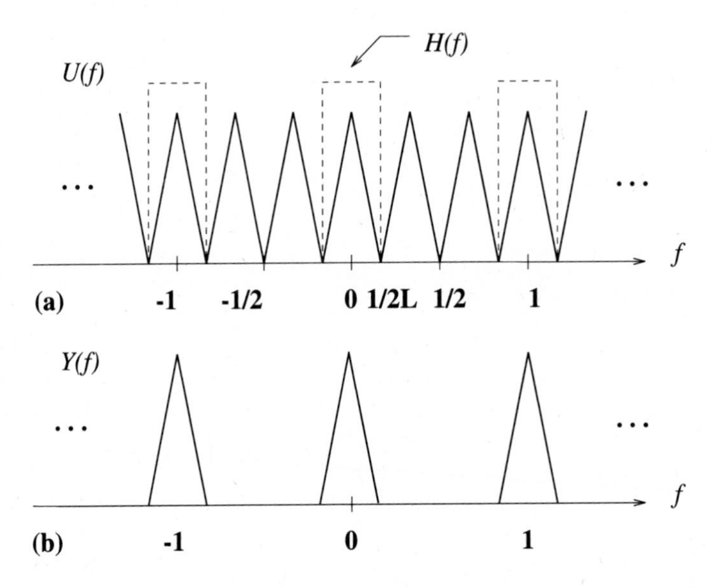

Figure 4.4: Ideal interpolation filter ($L = 3$): Spectra a) before and b) after filtering.

The impulse response of the ideal interpolation filter is a sinc function given by

$$h(n) = \frac{sin(\pi n/L)}{\pi n/L} \tag{4.3}$$

Thus, the interpolated signal samples are given by

$$y(n) = \sum_{k=-\infty}^{\infty} s(k) \frac{sin(\pi(n-k)/L)}{\pi(n-k)/L} \tag{4.4}$$

The impulse response of the ideal interpolation filter has the property that $h(0) = 1$, and $h(n) = 0$, for $n = \pm L, \pm 2L, \ldots$. Because of these zero-crossings, $y(n) = s(n)$

at the existing sample values, while assigning nonzero values for the zero samples in the upsampled signal.

Practical Interpolation Filters

The ideal interpolation filter is unrealizable since it is an infinite impulse response, noncausal filter; that is, its implementation requires infinite time delay. Thus, it is usually replaced by one of the following realizable approximations:

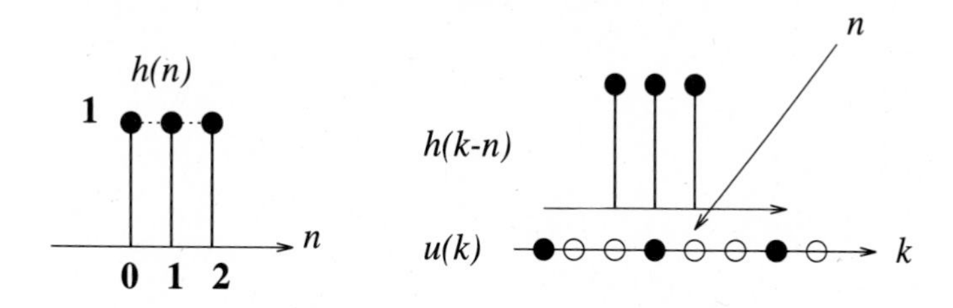

Figure 4.5: The zero-order hold filter for $L = 3$.

- Zero-order hold filter

Zero-order hold is the simplest interpolation filter. It corresponds to pixel replication. The impulse response and the implementation of the zero-order hold filter are shown in Figure 4.5. Note that the zero-order hold filter is a poor interpolation filter, since its frequency response is given by a sinc function, which has large sidelobes. These sidelobes cause leakage of energy from the replications in the frequency domain, which results in high-frequency artifacts in the interpolated signal.

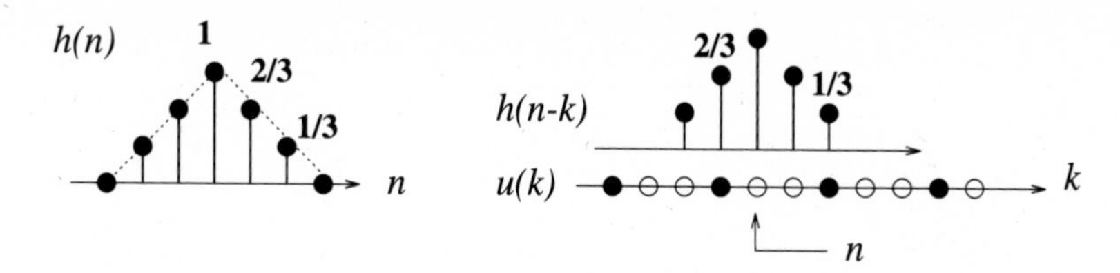

Figure 4.6: The linear interpolation filter for $L = 3$.

- Linear interpolation filter

Linear interpolation is achieved by a weighted sum of the neighboring pixels. The weights are inversely proportional to the distance of the pixel, to be interpolated, from its neighbors. The impulse response of the linear interpolation filter is shown in Figure 4.6 for $L = 3$. The frequency response of the linear interpolation filter is equal to the square of the sinc function. Thus, it has lower sidelobes than the zero-order hold filter.

- Cubic spline interpolation filter

The cubic spline interpolation filter approximates the impulse response of the ideal low-pass filter (4.3) by a piecewise cubic polynomial consisting of three cubics. A procedure for designing FIR cubic-spline interpolation filters is given by Keys [Key 81] (see Exercise 2). The impulse response of the cubic-spline filter, which has $4L + 1$ taps, is demonstrated in Figure 4.7. This filter also has the sample preserving property of the ideal low-pass filter. We note that the cubic spline filter approximates the unrealizable ideal low-pass filter better than a truncated sinc filter of the same length, because the frequency response of the truncated sinc filter suffers from ringing effects due to the well-known Gibbs phenomenon [Opp 89].

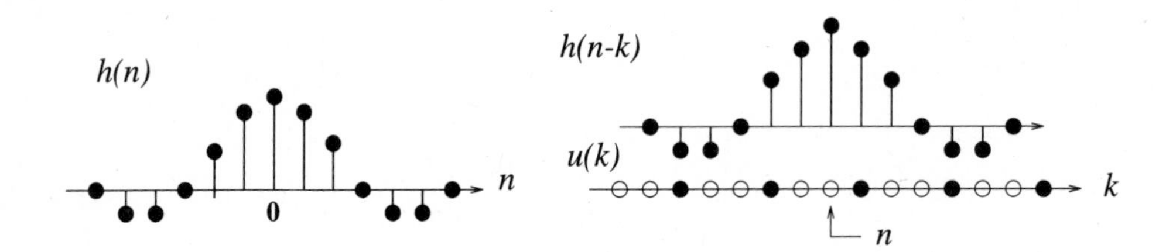

Figure 4.7: The cubic-spline filter for $L = 3$.

4.1.2 Decimation of 1-D Signals

The process of decimation by a factor of M can also be modeled in two steps: first, multiplication by an impulse train to replace $M - 1$ samples in between every Mth sample with zeros and then, discarding the zero samples to obtain a signal at the lower rate. We describe these steps in the following.

Given the input signal $s(n)$, define an intermediate signal $w(n)$ by

$$w(n) = s(n) \sum_{k=-\infty}^{\infty} \delta(n - kM) \tag{4.5}$$

Then, the signal decimated by a factor of M can be expressed as

$$y(n) = w(Mn) \tag{4.6}$$

The process of decimation is illustrated in Figure 4.8 for $M = 2$. The intermediate signal $w(n)$ is introduced to faciliate the frequency domain characterization of decimation.

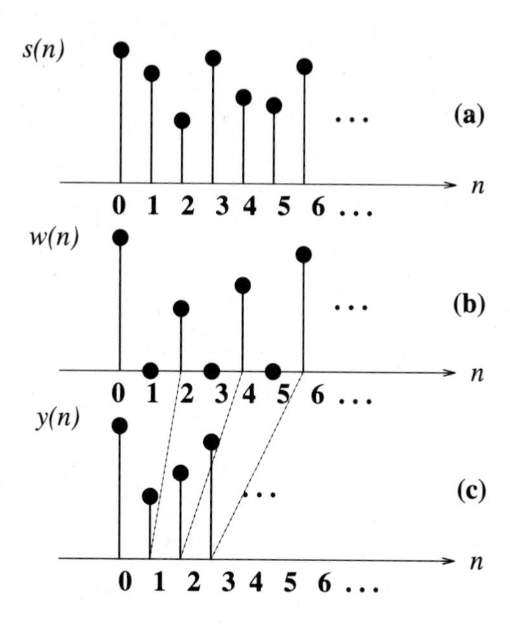

Figure 4.8: Decimation by $M = 2$: a) input signal, b) intermediate signal, and c) downsampled signal.

To analyze the spectrum of the decimated signal, we first compute the Fourier transform of the intermediate signal as

$$W(f) = \frac{1}{M} \sum_{k=0}^{M-1} S(f - \frac{k}{M}) \tag{4.7}$$

It can easily be seen that the spectrum of the intermediate signal consists of replications of the input signal spectrum. There are M replicas in the interval $(-1/2, 1/2)$ as shown in Figure 4.9. Then the spectrum of the decimated signal is given by

$$Y(f) = \sum_{n=-\infty}^{\infty} w(Mn) \exp\{-j2\pi fn\} = W(\frac{f}{M}) \tag{4.8}$$

which results in an expansion of the frequency axis for the spectrum of the intermediate signal. Note that in Figure 4.9 the bandwidth of the input signal is $1/4$; thus, decimation by two yields a full bandwidth signal.

If the bandwidth of the input signal is more than $1/(2M)$, then the replications will overlap and the decimated signal will suffer from aliasing. This is as expected, because the sampling rate of the decimated signal should not be allowed to fall below the Nyquist sampling rate. If an application mandates going below the Nyquist rate, then appropriate anti-alias filtering should be applied prior to the decimation. Ideal anti-alias filtering can be achieved by ideal low-pass filtering, as shown in Figure 4.10. For decimation by M, the cutoff frequency of the ideal anti-alias filter is $1/(2M)$.

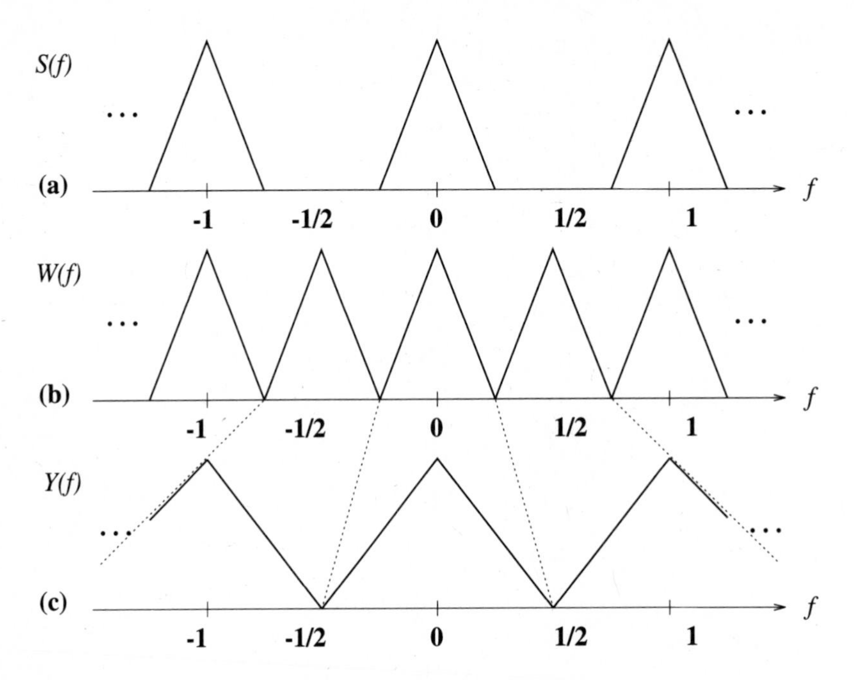

Figure 4.9: Spectra of the a) input signal, b) intermediate signal, and c) downsampled signal ($M = 2$).

The simplest realizable anti-alias filter is a box filter. Box filtering corresponds to averaging all pixels within a local window. Although they are a poor approximation of the ideal low-pass filter, box filters are usually preferred because of their simplicity.

4.1.3 Sampling Rate Change by a Rational Factor

The theory of decimation and interpolation (by an integer factor) extends to sampling rate change by a rational factor in a straightforward fashion. Sampling rate change by a rational factor, L/M, can be achieved by first interpolating the signal up by a factor of L and then decimating the result by a factor of M. The interpolation filter and the anti-alias filter of decimation can be merged into a single low-pass filter with the cutoff frequency $f_c = \min\{1/(2M), 1/(2L)\}$. A block diagram of a system for sampling rate change by a factor of L/M is depicted in Figure 4.11.

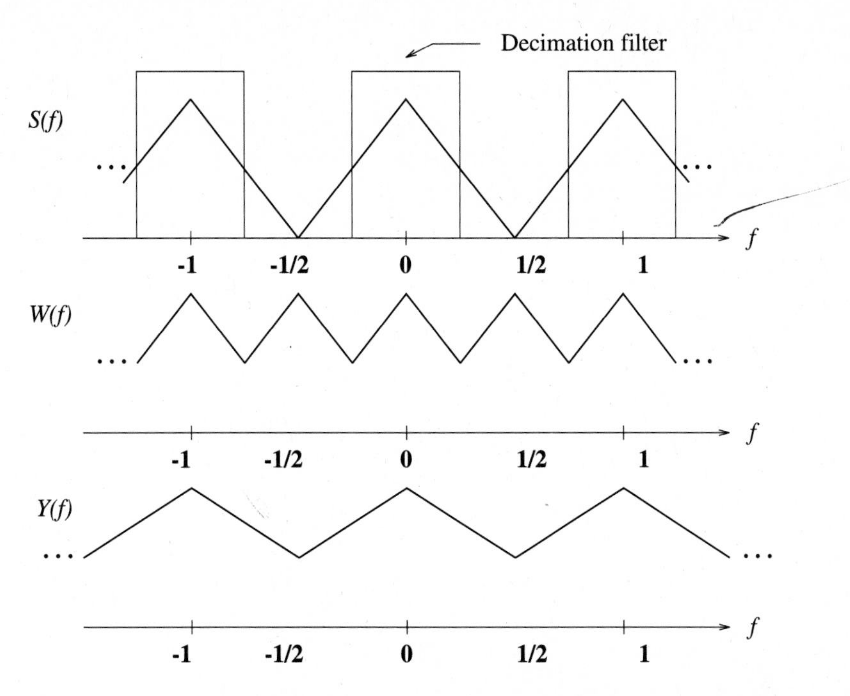

Figure 4.10: Anti-alias filtering for $M = 2$: Spectra of the a) input signal, b) intermediate signal, and c) downsampled signal with anti-alias filtering.

If $M > L$, then the sampling rate change problem is effectively a decimation problem; otherwise, it is an interpolation problem. In any case, the constraint that the values of the existing samples must be preserved needs to be incorporated into the low-pass filter design using one of the suggested interpolation filter design procedures. If a linear interpolation filter is to be used, the system given in Figure 4.11 can be simplified into a single filter which does not first require interpolation by a factor of L. This simplification may result in significant computational savings, especially for factors such as 9/2, where the numerator is large.

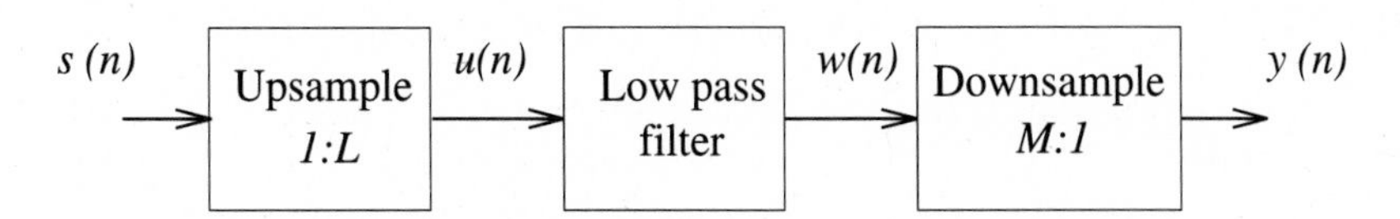

Figure 4.11: Sampling rate change by a factor of L/M.

4.2 Sampling Lattice Conversion

In this section, the theory of sampling rate change is extended to conversion from one 3-D sampling lattice to another. A system for sampling lattice conversion can be decomposed, as shown in Figure 4.12, where the input video is first upconverted to the lattice $\Lambda_1^3+\Lambda_2^3$. After appropriate filtering, it is downconverted to the output lattice Λ_2^3. In the following, we describe the components of this system in detail.

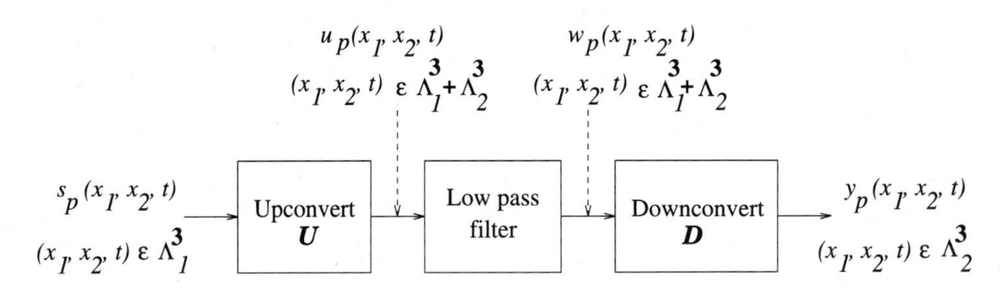

Figure 4.12: Decomposition of the system for sampling structure conversion.

We define the sum of two lattices as

$$\Lambda_1^3+\Lambda_2^3=\left\{\left[\begin{array}{c}\mathbf{x}_1\\ t_1\end{array}\right]+\left[\begin{array}{c}\mathbf{x}_2\\ t_2\end{array}\right] \mid (\mathbf{x}_1,t_1)\in\Lambda_1^3 \text{ and } (\mathbf{x}_2,t_2)\in\Lambda_2^3\right\} \tag{4.9}$$

Thus, the sum of two lattices can be found by adding each point in one lattice to every point of the other lattice. The intersection of two lattices is given by

$$\Lambda_1^3\bigcap\Lambda_2^3=\{(\mathbf{x},t)\mid(\mathbf{x},t)\in\Lambda_1^3 \text{ and } (\mathbf{x},t)\in\Lambda_2^3\} \tag{4.10}$$

The intersection $\Lambda_1^3\bigcap\Lambda_2^3$ is the largest lattice that is a sublattice of both Λ_1^3 and Λ_2^3, while the sum $\Lambda_1^3+\Lambda_2^3$ is the smallest lattice that contains both Λ_1^3 and Λ_2^3 as sublattices.

The upconversion from Λ_1^3 to $\Lambda_1^3+\Lambda_2^3$ is defined as follows:

$$u_p(\mathbf{x},t)=\mathcal{U}s_p(\mathbf{x},t)=\left\{\begin{array}{ll} s_p(\mathbf{x},t) & (\mathbf{x},t)\in\Lambda_1^3 \\ 0 & (\mathbf{x},t)\notin\Lambda_1^3,\quad(\mathbf{x},t)\in\Lambda_1^3+\Lambda_2^3\end{array}\right. \tag{4.11}$$

and downconversion from $\Lambda_1^3+\Lambda_2^3$ to Λ_2^3 as:

$$y_p(\mathbf{x},t)=\mathcal{D}w_p(\mathbf{x},t)=w_p(\mathbf{x},t),\quad(\mathbf{x},t)\in\Lambda_2^3 \tag{4.12}$$

The interpolation filter (anti-alias filter in the case of downsampling) is applied on the lattice $\Lambda_1^3+\Lambda_2^3$. By definition, this filter will be shift-invariant if the output is shifted by the vector $\mathbf{p}$ when the input is shifted by $\mathbf{p}$. Thus, we need $\mathbf{p}\in\Lambda_1^3\bigcap\Lambda_2^3$.

This condition is satisfied if $\Lambda_1^3 \bigcap \Lambda_2^3$ is a lattice, that is, $\mathbf{V}_1^{-1}\mathbf{V}_2$ is a matrix of integers. This requirement is the counterpart of L/M to be a rational number in the 1-D sampling rate change problems.

The linear, shift-invariant filtering operation on $\Lambda_1^3 + \Lambda_2^3$ can be expressed as

$$w_p(\mathbf{x}, t) = \sum_{(\mathbf{z},\tau)\in\Lambda_1^3+\Lambda_2^3} u_p(\mathbf{z}, \tau)\, h_p(\begin{bmatrix} \mathbf{x} \\ t \end{bmatrix} - \begin{bmatrix} \mathbf{z} \\ \tau \end{bmatrix}), \quad (\mathbf{x}, t) \in \Lambda_1^3 + \Lambda_2^3 \tag{4.13}$$

However, by the definition of the upsampling operator, $u_p(\mathbf{x}, t) = s_p(\mathbf{x}, t)$ for $(\mathbf{x}, t) \in \Lambda_1^3$ and zero otherwise; thus,

$$w_p(\mathbf{x}, t) = \sum_{(\mathbf{z},\tau)\in\Lambda_1^3} s_p(\mathbf{z}, \tau)\, h_p(\begin{bmatrix} \mathbf{x} \\ t \end{bmatrix} - \begin{bmatrix} \mathbf{z} \\ \tau \end{bmatrix}), \quad (\mathbf{x}, t) \in \Lambda_1^3 + \Lambda_2^3 \tag{4.14}$$

After the downsampling operation,

$$y_p(\mathbf{x}, t) = \sum_{(\mathbf{z},\tau)\in\Lambda_1^3} s_p(\mathbf{z}, \tau)\, h_p(\begin{bmatrix} \mathbf{x} \\ t \end{bmatrix} - \begin{bmatrix} \mathbf{z} \\ \tau \end{bmatrix}), \quad (\mathbf{x}, t) \in \Lambda_2^3 \tag{4.15}$$

The frequency response of the filter is periodic with the main period determined by the unit cell of $(\Lambda_1^3 + \Lambda_2^3)^*$. In order to avoid aliasing, the passband of the interpolation/anti-alias (low-pass) filter is restricted to the smaller of the Voronoi cells of Λ_1^{3*} and Λ_2^{3*}. Sampling lattice conversion is illustrated by means of the following examples.

Example: Conversion from Λ_1^2 to Λ_2^2 [Dub 85]

We consider a 2-D sampling lattice conversion from the input lattice Λ_1^2 to the output lattice Λ_2^2, which are shown in Figure 4.13 along with their sampling matrices, V_1 and V_2. The sampling densities of Λ_1^2 and Λ_2^2 are inversely proportional to the determinants of V_1 and V_2, given by

$$d(\Lambda_1^2) = 2\Delta x_1 \Delta x_2 \quad \text{and} \quad d(\Lambda_2^2) = 4\Delta x_1 \Delta x_2$$

Also shown in Figure 4.13 are the lattices $\Lambda_1^2 + \Lambda_2^2$ and $\Lambda_1^2 \bigcap \Lambda_2^2$, which are used to define the sampling conversion factor

$$Q = (\Lambda_1^2 + \Lambda_2^2 : \Lambda_1^2) = (\Lambda_2^2 : \Lambda_1^2 \bigcap \Lambda_2^2) = 2$$

Because we have a downconversion problem, by a factor of 2, anti-alias filtering must be performed on the lattice $\Lambda_1^2 + \Lambda_2^2$. The fundamental period of the filter frequency response is given by the unit cell of $(\Lambda_1^2 + \Lambda_2^2)^*$, which is indicated by the dotted lines in Figure 4.14. In order to avoid aliasing, the passband of the low-pass filter must be restricted to the Voronoi cell of Λ_2^{2*} which has an hexagonal shape.

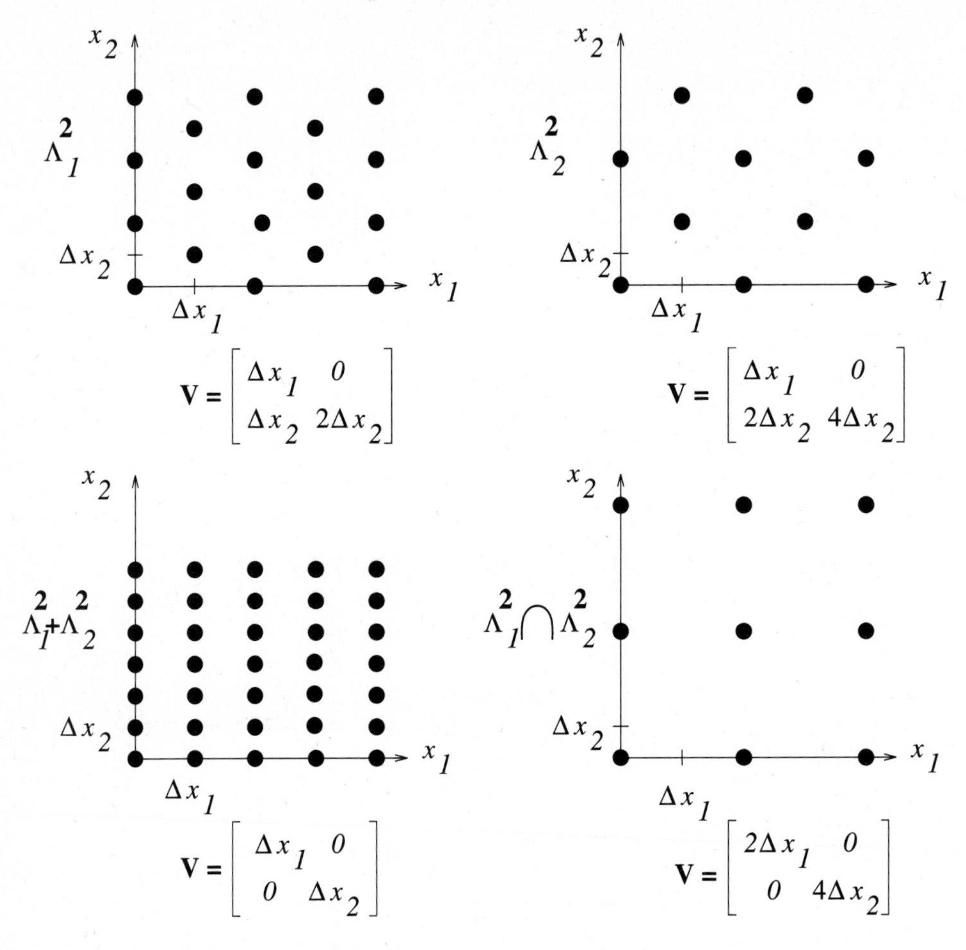

Figure 4.13: The lattices Λ_1^2, Λ_2^2, $\Lambda_1^2 + \Lambda_2^2$, and $\Lambda_1^2 \bigcap \Lambda_2^2$ [Dub 85] (©1985 IEEE).

Example: De-interlacing

De-interlacing refers to conversion from an interlaced sampling grid (input) to a progressive grid (output), shown in Figure 4.15 (a) and (b), respectively. The sampling matrices for the input and output grids are

$$\mathbf{V}_{in} = \begin{bmatrix} \Delta x_1 & 0 & 0 \\ 0 & 2\Delta x_2 & \Delta x_2 \\ 0 & 0 & \Delta t \end{bmatrix}$$

and

$$\mathbf{V}_{out} = \begin{bmatrix} \Delta x_1 & 0 & 0 \\ 0 & \Delta x_2 & 0 \\ 0 & 0 & \Delta t \end{bmatrix}$$

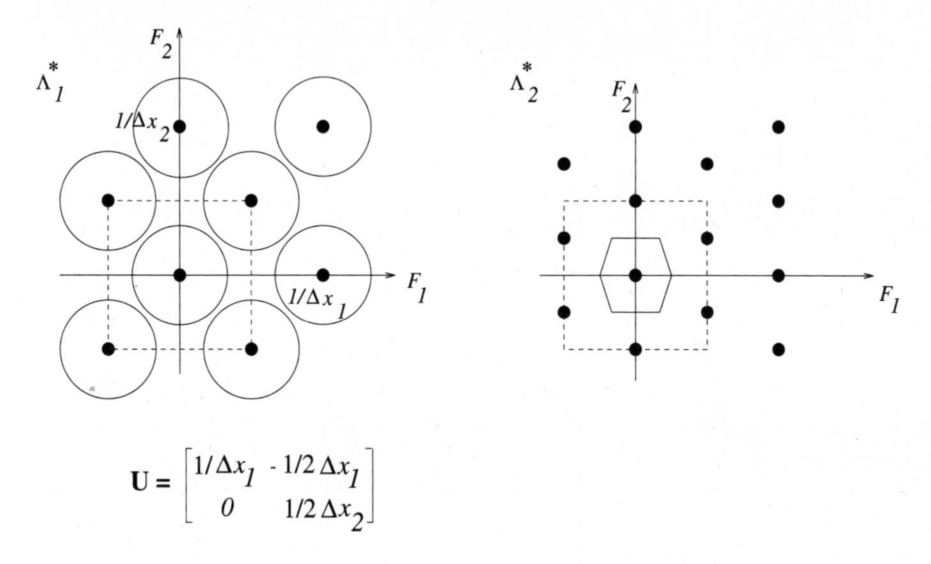

Figure 4.14: The spectrum of $s(\mathbf{x})$ with the periodicity matrix $\mathbf{U}$, and the support of the frequency response of the filter [Dub 85] (©1985 IEEE).

respectively. Note that $|det\mathbf{V}_{in}| = 2|det\mathbf{V}_{out}|$; hence, we have a spatio-temporal interpolation problem by a factor of 2. The interpolation can be achieved by zero-filling followed by ideal low-pass filtering. The pass-

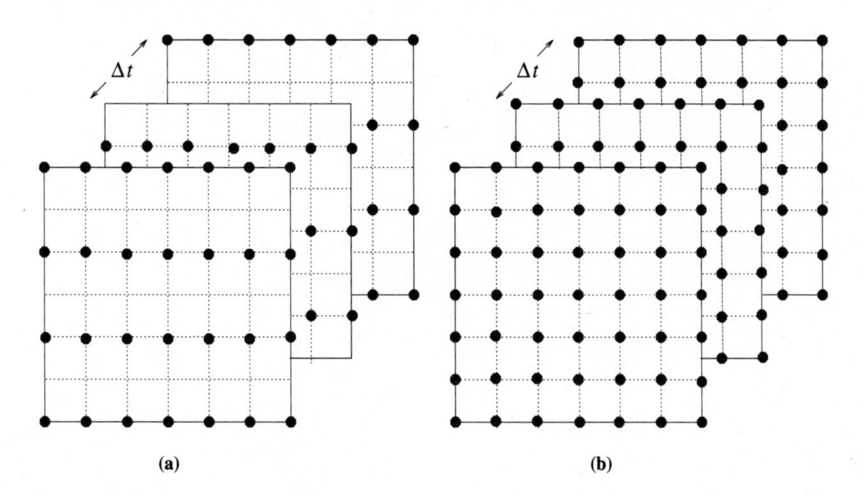

Figure 4.15: Interlaced to progressive conversion: a) input; b) output lattices.

band of the ideal low-pass filter should be restricted to the unit cell of the reciprocal lattice of the output sampling lattice, which is given by

$$\left(-\frac{1}{2\Delta x_1}, \frac{1}{2\Delta x_1}\right) \times \left(-\frac{1}{2\Delta x_2}, \frac{1}{2\Delta x_2}\right) \times \left(-\frac{1}{2\Delta t}, \frac{1}{2\Delta t}\right)$$

In spatio-temporal sampling structure down-conversion without motion compensation, there is a tradeoff between allowed aliasing errors and loss of resolution (blurring) due to anti-alias (low-pass) filtering prior to down-conversion. When anti-alias filtering has been used prior to down-conversion, the resolution that is lost cannot be recovered by subsequent interpolation. In Chapters 13 and 16, we present methods to incorporate interframe motion information to sampling structure conversion problems by means of motion compensation. Motion-compensated interpolation makes it possible to recover full-resolution frames by up-conversion of previously down converted frames if no anti-alias filtering has been applied in the down-conversion process.

4.3 Exercises

1. Find the frequency response of the zero-order hold and the linear interpolation filters shown in Figures 4.5 and 4.6, respectively. Compare them with that of the ideal (band-limited) interpolation filter.

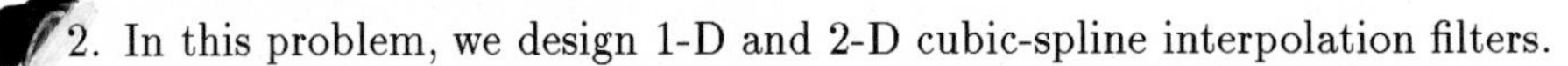

2. In this problem, we design 1-D and 2-D cubic-spline interpolation filters.

   a) We start by designing a continuous-time cubic spline reconstruction filter that approximates the ideal low-pass filter with the cutoff frequency $f_c = 1/2$. The impulse response $h(t)$ (shown below) of a filter that uses four neighboring samples at any time t can be expressed as

$$h(t) = \begin{cases} a_3|t|^3 + a_2|t|^2 + a_1|t| + a_0 & 0 \le |t| < 1 \\ b_3|t|^3 + b_2|t|^2 + b_1|t| + b_0 & 1 \le |t| < 2 \\ 0 & 2 \le |t| \end{cases} \tag{4.16}$$

   The design criteria are: $h(t) = 0$ for all integer t (i.e., original sample points), except $t = 0$ where $h(0) = 1$; and the slope of the impulse response $\frac{dh(t)}{dt}$ must be continuous across original sample points.

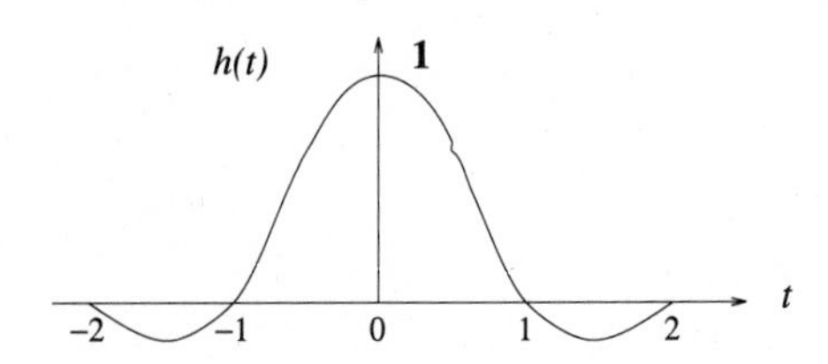

Show that the design criteria are met if the 8 unknown coefficients a_i, and b_i, $i = 0, \ldots, 3$ satisfy the following 7 equations:

$$a_0 = 1, \qquad a_1 = 0$$
$$a_3 + a_2 = -1, \qquad b_3 + b_2 + b_1 + b_0 = 0$$
$$8b_3 + 4b_2 + 2b_1 + b_0 = 0, \qquad 12b_3 + 4b_2 + b_1 = 0$$
$$3a_3 + 2a_2 = 3b_3 + 2b_2 + b_1$$

b) Let b_3 be a free parameter, and express all other parameters in terms of b_3. Determine a range of values for b_3 such that

$$\frac{d^2h(t)}{dt^2}|_{t=0} < 0 \quad \text{and} \quad \frac{d^2h(t)}{dt^2}|_{t=1} > 0$$

c) A digital cubic spline filter can be obtained by sampling $h(t)$. Show that for interpolation by a factor of L, the impulse response of the cubic-spline filter has $4L + 1$ taps. Determine the impulse response for $L = 2$ and $b_3 = -1$.

d) Design a 2-D separable cubic spline interpolation filter with the same parameters.

Bibliography

[Cro 83] R. E. Crochiere and L. R. Rabiner, *Multirate Digital Signal Processing*, Englewood Cliffs, NJ: Prentice Hall, 1983.

[Dub 85] E. Dubois, "The sampling and reconstruction of time-varying imagery with application in video systems," *Proc. IEEE*, vol. 73, no. 4, pp. 502–522, Apr. 1985.

[Key 81] R. G. Keys, "Cubic convolution interpolation for digital image processing," *IEEE Trans. Acoust. Speech and Sign. Proc.*, vol. ASSP-29, pp. 1153–1160, Dec. 1981.

[Lim 90] J. S. Lim, *Two-Dimensional Signal and Image Processing*, Englewood Cliffs, NJ: Prentice Hall, 1990.

[Opp 89] A. V. Oppenheim and R. W. Schafer, *Discrete-Time Signal Processing*, Englewood Cliffs, NJ: Prentice Hall, 1989.

Chapter 5

OPTICAL FLOW METHODS

Motion estimation, which may refer to image-plane motion (2-D motion) or object-motion (3-D motion) estimation, is one of the fundamental problems in digital video processing. Indeed, it has been the subject of much research effort [Hua 81, Hua 83, Agg 88, Sin 91, Fle 92, Sez 93]. This chapter has two goals: it provides a general introduction to the 2-D motion estimation problem; it also discusses specific algorithms based on the optical flow equation. Various other nonparametric approaches for 2-D motion estimation will be covered in Chapters 6-8. In Section 5.1, we emphasize the distinction between 2-D motion and apparent motion (optical flow or correspondence). The 2-D motion estimation problem is then formulated as an ill-posed problem in Section 5.2, and a brief overview of *a priori* motion field models is presented. Finally, we discuss a number of optical flow estimation methods based on the optical flow equation (also known as the differential methods) in Section 5.3. Besides being an integral component of motion-compensated filtering and compression, 2-D motion estimation is often the first step towards 3-D motion analysis, which will be studied in Chapters 9-12.

5.1 2-D Motion vs. Apparent Motion

Because time-varying images are 2-D projections of 3-D scenes, as described in Chapter 2, 2-D motion refers to the projection of the 3-D motion onto the image plane. We wish to estimate the 2-D motion (instantaneous velocity or displacement) field from time-varying images sampled on a lattice Λ^3. However, 2-D velocity or displacement fields may not always be observable for several reasons, which are cited below. Instead, what we observe is the so-called "apparent" motion (optical flow or correspondence) field. It is the aim of this section to clarify the distinction between 2-D velocity and optical flow, and 2-D displacement and correspondence fields, respectively.

5.1.1 2-D Motion

2-D motion, also called "projected motion," refers to the perspective or the orthographic projection of 3-D motion into the image plane. 3-D motion can be characterized in terms of either 3-D instantaneous velocity (hereafter velocity) or 3-D displacement of the object points. Expressions for the projections of the 3-D displacement and velocity vectors, under the assumption of rigid motion, into the image plane are derived in Chapters 9 and 10, respectively.

The concept of a 2-D displacement vector is illustrated in Figure 5.1. Let the object point $\mathbf{P}$ at time t move to $\mathbf{P}'$ at time t'. The perspective projection of the points $\mathbf{P}$ and $\mathbf{P}'$ to the image plane gives the respective image points $\mathbf{p}$ and $\mathbf{p}'$. Figure 5.2 depicts a 2-D view of the motion of the image point $\mathbf{p}$ at time t to $\mathbf{p}'$ at time t' as the perspective projection of the 3-D motion of the corresponding object points. Note that because of the projection operation, all 3-D displacement vectors whose tips lie on the dotted line would give the same 2-D displacement vector.

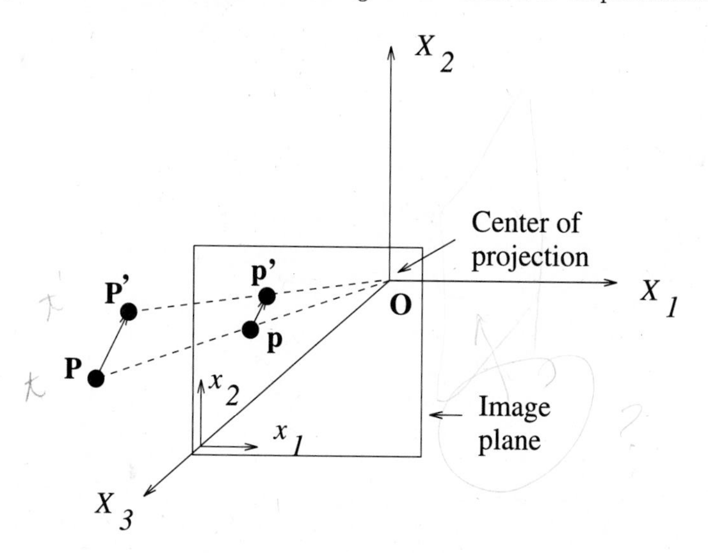

Figure 5.1: Three-dimensional versus two-dimensional motion.

The projected displacement between the times t and $t' = t + \ell\Delta t$, where ℓ is an integer and Δt is the temporal sampling interval, can be defined for all $(\mathbf{x}, t) \in \mathbf{R}^3$, resulting in a real-valued 2-D displacement vector function $\mathbf{d}_c(\mathbf{x}, t; \ell\Delta t)$ of the continuous spatio-temporal variables. The 2-D displacement vector field refers to a sampled representation of this function, given by

$$\mathbf{d}_p(\mathbf{x}, t; \ell\Delta t) = \mathbf{d}_c(\mathbf{x}, t; \ell\Delta t), \quad (\mathbf{x}, t) \in \Lambda^3, \tag{5.1}$$

or, equivalently,

$$\mathbf{d}(\mathbf{n}, k; \ell) = \mathbf{d}_p(\mathbf{x}, t; \ell\Delta t)|_{[x_1\ x_2\ t]^T = \mathbf{V}[n_1\ n_2\ k]^T}, \quad (\mathbf{n}, k) \in \mathbf{Z}^3, \tag{5.2}$$

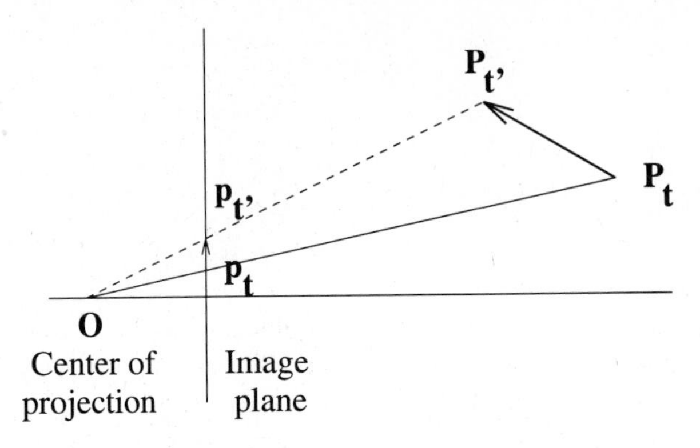

Figure 5.2: The projected motion.

where $\mathbf{V}$ is the sampling matrix of the lattice Λ^3. Thus, a 2-D displacement field is a collection of 2-D displacement vectors, $\mathbf{d}(\mathbf{x}, t; \ell\Delta t)$, where $(\mathbf{x}, t) \in \Lambda^3$.

The projected velocity function $\mathbf{v}_c(\mathbf{x}, t)$ at time t, and the 2-D velocity vector field $\mathbf{v}_p(\mathbf{x}, t) = v(\mathbf{n}, k)$, for $[x_1\ x_2\ t]^T = \mathbf{V}[n_1, n_2, k]^T \in \Lambda^3$ and $(\mathbf{n}, k) \in \mathbf{Z}^3$ can be similarly defined in terms of the 3-D instantaneous velocity $(\dot{X}_1, \dot{X}_2, \dot{X}_3)$, where the dot denotes a time derivative.

5.1.2 Correspondence and Optical Flow

The displacement of the image-plane coordinates $\mathbf{x}$ from time t to t', based on the variations of $s_c(\mathbf{x}, t)$, is called a correspondence vector. An optical flow vector is defined as the temporal rate of change of the image-plane coordinates, $(v_1, v_2) = (dx_1/dt, dx_2/dt)$, at a particular point $(\mathbf{x}, t) \in \mathbf{R}^3$ as determined by the spatio-temporal variations of the intensity pattern $s_c(\mathbf{x}, t)$. That is, it corresponds to the instantaneous pixel velocity vector. (Theoretically, the optical flow and correspondence vectors are identical in the limit $\Delta t = t' - t$ goes to zero, should we have access to the continuous video.) In practice, we define the correspondence (optical flow) field as a vector field of pixel displacements (velocities) based on the observable variations of the 2-D image intensity pattern on a spatio-temporal lattice Λ^3. The correspondence field and optical flow field are also known as the "apparent 2-D displacement" field and "apparent 2-D velocity" field, respectively.

The correspondence (optical flow) field is, in general, different from the 2-D displacement (2-D velocity) field due to [Ver 89]:

- Lack of sufficient spatial image gradient: There must be sufficient gray-level (color) variation within the moving region for the actual motion to be observable. An example of an unobservable motion is shown in Figure 5.3, where a circle with uniform intensity rotates about its center. This motion generates no optical flow, and thus is unobservable.

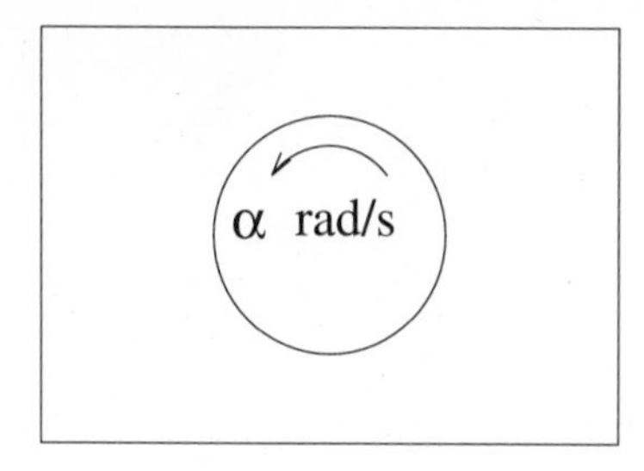

Figure 5.3: All projected motion does not generate optical flow.

- Changes in external illumination: An observable optical flow may not always correspond to an actual motion. For example, if the external illumination varies from frame to frame, as shown in Figure 5.4, then an optical flow will be observed even though there is no motion. Therefore, changes in the external illumination impair the estimation of the actual 2-D motion field.

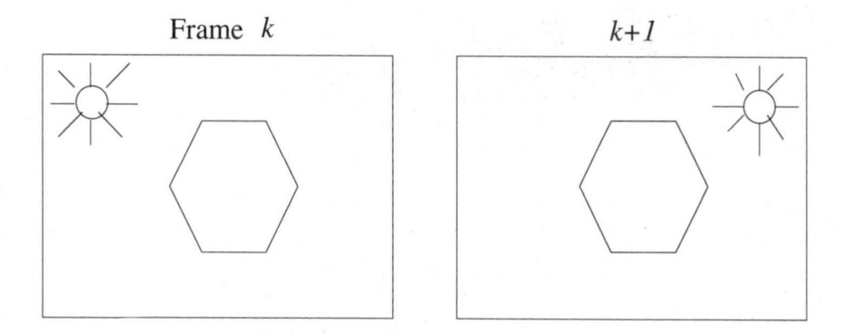

Figure 5.4: All optical flow does not correspond to projected motion.

In some cases, the shading may vary from frame to frame even if there is no change in the external illumination. For example, if an object rotates its surface normal changes, which results in a change in the shading. This change in shading may cause the intensity of the pixels along a motion trajectory to vary, which needs to be taken into account for 2-D motion estimation.

In conclusion, the 2-D displacement and velocity fields are projections of the respective 3-D fields into the image plane, whereas the correspondence and optical flow fields are the velocity and displacement functions perceived from the time-varying image intensity pattern. Since we can only observe optical flow and correspondence fields, we assume they are the same as the 2-D motion field in the remainder of this book.

5.2 2-D Motion Estimation

In this section, we first state the pixel correspondence and optical flow estimation problems. We then discuss the ill-posed nature of these problems, and introduce some *a priori* 2-D motion field models.

The 2-D motion estimation problem can be posed as either: i) the estimation of image-plane correspondence vectors $\mathbf{d}(\mathbf{x}, t; \ell\Delta t) = [d_1(\mathbf{x}, t; \ell\Delta t)\ d_2(\mathbf{x}, t; \ell\Delta t)]^T$ between the times t and $t + \ell\Delta t$, for all $(\mathbf{x}, t) \in \mathbf{\Lambda}^3$ and ℓ is an integer, or ii) the estimation of the optical flow vectors $\mathbf{v}(\mathbf{x}, t) = [v_1(\mathbf{x}, t)\ v_2(\mathbf{x}, t)]^T$ for all $(\mathbf{x}, t) \in \mathbf{\Lambda}^3$. Observe that the subscript "p" has been dropped for notational simplicity. The correspondence and optical flow vectors usually vary from pixel to pixel (space-varying motion), e.g., due to rotation of objects in the scene, and as a function of time, e.g., due to acceleration of objects.

The Correspondence Problem: The correspondence problem can be set up as a *forward* or *backward* motion estimation problem, depending on whether the motion vector is defined from time t to $t+\ell\Delta t$ or from t to $t-\ell\Delta t$, as depicted in Figure 5.5.

Forward Estimation: Given the spatio-temporal samples $s_p(\mathbf{x}, t)$ at times t and $t + \ell\Delta t$, which are related by

$$s_p(x_1, x_2, t) = s_c(x_1 + d_1(\mathbf{x}, t; \ell\Delta t), x_2 + d_2(\mathbf{x}, t; \ell\Delta t), t + \ell\Delta t) \tag{5.3}$$

or, equivalently,

$$s_k(x_1, x_2) = s_{k+\ell}(x_1 + d_1(\mathbf{x}), x_2 + d_2(\mathbf{x})), \quad \text{such that} \quad t = k\Delta t$$

find the real-valued correspondence vector $\mathbf{d}(\mathbf{x}) = [d_1(\mathbf{x})\ d_2(\mathbf{x})]^T$, where the temporal arguments of $\mathbf{d}(\mathbf{x})$ are dropped.

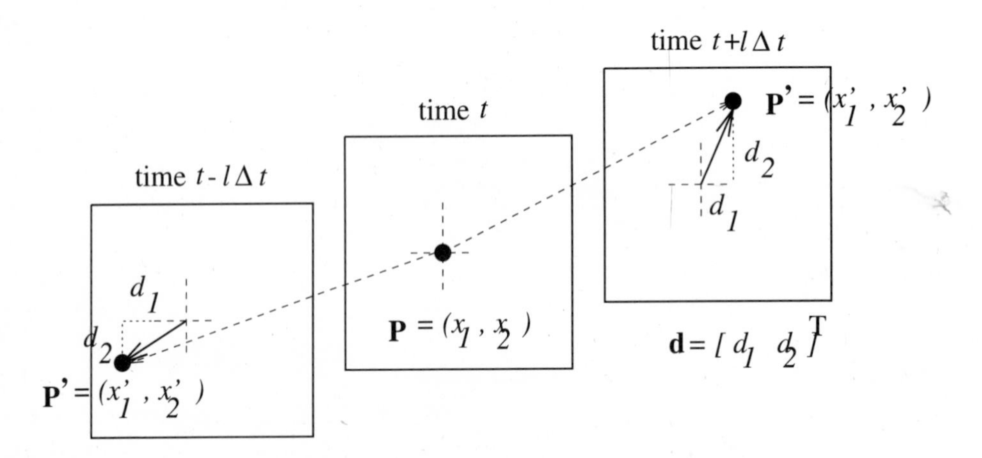

Figure 5.5: Forward and backward correspondence estimation.

Backward Estimation: If we define the correspondence vectors from time t to $t - \ell\Delta t$ then, the 2-D motion model becomes

$$s_k(x_1, x_2) = s_{k-\ell}(x_1 + d_1(\mathbf{x}), x_2 + d_2(\mathbf{x})), \quad \text{such that} \quad t = k\Delta t$$

Alternately, the motion vector can be defined from time $t - \ell\Delta t$ to t. Then we have

$$s_k(x_1, x_2) = s_{k-\ell}(x_1 - d_1(\mathbf{x}), x_2 - d_2(\mathbf{x})), \quad \text{such that} \quad t = k\Delta t$$

Although we discuss both types of motion estimation in this book, backward motion estimation is more convenient for forward motion compensation, which is commonly employed in predictive video compression. Observe that because $\mathbf{x} \pm \mathbf{d}(\mathbf{x})$ generally does not correspond to a lattice site, the right-hand sides of the expressions are given in terms of the continuous video $s_c(x_1, x_2, t)$, which is not available. Hence, most correspondence estimation methods incorporate some interpolation scheme. The correspondence problem also arises in stereo disparity estimation (Chapter 12), where we have a left-right pair instead of a temporal pair of images.

Image Registration: The registration problem is a special case of the correspondence problem, where the two frames are globally shifted with respect to each other, for example, multiple exposures of a static scene with a translating camera.

Optical Flow Estimation: Given the samples $s_p(x_1, x_2, t)$ on a 3-D lattice Λ^3, determine the 2-D velocity $\mathbf{v}(\mathbf{x}, t)$ for all $(\mathbf{x}, t) \in \Lambda^3$. Of course, estimation of optical flow and correspondence vectors from two frames are equivalent, with $\mathbf{d}(\mathbf{x}, t; \ell\Delta t) = \mathbf{v}(\mathbf{x}, t)\ell\Delta t$, assuming that the velocity remains constant during each time interval $\ell\Delta t$. Note that one needs to consider more than two frames at a time to estimate optical flow in the presence of acceleration.

2-D motion estimation, stated as either a correspondence or optical flow estimation problem, based only on two frames, is an "ill-posed" problem in the absence of any additional assumptions about the nature of the motion. A problem is called ill-posed if a unique solution does not exist, and/or solution(s) do(es) not continuously depend on the data [Ber 88]. 2-D motion estimation suffers from all of the existence, uniqueness, and continuity problems:

- Existence of a solution: No correspondence can be established for covered/uncovered background points. This is known as the *occlusion* problem.
- Uniqueness of the solution: If the components of the displacement (or velocity) at each pixel are treated as independent variables, then the number of unknowns is twice the number of observations (the elements of the frame difference). This leads to the so-called "aperture" problem.
- Continuity of the solution: Motion estimation is highly sensitive to the presence of observation noise in video images. A small amount of noise may result in a large deviation in the motion estimates.

The occlusion and aperture problems are described in detail in the following.

5.2.1 The Occlusion Problem

Occlusion refers to the covering/uncovering of a surface due to 3-D rotation and translation of an object which occupies only part of the field of view. The covered and uncovered background concepts are illustrated in Figure 5.6, where the object indicated by the solid lines translates in the x_1 direction from time t to t'. Let the index of the frames at time t and t' be k and $k+1$, respectively. The dotted region in the frame k indicates the background to be covered in frame $k+1$. Thus, it is not possible to find a correspondence for these pixels in frame $k+1$. The dotted region in frame $k+1$ indicates the background uncovered by the motion of the object. There is no correspondence for these pixels in frame k.

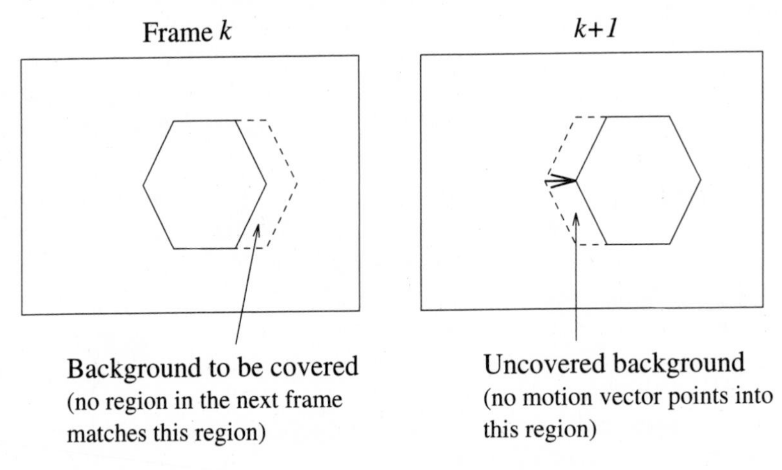

Figure 5.6: The covered/uncovered background problem.

5.2.2 The Aperture Problem

The aperture problem is a restatement of the fact that the solution to the 2-D motion estimation problem is not unique. If motion vectors at each pixel are considered as independent variables, then there are twice as many unknowns as there are equations, given by (5.3). The number of equations is equal to the number of pixels in the image, but for each pixel the motion vector has two components.

Theoretical analysis, which will be given in the next section, indicates that we can only determine motion that is orthogonal to the spatial image gradient, called the normal flow, at any pixel. The aperture problem is illustrated in Figure 5.7. Suppose we have a corner of an object moving in the x_2 direction (upward). If we estimate the motion based on a local window, indicated by Aperture 1, then it is not possible to determine whether the image moves upward or perpendicular to the edge. The motion in the direction perpendicular to the edge is called the normal flow.

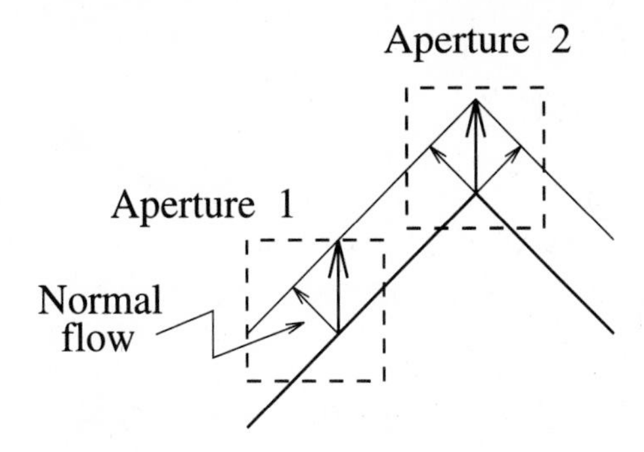

Figure 5.7: The aperture problem.

However, if we observe Aperture 2, then it is possible to estimate the correct motion, since the image has gradient in two perpendicular directions within this aperture. Thus, it is possible to overcome the aperture problem by estimating the motion based on a block of pixels that contain sufficient gray-level variation. Of course, implicit here is the assumption that all these pixels translate by the same motion vector. A less restrictive approach would be to represent the variation of the motion vectors from pixel to pixel by some parametric or nonparametric 2-D motion field models.

5.2.3 Two-Dimensional Motion Field Models

Because of the ill-posed nature of the problem, motion estimation algorithms need additional assumptions (models) about the structure of the 2-D motion field. We provide a brief overview of these models in the following.

Parametric Models

Parametric models aim to describe the orthographic or perspective projection of 3-D motion (displacement or velocity) of a surface into the image plane. In general, parametric 2-D motion field models depend on a representation of the 3-D surface. For example, a 2-D motion field resulting from 3-D rigid motion of a planar surface under orthographic projection can be described by a 6-parameter affine model, while under perspective projection it can be described by an 8-parameter nonlinear model [Ana 93]. There also exist more complicated models for quadratic surfaces [Agg 88]. We elaborate on these models in Chapters 9-12, where we discuss 3-D motion estimation.

A subclass of parametric models are the so-called quasi-parametric models, which treat the depth of each 3-D point as an independent unknown. Then the six 3-D motion parameters constrain the local image flow vector to lie along a specific line, while knowledge of the local depth value is required to determine the exact value of the motion vector [Ana 93]. These models may serve as constraints to regulate the estimation of the 2-D motion vectors, which lead to simultaneous 2-D and 3-D motion estimation formulations (see Chapter 11).

Nonparametric Models

The main drawback of the parametric models is that they are only applicable in case of 3-D rigid motion. Alternatively, nonparametric uniformity (smoothness) constraints can be imposed on the 2-D motion field without employing 3-D rigid motion models. The nonparametric constraints can be classified as deterministic versus stochastic smoothness models. The following is a brief preview of the nonparametric approaches covered in this book.

- *Optical flow equation (OFE) based methods:* Methods based on the OFE, studied in the rest of this chapter, attempt to provide an estimate of the optical flow field in terms of spatio-temporal image intensity gradients. With monochromatic images, the OFE needs to be used in conjunction with an appropriate spatio-temporal smoothness constraint, which requires that the displacement vector vary slowly over a neighborhood. With color images, the OFE can be imposed at each color band separately, which could possibly constrain the displacement vector in three different directions [Oht 90]. However, in most cases an appropriate smoothness constraint is still needed to obtain satisfactory results. Global smoothness constraints cause inaccurate motion estimation at the occlusion boundaries. More advanced directional smoothness constraints allow sudden discontinuities in the motion field.

- *Block motion model:* It is assumed that the image is composed of moving blocks. We discuss two approaches to determining the displacement of blocks from frame to frame: the phase-correlation and block-matching methods. In the phase-correlation approach, the linear term of the Fourier phase difference between two consecutive frames determines the motion estimate. Block matching searches for the location of the best-matching block of a fixed size in the next (and/or previous) frame(s) based on a distance criterion. The basic form of both methods apply only to translatory motion; however, generalized block matching can incorporate other spatial transformations. Block-based motion estimation is covered in Chapter 6.

- *Pel-recursive methods:* Pel-recursive methods are predictor-corrector type displacement estimators. The prediction can be taken as the value of the motion estimate at the previous pixel location or as a linear combination of motion estimates in a neighborhood of the current pixel. The update is based on gradient-based minimization of the displaced frame difference (DFD) at that pixel. The prediction step is generally considered as an implicit smoothness constraint. Extension of this approach to block-based estimation results in the so-called Wiener-type estimation strategies. Pel-recursive methods are presented in Chapter 7.

- *Bayesian Methods:* Bayesian methods utilize probabilistic smoothness constraints, usually in the form of a Gibbs random field to estimate the displacement field. Their main drawback is the extensive amount of computation that is required. A maximum *a posteriori* probability estimation method is developed in Chapter 8.

5.3 Methods Using the Optical Flow Equation

In this section, we first derive the optical flow equation (OFE). Then optical flow estimation methods using the OFE are discussed.

5.3.1 The Optical Flow Equation

Let $s_c(x_1, x_2, t)$ denote the continuous space-time intensity distribution. If the intensity remains constant along a motion trajectory, we have

$$\frac{ds_c(x_1, x_2, t)}{dt} = 0 \tag{5.4}$$

where x_1 and x_2 varies by t according to the motion trajectory. Equation (5.4) is a total derivative expression and denotes the rate of change of intensity along the motion trajectory. Using the chain rule of differentiation, it can be expressed as

$$\frac{\partial s_c(\mathbf{x};t)}{\partial x_1} v_1(\mathbf{x},t) + \frac{\partial s_c(\mathbf{x};t)}{\partial x_2} v_2(\mathbf{x},t) + \frac{\partial s_c(\mathbf{x};t)}{\partial t} = 0 \tag{5.5}$$

where $v_1(\mathbf{x},t) = dx_1/dt$ and $v_2(\mathbf{x},t) = dx_2/dt$ denote the components of the coordinate velocity vector in terms of the continuous spatial coordinates. The expression (5.5) is known as the optical flow equation or the optical flow constraint.

It can alternatively be expressed as

$$\langle\ \nabla s_c(\mathbf{x};t)\ ,\ \mathbf{v}(\mathbf{x},t)\ \rangle + \frac{\partial s_c(\mathbf{x};t)}{\partial t} = 0 \tag{5.6}$$

where $\nabla s_c(\mathbf{x};t) \doteq [\frac{\partial s_c(\mathbf{x};t)}{\partial x_1}\ \frac{\partial s_c(\mathbf{x};t)}{\partial x_2}]^T$ and $\langle\cdot,\cdot\rangle$ denotes vector inner product.

Naturally, the OFE (5.5) is not sufficient to uniquely specify the 2-D velocity (flow) field just like (5.3). The OFE yields one scalar equation in two unknowns, $v_1(\mathbf{x},t)$ and $v_2(\mathbf{x},t)$, at each site $(\mathbf{x},t)$. Inspection of (5.6) reveals that we can only estimate the component of the flow vector that is in the direction of the spatial image gradient $\frac{\nabla s_c(\mathbf{x};t)}{\|\nabla s_c(\mathbf{x};t)\|}$, called the normal flow $v_\perp(\mathbf{x},t)$, because the component that is orthogonal to the spatial image gradient disappears under the dot product.

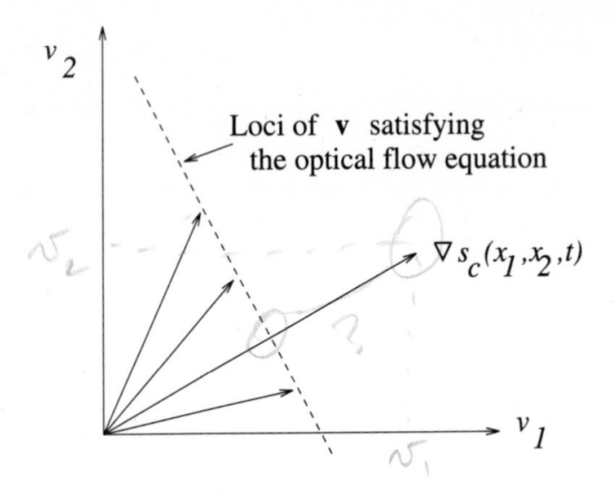

Figure 5.8: The normal flow.

This is illustrated in Figure 5.8, where all vectors whose tips lie on the dotted line satisfy (5.6). The normal flow at each site can be computed from (5.6) as

$$v_{\perp}(\mathbf{x},t) = \frac{-\frac{\partial s_c(\mathbf{x};t)}{\partial t}}{||\nabla s_c(\mathbf{x};t)||} \tag{5.7}$$

Thus, the OFE (5.5) imposes a constraint on the component of the flow vector that is in the direction of the spatial gradient of the image intensity at each site (pixel), which is consistent with the aperture problem. Observe that the OFE approach requires that first, the spatio-temporal image intensity be differentiable, and second, the partial derivatives of the intensity be available. In practice, optical flow estimation from two views can be shown to be equivalent to correspondence estimation under certain assumptions (see Exercise 2). In the following we present several approaches to estimate optical flow from estimates of normal flow.

5.3.2 Second-Order Differential Methods

In search of another constraint to determine both components of the flow vector at each pixel, several researchers [Nag 87, Ura 88] suggested the conservation of the spatial image gradient, $\nabla s_c(\mathbf{x};t)$, stated by

$$\frac{d\ \nabla s_c(\mathbf{x};t)}{dt} = 0 \tag{5.8}$$

An estimate of the flow field, which is obtained from (5.8), is given by

$$\begin{bmatrix} \hat{v}_1(\mathbf{x};t) \\ \hat{v}_2(\mathbf{x};t) \end{bmatrix} = \begin{bmatrix} \frac{\partial^2 s_c(\mathbf{x},t)}{\partial x_1^2} & \frac{\partial^2 s_c(\mathbf{x},t)}{\partial x_2 x_1} \\ \frac{\partial^2 s_c(\mathbf{x},t)}{\partial x_1 x_2} & \frac{\partial^2 s_c(\mathbf{x},t)}{\partial x_2^2} \end{bmatrix}^{-1} \begin{bmatrix} -\frac{\partial^2 s_c(\mathbf{x},t)}{\partial t \partial x_1} \\ -\frac{\partial^2 s_c(\mathbf{x},t)}{\partial t \partial x_2} \end{bmatrix} \tag{5.9}$$

However, the constraint (5.8) does not allow for some common motion such as rotation and zooming (see Exercise 6). Further, second-order partials cannot always be estimated with sufficient accuracy. This problem is addressed in Section 5.3.5. As a result, Equation (5.9) does not always yield reliable flow estimates.

5.3.3 Block Motion Model

Another approach to overcoming the aperture problem is to assume that the motion vector remains unchanged over a particular block of pixels, denoted by $\mathcal{B}$ (suggested by Lucas and Kanade [Luc 81]); that is,

$$\mathbf{v}(\mathbf{x},t) = \mathbf{v}(t) = [v_1(t)\; v_2(t)]^T, \quad \text{for } \mathbf{x} \in \mathcal{B} \tag{5.10}$$

Although such a model cannot handle rotational motion, it is possible to estimate a purely translational motion vector uniquely under this model provided that the block of pixels contain sufficient gray-level variation.

Let's define the error in the optical flow equation over the block of pixels $\mathcal{B}$ as

$$E = \sum_{\mathbf{x}\in\mathcal{B}} \left(\frac{\partial s_c(\mathbf{x},t)}{\partial x_1} v_1(t) + \frac{\partial s_c(\mathbf{x},t)}{\partial x_2} v_2(t) + \frac{\partial s_c(\mathbf{x},t)}{\partial t} \right)^2 \tag{5.11}$$

Computing the partials of the error E with respect to $v_1(t)$ and $v_2(t)$, respectively, and setting them equal to zero, we have

$$\sum_{\mathbf{x}\in\mathcal{B}} \left(\frac{\partial s_c(\mathbf{x},t)}{\partial x_1} \hat{v}_1(t) + \frac{\partial s_c(\mathbf{x},t)}{\partial x_2} \hat{v}_2(t) + \frac{\partial s_c(\mathbf{x},t)}{\partial t} \right) \frac{\partial s_c(\mathbf{x},t)}{\partial x_1} = 0$$

$$\sum_{\mathbf{x}\in\mathcal{B}} \left(\frac{\partial s_c(\mathbf{x},t)}{\partial x_1} \hat{v}_1(t) + \frac{\partial s_c(\mathbf{x},t)}{\partial x_2} \hat{v}_2(t) + \frac{\partial s_c(\mathbf{x},t)}{\partial t} \right) \frac{\partial s_c(\mathbf{x},t)}{\partial x_2} = 0$$

where $\hat{\cdot}$ denotes the estimate of the respective quantity. Solving these equations simultaneously, we have

$$\begin{bmatrix} \hat{v}_1(t) \\ \hat{v}_2(t) \end{bmatrix} = \begin{bmatrix} \sum_{\mathbf{x}\in\mathcal{B}} \frac{\partial s_c(\mathbf{x},t)}{\partial x_1}\frac{\partial s_c(\mathbf{x},t)}{\partial x_1} & \sum_{\mathbf{x}\in\mathcal{B}} \frac{\partial s_c(\mathbf{x},t)}{\partial x_1}\frac{\partial s_c(\mathbf{x},t)}{\partial x_2} \\ \sum_{\mathbf{x}\in\mathcal{B}} \frac{\partial s_c(\mathbf{x},t)}{\partial x_1}\frac{\partial s_c(\mathbf{x},t)}{\partial x_2} & \sum_{\mathbf{x}\in\mathcal{B}} \frac{\partial s_c(\mathbf{x},t)}{\partial x_2}\frac{\partial s_c(\mathbf{x},t)}{\partial x_2} \end{bmatrix}^{-1} \begin{bmatrix} -\sum_{\mathbf{x}\in\mathcal{B}} \frac{\partial s_c(\mathbf{x},t)}{\partial x_1}\frac{\partial s_c(\mathbf{x},t)}{\partial t} \\ -\sum_{\mathbf{x}\in\mathcal{B}} \frac{\partial s_c(\mathbf{x},t)}{\partial x_2}\frac{\partial s_c(\mathbf{x},t)}{\partial t} \end{bmatrix} \tag{5.12}$$

It is possible to increase the influence of the constraints towards the center of the block $\mathcal{B}$ by replacing all summations with weighted summations. A suitable weighting function may be in the form of a 2-D triangular window. Clearly, the accuracy of the flow estimates depends on the accuracy of the estimated spatial and temporal partial derivatives (see Section 5.3.5). We next discuss the method of Horn and Schunck [Hor 81], which imposes a less restrictive global smoothness constraint on the velocity field.

5.3.4 Horn and Schunck Method

Horn and Schunck seek a motion field that satisfies the OFE with the minimum pixel-to-pixel variation among the flow vectors. Let

$$\mathcal{E}_{of}(\mathbf{v}(\mathbf{x},t)) = \langle \nabla s_c(\mathbf{x};t) \, , \, \mathbf{v}(\mathbf{x},t) \rangle + \frac{\partial s_c(\mathbf{x};t)}{\partial t} \tag{5.13}$$

denote the error in the optical flow equation. Observe that the OFE is satisfied when $\mathcal{E}_{of}(\mathbf{v}(\mathbf{x},t))$ is equal to zero. In the presence of occlusion and noise, we aim to minimize the square of $\mathcal{E}_{of}(\mathbf{v}(\mathbf{x},t))$ in order to enforce the optical flow constraint.

The pixel-to-pixel variation of the velocity vectors can be quantified by the sum of the magnitude squares of the spatial gradients of the components of the velocity vector, given by

$$\begin{aligned}\mathcal{E}_s^2(\mathbf{v}(\mathbf{x},t)) &= ||\nabla v_1(\mathbf{x},t)||^2 + ||\nabla v_2(\mathbf{x},t)||^2 \\ &= (\frac{\partial v_1}{\partial x_1})^2 + (\frac{\partial v_1}{\partial x_2})^2 + (\frac{\partial v_2}{\partial x_1})^2 + (\frac{\partial v_2}{\partial x_2})^2\end{aligned} \tag{5.14}$$

where we assume that the spatial and temporal coordinates are continuous variables. It can easily be verified that the smoother the velocity field, the smaller $\mathcal{E}_s^2(\mathbf{v}(\mathbf{x},t))$.

Then the Horn and Schunck method minimizes a weighted sum of the error in the OFE and a measure of the pixel-to-pixel variation of the velocity field

$$\min_{\mathbf{v}(\mathbf{x},t)} \int_{\mathcal{A}} (\mathcal{E}_{of}^2(\mathbf{v}) + \alpha^2 \mathcal{E}_s^2(\mathbf{v})) d\mathbf{x} \tag{5.15}$$

to estimate the velocity vector at each point $\mathbf{x}$, where $\mathcal{A}$ denotes the continuous image support. The parameter α^2, usually selected heuristically, controls the strength of the smoothness constraint. Larger values of α^2 increase the influence of the constraint.

The minimization of the functional (5.15), using the calculus of variations, requires solving the two equations

$$\begin{aligned}(\frac{\partial s_c}{\partial x_1})^2 \hat{v}_1(\mathbf{x},t) + \frac{\partial s_c}{\partial x_1}\frac{\partial s_c}{\partial x_2}\hat{v}_2(\mathbf{x},t) &= \alpha^2 \nabla^2 \hat{v}_1(\mathbf{x},t) - \frac{\partial s_c}{\partial x_1}\frac{\partial s_c}{\partial t} \\ \frac{\partial s_c}{\partial x_1}\frac{\partial s_c}{\partial x_2}\hat{v}_1(\mathbf{x},t) + (\frac{\partial s_c}{\partial x_2})^2 \hat{v}_2(\mathbf{x},t) &= \alpha^2 \nabla^2 \hat{v}_2(\mathbf{x},t) - \frac{\partial s_c}{\partial x_2}\frac{\partial s_c}{\partial t}\end{aligned} \tag{5.16}$$

simultaneously, where ∇^2 denotes the Laplacian and $\hat{\cdot}$ denotes the estimate of the respective quantity. In the implementation of Horn and Schunck [Hor 81], the Laplacians of the velocity components have been approximated by FIR highpass filters to arrive at the Gauss-Seidel iteration

$$\begin{aligned}\hat{v}_1^{(n+1)}(\mathbf{x},t) &= \bar{v}_1^{(n)}(\mathbf{x},t) - \frac{\partial s_c}{\partial x_1} \frac{\frac{\partial s_c}{\partial x_1}\bar{v}_1^{(n)}(\mathbf{x},t) + \frac{\partial s_c}{\partial x_2}\bar{v}_2^{(n)}(\mathbf{x},t) + \frac{\partial s_c}{\partial t}}{\alpha^2 + (\frac{\partial s_c}{\partial x_1})^2 + (\frac{\partial s_c}{\partial x_2})^2} \\ \hat{v}_2^{(n+1)}(\mathbf{x},t) &= \bar{v}_2^{(n)}(\mathbf{x},t) - \frac{\partial s_c}{\partial x_2} \frac{\frac{\partial s_c}{\partial x_1}\bar{v}_1^{(n)}(\mathbf{x},t) + \frac{\partial s_c}{\partial x_2}\bar{v}_2^{(n)}(\mathbf{x},t) + \frac{\partial s_c}{\partial t}}{\alpha^2 + (\frac{\partial s_c}{\partial x_1})^2 + (\frac{\partial s_c}{\partial x_2})^2}\end{aligned} \tag{5.17}$$

where n is the iteration counter, the overbar denotes weighted local averaging (excluding the present pixel), and all partials are evaluated at the point $(\mathbf{x}, t)$. The reader is referred to [Hor 81] for the derivation of this iterative estimator. The initial estimates of the velocities $v_1^{(0)}(\mathbf{x}, t)$ and $v_2^{(0)}(\mathbf{x}, t)$ are usually taken as zero. The above formulation assumes a continuous spatio-temporal intensity distribution. In computer implementation, all spatial and temporal image gradients need to be estimated numerically from the observed image samples, which will be discussed in the next subsection.

5.3.5 Estimation of the Gradients

We discuss two gradient estimation methods. The first method makes use of finite differences, and the second is based on polynomial fitting.

Gradient Estimation Using Finite Differences

One approach to estimating the partial derivatives from a discrete image $s(n_1, n_2, k)$ is to approximate them by the respective forward or backward finite differences. In order to obtain more robust estimates of the partials, we can compute the average of the forward and backward finite differences, called the average difference. Furthermore, we can compute a local average of the average differences to eliminate the effects of observation noise. Horn and Schunck [Hor 81] proposed averaging four finite differences to obtain

$$\begin{aligned}
\frac{\partial s_c(x_1, x_2, t)}{\partial x_1} \approx\; & \frac{1}{4}\{\, s(n_1+1, n_2, k) - s(n_1, n_2, k) + s(n_1+1, n_2+1, k) \\
& -s(n_1, n_2+1, k) + s(n_1+1, n_2, k+1) - s(n_1, n_2, k+1) \\
& +s(n_1+1, n_2+1, k+1) - s(n_1, n_2+1, k+1) \,\} \\
\frac{\partial s_c(x_1, x_2, t)}{\partial x_2} \approx\; & \frac{1}{4}\{\, s(n_1, n_2+1, k) - s(n_1, n_2, k) + s(n_1+1, n_2+1, k) \\
& -s(n_1+1, n_2, k) + s(n_1, n_2+1, k+1) - s(n_1, n_2, k+1) \\
& +s(n_1+1, n_2+1, k+1) - s(n_1+1, n_2, k+1) \,\} \\
\frac{\partial s_c(x_1, x_2, t)}{\partial t} \approx\; & \frac{1}{4}\{\, s(n_1, n_2, k+1) - s(n_1, n_2, k) + s(n_1+1, n_2, k+1) \\
& -s(n_1+1, n_2, k) + s(n_1, n_2+1, k+1) - s(n_1, n_2+1, k) \\
& +s(n_1+1, n_2+1, k+1) - s(n_1+1, n_2+1, k) \,\} \qquad (5.18)
\end{aligned}$$

Various other averaging strategies exist for estimating partials using finite differences [Lim 90]. Spatial and temporal presmoothing of video with Gaussian kernels usually helps gradient estimation.

Gradient Estimation by Local Polynomial Fitting

An alternative approach is to approximate $s_c(x_1, x_2, t)$ locally by a linear combination of some low-order polynomials in x_1, x_2 and t; that is,

$$s_c(x_1, x_2, t) \approx \sum_{i=0}^{N-1} a_i \phi_i(x_1, x_2, t) \tag{5.19}$$

where $\phi_i(x_1, x_2, t)$ are the basis polynomials, N is the number of basis functions used in the polynomial approximation, and a_i are the coefficients of the linear combination. Here, we will set N equal to 9, with the following choice of the basis functions:

$$\phi_i(x_1, x_2, t) = 1, x_1, x_2, t, x_1^2, x_2^2, x_1x_2, x_1t, x_2t \tag{5.20}$$

which are suggested by Lim [Lim 90]. Then, Equation (5.19) becomes,

$$s_c(x_1, x_2, t) \approx a_0 + a_1x_1 + a_2x_2 + a_3t + a_4x_1^2 + a_5x_2^2 + a_6x_1x_2 + a_7x_1t + a_8x_2t \tag{5.21}$$

The coefficients a_i, $i = 0, \ldots, 8$, are estimated by using the least squares method, which minimizes the error function

$$e^2 = \sum_{n_1}\sum_{n_2}\sum_{k} \left(s(n_1, n_2, k) - \sum_{i=0}^{N-1} a_i \phi_i(x_1, x_2, t)|_{[x_1\ x_2\ t]^T = \mathbf{V}[n_1\ n_2\ k]^T} \right)^2 \tag{5.22}$$

with respect to these coefficients. The summation over (n_1, n_2, k) is carried within a local neighborhood of the pixel for which the polynomial approximation is made. A typical case involves 50 pixels, 5×5 spatial windows in two consecutive frames.

Once the coefficients a_i are estimated, the components of the gradient can be found by simple differentiation,

$$\frac{\partial s_c(x_1, x_2, t)}{\partial x_1} \approx a_1 + 2a_4x_1 + a_6x_2 + a_7t|_{x_1=x_2=t=0} = a_1 \tag{5.23}$$

$$\frac{\partial s_c(x_1, x_2, t)}{\partial x_2} \approx a_2 + 2a_5x_2 + a_6x_1 + a_8t|_{x_1=x_2=t=0} = a_2 \tag{5.24}$$

$$\frac{\partial s_c(x_1, x_2, t)}{\partial t} \approx a_3 + a_7x_1 + a_8x_2|_{x_1=x_2=t=0} = a_3 \tag{5.25}$$

Similarly, the second-order and mixed partials can be easily estimated in terms of the coefficients a_4 through a_8.

5.3.6 Adaptive Methods

The Horn-Schunck method imposes the optical flow and smoothness constraints globally over the entire image, or over a motion estimation window. This has two undesired effects:

i) The smoothness constraint does not hold in the direction perpendicular to an occlusion boundary. Thus, a global smoothness constraint blurs "motion edges." For example, if an object moves against a stationary background, there is a sudden change in the motion field at the boundary of the object. Motion edges can be preserved by imposing the smoothness constraint only in the directions along which the pixel intensities do not significantly change. This is the basic concept of the so-called directional or oriented smoothness constraint.
ii) The nonadaptive method also enforces the optical flow constraint at the occlusion regions, where it should be turned off. This can be achieved by adaptively varying α to control the relative strengths of the optical flow and smoothness constraints. For example, at occlusion regions, such as the dotted regions shown in Figure 5.6, the optical flow constraint can be completely turned off, and the smoothness constraint can be fully on.

Several researchers proposed to impose the smoothness constraint along the boundaries but not perpendicular to the occlusion boundaries. Hildreth [Hil 84] minimized the criterion function of Horn and Schunck given by (5.15) along object contours. Nagel and Enkelman [Nag 86, Enk 88] introduced the concept of directional smoothness, which suppresses the smoothness constraint in the direction of the spatial image gradient. Fogel [Fog 91] used directional smoothness constraints with adaptive weighting in a hierarchical formulation. Note that adaptive weighting methods require strategies to detect moving object (occlusion) boundaries. Recently, Snyder [Sny 91] proposed a general formulation of the smoothness constraint that includes some of the above as special cases.

Directional-Smoothness Constraint

The directional smoothness constraint can be expressed as

$$\mathcal{E}_{ds}^2(\mathbf{v}(\mathbf{x},t)) = (\nabla v_1)^T \mathbf{W} (\nabla v_1) + (\nabla v_2)^T \mathbf{W} (\nabla v_2) \tag{5.26}$$

where $\mathbf{W}$ is a weight matrix to penalize variations in the motion field depending on the spatial changes in gray-level content of the video. Various alternatives for the weight matrix $\mathbf{W}$ exist [Nag 86, Nag 87, Enk 88]. For example, $\mathbf{W}$ can be chosen as

$$\mathbf{W} = \frac{\mathbf{F} + \delta \mathbf{I}}{\text{trace}(\mathbf{F} + \delta \mathbf{I})} \tag{5.27}$$

where $\mathbf{I}$ is the identity matrix representing a global smoothness term to ensure a nonzero weight matrix at spatially uniform regions, δ is a scalar, and

$$\mathbf{F} = \begin{bmatrix} (\frac{\partial s_c}{\partial x_1})^2 + b^2((\frac{\partial^2 s_c}{\partial x_1^2})^2 + (\frac{\partial^2 s_c}{\partial x_1 \partial x_2})^2) & \frac{\partial s_c}{\partial x_1}\frac{\partial s_c}{\partial x_2} + b^2 \frac{\partial^2 s_c}{\partial x_1 \partial x_2}(\frac{\partial^2 s_c}{\partial x_1^2} + \frac{\partial^2 s_c}{\partial x_2^2}) \\ \frac{\partial s_c}{\partial x_1}\frac{\partial s_c}{\partial x_2} + b^2 \frac{\partial^2 s_c}{\partial x_1 \partial x_2}(\frac{\partial^2 s_c}{\partial x_1^2} + \frac{\partial^2 s_c}{\partial x_2^2}) & (\frac{\partial s_c}{\partial x_2})^2 + b^2((\frac{\partial^2 s_c}{\partial x_1 \partial x_2})^2 + (\frac{\partial^2 s_c}{\partial x_2^2})^2) \end{bmatrix}^{-1}$$

with b^2 a constant.

Then, the directional-smoothness method minimizes the criterion function

$$\min_{\mathbf{v}(\mathbf{x},t)} \int_{\mathcal{A}} (\mathcal{E}_{of}^2(\mathbf{v}) + \alpha^2 \mathcal{E}_{ds}^2(\mathbf{v}))d\mathbf{x} \tag{5.28}$$

to find an estimate of the motion field, where $\mathcal{A}$ denotes the image support and α^2 is the smoothness parameter. Observe that the method of Horn and Schunck (5.15) is a special case of this formulation with $\delta = 1$ and $\mathbf{F} = \mathbf{0}$. A Gauss-Seidel iteration to minimize (5.28) has been described in [Enk 88], where the update term in each iteration is computed by means of a linear algorithm. The performance of the directional-smoothness method depends on how accurately the required second and mixed partials of image intensity can be estimated.

Hierarchical Approach

Fogel [Fog 91] used the concepts of directional smoothness and adaptive weighting in an elegant hierarchical formulation. A multiresolution representation $s_c^\alpha(x_1, x_2, t)$ of the video was defined as

$$s_c^\alpha(x_1, x_2, t) \doteq \int_{\mathcal{A}} s_c(x_1, x_2, t) h(\frac{\mu - x_1}{\alpha}, \frac{\eta - x_2}{\alpha}) d\mu d\eta \tag{5.29}$$

where α is the resolution parameter, $\mathcal{A}$ denotes the image support, and

$$h(x_1, x_2) = \begin{cases} A \exp\{\frac{-(C^2x_1^2 + D^2x_2^2)}{B^2 - (C^2x_1^2 + D^2x_2^2)}\} & C^2x_1^2 + D^2x_2^2 < B^2 \\ 0 & \text{otherwise} \end{cases} \tag{5.30}$$

denotes a low pass filter with parameters A, B, C and D. It can readily be seen that the spatial resolution of $s_c^\alpha(x_1, x_2, t)$ decreases as α increases. Observe that spatial partial derivatives of $s_c^\alpha(x_1, x_2, t)$ can be computed by correlating $s_c(x_1, x_2, t)$ with the partial derivatives of $h(x_1, x_2)$ which can be computed analytically. A nonlinear optimization problem was solved at each resolution level using a Quasi-Newton method. The estimate obtained at one resolution level was used as the initial estimate at the next higher resolution level. Interested readers are referred to [Fog 91] for implementational details.

5.4 Examples

We compare the results of three representative methods of increasing complexity: a simple Lucas-Kanade (L-K) type estimator given by (5.12) based on the block motion model, the Horn-Schunck (H-S) method (5.17) imposing a global smoothness constraint, and the directional-smoothness method of Nagel. In all cases, the spatial and temporal gradients have been approximated by both average finite differences and polynomial fitting as discussed in Section 5.3.5. The images are spatially presmoothed by a 5×5 Gaussian kernel with the variance 2.5 pixels.

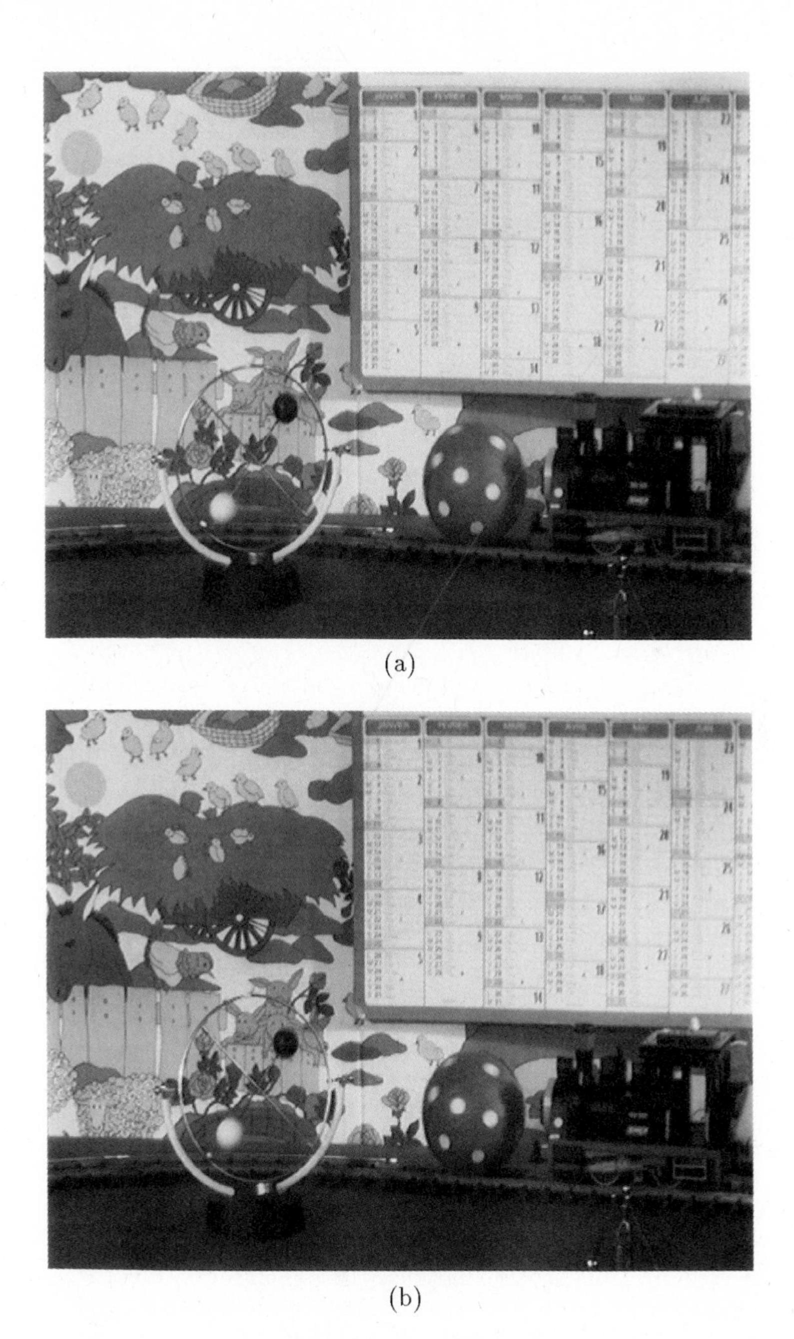

(a)

(b)

Figure 5.9: a) First and b) second frames of the Mobile and Calendar sequence. (source: CCETT, Cesson Sevigne Cedex, France.)

Figure 5.10: Absolute value of the frame difference. (Courtesy Gozde Bozdagi)

Our implementation of the L-K method considers 11×11 blocks with no weighting. In the H-S algorithm, we set $\alpha^2 = 625$, and allowed for 20 to 150 iterations. The parameters of the Nagel algorithm were set to $\alpha^2 = 25$ and $\delta = 5$ with 20 iterations.

These methods have been applied to estimate the motion between the seventh and eighth frames of a progressive video, known as the "Mobile and Calendar" sequence, shown in Figure 5.9 (a) and (b), respectively. Figure 5.10 (a) shows the absolute value of the frame difference (multiplied by 3), without any motion compensation, to indicate the amount of motion present. The lighter pixels are those whose intensity has changed with respect to the previous frame due to motion. Indeed, the scene contains multiple motions: the train is moving forward (from right to left), pushing the ball in front of it; there is a small ball in the foreground spinning around a circular ring; the background moves toward the right due to camera pan; and the calendar moves up and down. The motion fields estimated by the L-K and the H-S methods are depicted in Figure 5.11 (a) and (b), respectively. It can be seen that the estimated fields capture most of the actual motion.

We evaluate the goodness of the motion estimates on the basis of the peak signal-to-noise ratio (PSNR) of the resulting displaced frame difference (DFD) between the seventh and eighth frames, defined by

$$PSNR = 10 \log_{10} \frac{255 \times 255}{\sum [s_8(n_1, n_2) - s_7(n_1 + d_1(n_1, n_2), n_2 + d_2(n_1, n_2))]^2} \qquad (5.31)$$

where d_1 and d_2 are the components of the motion estimates at each pixel. We also computed the entropy of the estimated 2-D motion field, given by

$$H = -\sum_{d_1} P(d_1)\ log_2 P(d_1) - \sum_{d_2} P(d_2)\ log_2 P(d_2) \tag{5.32}$$

where $P(d_1)$ and $P(d_2)$ denote the relative frequency of occurence of the horizontal and vertical components of the motion vector $\mathbf{d}$. The entropy, besides serving as a measure of smoothness of the motion field, is especially of interest in motion-compensated video compression, where one wishes to minimize both the entropy of the motion field (for cheaper transmission) and the energy of the DFD. The PSNR and the entropy of all methods are listed in Table 5.1 after 20 iterations (for H-S and Nagel algorithms). In the case of the H-S method, the PSNR increases to 32.23 dB after 150 iterations using average finite differences. Our experiments indicate that the Nagel algorithm is not as robust as the L-K and H-S algorithms, and may diverge if some stopping criterion is not employed. This may be due to inaccuracies in the estimated second and mixed partials of the spatial image intensity.

Table 5.1: Comparison of the differential methods. (Courtesy Gozde Bozdagi)

Method	PSNR (dB)		Entropy (bits)	
	Polynomial	Differences	Polynomial	Differences
Frame-Difference	23.45	-	-	-
Lucas-Kanade	30.89	32.09	6.44	6.82
Horn-Schunck	28.14	30.71	4.22	5.04
Nagel	29.08	31.84	5.83	5.95

The reader is alerted that the mean absolute DFD provides a measure of the goodness of the pixel correspondence estimates. However, it does not provide insight about how well the estimates correlate with the projected 3-D displacement vectors. Recall that the optical flow equation enables estimation of the normal flow vectors at each pixel, rather than the actual projected flow vectors.

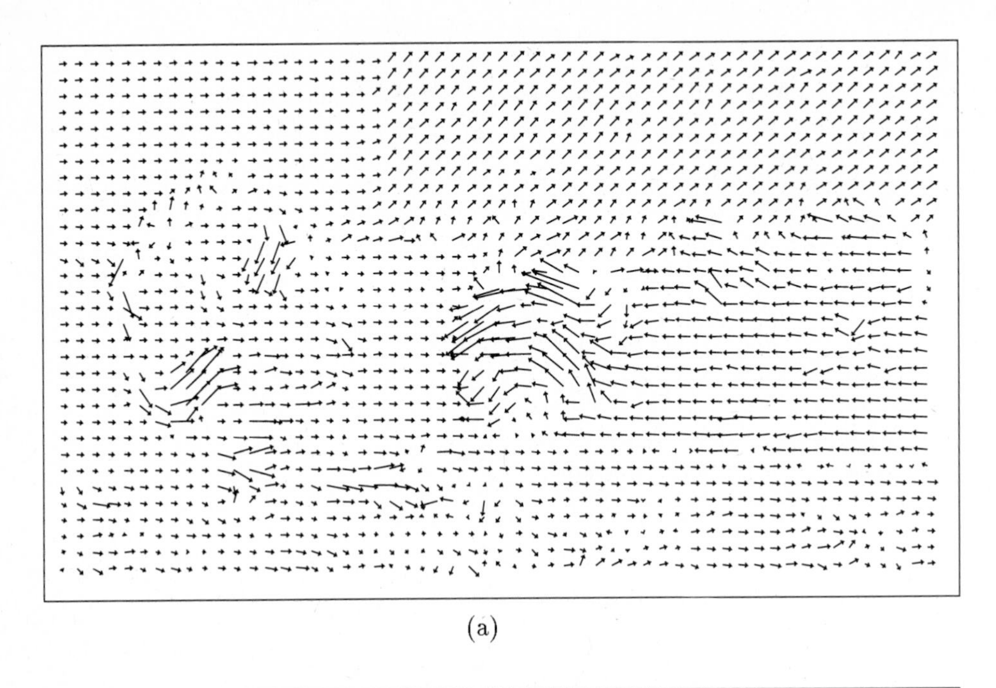

(a)

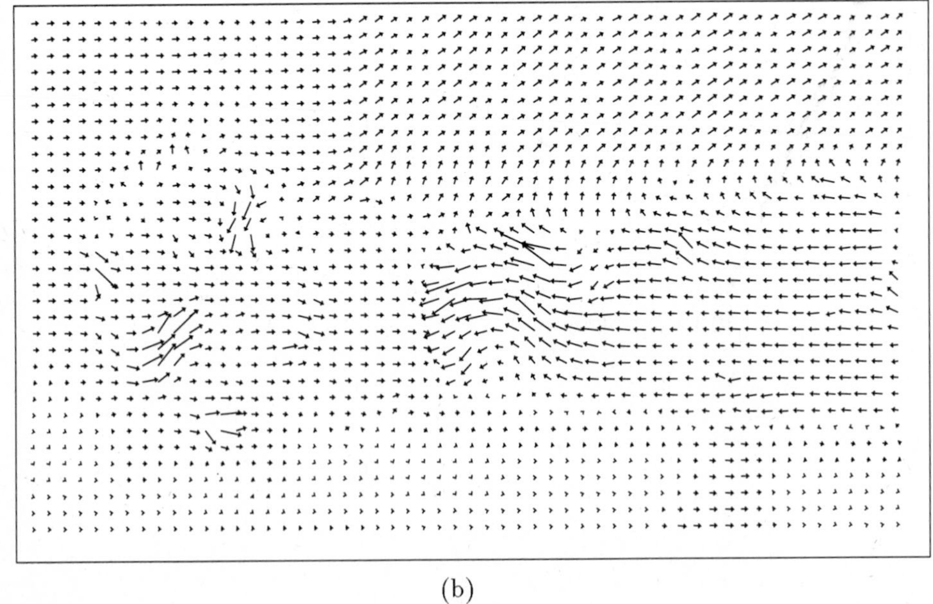

(b)

Figure 5.11: Motion field obtained by a) the Lucas-Kanade method and b) the Horn-Schunck method. (Courtesy Gozde Bozdagi and Mehmet Ozkan)

5.5 Exercises

1. For a color image, the optical flow equation (5.5) can be written for each of the R, G, and B channels separately. State the conditions on the (R,G,B) intensities so that we have at least two linearly independent equations at each pixel. How valid are these conditions for general color images?

2. State the conditions on spatio-temporal image intensity and the velocity under which the optical flow equation can be used for displacement estimation. Why do we need the small motion assumption?

3. What are the conditions for the existence of normal flow (5.7)? Can we always recover optical flow from the normal flow? Discuss the relationship between the spatial image gradients and the aperture problem.

4. Suggest methods to detect occlusion.

5. Derive (5.9) from (5.8).

6. Show that the constraint (5.8) does not hold when there is rotation or zoom.

7. Find the least squares estimates of a_1, a_2 and a_3 in (5.23)-(5.25).

Bibliography

[Agg 88] J. K. Aggarwal and N. Nandhakumar, "On the computation of motion from sequences of images," *Proc. IEEE*, vol. 76, pp. 917–935, Aug. 1988.

[Ana 93] P. Anandan, J. R. Bergen, K. J. Hanna, and R. Hingorani, "Hierarchical model-based motion estimation," in *Motion Analysis and Image Sequence Processing*, M. I. Sezan and R. L. Lagendijk, eds., Norwell, MA: Kluwer, 1993.

[Bar 94] J. L. Barron, D. J. Fleet, and S. S. Beauchemin, "Systems and experiment: Performance of optical flow techniques," *Int. J. Comp. Vision*, vol. 12:1, pp. 43–77, 1994.

[Ber 88] M. Bertero, T. A. Poggio and V. Torre, "Ill-posed problems in early vision," *Proc. IEEE*, vol. 76, pp. 869–889, August 1988.

[Enk 88] W. Enkelmann, "Investigations of multigrid algorithms for the estimation of optical flow fields in image sequences," *Comp. Vis. Graph Image Proc.*, vol. 43, pp. 150–177, 1988.

[Fle 92] D. J. Fleet, *Measurement of Image Velocity*, Norwell, MA: Kluwer, 1992.

[Fog 91] S. V. Fogel, "Estimation of velocity vector fields from time-varying image sequences," *CVGIP: Image Understanding*, vol. 53, pp. 253–287, 1991.

[Hil 84] E. C. Hildreth, "Computations underlying the measurement of visual motion," *Artif. Intel.*, vol. 23, pp. 309–354, 1984.

[Hor 81] B. K. P. Horn and B. G. Schunck, "Determining optical flow," *Artif. Intell.*, vol. 17, pp. 185–203, 1981.

[Hua 81] T. S. Huang, ed., *Image Sequence Analysis*, Springer Verlag, 1981.

[Hua 83] T. S. Huang, ed., *Image Sequence Processing and Dynamic Scene Analysis*, Berlin, Germany: Springer-Verlag, 1983.

[Lim 90] J. S. Lim, *Two-Dimensional Signal and Image Processing*, Englewood Cliffs, NJ: Prentice Hall, 1990.

[Luc 81] B. D. Lucas and T. Kanade, "An iterative image registration technique with an application to stereo vision," *Proc. DARPA Image Understanding Workshop*, pp. 121–130, 1981.

[Nag 86] H. H. Nagel and W. Enkelmann, "An investigation of smoothness constraints for the estimation of displacement vector fields from image sequences," *IEEE Trans. Patt. Anal. Mach. Intel.*, vol. 8, pp. 565–593, 1986.

[Nag 87] H. H. Nagel, "On the estimation of optical flow: Relations between different approaches and some new results," *Artificial Intelligence*, vol. 33, pp. 299–324, 1987.

[Oht 90] N. Ohta, "Optical flow detection by color images," *NEC Res. and Dev.*, no. 97, pp. 78–84, Apr. 1990.

[Sez 93] M. I. Sezan and R. L. Lagendijk, eds., *Motion Analysis and Image Sequence Processing*, Norwell, MA: Kluwer, 1993.

[Sin 91] A. Singh, *Optic Flow Computation*, Los Alamitos, CA: IEEE Computer Soc. Press, 1991.

[Sny 91] M. A. Snyder, "On the mathematical foundations of smoothness constraints for the determination of optical flow and for surface reconstruction," *IEEE Trans. Patt. Anal. Mach. Intel.*, vol. 13, pp. 1105–1114, 1991.

[Ura 88] S. Uras, F. Girosi, A. Verri, and V. Torre, "A computational approach to motion perception," *Biol. Cybern.*, vol. 60, pp. 79–97, 1988.

[Ver 89] A. Verri and T. Poggio, "Motion field and optical flow: Qualitative properties," *IEEE Trans. Patt. Anal. Mach. Intel.*, vol. PAMI-11, pp. 490–498, May 1989.

Chapter 6

BLOCK-BASED METHODS

Block-based motion estimation and compensation are among the most popular approaches. Block-based motion compensation has been adopted in the international standards for digital video compression, such as H.261 and MPEG 1-2. Although these standards do not specify a particular motion estimation method, block-based motion estimation becomes a natural choice. Block-based motion estimation is also widely used in several other digital video applications, including motion-compensated filtering for standards conversion.

We start with a brief introduction of the block-motion models in Section 6.1. A simple block-based motion estimation scheme, based on the translatory block model, was already discussed in Chapter 5, in the context of estimation using the optical flow equation. In this chapter, we present two other translatory-block-based motion estimation strategies. The first approach, discussed in Section 6.2, is a spatial frequency domain technique, called the phase-correlation method. The second, presented in Section 6.3, is a spatial domain search approach, called the block-matching method. Both methods can be implemented hierarchically, using a multiresolution description of the video. Hierarchical block-motion estimation is addressed in Section 6.4. Finally, in Section 6.5, possible generalizations of the block-matching framework are discussed, including 2-D deformable block motion estimation, in order to overcome some shortcomings of the translatory block-motion model.

6.1 Block-Motion Models

The block-motion model assumes that the image is composed of moving blocks. We consider two types of block motion: i) simple 2-D translation, and ii) various 2-D deformations of the blocks.

6.1.1 Translational Block Motion

The simplest form of this model is that of translatory blocks, restricting the motion of each block to a pure translation. Then an $N \times N$ block $\mathcal{B}$ in frame k centered about the pixel $\mathbf{n} = (n_1, n_2)$ is modeled as a globally shifted version of a same-size block in frame $k + \ell$, for an integer ℓ. That is,

$$s(n_1, n_2, k) = s_c(x_1 + d_1, x_2 + d_2, t + \ell\Delta t)|_{\begin{bmatrix} \mathbf{x} \\ t \end{bmatrix} = \mathbf{V} \begin{bmatrix} \mathbf{n} \\ k \end{bmatrix}} \quad (6.1)$$

for all $(n_1, n_2) \in \mathcal{B}$, where d_1 and d_2 are the components of the displacement (translation) vector for block $\mathcal{B}$. Recall from the previous chapter that the right-hand side of (6.1) is given in terms of the continuous time-varying image $s_c(x_1, x_2, t)$, because d_1 and d_2 are real-valued. Assuming the values of d_1 and d_2 are quantized to the nearest integer, the model (6.1) can be simplified as

$$s(n_1, n_2, k) = s(n_1 + d_1, n_2 + d_2, k + \ell) \quad (6.2)$$

Observe that it is possible to obtain $1/2^L$ pixel accuracy in the motion estimates, using either the phase-correlation or the block-matching methods, if the frames k and $k + \ell$ in (6.2) are interpolated by a factor of L.

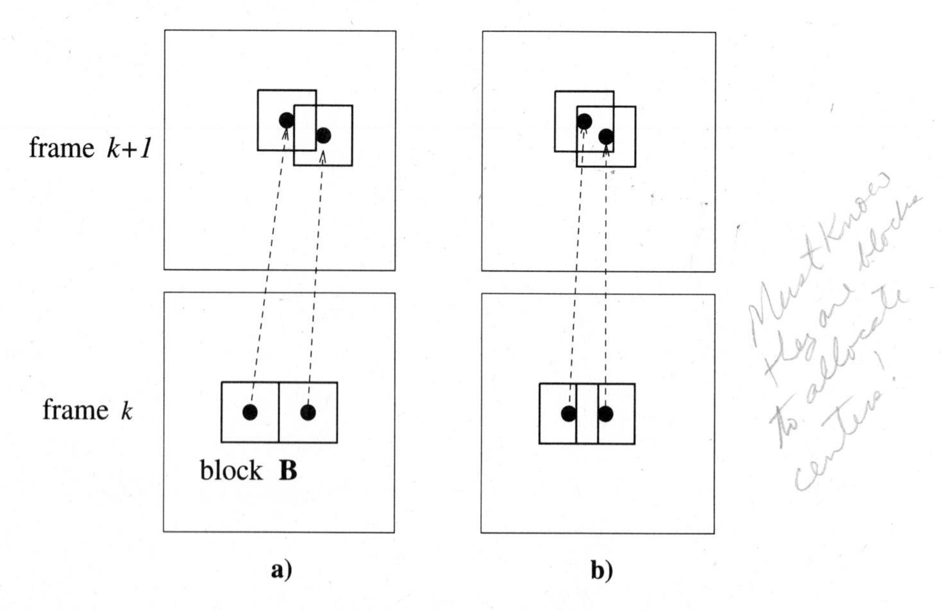

Figure 6.1: Block-motion models: a) nonoverlapping and b) overlapping blocks.

In the model (6.1), the blocks $\mathcal{B}$ may be nonoverlapping or overlapping as shown in Figure 6.1 (a) and (b), respectively. In the nonoverlapping case, the entire block is assigned a single motion vector. Hence, motion compensation can be achieved by copying the gray-scale or color information from the corresponding block in the frame $k+1$ on a pixel-by-pixel basis. In the case of overlapping blocks, we can either compute the average of the motion vectors within the overlapping regions, or select one of the estimated motion vectors. Motion compensation in the case of overlapping blocks was discussed in [Sul 93], where a multihypothesis expectation approach was proposed.

The popularity of motion compensation and estimation based on the model of translational blocks originates from:

- low overhead requirements to represent the motion field, since one motion vector is needed per block, and
- ready availability of low-cost VLSI implementations.

However, motion compensation using translational blocks i) fails for zoom, rotational motion, and under local deformations, and ii) results in serious blocking artifacts, especially for very-low-bitrate applications, because the boundaries of objects do not generally agree with block boundaries, and adjacent blocks may be assigned substantially different motion vectors.

6.1.2 Generalized/Deformable Block Motion

In order to generalize the translational block model (6.1), note that it can be characterized by a simple frame-to-frame pixel coordinate (spatial) transformation of the form

$$
\begin{aligned}
x_1' &= x_1 + d_1 \\
x_2' &= x_2 + d_2
\end{aligned}
\tag{6.3}
$$

where (x_1', x_2') denotes the coordinates of a point in the frame $k+l$. The spatial transformation (6.3) can be generalized to include affine coordinate transformations, given by

$$
\begin{aligned}
x_1' &= a_1 x_1 + a_2 x_2 + d_1 \\
x_2' &= a_3 x_1 + a_4 x_2 + d_2
\end{aligned}
\tag{6.4}
$$

The affine transformation (6.4) can handle rotation of blocks as well as 2-D deformation of squares (rectangles) into parallelograms, as depicted in Figure 6.2. Other spatial transformations include the perspective and bilinear coordinate transformations. The perspective transformation is given by

$$
\begin{aligned}
x_1' &= \frac{a_1 x_1 + a_2 x_2 + a_3}{a_7 x_1 + a_8 x_2 + 1} \\
x_2' &= \frac{a_4 x_1 + a_5 x_2 + a_6}{a_7 x_1 + a_8 x_2 + 1}
\end{aligned}
\tag{6.5}
$$

whereas the bilinear transformation can be expressed as

bilinear

$$\begin{aligned} x_1' &= a_1 x_1 + a_2 x_2 + a_3 x_1 x_2 + a_4 \\ x_2' &= a_5 x_1 + a_6 x_2 + a_7 x_1 x_2 + a_8 \end{aligned} \tag{6.6}$$

We will see in Chapter 9 that the affine and perspective coordinate transformations correspond to the orthographic and perspective projection of the 3-D rigid motion of a planar surface, respectively. However, the bilinear transformation is not related to any physical 3-D motion. Relevant algebraic and geometric properties of these transformations have been developed in [Wol 90].

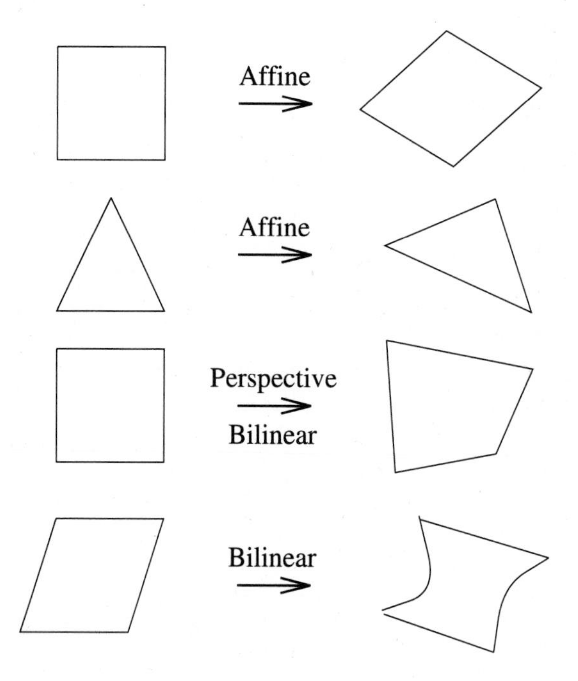

Figure 6.2: Examples of spatial transformations.

While the basic phase-correlation and block-matching methods are based on the translational model (6.3), generalizations of the block-matching method to track 2-D deformable motion based on the spatial transformations depicted in Figure 6.2 will be addressed in Section 6.5. We note that various block-motion models, including (6.4) and (6.5), fall under the category of parametric motion models (discussed in Chapter 5), and may be considered as local regularization constraints on arbitrary displacement fields to overcome the aperture problem, in conjunction with the optical-flow-equation-based methods (see Section 5.3.3), or pel-recursive methods (which are described in Chapter 7).

6.2 Phase-Correlation Method

Taking the 2-D Fourier transform of both sides of the discrete motion model (6.2), with $\ell = 1$, over a block $\mathcal{B}$ yields

$$S_k(f_1, f_2) = S_{k+1}(f_1, f_2) \exp\{j2\pi(d_1 f_1 + d_2 f_2)\} \tag{6.7}$$

where $S_k(f_1, f_2)$ denotes the 2-D Fourier transform of the frame k with respect to the spatial variables x_1 and x_2. It follows that, in the case of translational motion, the difference of the 2-D Fourier phases of the respective blocks,

$$\arg\{S(f_1, f_2, k)\} - \arg\{S(f_1, f_2, k+1)\} = 2\pi(d_1 f_1 + d_2 f_2) \tag{6.8}$$

defines a plane in the variables (f_1, f_2). Then the interframe motion vector can be estimated from the orientation of the plane (6.8). This seemingly straightforward approach runs into two important problems: i) estimation of the orientation of the plane in general requires 2-D phase unwrapping, which is not trivial by any means; and ii) it is not usually easy to identify the motion vectors for more than one moving object within a block. The phase-correlation method alleviates both problems [Tho 87]. Other frequency-domain motion estimation methods include those based on 3-D spatio-temporal frequency-domain analysis using Wigner distributions [Jac 87] or a set of Gabor filters [Hee 87].

The phase-correlation method estimates the relative shift between two image blocks by means of a normalized cross-correlation function computed in the 2-D spatial Fourier domain. It is also based on the principle that a relative shift in the spatial domain results in a linear phase term in the Fourier domain. In the following, we first show the derivation of the phase-correlation function, and then discuss some issues related to its implementation. Although an extension of the phase-correlation method to include rotational motion was also suggested [Cas 87], that will not be covered here.

6.2.1 The Phase-Correlation Function

The cross-correlation function between the frames k and $k+1$ is defined as

$$c_{k,k+1}(n_1, n_2) = s(n_1, n_2, k+1) ** s(-n_1, -n_2, k) \tag{6.9}$$

where $**$ denotes the 2-D convolution operation. Taking the Fourier transform of both sides, we obtain the complex-valued cross-power spectrum expression

$$C_{k,k+1}(f_1, f_2) = S_{k+1}(f_1, f_2) S_k^*(f_1, f_2) \tag{6.10}$$

Normalizing $C_{k,k+1}(f_1, f_2)$ by its magnitude gives the phase of the cross-power spectrum

$$\tilde{C}_{k,k+1}(f_1, f_2) = \frac{S_{k+1}(f_1, f_2) S_k^*(f_1, f_2)}{|S_{k+1}(f_1, f_2) S_k^*(f_1, f_2)|} \tag{6.11}$$

Assuming translational motion, we substitute (6.7) into (6.11) to obtain

$$\tilde{C}_{k,k+1}(f_1, f_2) = \exp\{-j2\pi(f_1 d_1 + f_2 d_2)\} \tag{6.12}$$

Taking the inverse 2-D Fourier transform of this expression yields the phase-correlation function

$$\tilde{c}_{k,k+1}(n_1, n_2) = \delta(n_1 - d_1, n_2 - d_2) \tag{6.13}$$

We observe that the phase-correlation function consists of an impulse whose location yields the displacement vector.

6.2.2 Implementation Issues

Implementation of the phase-correlation method in the computer requires replacing the 2-D Fourier transforms by the 2-D DFT, resulting in the following algorithm:

1. Compute the 2-D DFT of the respective blocks from the kth and $k+1$th frames.

2. Compute the phase of the cross-power spectrum as in (6.11).

3. Compute the 2-D inverse DFT of $\tilde{C}_{k,k+1}(f_1, f_2)$ to obtain the phase-correlation function $\tilde{c}_{k,k+1}(n_1, n_2)$.

4. Detect the location of the peak(s) in the phase-correlation function.

Ideally, we expect to observe a single impulse in the phase-correlation function indicating the relative displacement between the two blocks. In practice, a number of factors contribute to the degeneration of the phase-correlation function to contain one or more peaks. They are the use of the 2-D DFT instead of the 2-D Fourier transform, the presence of more than one moving object within the block, and the presence of observation noise.

The use of the 2-D DFT instead of the 2-D Fourier transform has a number of consequences:

- Boundary effects: In order to obtain a perfect impulse, the shift must be cyclic. Since things disappearing at one end of the window generally do not reappear at the other end, the impulses degenerate into peaks. Further, it is well known that the 2-D DFT assumes periodicity in both directions. Discontinuities from left to right boundaries, and from top to bottom, may introduce spurious peaks.

- Spectral leakage due to noninteger motion vectors: In order to observe a perfect impulse, the components of the displacement vector must correspond to an integer multiple of the fundamental frequency. Otherwise, the impulse will degenerate into a peak due to the well-known spectral leakage phenomenon.

- Range of displacement estimates: Since the 2-D DFT is periodic with the block size (N_1, N_2), the displacement estimates need to be unwrapped as

$$\hat{d}_i = \begin{cases} d_i & \text{if } |d_i| \leq N_i/2,\ N_i \text{ even, or } |d_i| \leq (N_i - 1)/2,\ N_i \text{ odd} \\ d_i - N_i & \text{otherwise} \end{cases} \tag{6.14}$$

to accommodate negative displacements. Thus, the range of estimates is $[-N_i/2+1, N_i/2]$ for N_i even. For example, to estimate displacements within a range [-31,32], the block size should be at least 64 × 64.

The block size is one of the most important parameters in any block-based motion estimation algorithm. Selection of the block size usually involves a tradeoff between two conflicting requirements. The window must be large enough in order to be able to estimate large displacement vectors. On the other hand, it should be small enough so that the displacement vector remains constant within the window. These two contradicting requirements can usually be addressed by hierarchical methods [Erk 93]. Hierarchical methods will be treated for the case of block matching that also faces the same tradeoff in window size selection.

The phase-correlation method has some desirable properties:

Frame-to-frame intensity changes: The phase-correlation method is relatively insensitive to changes in illumination, because shifts in the mean value or multiplication by a constant do not affect the Fourier phase. Since the phase-correlation function is normalized with the Fourier magnitude, the method is also insensitive to any other Fourier-magnitude-only degradation.

Multiple moving objects: It is of interest to know what happens when multiple moving objects with different velocities are present within a single window. Experiments indicate that multiple peaks are observed, each indicating the movement of a particular object [Tho 87]. Detection of the significant peaks then generates a list of candidate displacement vectors for each pixel within the block. An additional search is required to find which displacement vector belongs to which pixel within the block. This can be verified by testing the magnitude of the displaced frame difference with each of the candidate vectors.

6.3 Block-Matching Method

Block matching can be considered as the most popular method for practical motion estimation due to its lesser hardware complexity [Jai 81, Ghr 90]. As a result it is widely available in VLSI, and almost all H.261 and MPEG 1-2 codecs are utilizing block matching for motion estimation. In block matching, the best motion vector estimate is found by a pixel-domain search procedure.

The basic idea of block matching is depicted in Figure 6.3, where the displacement for a pixel (n_1, n_2) in frame k (the present frame) is determined by considering an $N_1 \times N_2$ block centered about (n_1, n_2), and searching frame $k+1$ (the search frame) for the location of the best-matching block of the same size. The search is

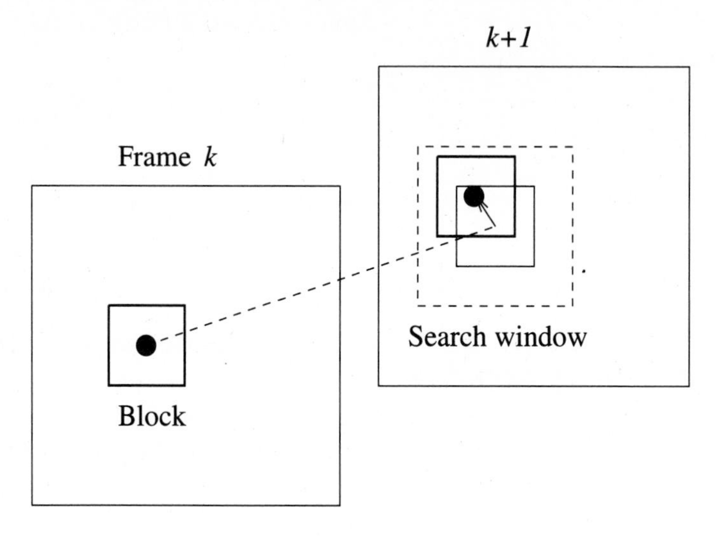

Figure 6.3: Block matching.

usually limited to an $N_1 + 2M_1 \times N_2 + 2M_2$ region called the search window for computational reasons.

Block-matching algorithms differ in:

- the matching criteria (e.g., maximum cross-correlation, minimum error)
- the search strategy (e.g., three-step search, cross search), and
- the determination of block size (e.g., hierarchical, adaptive)

We discuss some of the popular options in the following.

6.3.1 Matching Criteria

The matching of the blocks can be quantified according to various criteria including the maximum cross-correlation (similar to the phase-correlation function), the minimum mean square error (MSE), the minimum mean absolute difference (MAD), and maximum matching pel count (MPC).

In the minimum MSE criterion, we evaluate the MSE, defined as

$$MSE(d_1, d_2) = \frac{1}{N_1 N_2} \sum_{(n_1,n_2)\in\mathcal{B}} [s(n_1, n_2, k) - s(n_1 + d_1, n_2 + d_2, k + 1)]^2 \quad (6.15)$$

where $\mathcal{B}$ denotes an $N_1 \times N_2$ block, for a set of candidate motion vectors (d_1, d_2). The estimate of the motion vector is taken to be the value of (d_1, d_2) which minimizes the MSE. That is,

$$[\hat{d}_1 \ \hat{d}_2]^T = \arg \min_{(d_1,d_2)} MSE(d_1, d_2) \tag{6.16}$$

Minimizing the MSE criterion can be viewed as imposing the optical flow constraint on all pixels of the block. In fact, expressing the displaced frame difference,

$$dfd_{k,k+1}(n_1, n_2) = s(n_1, n_2, k) - s(n_1 + d_1, n_2 + d_2, k + 1) \tag{6.17}$$

in terms of a first order Taylor series (see Section 7.1), it can be easily seen that minimizing the MSE (6.15) is equivalent to minimizing $\mathcal{E}_{of}(\mathbf{v}(\mathbf{x}, t))$ given by (5.13) in the Horn and Schunck method. However, the minimum MSE criterion is not commonly used in VLSI implementations because it is difficult to realize the square operation in hardware.

Instead, the minimum MAD criterion, defined as

$$MAD(d_1, d_2) = \frac{1}{N_1 N_2} \sum_{(n_1,n_2) \in \mathcal{B}} |s(n_1, n_2, k) - s(n_1 + d_1, n_2 + d_2, k + 1)| \tag{6.18}$$

is the most popular choice for VLSI implementations. Then the displacement estimate is given by

$$[\hat{d}_1 \ \hat{d}_2]^T = \arg \min_{(d_1,d_2)} MAD(d_1, d_2) \tag{6.19}$$

It is well-known that the performance of the MAD criterion deteriorates as the search area becomes larger due to the presence of several local minima.

Another alternative is the maximum matching pel count (MPC) criterion. In this approach, each pel within the block $\mathcal{B}$ is classified as either a matching pel or a mismatching pel according to

$$T(n_1, n_2; d_1, d_2) = \begin{cases} 1 & \text{if } |s(n_1, n_2, k) - s(n_1 + d_1, n_2 + d_2, k + 1)| \leq t \\ 0 & \text{otherwise} \end{cases} \tag{6.20}$$

where t is a predetermined threshold. Then the number of matching pels within the block is given by

$$MPC(d_1, d_2) = \sum_{(n_1,n_2) \in \mathcal{B}} T(n_1, n_2; d_1, d_2) \tag{6.21}$$

and

$$[\hat{d}_1, \hat{d}_2]^T = \arg \max_{(d_1,d_2)} MPC(d_1, d_2) \tag{6.22}$$

That is, the motion estimate is the value of (d_1, d_2) which gives the highest number of matching pels. The MPC criterion requires a threshold comparator, and a $\log_2(N_1 \times N_2)$ counter [Gha 90].

6.3.2 Search Procedures

Finding the best-matching block requires optimizing the matching criterion over all possible candidate displacement vectors at each pixel (n_1, n_2). This can be accomplished by the so-called "full search," which evaluates the matching criterion for all values of (d_1, d_2) at each pixel, and is extremely time-consuming.

As a first measure to reduce the computational burden, we usually limit the search area to a

$$-M_1 \leq d_1 \leq M_1 \quad \text{and} \quad -M_2 \leq d_2 \leq M_2$$

"search window" centered about each pixel for which a motion vector will be estimated, where M_1 and M_2 are predetermined integers. A search window is shown in Figure 6.3. Another commonly employed practice to lower the computational load is to estimate motion vectors on a sparse grid of pixels, e.g., once every eight pixels and eight lines using a 16×16 block, and to then interpolate the motion field to estimate the remaining motion vectors.

In most cases, however, search strategies faster than the full search are utilized, although they lead to suboptimal solutions. Some examples of faster search algorithms include the

- three-step search and,
- cross-search.

These faster search algorithms evaluate the criterion function only at a predetermined subset of the candidate motion vector locations. Let's note here that the expected accuracy of motion estimates varies according to the application. In motion-compensated compression, all we seek is a matching block, in terms of some metric, even if the match does not correlate well with the actual projected motion. It is for this reason that faster search algorithms serve video compression applications reasonably well.

Three-Step Search

We explain the three-step search procedure with the help of Figure 6.4, where only the search frame is depicted with the search window parameters $M_1 = M_2 = 7$. The "0" marks the pixel in the search frame that is just behind the present pixel. In the first step, the criterion function is evaluated at nine points, the pixel "0" and the pixels marked as "1." If the lowest MSE or MAD is found at the pixel "0," then we have "no motion." In the second step, the criterion function is evaluated at 8 points that are marked as "2" centered about the pixel chosen as the best match in the first stage (denoted by a circled "1"). Note that in the initial step, the search pixels are the corners of the search window, and then at each step we halve the distance of the search pixels from the new center to obtain finer-resolution estimates. The motion estimate is obtained after the third step, where the search pixels are all 1 pixel away from the center.

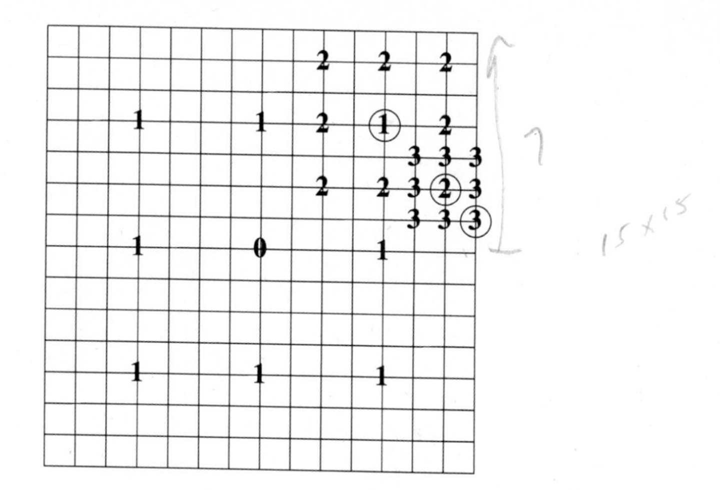

Figure 6.4: Three-step search.

Additional steps may be incorporated into the procedure if we wish to obtain subpixel accuracy in the motion estimates. Note that the search frame needs to be interpolated to evaluate the criterion function at subpixel locations. Generalization of this procedure to other search window parameters yields the so-called "n-step search" or "log-D search" procedures.

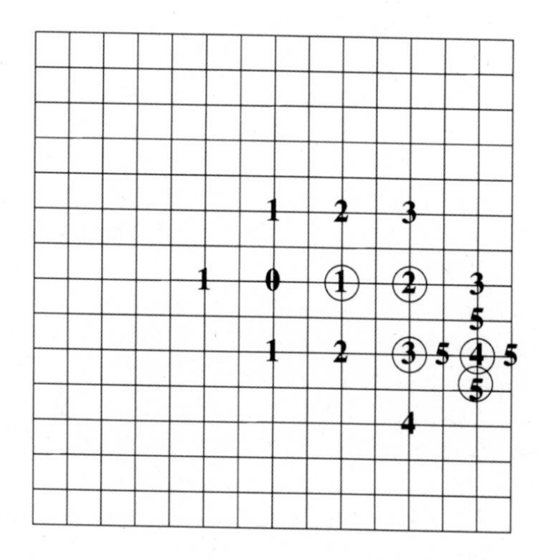

Figure 6.5: Cross-search

Cross-Search

The cross-search method is another logarithmic search strategy, where at each step there are four search locations which are the end points of an (x)-shape cross or a (+)-shape cross [Gha 90]. The case of a (+)-shape cross is depicted in Figure 6.5.

The distance between the search points is reduced if the best match is at the center of the cross or at the boundary of the search window. Several variations of these search strategies exist in the literature [Ghr 90]. Recently, more efficient search algorithms that utilize the spatial and temporal correlations between the motion vectors across the blocks have been proposed [Liu 93]. Block matching has also been generalized to arbitrary shape matching, such as matching contours and curvilinear shapes using heuristic search strategies [Cha 91, Dav 83].

The selection of an appropriate block size is essential for any block-based motion estimation algorithm. There are conflicting requirements on the size of the search blocks. If the blocks are too small, a match may be established between blocks containing similar gray-level patterns which are unrelated in the sense of motion. On the other hand, if the blocks are too large, then actual motion vectors may vary within a block, violating the assumption of a single motion vector per block. Hierarchical block matching, discussed in the next section, addresses these conflicting requirements.

6.4 Hierarchical Motion Estimation

Hierarchical (multiresolution) representations of images (frames of a sequence) in the form of a Laplacian pyramid or wavelet transform may be used with both the phase-correlation and block-matching methods for improved motion estimation. A pyramid representation of a single frame is depicted in Figure 6.6 where the full-resolution image (layer-1) is shown at the bottom. The images in the upper levels are lower and lower resolution images obtained by appropriate low-pass filtering and subsampling. In the following, we only discuss the hierarchical block-matching method. A hierarchical implementation of the phase-correlation method follows the same principles.

The basic idea of hierarchical block-matching is to perform motion estimation at each level successively, starting with the lowest resolution level [Bie 88]. The lower resolution levels serve to determine a rough estimate of the displacement using relatively larger blocks. Note that the "relative size of the block" can be measured as the size of the block normalized by the size of the image at a particular resolution level. The estimate of the displacement vector at a lower resolution level is then passed onto the next higher resolution level as an initial estimate. The higher resolution levels serve to fine-tune the displacement vector estimate. At higher resolution levels, relatively smaller window sizes can be used since we start with a good initial estimate.

In practice, we may skip the subsampling step. Then the pyramid contains images that are all the same size but successively more blurred as we go to the

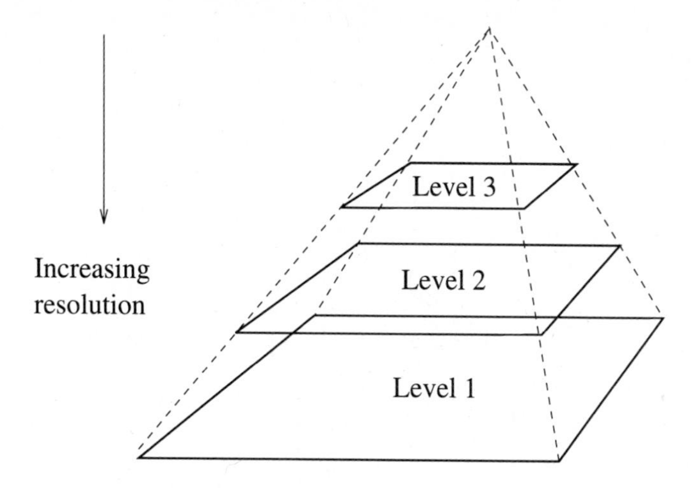

Figure 6.6: Hierarchical image representation.

lower resolution levels. Hierarchical block-matching in such a case is illustrated in Figure 6.7, where the larger blocks are applied to more blurred versions of the image. For simplicity, the low-pass filtering (blurring) may be performed by a box filter which replaces each pixel by a local mean. A typical set of parameters for

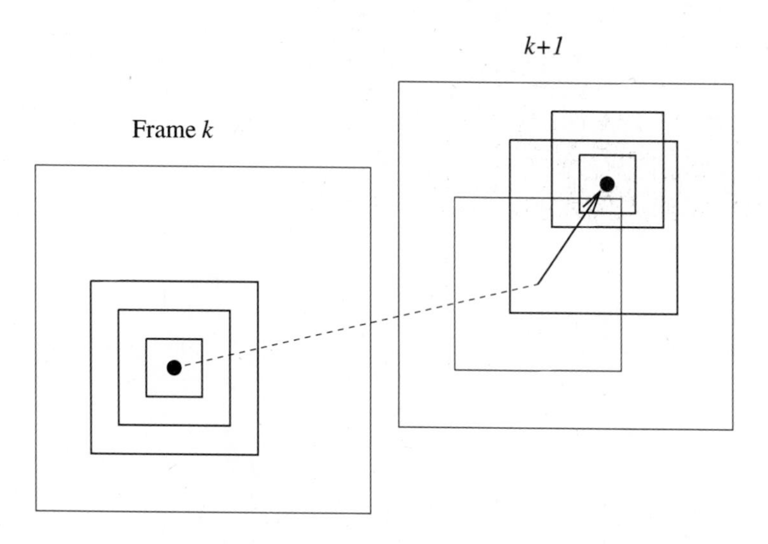

Figure 6.7: Hierarchical block-matching.

Table 6.1: Typical set of parameters for 5-Level hierarchical block-matching.

LEVEL:	5	4	3	2	1
Filter Size	10	10	5	5	3
Maximum Displacement	±31	±15	±7	±3	±1
Block Size	64	32	16	8	4

5-level hierarchical block-matching (with no subsampling) is shown in Table 6.1. Here, the filter size refers to the size of a square window used in computing the local mean.

Figure 6.8 illustrates hierarchical block-matching with 2 levels, where the maximum allowed displacement $M = 7$ for level 2 and $M = 3$ for level 1. The best estimate at the lower resolution level (level 2) is indicated by the circled "3." The center of the search area in level 1 (denoted by "0") corresponds to the best estimate from the second level. The estimates in the second and first levels are $[7,1]^T$ and $[3,1]^T$, respectively, resulting in an overall estimate of $[10,2]^T$. Hierarchical block-matching can also be performed with subpixel accuracy by incorporating appropriate interpolation.

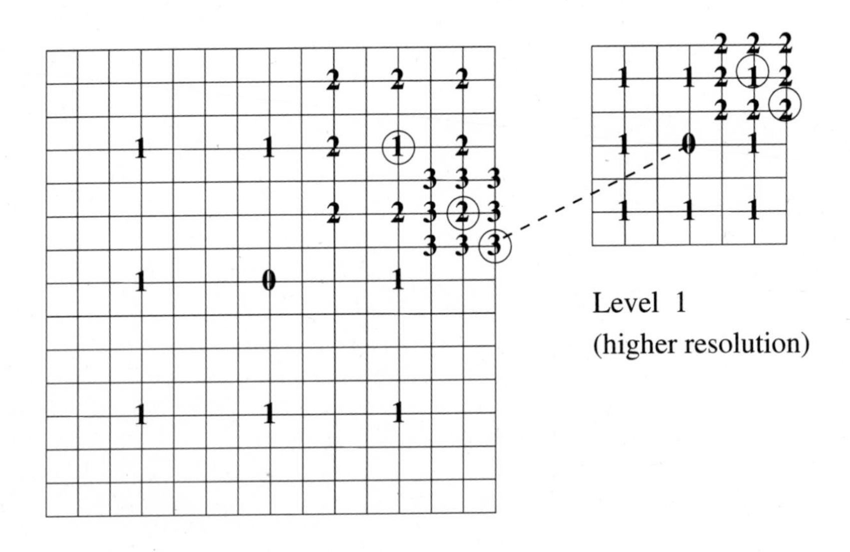

Figure 6.8: Example of hierarchical block-matching with 2 levels.

6.5 Generalized Block-Motion Estimation

While motion estimation based on the translatory block model is simple, it deals poorly with rotations and deformations of blocks from frame to frame, as well as discontinuities in the motion field. We discuss two approaches in the following for improved motion tracking and compensation using block-based methods: a postprocessing approach, and a generalized block-matching approach using spatial transformations.

6.5.1 Postprocessing for Improved Motion Compensation

Block-based motion representation and compensation have been adopted in international standards for video compression, such as H.261, MPEG-1, and MPEG-2, where a single motion vector is estimated for each 16×16 square block, in order to limit the number of motion vectors that need to be transmitted. However, this low-resolution representation of the motion field results in inaccurate compensation and visible blocking artifacts, especially at the borders of the blocks.

To this effect, Orchard [Orc 93] proposed a postprocessing method to reconstruct a higher-resolution motion field based on image segmentation. In this method, a single motion vector is estimated per block, as usual, in the first pass. Next, image blocks are segmented into K regions such that each region is represented by a single motion vector. To avoid storage/transmission of additional motion vectors, the K candidate motion vectors are selected from a set of already estimated motion vectors for the neighboring blocks. Then, the motion vector that minimizes the DFD at each pixel of the block is selected among the set of K candidate vectors. A predictive MAP segmentation scheme has been proposed to avoid transmitting overhead information about the boundaries of these segments, where the decoder/receiver can duplicate the segmentation process. As a result, this method allows for motion compensation using a pixel-resolution motion field, while transmitting the same amount of motion information as classical block-based methods.

6.5.2 Deformable Block Matching

In deformable (or generalized) block matching, the current frame is divided into triangular, rectangular, or arbitrary quadrilateral patches. We then search for the best matching triangle or quadrilateral in the search frame under a given spatial transformation. This is illustrated in Figure 6.9. The choice of patch shape and the spatial transformation are mutually related. For example, triangular patches offer sufficient degrees of freedom (we have two equations per node) with the affine transformation, which has only six independent parameters. Perspective and bilinear transformations have eight free parameters. Hence, they are suitable for use with rectangular or quadrilateral patches. Note that using affine transformation with quadrilateral patches results in an overdetermined motion estimation problem in that affine transformation preserves parallel lines (see Figure 6.2).

The spatial transformations (6.4), (6.5), and (6.6) clearly provide superior motion tracking and rendition, especially in the presence of rotation and zooming, compared to the translational model (6.3). However, the complexity of the motion estimation increases significantly. If a search-based motion estimation strategy is adopted, we now have to perform a search in a 6- or 8-D parameter space (affine or perspective/bilinear transformation) instead of a 2-D space (the x_1 and x_2 components of the translation vector). Several generalized motion estimation schemes have been proposed, including a full-search method [Sef 93], a faster hexagonal search [Nak 94], and a global spline surface-fitting method using a fixed number of point correspondences [Flu92].

The full-search method can be summarized as follows [Sef 93]:

1. Segment the current frame into rectangular (triangular) blocks.
2. Perturb the coordinates of the corners of a matching quadrilateral (triangle) in the search frame starting from an initial guess.
3. For each quadrilateral (triangle), find the parameters of a prespecified spatial transformation that maps this quadrilateral (triangle) onto the rectangular (triangular) block in the current frame using the coordinates of the four (three) matching corners.
4. Find the coordinates of each corresponding pixel within the quadrilateral (triangle) using the computed spatial transformation, and calculate the MSE between the given block and the matching patch.
5. Choose the spatial transformation that yields the smallest MSE or MAD.

In order to reduce the computational load imposed by generalized block-matching, generalized block-matching is only used for those blocks where standard block-matching is not satisfactory. The displaced frame difference resulting from standard block matching can be used as a decision criterion. The reader is referred to [Flu92] and [Nak 94] for other generalized motion estimation strategies.

The generalized block-based methods aim to track the motion of all pixels within a triangular or quadrilateral patch using pixel correspondences established at the corners of the patch. Thus, it is essential that the current frame is segmented into triangular or rectangular patches such that each patch contains pixels only from a single moving object. Otherwise, local motion within these patches cannot be tracked using a single spatial transformation. This observation leads us to feature- and/or motion-based segmentation of the current frame into adaptive meshes (also known as irregular-shaped meshes). Regular and adaptive mesh models are depicted in Figure 6.10 for comparison.

There are, in general, two segmentation strategies: i) Feature-based segmentation: The edges of the patches are expected to align sufficiently closely with gray-level and/or motion edges. Ideally, adaptive mesh fitting and motion tracking should be performed simultaneously, since it is not generally possible to isolate motion rendition errors due to adaptive mesh misfit and motion estimation. A simultaneous optimization method was recently proposed by Wang *et al.* [Wan 94].

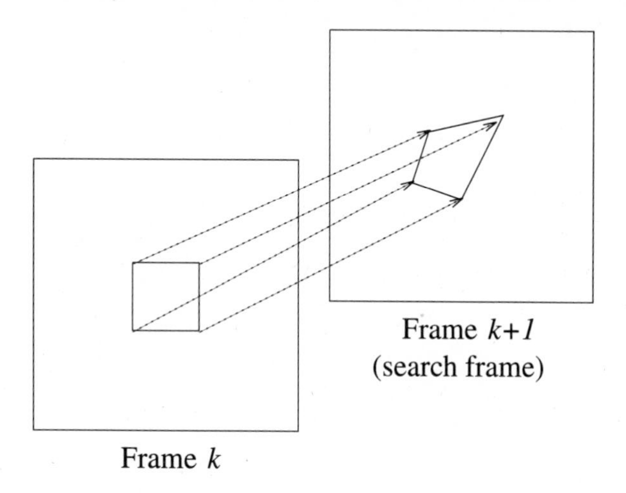

Figure 6.9: Generalized block-matching.

ii) Hierarchical segmentation: It is proposed to start with an initial coarse mesh, and perform a synthesis of the present frame based on this mesh. Then this mesh is refined by successive partitioning of the patches where the initial synthesis error is above a threshold. A method based on the hierarchical approach was proposed by Flusser [Flu92].

Fitting and tracking of adaptive mesh models is an active research area at present. Alternative strategies for motion-based segmentation (although not necessarily into mesh structures) and parametric motion model estimation will be discussed in Chapter 11.

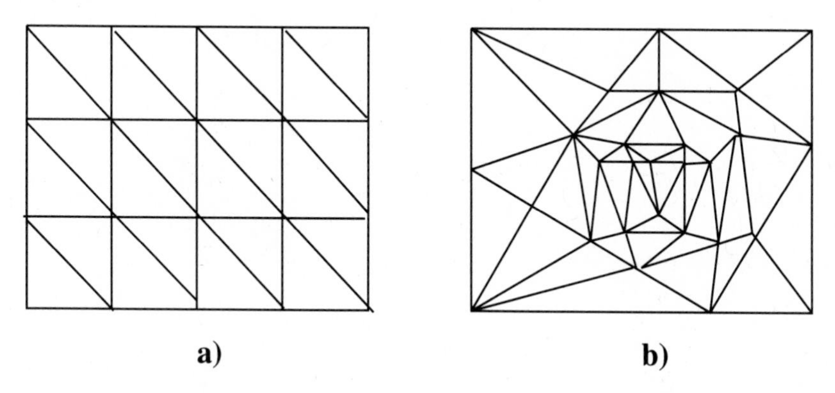

Figure 6.10: a) Regular and b) adaptive mesh models fitted to frame k.

6.6 Examples

We have evaluated four algorithms, phase correlation, block matching (BM), hierarchical BM, and the generalized BM, using the same two frames of the Mobile and Calendar sequence shown in Figure 5.9 (a) and (b).

An example of the phase-correlation function, computed on a 16×16 block on the calendar, using a 32×32 DFT with appropriate zero padding, which allows for a maximum of ± 8 for each component of the motion vector, is plotted in Figure 6.11 (a). The impulsive nature of the function can be easily observed. Figure 6.11 (b) depicts the motion estimates obtained by applying the phase-correlation method with the same parameters centered about each pixel. In our implementation of the phase-correlation method, the motion vector corresponding to the highest peak for each block has been retained. We note that improved results could be obtained by postprocessing of the motion estimates. For example, we may consider two or three highest peaks for each block and then select the vector yielding the smallest displaced frame difference at each pixel.

The BM and 3-level hierarchical BM (HBM) methods use the mean-squared error measure and the three-step search algorithm. The resulting motion fields are shown in Figure 6.12 (a) and (b), respectively. The BM algorithm uses 16×16 blocks and 3×3 blurring for direct comparison with the phase-correlation method. In the case of the HBM, the parameter values given in Table 6.1 (for the first three levels) have been employed, skipping the subsampling step. The generalized BM algorithm has been applied only to those pixels where the displaced frame difference (DFD) resulting from the 3-level HBM is above a threshold. The PSNR of the DFD and the entropy of the estimated motion fields are shown in Table 6.2 for all four methods. The 3-level HBM is prefered over the generalized BM algorithm, since the latter generally requires an order of magnitude more computation.

Table 6.2: Comparison of the block-based methods. (Courtesy Gozde Bozdagi)

Method	PSNR (dB)	Entropy (bits)
Frame-difference	23.45	-
Phase correlation	32.70	5.64
Block matching (BM)	29.76	4.62
Hierarchical BM	36.29	7.32
Generalized BM	36.67	7.31

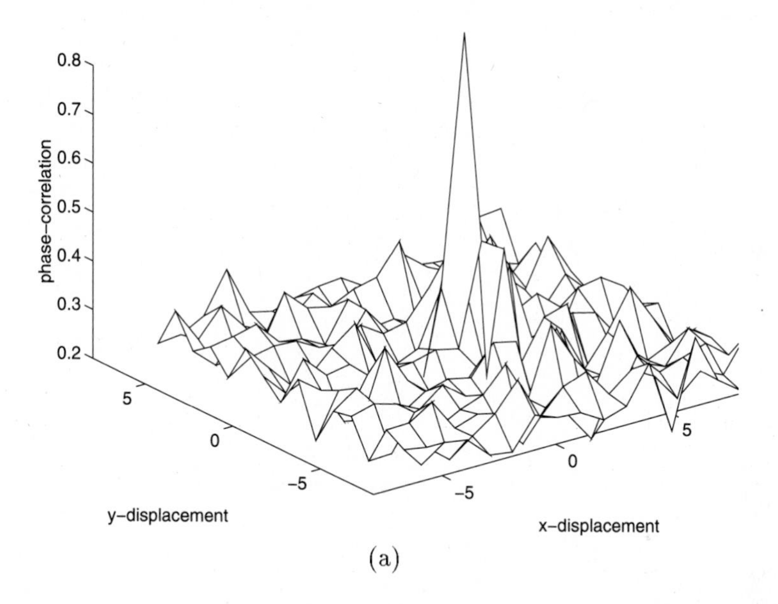

(a)

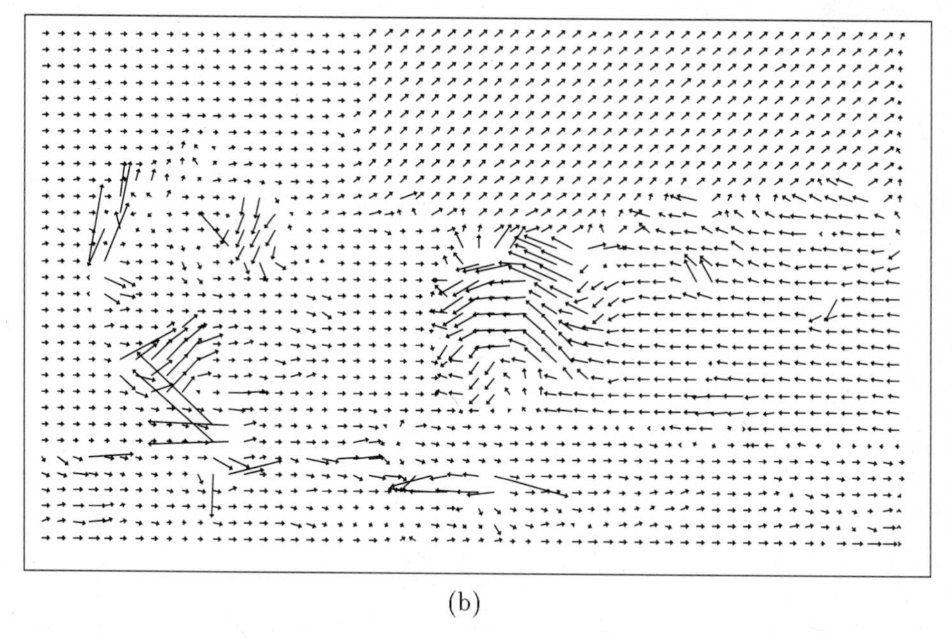

(b)

Figure 6.11: a) The phase-correlation function, and b) the motion field obtained by the phase-correlation method. (Courtesy Gozde Bozdagi)

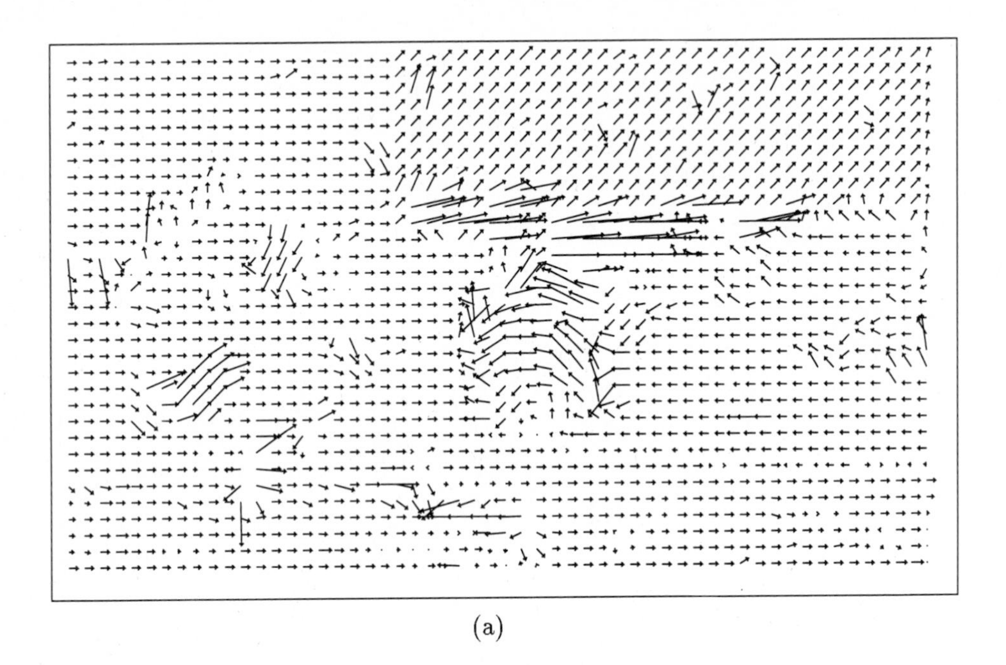

(a)

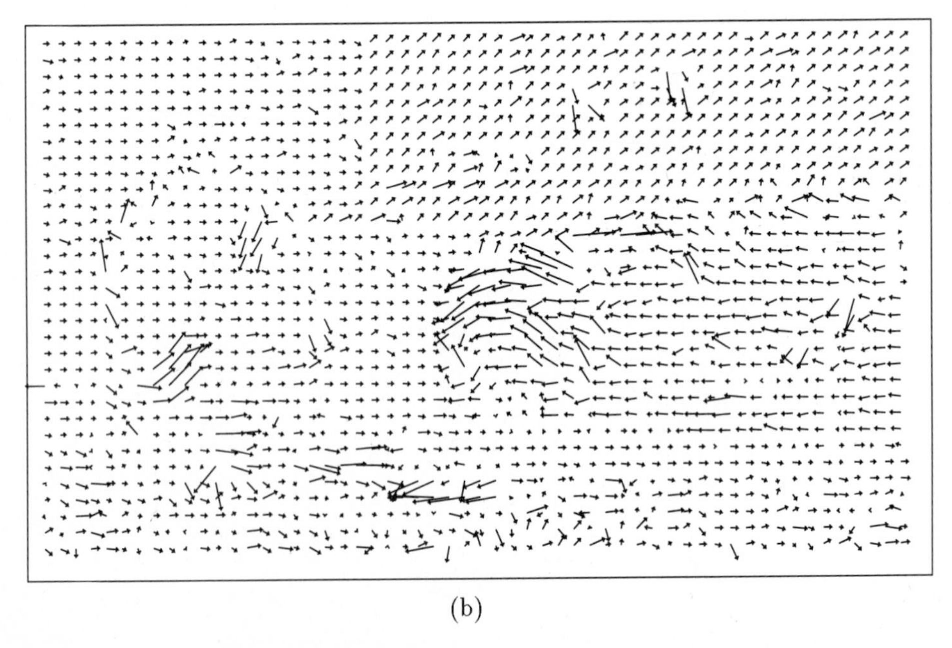

(b)

Figure 6.12: The motion field obtained by a) the block matching and b) the hierarchical block-matching methods. (Courtesy Mehmet Ozkan and Gozde Bozdagi)

6.7 Exercises

1. How do you deal with the boundary effects in the phase-correlation method? Can we use the discrete cosine transform (DCT) instead of the DFT?
2. Suggest a model to quantify the spectral leakage due to subpixel motion in the phase-correlation method.
3. State the symmetry properties of the DFT for N even and odd, respectively. Verify Equation (6.14).
4. Discuss the aperture problem for the cases of i) single pixel matching, ii) line matching, iii) curve matching, and iv) corner matching.
5. Compare the computational complexity of full search versus i) the three-step search, and ii) search using integral projections [Kim 92].
6. Propose a method which uses the optical flow equation (5.5) for deformable block matching with the affine model. (Hint: Review Section 5.3.3.)

Bibliography

[Bie 88] M. Bierling, "Displacement estimation by hierarchical block-matching," *Proc. Visual Comm. and Image Proc.*, SPIE vol. 1001, pp. 942–951, 1988.

[Cas 87] E. De Castro and C. Morandi, "Registration of translated and rotated images using finite Fourier transforms," *IEEE Trans. Patt. Anal. Mach. Intel.*, vol. 9, no. 5, pp. 700–703, Sep. 1987.

[Cha 91] S. Chaudhury, S. Subramanian, and G. Parthasarathy, "Heuristic search approach to shape matching in image sequences," *IEE Proc.-E*, vol. 138, no. 2, pp. 97–105, 1991.

[Dav 83] L. S. Davis, Z. Wu, and H. Sun, "Contour-based motion estimation," *Comput. Vis. Graphics Image Proc.*, vol. 23, pp. 313–326, 1983.

[Erk 93] Y. M. Erkam, M. I. Sezan, and A. T. Erdem, "A hierarchical phase-correlation method for motion estimation," *Proc. Conf. on Info. Scien. and Systems*, Baltimore MD, Mar. 1993, pp. 419–424.

[Flu92] J. Flusser, "An Adaptive Method for Image Registration," *Patt. Recog.*, vol. 25, pp. 45–54, 1992.

[Gha 90] M. Ghanbari, "The cross-search algorithm for motion estimation," *IEEE Trans. Commun.*, vol. 38, pp. 950–953, 1990.

[Ghr 90] H. Gharavi and M. Mills, "Block-matching motion estimation algorithms: New results," *IEEE Trans. Circ. and Syst.*, vol. 37, pp. 649–651, 1990.

[Hee 87] D. J. Heeger, "Model for the extraction of image flow," *J. Opt. Soc. Am. A*, vol. 4, no. 8, pp. 1455–1471, Aug. 1987.

[Jac 87] L. Jacobson and H. Wechsler, "Derivation of optical flow using a spatiotemporal frequency approach," *Comp. Vision Graph. Image Proc.*, vol. 38, pp. 57–61, 1987.

[Jai 81] J. R. Jain and A. K. Jain, "Displacement measurement and its application in interframe image coding," *IEEE Trans. Commun.*, vol. 29, pp. 1799–1808, 1981.

[Kim 92] J.-S. Kim and R.-H. Park, "A fast feature-based block-matching algorithm using integral projections," *IEEE J. Selected Areas in Comm.*, vol. 10, pp. 968–971, June 1992.

[Liu 93] B. Liu and A. Zaccarin, "New fast algorithms for the estimation of block motion vectors," *IEEE Trans. Circ. and Syst. Video Tech.*, vol. 3, no. 2, pp. 148–157, Apr. 1993.

[Nak 94] Y. Nakaya and H. Harashima, "Motion compensation based on spatial transformations," *IEEE Trans. CAS Video Tech.*, vol. 4, pp. 339–356, June 1994.

[Orc 93] M. Orchard, "Predictive motion-field segmentation for image sequence coding," *IEEE Trans. Circ. and Syst: Video Tech.*, vol. 3, pp. 54–70, Feb. 1993.

[Sef 93] V. Seferidis and M. Ghanbari, "General approach to block-matching motion estimation," *Optical Engineering*, vol. 32, pp. 1464–1474, July 1993.

[Sul 93] G. Sullivan, "Multi-hypothesis motion compensation for low bit-rate video coding," *Proc. IEEE Int. Conf. ASSP, Minneapolis, MN*, vol. 5, pp. 437–440, 1993.

[Tho 87] G. A. Thomas and B. A. Hons, "Television motion measurement for DATV and other applications," Tech. Rep. BBC-RD-1987-11, 1987.

[Wan 94] Y. Wang and O. Lee, "Active mesh: A feature seeking and tracking image sequence representation scheme," *IEEE Trans. Image Proc.*, vol. 3, pp. 610–624, Sep. 1994.

[Wol 90] G. Wolberg, *Digital Image Warping*, Los Alamitos, CA: IEEE Comp. Soc. Press, 1990.

Chapter 7

PEL-RECURSIVE METHODS

All motion estimation methods, in one form or another, employ the optical flow constraint accompanied by some smoothness constraint. Pel-recursive methods are predictor-corrector-type estimators, of the form

$$\hat{\mathbf{d}}_a(\mathbf{x}, t; \Delta t) = \hat{\mathbf{d}}_b(\mathbf{x}, t; \Delta t) + \mathbf{u}(\mathbf{x}, t; \Delta t) \tag{7.1}$$

where $\hat{\mathbf{d}}_a(\mathbf{x}, t; \Delta t)$ denotes the estimated motion vector at the location $\mathbf{x}$ and time t, $\hat{\mathbf{d}}_b(\mathbf{x}, t; \Delta t)$ denotes the predicted motion estimate, and $\mathbf{u}(\mathbf{x}, t; \Delta t)$ is the update term. The subscripts "a" and "b" denote after and before the update at the pel location $(\mathbf{x}, t)$. The prediction step, at each pixel, imposes a local smoothness constraint on the estimates, and the update step enforces the optical flow constraint.

The estimator (7.1) is usually employed in a recursive fashion, by performing one or more iterations at $(\mathbf{x}, t)$ and then proceeding to the next pixel in the direction of the scan; hence the name pel-recursive. Early pel-recursive approaches focused on ease of hardware implementation and real-time operation, and thus employed simple prediction and update equations [Rob 83]. Generally, the best available estimate at the previous pel was taken as the predicted estimate for the next pel, followed by a single gradient-based update to minimize the square of the displaced frame difference at that pel. Later, more sophisticated prediction and update schemes that require more computation were proposed [Wal 84, Bie 87, Dri 91, Bor 91].

We start this chapter with a detailed discussion of the relationship between the minimization of the displaced frame difference and the optical flow constraint in Section 7.1. We emphasize that the update step, which minimizes the displaced frame difference at the particular pixel location, indeed enforces the optical flow equation (constraint) at that pixel. Section 7.2 provides an overview of some gradient-based minimization methods that are an integral part of basic pel-recursive methods. Basic pel-recursive methods are presented in Section 7.3. An extension of these methods, called Wiener-based estimation, is covered in Section 7.4.

7.1 Displaced Frame Difference

The fundamental principle in almost all motion estimation methods, known as the optical flow constraint, is that the image intensity remains unchanged from frame to frame along the true motion path (or changes in a known or predictable fashion). The optical flow constraint may be employed in the form of the optical flow equation (5.5), as in Chapter 5, or may be imposed by minimizing the displaced frame difference (6.15), as in block-matching and pel-recursive methods. This section provides a detailed description of the relationship between the minimization of the displaced frame difference (DFD) and the optical flow equation (OFE).

Let the DFD between the time instances t and $t' = t + \Delta t$ be defined by

$$dfd(\mathbf{x}, \mathbf{d}) \doteq s_c(\mathbf{x} + \mathbf{d}(\mathbf{x}, t; \Delta t); t + \Delta t) - s_c(\mathbf{x}, t) \tag{7.2}$$

where $s_c(x_1, x_2, t)$ denotes the time-varying image distribution, and

$$\mathbf{d}(\mathbf{x}, t; \Delta t) \doteq \mathbf{d}(\mathbf{x}) = [d_1(\mathbf{x}) \;\; d_2(\mathbf{x})]^T$$

denotes the displacement vector field between the times t and $t + \Delta t$. We observe that i) if the components of $\mathbf{d}(\mathbf{x})$ assume noninteger values, interpolation is required to compute the DFD at each pixel location; and ii) if $\mathbf{d}(\mathbf{x})$ were equal to the true displacement vector at site $\mathbf{x}$ and there were no interpolation errors, the DFD attains the value of zero at that site under the optical flow constraint.

Next, we expand $s_c(\mathbf{x} + \mathbf{d}(\mathbf{x}), t + \Delta t)$ into a Taylor series about $(\mathbf{x}; t)$, for $\mathbf{d}(\mathbf{x})$ and Δt small, as

$$\begin{aligned} s_c(x_1 + d_1(\mathbf{x}), x_2 + d_2(\mathbf{x}); t + \Delta t) &= s_c(\mathbf{x}; t) + d_1(\mathbf{x}) \frac{\partial s_c(\mathbf{x}; t)}{\partial x_1} \\ &+ d_2(\mathbf{x}) \frac{\partial s_c(\mathbf{x}; t)}{\partial x_2} + \Delta t \frac{\partial s_c(\mathbf{x}; t)}{\partial t} + h.o.t. \end{aligned} \tag{7.3}$$

Substituting (7.3) into (7.2), and neglecting the higher-order terms ($h.o.t.$),

$$dfd(\mathbf{x}, \mathbf{d}) = \frac{\partial s_c(\mathbf{x}; t)}{\partial x_1} d_1(\mathbf{x}) + \frac{\partial s_c(\mathbf{x}; t)}{\partial x_2} d_2(\mathbf{x}) + \Delta t \frac{\partial s_c(\mathbf{x}; t)}{\partial t} \tag{7.4}$$

We investigate the relationship between the DFD and OFE in two cases:

1. *Limit Δt approaches 0*: Setting $dfd(\mathbf{x}, \mathbf{d}) = 0$, dividing both sides of (7.4) by Δt, and taking the limit as Δt approaches 0, we obtain the OFE

$$\frac{\partial s_c(\mathbf{x}; t)}{\partial x_1} v_1(\mathbf{x}, t) + \frac{\partial s_c(\mathbf{x}; t)}{\partial x_2} v_2(\mathbf{x}, t) + \frac{\partial s_c(\mathbf{x}; t)}{\partial t} = 0 \tag{7.5}$$

   where $\mathbf{v}(\mathbf{x}, t) = [v_1(\mathbf{x}, t) \; v_2(\mathbf{x}, t)]^T$ denotes the velocity vector at time t. That is, velocity estimation using the OFE and displacement estimation by setting the DFD equal to zero are equivalent in the limit Δt goes to zero.

2. *For Δt finite*: An estimate of the displacement vector $\hat{\mathbf{d}}(\mathbf{x})$ between any two frames that are Δt apart can be obtained from (7.4) in a number of ways:

   (a) Search for $\hat{\mathbf{d}}(\mathbf{x})$ which would set the left-hand side of (7.4), given by (7.2), to zero over a block of pixels (block-matching strategy).

   (b) Compute $\hat{\mathbf{d}}(\mathbf{x})$ which would set the left-hand side of (7.4) to zero on a pixel-by-pixel basis using a gradient-based optimization scheme (pel-recursive strategy).

   (c) Set $\Delta t = 1$ and $dfd(\mathbf{x}, \hat{\mathbf{d}}) = 0$; solve for $\hat{\mathbf{d}}(\mathbf{x})$ using a set of linear equations obtained from the right-hand side of (7.4) using a block of pixels.

   All three approaches can be shown to be identical if i) local variation of the spatio-temporal image intensity is linear, and ii) velocity is constant within the time interval Δt; that is,

$$\hat{d}_1(\mathbf{x}) = \hat{v}_1(\mathbf{x}, t)\Delta t \quad \text{and} \quad \hat{d}_2(\mathbf{x}) = \hat{v}_2(\mathbf{x}, t)\Delta t$$

In practice, the DFD, $dfd(\mathbf{x}, \mathbf{d})$, hardly ever becomes exactly zero for any value of $\mathbf{d}(\mathbf{x})$, because: i) there is observation noise, ii) there is occlusion (covered/uncovered background problem), iii) errors are introduced by the interpolation step in the case of noninteger displacement vectors, and iv) scene illumination may vary from frame to frame. Therefore, we generally aim to minimize the absolute value or the square of the dfd (7.2) or the left-hand side of the OFE (7.5) to estimate the 2-D motion field. The pel-recursive methods presented in this chapter employ gradient-based optimization techniques to minimize the square of the dfd (as opposed to a search method used in block matching) with an implicit smoothness constraint in the prediction step.

7.2 Gradient-Based Optimization

The most straightforward way to minimize a function $f(u_1, \ldots, u_n)$ of several unknowns is to calculate its partials with respect to each unknown, set them equal to zero, and solve the resulting equations

$$\begin{aligned} \frac{\partial f(\mathbf{u})}{\partial u_1} &= 0 \\ &\vdots \\ \frac{\partial f(\mathbf{u})}{\partial u_n} &= 0 \end{aligned} \tag{7.6}$$

simultaneously for $u_1, \ldots, u_n$. This set of simultaneous equations can be expressed as a vector equation,

$$\nabla_{\mathbf{u}} f(\mathbf{u}) = \mathbf{0} \tag{7.7}$$

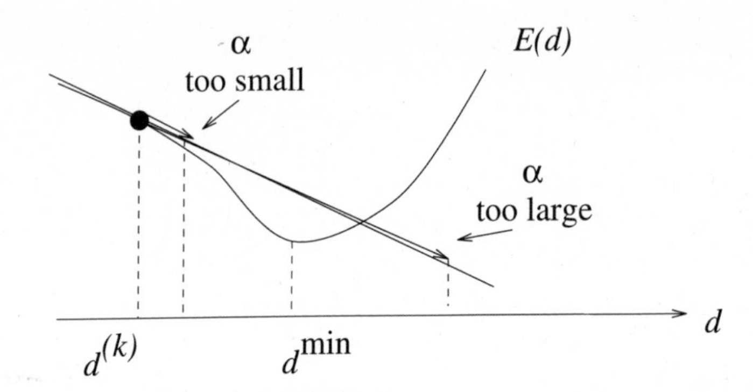

Figure 7.1: An illustration of the gradient descent method.

where $\nabla_{\mathbf{u}}$ is the gradient operator with respect to the unknown vector $\mathbf{u}$. Because it is difficult to define a closed-form criterion function $f(u_1, \ldots, u_n)$ for motion estimation, and/or to solve the set of equations (7.7) in closed form, we resort to iterative (numerical) methods. For example, the DFD is a function of pixel intensities which cannot be expressed in closed form.

7.2.1 Steepest-Descent Method

Steepest descent is probably the simplest numerical optimization method. It updates the present estimate of the location of the minimum in the direction of the negative gradient, called the steepest-descent direction. Recall that the gradient vector points in the direction of the maximum. That is, in one dimension (function of a single variable), its sign will be positive on an "uphill" slope. Thus, the direction of steepest descent is just the opposite direction, which is illustrated in Figure 7.1.

In order to get closer to the minimum, we update our current estimate as

$$\mathbf{u}^{(k+1)} = \mathbf{u}^{(k)} - \alpha \nabla_{\mathbf{u}} f(\mathbf{u})|_{\mathbf{u}^{(k)}} \tag{7.8}$$

where α is some positive scalar, known as the step size. The step size is critical for the convergence of the iterations, because if α is too small, we move by a very small amount each time, and the iterations will take too long to converge. On the other hand, if it is too large the algorithm may become unstable and oscillate about the minimum. In the method of steepest descent, the step size is usually chosen heuristically.

7.2.2 Newton-Raphson Method (one step)

The optimum value for the step size α can be estimated using the well-known Newton-Raphson method for root finding. Here, the derivation for the case of a function of a single variable is shown for simplicity. In one dimension, we would like to find a root of $f'(u)$. To this effect, we expand $f'(u)$ in a Taylor series about the point $u^{(k)}$ to obtain

$$f'(u^{(k+1)}) = f'(u^{(k)}) + (u^{(k+1)} - u^{(k)})f''(u^{(k)}) \tag{7.9}$$

Since we wish $u^{(k+1)}$ to be a zero of $f'(u)$, we set

$$f'(u^{(k)}) + (u^{(k+1)} - u^{(k)})f''(u^{(k)}) = 0 \tag{7.10}$$

Solving (7.10) for $u^{(k+1)}$, we have

$$u^{(k+1)} = u^{(k)} - \frac{f'(u^{(k)})}{f''(u^{(k)})} \tag{7.11}$$

This result can be generalized for the case of a function of several unknowns as

$$\mathbf{u}^{(k+1)} = \mathbf{u}^{(k)} - \mathbf{H}^{-1} \nabla_{\mathbf{u}} f(\mathbf{u})|_{\mathbf{u}^{(k)}} \tag{7.12}$$

where $\mathbf{H}$ is the Hessian matrix

$$\mathbf{H}_{ij} = \left[\frac{\partial^2 f(\mathbf{u})}{\partial u_i \partial u_j} \right]$$

The Newton-Raphson method finds an analytical expression for the step-size parameter in terms of the second-order partials of the criterion function. When a closed-form criterion function is not available, the Hessian matrix can be estimated by using numerical methods [Fle 87].

7.2.3 Local vs. Global Minima

The gradient descent approach suffers from a serious drawback: the solution depends on the initial point. If we start in a "valley," it will be stuck at the bottom of that valley, even if it is a "local" minimum. Because the gradient vector is zero or near zero, at or around a local minimum, the updates become too small for the method to move out of a local minimum. One solution to this problem is to initialize the algorithm at several different starting points, and then pick the solution that gives the smallest value of the criterion function.

More sophisticated optimization methods, such as simulated annealing, exist in the literature to reach the global minimum regardless of the starting point. However, these methods usually require significantly more processing time. We will discuss simulated annealing techniques in detail in the next chapter.

7.3 Steepest-Descent-Based Algorithms

Pel-recursive motion estimation is usually preceeded by a change detection stage, where the frame difference at each pixel is tested against a threshold. Estimation is performed only at those pixels belonging to the changed region. The steepest-descent-based pel-recursive algorithms estimate the update term $\mathbf{u}(\mathbf{x}, t; \Delta t)$ in (7.1), at each pel in the changed region, by minimizing a positive-definite function E of the frame difference dfd with respect to $\mathbf{d}$. The function E needs to be positive definite, such as the square function, so that the minimum occurs when the dfd is zero. The dfd converges to zero locally, when $\hat{\mathbf{d}}_a(\mathbf{x}, t; \Delta t)$ converges to the actual displacement. Therefore, the update step corresponds to imposing the optical flow constraint locally. In the following, we present some variations on the basic steepest-descent-based pel-recursive estimation scheme.

7.3.1 Netravali-Robbins Algorithm

The Netravali-Robbins algorithm finds an estimate of the displacement vector, which minimizes the square of the DFD at each pixel, using a gradient descent method. Then the criterion function to be minimized is given by

$$E(\mathbf{x}; \mathbf{d}) = [dfd(\mathbf{x}, \mathbf{d})]^2 \tag{7.13}$$

From Section 7.2, minimization of $E(\mathbf{x}; \mathbf{d})$ with respect to $\mathbf{d}$, at pixel $\mathbf{x}$, by the steepest descent method yields the iteration

$$\begin{aligned} \hat{\mathbf{d}}^{i+1}(\mathbf{x}) &= \hat{\mathbf{d}}^{i}(\mathbf{x}) - (1/2)\epsilon \, \nabla_{\mathbf{d}} [dfd(\mathbf{x}, \mathbf{d})|_{\mathbf{d}=\hat{\mathbf{d}}^i}]^2 \\ &= \hat{\mathbf{d}}^{i}(\mathbf{x}) - \epsilon \; dfd(\mathbf{x}, \hat{\mathbf{d}}^i) \, \nabla_{\mathbf{d}} dfd(\mathbf{x}, \mathbf{d})|_{\mathbf{d}=\hat{\mathbf{d}}^i} \end{aligned} \tag{7.14}$$

where ∇ is the gradient with respect to $\mathbf{d}$, and ϵ is the step size. Recall that the negative gradient points to the direction of steepest descent.

We now discuss the evaluation of $\nabla_{\mathbf{d}} dfd(\mathbf{x}, \mathbf{d})$. From (7.2), we can write

$$dfd(\mathbf{x}, \mathbf{d}) - dfd(\mathbf{x}, \hat{\mathbf{d}}^i) = s_c(\mathbf{x} + \mathbf{d}, t + \Delta t) - s_c(\mathbf{x} + \hat{\mathbf{d}}^i, t + \Delta t) \tag{7.15}$$

Now, expanding the intensity $s_c(\mathbf{x} + \mathbf{d}, t + \Delta t)$ at an arbitrary point $\mathbf{x} + \mathbf{d}$ into a Taylor series about $\mathbf{x} + \hat{\mathbf{d}}^i$, we have

$$\begin{aligned} s_c(\mathbf{x} + \mathbf{d}, t + \Delta t) &= s_c(\mathbf{x} + \hat{\mathbf{d}}^i, t + \Delta t) + \\ & \quad (\mathbf{d} - \hat{\mathbf{d}}^i)^T \nabla_{\mathbf{x}} s_c(\mathbf{x} - \mathbf{d}; t - \Delta t)|_{\mathbf{d}=\hat{\mathbf{d}}^i} + o(\mathbf{x}, \hat{\mathbf{d}}^i) \end{aligned} \tag{7.16}$$

where $o(\mathbf{x}, \hat{\mathbf{d}}^i)$ denotes the higher-order terms in the series. Substituting the Taylor series expansion into (7.15), we obtain the linearized DFD expression

$$dfd(\mathbf{x}, \mathbf{d}) = dfd(\mathbf{x}, \hat{\mathbf{d}}^i) + \nabla_{\mathbf{x}}^T s_c(\mathbf{x} - \hat{\mathbf{d}}^i; t - \Delta t)(\mathbf{d} - \hat{\mathbf{d}}^i) + o(\mathbf{x}, \hat{\mathbf{d}}^i) \tag{7.17}$$

where $\nabla_{\mathbf{x}} s_c(\mathbf{x} - \hat{\mathbf{d}}^i; t - \Delta t) \doteq \nabla_{\mathbf{x}} s_c(\mathbf{x} - \mathbf{d}; t - \Delta t)|_{\mathbf{d}=\hat{\mathbf{d}}^i}$.

Using (7.17) and ignoring the higher-order terms, we can express the gradient of the DFD with respect to $\mathbf{d}$ in terms of the spatial gradient of image intensity as

$$\nabla_{\mathbf{d}} dfd(\mathbf{x}, \mathbf{d})|_{\mathbf{d}=\hat{\mathbf{d}}^i} = \nabla_{\mathbf{x}} s_c(\mathbf{x} - \hat{\mathbf{d}}^i; t - \Delta t) \tag{7.18}$$

and the pel-recursive estimator becomes

$$\hat{\mathbf{d}}^{i+1}(\mathbf{x}) = \hat{\mathbf{d}}^{i}(\mathbf{x}) - \epsilon \; dfd(\mathbf{x}, \hat{\mathbf{d}}^{i}) \; \nabla_{\mathbf{x}} s_c(\mathbf{x} - \hat{\mathbf{d}}^{i}; t - \Delta t) \tag{7.19}$$

In (7.19), the first and second terms are the prediction and update terms, respectively. Note that the evaluation of the frame difference $dfd(\mathbf{x}, \hat{\mathbf{d}}^{i})$ and the spatial gradient vector may require interpolation of the intensity value for noninteger displacement estimates.

The aperture problem is also apparent in the pel-recursive algorithms. The update term is a vector along the spatial gradient of the image intensity. Clearly, no correction is performed in the direction perpendicular to the gradient vector.

In an attempt to further simplify the structure of the estimator, Netravali and Robbins also proposed the modified estimation formula

$$\hat{\mathbf{d}}^{i+1}(\mathbf{x}) = \hat{\mathbf{d}}^{i}(\mathbf{x}) - \epsilon \; sgn\{dfd(\mathbf{x}, \hat{\mathbf{d}}^{i})\} sgn\{\nabla_{\mathbf{x}} s_c(\mathbf{x} - \hat{\mathbf{d}}^{i}; t - \Delta t)\} \tag{7.20}$$

where the update term takes one of the three values, $\pm\epsilon$ and zero. In this case, the motion estimates are updated only in 0, 45, 90, 135, ... degrees directions.

The convergence and the rate of convergence of the Netravali-Robbins algorithm depend on the choice of the step size parameter ϵ. For example, if $\epsilon = 1/16$, then at least 32 iterations are required to estimate a displacement by 2 pixels. On the other hand, a large choice for the step size may cause an oscillatory behavior. Several strategies can be advanced to facilitate faster convergence of the algorithm.

7.3.2 Walker-Rao Algorithm

Walker and Rao [Wal 84] proposed an adaptive step size motivated by the following observations: i) In the neighborhood of an edge where $|\nabla s_c(x_1, x_2, t)|$ is large, ϵ should be small if the DFD is small, so that we do not make an unduly large update. Furthermore, in the neighborhood of an edge, the accuracy of motion estimation is vitally important, which also necessitates a small step size. ii) In uniform image areas where $|\nabla s_c(x_1, x_2, t)|$ is small, we need a large step size when the DFD is large. Both of these requirements can be met by a step size of the form

$$\epsilon = \frac{1}{2 \; ||\nabla_{\mathbf{x}} s_c(\mathbf{x} - \hat{\mathbf{d}}^{i}; t - \Delta t)||^2} \tag{7.21}$$

In addition, Walker and Rao have introduced the following heuristic rules:

1. If the DFD is less than a threshold, the update term is set equal to zero.

2. If the DFD exceeds the threshold, but the magnitude of the spatial image gradient is zero, then the update term is again set equal to zero.

3. If the absolute value of the update term (for each component) is less than 1/16, then it is set equal to $\pm 1/16$.

4. If the absolute value of the update term (for each component) is more than 2, then it is set equal to ± 2.

Caffario and Rocca have also developed a similar step-size expression [Caf 83]

$$\epsilon = \frac{1}{||\nabla_{\mathbf{x}} s_c(\mathbf{x} - \hat{\mathbf{d}}^i; t - \Delta t)||^2 + \eta^2} \tag{7.22}$$

which includes a bias term η^2 to avoid division by zero in areas of constant intensity where the spatial gradient is almost zero. A typical value for $\eta^2 = 100$.

Experimental results indicate that using an adaptive step size greatly improves the convergence of the Netravali-Robbins algorithm. It has been found that five iterations were sufficient to achieve satisfactory results in most cases.

7.3.3 Extension to the Block Motion Model

It is possible to impose a stronger regularity constraint on the estimates at each pixel $\mathbf{x}$, by assuming that the displacement remains constant locally over a sliding support $\mathcal{B}$ about each pixel. We can, then, minimize the DFD over the support $\mathcal{B}$, defined as

$$E(\mathbf{x}, \mathbf{d}) = \sum_{\mathbf{x}_B \in \mathcal{B}} [dfd(\mathbf{x}_B, \mathbf{d})]^2 \tag{7.23}$$

as opposed to on a pixel-by-pixel basis. Notice that the support $\mathcal{B}$ needs to be "causal" (in the sense of recursive computability) in order to preserve the pel-recursive nature of the algorithm. A typical causal support with $N = 7$ pixels is shown in Figure 7.2.

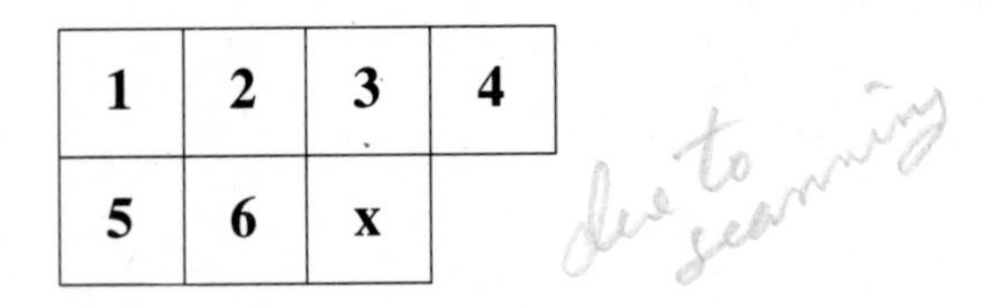

Figure 7.2: A causal support $\mathcal{B}$ for $N = 7$.

Following steps similar to those in the derivation of the pixel-by-pixel algorithm, steepest-descent minimization of this criterion function yields the iteration

$$\hat{\mathbf{d}}^{i+1}(\mathbf{x}) = \hat{\mathbf{d}}^i(\mathbf{x}) - \epsilon \sum_{\mathbf{x}_B \in \mathcal{B}} dfd(\mathbf{x}_B, \hat{\mathbf{d}}^i(\mathbf{x}))\ \nabla_{\mathbf{x}}\ s_c(\mathbf{x}_B - \hat{\mathbf{d}}^i(\mathbf{x}); t - \Delta t) \tag{7.24}$$

Observe that this formulation is equivalent to that of block matching, except for the shape of the support $\mathcal{B}$. Here, the solution is sought for using the steepest-descent minimization rather than a search strategy.

7.4 Wiener-Estimation-Based Algorithms

The Wiener-estimation-based method is an extension of the Netravali-Robbins algorithm in the case of block motion, where the higher-order terms in the linearized DFD expression (7.17) are not ignored. Instead, a linear least squares error (LLSE) or linear minimum mean square error (LMMSE) estimate of the update term

$$\mathbf{u}^i(\mathbf{x}) = \mathbf{d}(\mathbf{x}) - \hat{\mathbf{d}}^i(\mathbf{x}) \tag{7.25}$$

where $\mathbf{d}(\mathbf{x})$ denotes the true displacement vector, is derived based on a neighborhood $\mathcal{B}$ of a pel $\mathbf{x}$. In the following, we provide the derivation of the Wiener estimator for the case of N observations with a common displacement vector $\mathbf{d}(\mathbf{x})$.

Writing the linearized DFD (7.17), with $dfd(\mathbf{x}_B, \mathbf{d}) = 0$, at all N pixels $\mathbf{x}_B$ within the support $\mathcal{B}$, we have N equations in the two unknowns (the components of $\mathbf{u}^i$) given by

$$\begin{aligned}
-dfd(\mathbf{x}_B(1), \hat{\mathbf{d}}^i(\mathbf{x})) &= \nabla^T s_c(\mathbf{x}_B(1) - \hat{\mathbf{d}}^i(\mathbf{x}), t - \Delta t)\mathbf{u}^i + o(\mathbf{x}_B(1), \hat{\mathbf{d}}^i(\mathbf{x})) \\
-dfd(\mathbf{x}_B(2), \hat{\mathbf{d}}^i(\mathbf{x})) &= \nabla^T s_c(\mathbf{x}_B(2) - \hat{\mathbf{d}}^i(\mathbf{x}), t - \Delta t)\mathbf{u}^i + o(\mathbf{x}_B(2), \hat{\mathbf{d}}^i(\mathbf{x})) \\
\vdots &= \vdots \\
-dfd(\mathbf{x}_B(N), \hat{\mathbf{d}}^i(\mathbf{x})) &= \nabla^T s_c(\mathbf{x}_B(N) - \hat{\mathbf{d}}^i(\mathbf{x}), t - \Delta t)\mathbf{u}^i + o(\mathbf{x}_B(N), \hat{\mathbf{d}}^i(\mathbf{x}))
\end{aligned}$$

where $\mathbf{x}_B(1), \ldots, \mathbf{x}_B(N)$ denote an ordering of the pixels within the support $\mathcal{B}$ (shown in Figure 7.2). These equations can be expressed in vector-matrix form as

$$\mathbf{z} = \mathbf{\Phi}\mathbf{u}(\mathbf{x}) + \mathbf{n} \tag{7.26}$$

where

$$\mathbf{z} = \begin{bmatrix} -dfd(\mathbf{x}_B(1), \hat{\mathbf{d}}^i(\mathbf{x})) \\ -dfd(\mathbf{x}_B(2), \hat{\mathbf{d}}^i(\mathbf{x})) \\ \vdots \\ -dfd(\mathbf{x}_B(N), \hat{\mathbf{d}}^i(\mathbf{x})) \end{bmatrix}$$

$$\mathbf{\Phi} = \begin{bmatrix} \frac{\partial s_c(\mathbf{x}_B(1) - \hat{\mathbf{d}}^i, t-\Delta t)}{\partial x_1} & \frac{\partial s_c(\mathbf{x}_B(1) - \hat{\mathbf{d}}^i, t-\Delta t)}{\partial x_2} \\ \frac{\partial s_c(\mathbf{x}_B(2) - \hat{\mathbf{d}}^i, t-\Delta t)}{\partial x_1} & \frac{\partial s_c(\mathbf{x}_B(2) - \hat{\mathbf{d}}^i, t-\Delta t)}{\partial x_2} \\ \vdots & \\ \frac{\partial s_c(\mathbf{x}_B(N) - \hat{\mathbf{d}}^i, t-\Delta t)}{\partial x_1} & \frac{\partial s_c(\mathbf{x}_B(N) - \hat{\mathbf{d}}^i, t-\Delta t)}{\partial x_2} \end{bmatrix}$$

and

$$\mathbf{n} = \begin{bmatrix} o(\mathbf{x}_B(1), \hat{\mathbf{d}}^i) \\ o(\mathbf{x}_B(2), \hat{\mathbf{d}}^i) \\ \vdots \\ o(\mathbf{x}_B(N), \hat{\mathbf{d}}^i) \end{bmatrix}$$

Assuming that the update term $\mathbf{u}(\mathbf{x})$ and the truncation error $\mathbf{n}$ are uncorrelated random vectors, and using the principle of orthogonality, the LMMSE estimate of the update term is given by [Bie 87]

$$\hat{\mathbf{u}}(\mathbf{x}) = [\mathbf{\Phi}^T \mathbf{R}_n^{-1} \mathbf{\Phi} + \mathbf{R}_u^{-1}]^{-1} \mathbf{\Phi}^T \mathbf{R}_n^{-1} \mathbf{z} \tag{7.27}$$

Note that the solution requires the knowledge of the covariance matrices of both the update term $\mathbf{R}_u$ and the linearization error $\mathbf{R}_n$. In the absence of exact knowledge of these quantities, we will make the simplifying assumptions that both vectors have zero mean value, and their components are uncorrelated among each other. That is, we have $\mathbf{R}_u = \sigma_u^2 \mathbf{I}$ and $\mathbf{R}_n = \sigma_n^2 \mathbf{I}$, where $\mathbf{I}$ is a 2×2 identity matrix, and σ_u^2 and σ_n^2 are the variances of the components of the two vectors, respectively. Then the LMMSE estimate (7.27) simplifies to

$$\hat{\mathbf{u}}(\mathbf{x}) = [\mathbf{\Phi}^T \mathbf{\Phi} + \mu \mathbf{I}]^{-1} \mathbf{\Phi}^T \mathbf{z} \tag{7.28}$$

where $\mu = \sigma_n^2 / \sigma_u^2$ is called the damping parameter. Equation (7.28) gives the least squares estimate of the update term. The assumptions that are used to arrive at the simplified estimator are not, in general, true; for example, the linearization error is not uncorrelated with the update term, and the updates and the linearization errors at each pixel are not uncorrelated with each other. However, experimental results indicate improved performance compared with other pel-recursive estimators [Bie 87].

Having obtained an estimate of the update term, the Wiener-based pel-recursive estimator can be written as

$$\hat{\mathbf{d}}^{i+1}(\mathbf{x}) = \hat{\mathbf{d}}^{i}(\mathbf{x}) + [\mathbf{\Phi}^T \mathbf{\Phi} + \mu \mathbf{I}]^{-1} \mathbf{\Phi}^T \mathbf{z} \tag{7.29}$$

It has been pointed out that the Wiener-based estimator is related to the Walker-Rao and Caffario-Rocca algorithms [Bie 87]. This can easily be seen by writing (7.29) for the special case of $N = 1$ as

$$\hat{\mathbf{d}}^{i+1}(\mathbf{x}) = \hat{\mathbf{d}}^{i}(\mathbf{x}) - \frac{dfd(\mathbf{x}, \hat{\mathbf{d}}^i) \nabla^T s_c(\mathbf{x} - \hat{\mathbf{d}}^i, t - \Delta t)}{|\nabla^T s_c(\mathbf{x} - \hat{\mathbf{d}}^i, t - \Delta t)|^2 + \mu} \tag{7.30}$$

The so-called simplified Caffario-Rocca algorithm results when $\mu = 100$. We obtain the Walker-Rao algorithm when we set $\mu = 0$ and multiply the update term by $1/2$. The convergence properties of the Wiener-based estimators have been analyzed in [Bor 91].

The Wiener-estimation-based scheme presented in this section employs the best motion vector estimate from the previous iteration as the prediction for the next iteration. An improved algorithm with a motion-compensated spatio-temporal vector predictor was proposed in [Dri 91]. Pel-recursive algorithms have also been extended to include rotational motion [Bie 88]. As a final remark, we note that all pel-recursive algorithms can be applied hierarchically, using a multiresolution representation of the images to obtain improved results. Pel-recursive algorithms have recently evolved into Bayesian motion estimation methods, employing stochastic motion-field models, which will be introduced in the next chapter.

7.5 Examples

We have applied the Walker-Rao algorithm, given by (7.19), (7.22), and the heuristic rules stated in Section 7.3.2, and the Wiener-based method given by (7.29) to the same two frames of the Mobile and Calendar sequence that we used in Chapters 5 and 6. We generated two sets of results with the Walker-Rao algorithm, where we allowed 2 and 10 iterations at each pixel, respectively. In both cases, we have set the threshold on the DFD (see heuristic rules 1 and 2) equal to 3, and limited the maximum displacement to ± 10 as suggested by [Wal 84]. In the Wiener-estimation-based approach, we employed the support shown in Figure 7.2 with $N = 7$, allowed 2 iterations at each pixel, and set the damping parameter $\mu = 100$.

The effectiveness of the methods is evaluated visually, by inspection of the motion vector fields shown in Figure 7.3 (for the case of two iterations/pixel, I=2), and numerically, by comparing the PSNR and entropy values tabulated in Table 7.1. A few observations about the estimated motion fields are in order: i) Inspection of the upper left corner of the Walker-Rao estimate, shown in Figure 7.3 (a) indicates that the displacement estimates converge to the correct background motion after processing about 10 pixels. ii) In the Walker-Rao estimate, there are many outlier vectors in the regions of rapidly changing motion. iii) The Wiener estimate, shown in Figure 7.3 (b), contains far fewer outliers but gives zero motion vectors in the uniform areas (see flat areas in the background, the foreground, and the calendar). This may be overcome by using a larger support $\mathcal{B}$. In pel-recursive estimation, propagation of erroneous motion estimates may be overcome by resetting the predicted estimate to the zero vector, if the frame difference is smaller than the DFD obtained by the actual predicted estimate at that pixel. However, we have not implemented this option in our software.

In terms of motion compensation, the Wiener-based method is about 0.8 dB better than the Walker-Rao method if we allow 2 iterations/pixel. However, if we allow 10 iterations/pixel, the performance of the Wiener-based method remains about the same, and both methods perform similarly. Observe that the entropy of the Wiener-based motion field estimate is smaller than half of both Walker-Rao estimates, for I=2 and I=10, by virtue of its regularity. The two Walker-Rao motion field estimates are visually very close, as indicated by their similar entropies.

Table 7.1: Comparison of the pel-recursive methods. (Courtesy Gozde Bozdagi)

Method	PSNR (dB)	Entropy (bits)
Frame-Difference	23.45	-
Walker-Rao (I=2)	29.01	8.51
Wiener (I=2)	29.82	4.49
Walker-Rao (I=10)	29.92	8.54

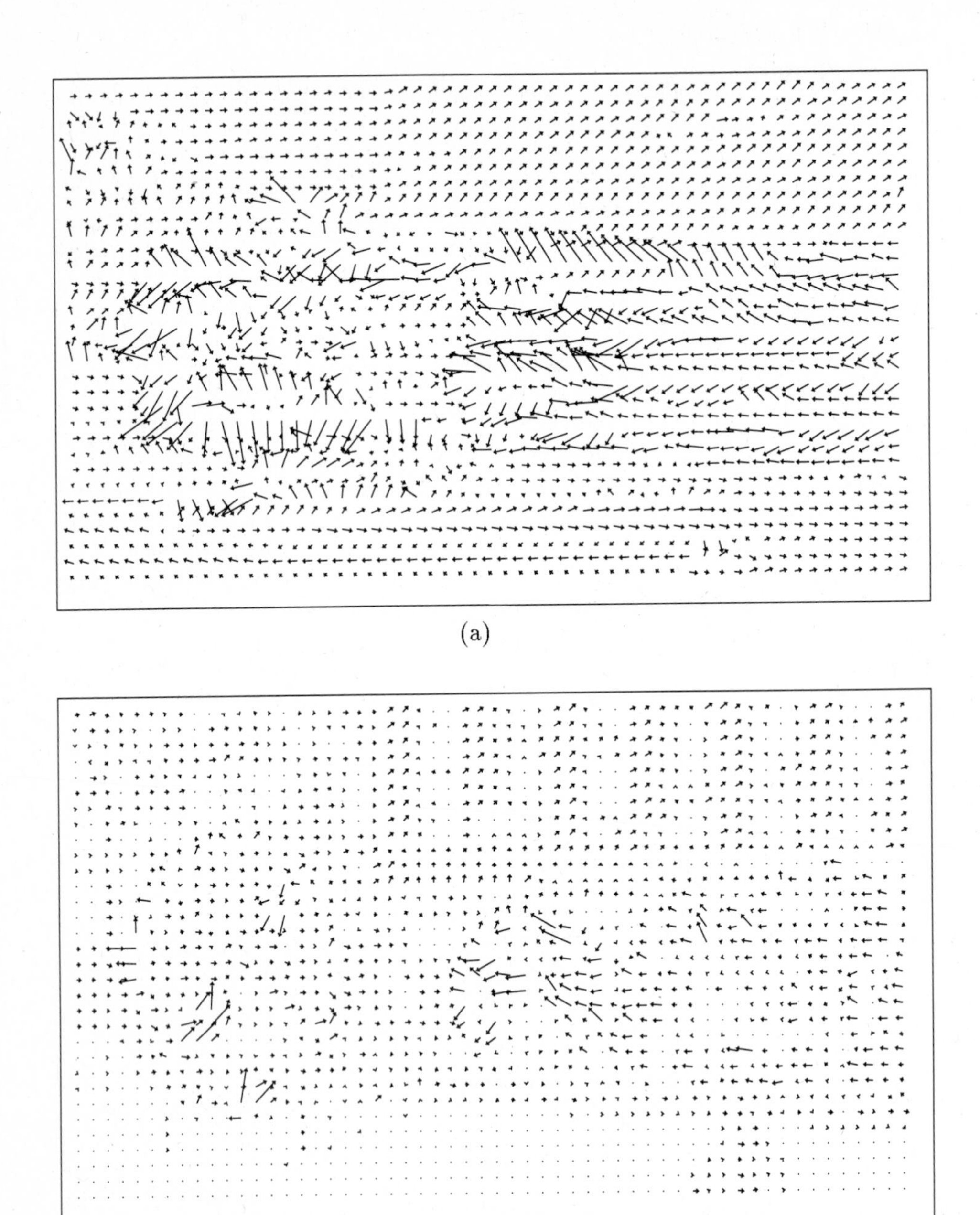

Figure 7.3: Motion field obtained by a) the Walker-Rao method and b) the Wiener-based method. (Courtesy Gozde Bozdagi and Mehmet Ozkan)

7.6 Exercises

1. Derive (7.12). How would you compute the optimal step size for the Netravali-Robbins algorithm?
2. Derive an explicit expression for (7.24). Comment on the relationship between block matching and the algorithm given by (7.24).
3. Comment on the validity of the assumptions made to arrive at the Wiener-based estimator (7.29).
4. Derive (7.30) from (7.29).

Bibliography

[Bie 87] J. Biemond, L. Looijenga, D. E. Boekee, and R. H. J. M. Plompen, "A pel-recursive Wiener-based displacement estimation algorithm," *Sign. Proc.*, vol. 13, pp. 399–412, Dec. 1987.

[Bie 88] J. Biemond, J. N. Driessen, A. M. Geurtz, and D. E. Boekee, "A pel-recursive Wiener based algorithm for the simultaneous estimation of rotation and translation," *Proc. SPIE Conf. Visual Commun. Image Proc.*, pp. 917–924, Cambridge MA, 1988.

[Bor 91] L. Boroczky, *Pel-Recursive Motion Estimation for Image Coding*, Ph.D. thesis, Delft University of Technology, 1991.

[Caf 83] C. Cafforio and F. Rocca, "The differential method for image motion estimation," in *Image Sequence Processing and Dynamic Scene Analysis*, T. S. Huang, ed., pp. 104–124, Berlin, Germany: Springer-Verlag, 1983.

[Dri 91] J. N. Driessen, L. Boroczky, and J. Biemond, "Pel-recursive motion field estimation from image sequences," *J. Vis. Comm. Image Rep.*, vol. 2, no. 3, pp. 259–280, 1991.

[Fle 87] R. Fletcher, *Practical Methods of Optimization*, Vol. 1, 2nd ed., Chichester, UK: John Wiley and Sons, 1987.

[Rob 83] J. D. Robbins and A. N. Netravali, "Recursive motion compensation: A review," in *Image Sequence Processing and Dynamic Scene Analysis*, T. S. Huang, ed., pp. 76–103, Berlin, Germany: Springer-Verlag, 1983.

[Wal 84] D. R. Walker and K. R. Rao, "Improved pel-recursive motion compensation," *IEEE Trans. Commun.*, vol. COM-32, pp. 1128–1134, Oct. 1984.

Chapter 8

BAYESIAN METHODS

In this chapter, 2-D motion estimation is formulated and solved as a Bayesian estimation problem. In the previous chapters, where we presented deterministic formulations of the problem, we minimized either the error in the optical flow equation or a function of the displaced frame difference (DFD). Here, the deviation of the DFD from zero is modeled by a random process that is exponentially distributed. Furthermore, a stochastic smoothness constraint is introduced by modeling the 2-D motion vector field in terms of a Gibbs distribution. The reader is referred to Appendix A for a brief review of the definitions and properties of Markov and Gibbs random fields. The clique potentials of the underlying Gibbsian distribution are selected to assign a higher *a priori* probability to slowly varying motion fields. In order to formulate directional smoothness constraints, more structured Gibbs random fields (GRF) with line processes are also introduced.

Since Bayesian estimation requires global optimization of a cost function, we study a number of optimization methods, including simulated annealing (SA), iterated conditional modes (ICM), mean field annealing (MFA), and highest confidence first (HCF) in Section 8.1. Section 8.2 provides the basic formulation of the maximum *a posteriori* probability (MAP) estimation problem. Extensions of the basic formulation to deal with motion discontinuities and occlusion areas are discussed in Section 8.3. It will be seen that block/pel matching and Horn-Schunck algorithms form special cases of the MAP estimator under certain assumptions.

8.1 Optimization Methods

Many motion estimation/segmentation problems require the minimization of a non-convex criterion function $E(\mathbf{u})$, where $\mathbf{u}$ is some N-dimensional unknown vector. Then, the motion estimation/segmentation problem can be posed so as to find

$$\hat{\mathbf{u}} = \arg\{\min_{\mathbf{u}} E(\mathbf{u})\} \tag{8.1}$$

This minimization is exceedingly difficult due to large dimensionality of the unknown vector and the presence of local minima. With nonconvex criterion functions, gradient descent methods, discussed in Chapter 7, generally cannot reach the global minimum, because they get trapped in the nearest local minimum.

In this section, we present two simulated (stochastic) annealing algorithms, the Metropolis algorithm [Met 53] and the Gibbs sampler [Gem 84], which are capable of finding the global minimum; and three deterministic algorithms, the iterative conditional modes (ICM) algorithm [Bes 74], the mean field annealing algorithm [Bil 91a], and the highest confidence first (HCF) algorithm [Cho 90], to obtain faster convergence. For a detailed survey of popular annealing procedures the reader is referred to [Kir 83, Laa 87].

8.1.1 Simulated Annealing

Simulated annealing (SA) refers to a class of stochastic relaxation algorithms known as Monte Carlo methods. They are essentially prescriptions for a partially random search of the solution space. At each step of the algorithm, the previous solution is subjected to a random perturbation. Unlike deterministic gradient-based iterative algorithms which always move in the direction of decreasing criterion function, simulated annealing permits, on a random basis, changes that increase the criterion function. This is because an uphill move is sometimes necessary in order to prevent the solution from settling in a local minimum.

The probability of accepting uphill moves is controlled by a temperature parameter. The simulated annealing process starts by first "melting" the system at a high enough temperature that almost all random moves are accepted. Then the temperature is lowered slowly according to a "cooling" regime. At each temperature, the simulation must proceed long enough for the system to reach a "steady-state." The sequence of temperatures and the number of perturbations at each temperature constitute the "annealing schedule." The convergence of the procedure is strongly related to the annealing schedule. In their pioneering work, Geman and Geman [Gem 84] proposed the following temperature schedule:

$$T = \frac{\tau}{\ln(i+1)}, \quad i = 1, \ldots \tag{8.2}$$

where τ is a constant and i is the iteration cycle. This schedule is overly conservative but guarantees reaching the global minimum. Schedules that lower the temperature at a faster rate have also been shown to work (without a proof of convergence).

The process of generating random perturbations is referred to as sampling the solution space. In the following, we present two algorithms that differ in the way they sample the solution space.

The Metropolis Algorithm

In Metropolis sampling, at each step of the algorithm a new candidate solution is generated at random. If this new solution decreases the criterion function, it is always accepted; otherwise, it is accepted according to an exponential probability distribution. The probability P of accepting the new solution is then given by

$$P = \begin{cases} \exp(-\Delta E/T) & \text{if } \Delta E > 0 \\ 1 & \text{if } \Delta E \leq 0 \end{cases}$$

where ΔE is the change in the criterion function due to the perturbation, and T is the temperature parameter. If T is relatively large, the probability of accepting a positive energy change is higher than when T is small for a given ΔE. We provide a summary of the Metropolis algorithm in the following [Met 53]:

1. Set $i = 0$ and $T = T_{max}$. Choose an initial $\mathbf{u}^{(0)}$ at random.
2. Generate a new candidate solution $\mathbf{u}^{(i+1)}$ at random.
3. Compute $\Delta E = E(\mathbf{u}^{(i+1)}) - E(\mathbf{u}^{(i)})$.
4. Compute P from

$$P = \begin{cases} \exp(-\Delta E/T) & \text{if } \Delta E > 0 \\ 1 & \text{if } \Delta E \leq 0 \end{cases}$$

5. If $P = 1$, accept the perturbation; otherwise, draw a random number that is uniformly distributed between 0 and 1. If the number drawn is less than P, accept the perturbation.

6. Set $i = i + 1$. If $i \leq I_{max}$, where I_{max} is predetermined, go to 2.

7. Set $i = 0$, and $\mathbf{u}^{(0)} = \mathbf{u}^{(I_{max})}$. Reduce T according to a temperature schedule. If $T > T_{min}$, go to 2; otherwise, terminate.

Because the candidate solutions are generated by random perturbations, the algorithm typically requires a large number of iterations for convergence. Thus, the computational load of simulated annealing is significant, especially when the set of allowable values Γ (defined in Appendix A for $\mathbf{u}$ discrete) contains a large number of values or $\mathbf{u}$ is a continuous variable. Also, the computational load increases with the number of components in the unknown vector.

The Gibbs Sampler

Let's assume that $\mathbf{u}$ is a random vector composed of lexicographic ordering of the elements of a scalar GRF $u(\mathbf{x})$. In Gibbs sampling, the perturbations are generated according to local conditional probability density functions (pdf) derived from the given Gibbsian distribution, according to (A.5) in Appendix A, rather than making totally random perturbations and then deciding whether or not to accept them. The Gibbs sampler method is summarized in the following.

1. Set $T = T_{max}$. Choose an initial $\mathbf{u}$ at random.

2. Visit each site $\mathbf{x}$ to perturb the value of $\mathbf{u}$ at that site as follows:

a) At site $\mathbf{x}$, first compute the conditional probability of $u(\mathbf{x})$ to take each of the allowed values from the set Γ, given the present values of its neighbors using (A.5) in Appendix A. This step is illustrated for a scalar $u(\mathbf{x})$ by an example below.

Example: Computation of local conditional probabilities

Let $\Gamma = \{0, 1, 2, 3\}$. Given a 3×3 binary GRF, we wish to compute the conditional probability of the element marked with "x" in Figure 8.1 being equal to "0," "1," "2," or "3" given the values of its neighbors. If we define

$$P(\gamma) = P(u(\mathbf{x}_i) = \gamma \mid u(\mathbf{x}_j), \mathbf{x}_j \in N_{\mathbf{x}_i}), \quad \text{for all } \gamma \in \Gamma$$

to denote the joint probabilities of possible configurations, then, using (A.4), we have

$$P(u(\mathbf{x}_i) = 0 \mid u(\mathbf{x}_j), \mathbf{x}_j \in N_{\mathbf{x}_i}) = \frac{P(0)}{P(0) + P(1) + P(2) + P(3)}$$

$$P(u(\mathbf{x}_i) = 1 \mid u(\mathbf{x}_j), \mathbf{x}_j \in N_{\mathbf{x}_i}) = \frac{P(1)}{P(0) + P(1) + P(2) + P(3)}$$

$$P(u(\mathbf{x}_i) = 2 \mid u(\mathbf{x}_j), \mathbf{x}_j \in N_{\mathbf{x}_i}) = \frac{P(2)}{P(0) + P(1) + P(2) + P(3)}$$

and

$$P(u(\mathbf{x}_i) = 3 \mid u(\mathbf{x}_j), \mathbf{x}_j \in N_{\mathbf{x}_i}) = \frac{P(3)}{P(0) + P(1) + P(2) + P(3)}$$

The evaluation of $P(\gamma)$ for all $\gamma \in \Gamma$, using the 4-pixel neighborhood system shown in Figure A.1 and the 2-pixel clique potential specified by (8.11), is left as an exercise. You may assume that the *a priori* probabilities of a "0" and a "1" are equal.

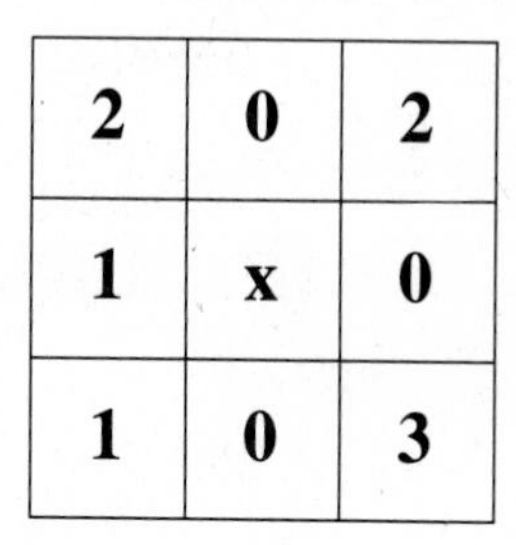

Figure 8.1: Illustration of local probability computation.

b) Once the probabilities for all elements of the set Γ are computed, draw the new value of $u(\mathbf{x})$ from this distribution. To clarify the meaning of "draw," again an example is provided.

Example: Suppose that $\Gamma = \{0, 1, 2, 3\}$, and it was found that

$$
\begin{aligned}
P(u(\mathbf{x}_i) = 0 \mid u(\mathbf{x}_j),\ \mathbf{x}_j \in \mathcal{N}_{\mathbf{x}_i}) &= 0.2 \\
P(u(\mathbf{x}_i) = 1 \mid u(\mathbf{x}_j),\ \mathbf{x}_j \in \mathcal{N}_{\mathbf{x}_i}) &= 0.1 \\
P(u(\mathbf{x}_i) = 2 \mid u(\mathbf{x}_j),\ \mathbf{x}_j \in \mathcal{N}_{\mathbf{x}_i}) &= 0.4 \\
P(u(\mathbf{x}_i) = 3 \mid u(\mathbf{x}_j),\ \mathbf{x}_j \in \mathcal{N}_{\mathbf{x}_i}) &= 0.3
\end{aligned}
$$

Then a random number R, uniformly distributed between 0 and 1, is generated, and the value of $u(\mathbf{x}_i)$ is decided as follows: if $0 \leq R \leq 0.2$ then $u(\mathbf{x}_i) = 0$ if $0.2 < R \leq 0.3$ then $u(\mathbf{x}_i) = 1$ if $0.3 < R \leq 0.7$ then $u(\mathbf{x}_i) = 2$ if $0.7 < R \leq 1$ then $u(\mathbf{x}_i) = 3$.

3. Repeat step 2 a sufficient number of times at a given temperature, then lower the temperature, and go to 2. Note that the conditional probabilities depend on the temperature parameter.

Perturbations through Gibbs sampling lead to very interesting properties that have been shown by Geman and Geman [Gem 84]:
(i) For any initial estimate, Gibbs sampling will yield a distribution that is asymptotically Gibbsian, with the same properties as the Gibbs distribution used to generate it. This result can be used to simulate a Gibbs random field.
(ii) For the particular temperature schedule (8.2), the global optimum will be reached. However, in practice, convergence with this schedule may be too slow.

8.1.2 Iterated Conditional Modes

Iterated conditional modes (ICM), also known as the greedy algorithm, is a deterministic procedure which aims to reduce the computational load of the stochastic annealing methods. It can be posed as special cases of both the Metropolis and Gibbs sampler algorithms.

ICM can best be conceptualized as the "instant freezing" case of the Metropolis algorithm, that is, when the temperature T is set equal to zero for all iterations. Then the probability of accepting perturbations that increase the value of the cost function is always 0 (refer to step 4 of the Metropolis algorithm). Alternatively, it has been shown that ICM converges to the solution that maximizes the local conditional probabilities given by (A.5) at each site. Hence, it can be implemented as in Gibbs sampling, but by choosing the value at each site that gives the maximum local conditional probability rather than drawing a value based on the conditional probability distribution.

ICM convergence much faster than the stochastic SA algorithms. However, because ICM only allows those perturbations yielding negative ΔE, it is likely to get trapped in a local minimum, much like gradient-descent algorithms. Thus, it is critical to initialize ICM with a reasonably good initial estimate. The use of ICM has been reported for image restoration [Bes 74] and image segmentation [Pap 92].

8.1.3 Mean Field Annealing

Mean field annealing is based on the "mean field approximation" (MFA) idea in statistical mechanics. MFA allows replacing each random variable (random field evaluated at a particular site) by the mean of its marginal probability distribution at a given temperature. Then mean field annealing is concerned about the estimation of these means at each site. Because the estimation of each mean is dependent on the means of the neighboring sites, this estimation is performed using an annealing schedule. The algorithm for annealing the mean field is similar to SA except that stochastic relaxation at each temperature is replaced by a deterministic relaxation to minimize the so-called mean field error, usually using a gradient-descent algorithm.

Historically, the mean field algorithms were limited to Ising-type models described by a criterion function involving a binary vector. However, it has recently been extended to a wider class of problems, including those with continuous variables [Bil 91b]. Experiments suggest that the MFA is valid for MRFs with local interactions over small regions. Thus, computations of the means and the mean field error are often based on Gibbsian distributions. It has been claimed that mean field annealing converges to an acceptable solution approximately 50 times faster than SA. The implementation of MFA is not unique. Covering all seemingly different implementations of the mean field annealing [Orl 85, Bil 92, Abd 92, Zha 93] is beyond the scope of this book.

8.1.4 Highest Confidence First

The highest confidence first (HCF) algorithm proposed by Chou and Brown [Cho 90] is a deterministic, noniterative algorithm. It is guaranteed to reach a local minimum of the potential function after a finite number of steps.

In the case of a discrete-valued GRF, the minimization is performed on a site-by-site basis according to the following rules: 1) Sites with reliable data can be labeled without using the prior probability model. 2) Sites where the data is unreliable should rely on neighborhood interaction for label assignment. 3) Sites with unreliable data should not affect sites with reliable data through neighborhood interaction. Guided by these principles, a scheme that determines a particular order for assigning labels and systematically increases neighborhood interaction is designed. Initially, all sites are labeled "uncommitted." Once a label is assigned to an uncommitted site, the site is committed and cannot return to the uncommitted state. However, the label of a committed site can be changed through another assignment. A "stability" measure is calculated for each site based on the local conditional *a posteriori* probability of the labels at that site, to determine the order in which the sites are to be visited. The procedure terminates when the criterion function can no longer be decreased by reassignment of the labels.

Among the deterministic methods, HCF is simpler and more robust than MFA, and more accurate than the ICM. Extensions of HCF for the case of continuous variables also exist.

8.2 Basics of MAP Motion Estimation

In this section, 2-D motion estimation is formulated as a maximum *a posteriori* probability (MAP) estimation problem. The MAP formulation requires two pdf models: the conditional pdf of the observed image intensity given the motion field, called the likelihood model or the observation model, and the *a priori* pdf of the motion vectors, called the motion field model. The basic formulation assumes a Gaussian observation model to impose consistency with the observations, and a Gibbsian motion field model to impose a probabilistic global smoothness constraint.

In order to use a compact notation, we let the vector $\mathbf{s}_k$ denote a lexicographic ordering of the ideal pixel intensities $s_k(\mathbf{x})$ in the kth picture (frame or field), such that $(\mathbf{x}, t) \in \Lambda^3$ for $t = k\Delta t$. If $\mathbf{d}(\mathbf{x}) = (d_1(\mathbf{x}), d_2(\mathbf{x}))$ denotes the displacement vector from frame/field $k-1$ to k at the site $\mathbf{x} = (x_1, x_2)$, then we let $\mathbf{d}_1$ and $\mathbf{d}_2$ denote a lexicographic ordering of $d_1(\mathbf{x})$ and $d_2(\mathbf{x})$, respectively. Hence, ignoring covered/uncovered background regions and intensity variations due to changes in the illumination and shading, we have

$$s_k(\mathbf{x}) = s_{k-1}(\mathbf{x} - \mathbf{d}(\mathbf{x})) \doteq s_{k-1}(x_1 - d_1(\mathbf{x}), x_2 - d_2(\mathbf{x})) \qquad (8.3)$$

which is a restatement of the optical flow constraint.

In general, we can only observe video that is corrupted by additive noise, given by

$$g_k(\mathbf{x}) = s_k(\mathbf{x}) + v_k(\mathbf{x}) \qquad (8.4)$$

and need to estimate 2-D motion from the noisy observations. Then the basic MAP problem can be stated as: given two frames $\mathbf{g}_k$ and $\mathbf{g}_{k-1}$, find

$$(\hat{\mathbf{d}}_1, \hat{\mathbf{d}}_2) = \arg\max_{\mathbf{d}_1, \mathbf{d}_2} p(\mathbf{d}_1, \mathbf{d}_2 | \mathbf{g}_k, \mathbf{g}_{k-1}) \qquad (8.5)$$

where $p(\mathbf{d}_1, \mathbf{d}_2 | \mathbf{g}_k, \mathbf{g}_{k-1})$ denotes the *a posteriori* pdf of the motion field given the two frames. Using the Bayes theorem,

$$p(\mathbf{d}_1, \mathbf{d}_2 | \mathbf{g}_k, \mathbf{g}_{k-1}) = \frac{p(\mathbf{g}_k | \mathbf{d}_1, \mathbf{d}_2, \mathbf{g}_{k-1}) p(\mathbf{d}_1, \mathbf{d}_2 | \mathbf{g}_{k-1})}{p(\mathbf{g}_k | \mathbf{g}_{k-1})} \qquad (8.6)$$

where $p(\mathbf{g}_k | \mathbf{d}_1, \mathbf{d}_2, \mathbf{g}_{k-1})$ is the conditional probability, or the "consistency (likelihood) measure," that measures how well the motion fields $\mathbf{d}_1$ and $\mathbf{d}_2$ explain the observations $\mathbf{g}_k$ through (8.3) given $\mathbf{g}_{k-1}$; and $p(\mathbf{d}_1, \mathbf{d}_2 | \mathbf{g}_{k-1})$ is the *a priori* pdf of the motion field that reflects our prior knowledge about the actual motion field. Since the denominator is not a function of $\mathbf{d}_1$ and $\mathbf{d}_2$, it is a constant for the purposes of motion estimation. Then the MAP estimate (8.5) can be expressed as

$$(\hat{\mathbf{d}}_1, \hat{\mathbf{d}}_2) = \arg\max_{\mathbf{d}_1, \mathbf{d}_2} p(\mathbf{g}_k | \mathbf{d}_1, \mathbf{d}_2, \mathbf{g}_{k-1}) p(\mathbf{d}_1, \mathbf{d}_2 | \mathbf{g}_{k-1}) \qquad (8.7)$$

or as

$$(\hat{\mathbf{d}}_1, \hat{\mathbf{d}}_2) = \arg\max_{\mathbf{d}_1, \mathbf{d}_2} p(\mathbf{g}_{k-1} | \mathbf{d}_1, \mathbf{d}_2, \mathbf{g}_k) p(\mathbf{d}_1, \mathbf{d}_2 | \mathbf{g}_k)$$

Next, we develop models for the conditional and the prior pdfs.

8.2.1 The Likelihood Model

Based on the models (8.3) and (8.4), the change in the intensity of a pixel along the true motion trajectory is due to observation noise. Assuming that the observation noise is white, Gaussian with zero mean and variance σ^2, the conditional pdf $p(\mathbf{g}_k|\mathbf{d}_1, \mathbf{d}_2, \mathbf{g}_{k-1})$ can be modeled as

$$p(\mathbf{g}_k|\mathbf{d}_1, \mathbf{d}_2, \mathbf{g}_{k-1}) = (2\pi\sigma^2)^{-\frac{1}{2d(\Lambda)}} \exp\left\{-\sum_{\mathbf{x}\in\Lambda} \frac{[g_k(\mathbf{x}) - g_{k-1}(\mathbf{x} - \mathbf{d}(\mathbf{x}))]^2}{2\sigma^2}\right\} \quad (8.8)$$

where $d(\Lambda)$ denotes the determinant of Λ which gives the reciprocal of the sampling density. The conditional pdf (8.8) gives the likelihood of observing the intensity $\mathbf{g}_k$ given the true motion field, $\mathbf{d}_1$ and $\mathbf{d}_2$, and the intensity vector of the previous frame $\mathbf{g}_{k-1}$.

8.2.2 The Prior Model

The motion field is assumed to be a realization of a continuous-valued GRF, where the clique potential functions are chosen to impose a local smoothness constraint on pixel-to-pixel variations of the motion vectors. Thus, the joint *a priori* pdf of the motion vector field can be expressed as

$$p(\mathbf{d}_1, \mathbf{d}_2|\mathbf{g}_{k-1}) = \frac{1}{Q_d} \exp\{-U_d(\mathbf{d}_1, \mathbf{d}_2|\mathbf{g}_{k-1})\} \quad (8.9)$$

where Q_d is the partition function, and

$$U_d(\mathbf{d}_1, \mathbf{d}_2|\mathbf{g}_{k-1}) = \lambda_d \sum_{c\in\mathcal{C}_d} V_d^c(\mathbf{d}_1, \mathbf{d}_2|\mathbf{g}_{k-1})$$

Here $\mathcal{C}_d$ denotes the set of all cliques for the displacement field, $V_d^c(.)$ represents the clique potential function for $c \in \mathcal{C}_d$, and λ_d is a positive constant. The clique potentials will be chosen to assign smaller probability to configurations where the motion vector varies significantly from pixel to pixel. This is demonstrated by two examples in the following.

Example: The case of a continuous-valued GRF

This example demonstrates that a spatial smoothness constraint can be formulated as an *a priori* pdf in the form of a Gibbs distribution. Let us employ a four-point neighborhood system, depicted in Figure A.1, with two-pixel cliques (see Appendix A). For continuous-valued GRF, a suitable potential function for the two-pixel cliques may be

$$V_d^{c_2}(\mathbf{d}(\mathbf{x}_i), \mathbf{d}(\mathbf{x}_j)) = ||\mathbf{d}(\mathbf{x}_i) - \mathbf{d}(\mathbf{x}_j)||^2 \quad (8.10)$$

where $\mathbf{x}_i$ and $\mathbf{x}_j$ denote the elements of any two-pixel clique c_2, and $||.||$ is the Euclidian distance. In Equation 8.9, $V_d^{c_2}(\mathbf{d}(\mathbf{x}_i), \mathbf{d}(\mathbf{x}_j))$ needs to be summed over all two-pixel cliques. Clearly, a spatial configuration of motion vectors with larger potential would have a smaller *a priori* probability.

Example: The case of a discrete-valued GRF

If the motion vectors are quantized, say, to 0.5 pixel accuracy, then we have a discrete-valued GRF. The reader is reminded that the definition of the pdf (8.9) needs to be modified to include a Dirac delta function in this case (see (A.2)). Suppose that a discrete-valued GRF $\mathbf{z}$ is defined over the 4×4 lattice, shown in Figure 8.2 (a). Figure 8.2 (b) and (c) show two realizations of 4×4 binary images.

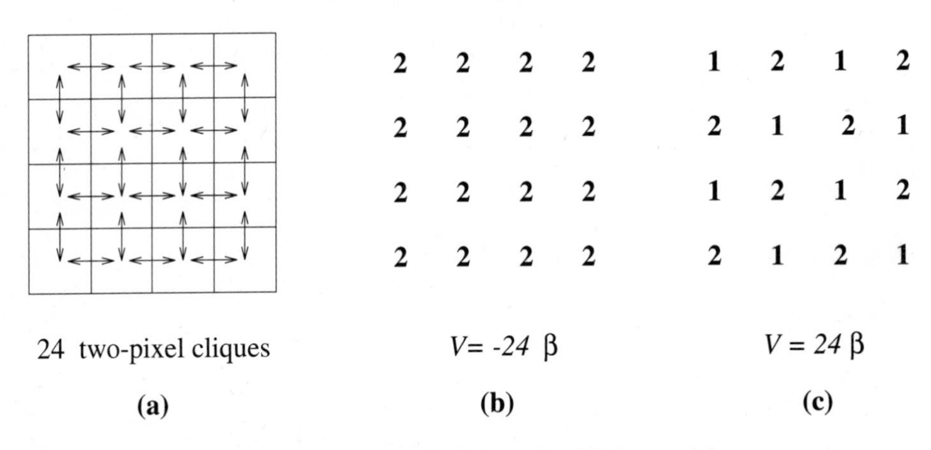

Figure 8.2: Demonstration of a Gibbs model.

Let the two-pixel clique potential be defined as

$$V_C(z(\mathbf{x}_i), z(\mathbf{x}_j)) = \begin{cases} -\beta & \text{if } z(\mathbf{x}_i) = z(\mathbf{x}_j) \\ +\beta & \text{otherwise} \end{cases} \tag{8.11}$$

where β is a positive number.

There are a total of 24 two-pixel cliques in a 4×4 image (shown by double arrows). It can be easily verified, by summing all clique potentials, that the configurations shown in Figures 8.2 (b) and (c) have the Gibbs potentials -24β and $+24\beta$, respectively. Clearly, with the choice of the clique potential function (8.11), the spatially smooth configuration in Figure 8.2 (b) has a higher *a priori* probability.

The basic formulation of the MAP motion estimation problem can be obtained by substituting the likelihood model (8.8) and the a priori model (8.9) into (8.7). Simplification of the resulting expression indicates that maximization of (8.7) is equivalent to minimization of a weighted sum of the square of the displaced frame difference and a global smoothness constraint term. The MAP estimation is usually implemented by minimizing the corresponding criterion function using simulated annealing.

A practical problem with the basic formulation is that it imposes a global smoothness constraint over the entire motion field, resulting in the blurring of optical flow boundaries. This blurring also adversely affects the motion estimates elsewhere in the image. Extensions of the basic approach to address this problem are presented in the following.

8.3 MAP Motion Estimation Algorithms

Three different approaches to account for the presence of optical flow boundaries and occlusion regions are discussed. First, a formulation which utilizes more structured motion field models, including an occlusion field and a discontinuity (line) field, is introduced. While the formulation with the discontinuity models is an elegant one, it suffers from a heavy computational burden because the discontinuity models introduce many more unknowns. To this effect, we also present two other noteworthy algorithms, namely the Local Outlier Rejection method proposed by Iu [Iu 93] and the region-labeling method proposed by Stiller [Sti 93].

8.3.1 Formulation with Discontinuity Models

We introduce two binary auxilary MRFs, called the occlusion field $\mathbf{o}$ and the line field $\mathbf{l}$, to improve the accuracy of motion estimation. The *a priori* pdf (8.9) penalizes any discontinuity in the motion field. In order to avoid oversmoothing actual optical flow boundaries, the line field is introduced, which marks the location of all allowed discontinuities in the motion field. The line field, $\mathbf{l}$, has sites between every pair of pixel sites in the horizontal and vertical directions. Figure 8.3 (a) illustrates a 4-line clique of a line field, composed of horizontal and vertical lines indicating possible discontinuities in the vertical and horizontal directions, respectively. The state of each line site can be either *ON* ($l = 1$) or *OFF* ($l = 0$), expressing the presence or absence of a discontinuity in the respective direction. The actual state of each line field site is *a priori* unknown and needs to be estimated along with the motion vector field.

While the line field is defined to improve the *a priori* motion field model, the occlusion field, $\mathbf{o}$, is used to improve the likelihood model. The occlusion field, which occupies the same lattice as the pixel sites, is an indicator function of occlusion pixels (refer to Section 5.2.1 for a discussion of occlusion) defined by

$$o(\mathbf{x}) = \begin{cases} 0 & \text{if } \mathbf{d}(x_1, x_2) \text{ is well defined} \\ 1 & \text{if } \mathbf{x} = (x_1, x_2) \text{ is an occlusion point} \end{cases} \tag{8.12}$$

Because the pixels where $o(\mathbf{x})$ is ON are also *a priori* unknown, the occlusion field needs to be estimated along with the motion vector field and the line field.

With the introduction of the occlusion and line fields, the MAP motion estimation problem can be restated as: given the two frames, $\mathbf{g}_k$ and $\mathbf{g}_{k-1}$, find

$$\{\hat{\mathbf{d}}_1, \hat{\mathbf{d}}_2, \hat{\mathbf{o}}, \hat{\mathbf{l}}\} = \arg\max_{\mathbf{d}_1,\mathbf{d}_2,\mathbf{o},\mathbf{l}} p(\mathbf{d}_1, \mathbf{d}_2, \mathbf{o}, \mathbf{l} | \mathbf{g}_k, \mathbf{g}_{k-1}) \tag{8.13}$$

Using the Bayes rule to factor the *a posteriori* probability and following the same steps as in the basic formulation, the MAP estimate can be expressed as

$$\begin{aligned}\{\hat{\mathbf{d}}_1, \mathbf{d}_2, \hat{\mathbf{o}}, \hat{\mathbf{l}}\} = \arg\max_{\mathbf{d}_1,\mathbf{d}_2,\mathbf{o},\mathbf{l}} & \; p(\mathbf{g}_k | \mathbf{d}_1, \mathbf{d}_2, \mathbf{o}, \mathbf{l}, \mathbf{g}_{k-1}) \\ & p(\mathbf{d}_1, \mathbf{d}_2 | \mathbf{o}, \mathbf{l}, \mathbf{g}_{k-1}) p(\mathbf{o} | \mathbf{l}, \mathbf{g}_{k-1}) p(\mathbf{l} | \mathbf{g}_{k-1})\end{aligned} \tag{8.14}$$

where the first term is the improved likelihood model, the second, third, and fourth terms are the displacement field, the occlusion field, and the discontinuity field models, respectively. Next, we develop expressions for these models.

The Likelihood Model

The conditional probability model (8.8) fails at the occlusion areas, since the displaced frame difference in the occlusion areas cannot be accurately modeled as white, Gaussian noise. The modeling error at the occlusion points can be avoided by modifying (8.8) based on the knowledge of the occlusion field as

$$\begin{aligned} & p(\mathbf{g}_k | \mathbf{d}_1, \mathbf{d}_2, \mathbf{o}, \mathbf{l}, \mathbf{g}_{k-1}) = \\ & \quad (2\pi\sigma^2)^{-\frac{1}{2N}} \exp\left\{ - \sum_{\mathbf{x} \in \Lambda} \frac{(1 - o(\mathbf{x}))[g_k(\mathbf{x}) - g_{k-1}(\mathbf{x} - \mathbf{d}(\mathbf{x}))]^2}{2\sigma^2} \right\} \end{aligned} \tag{8.15}$$

where the contributions of the pixels in the occluded areas are discarded, and N is the number of sites that do not experience occlusion. Note that this improved likelihood model does not depend on the line field explicitly. The pdf (8.15) can be expressed more compactly in terms of a "potential function" as

$$p(\mathbf{g}_k | \mathbf{d}_1, \mathbf{d}_2, \mathbf{o}, \mathbf{l}, \mathbf{g}_{k-1}) = \exp\{-U_g(\mathbf{g}_k | \mathbf{d}_1, \mathbf{d}_2, \mathbf{o}, \mathbf{g}_{k-1})\}$$

where

$$\begin{aligned} U_g(\mathbf{g}_k | \mathbf{d}_1, \mathbf{d}_2, \mathbf{o}, \mathbf{g}_{k-1}) = & \frac{\log(2\pi\sigma^2)}{2\,N} \\ & + \frac{1}{2\sigma^2} \sum_{\mathbf{x} \in \Lambda} (1 - o(\mathbf{x}))[g_k(\mathbf{x}) - g_{k-1}(\mathbf{x} - \mathbf{d}(\mathbf{x}))]^2 \end{aligned} \tag{8.16}$$

We must assign an appropriate penalty for the use of the occlusion state "ON." Otherwise, the displaced frame difference can be made arbitrarily small by using more occlusion states. This penalty is imposed by the occlusion model discussed below.

The Motion Field Model

We incorporate the line field model to the prior motion vector field model (8.9) in order not to unduly penalize optical flow boundaries. The improved *a priori* motion field model can be expressed as

$$p(\mathbf{d}_1, \mathbf{d}_2|\mathbf{o}, \mathbf{l}, \mathbf{g}_{k-1}) = \frac{1}{Q_d} \exp\{-U_d(\mathbf{d}_1, \mathbf{d}_2|\mathbf{o}, \mathbf{l}, \mathbf{g}_{k-1})/\beta_d\} \tag{8.17}$$

where Q_d is the partition function, and

$$U_d(\mathbf{d}_1, \mathbf{d}_2|\mathbf{o}, \mathbf{l}, \mathbf{g}_{k-1}) = \sum_{c \in \mathcal{C}_d} V_d^c(\mathbf{d}_1, \mathbf{d}_2|\mathbf{o}, \mathbf{l}, \mathbf{g}_{k-1}) \tag{8.18}$$

Here $\mathcal{C}_d$ denotes the set of all cliques for the displacement field, and $V_d^c(.)$ represents the clique potential function for $c \in \mathcal{C}_d$.

Typically, the dependency of the clique potentials on $\mathbf{o}$ and $\mathbf{g}_{k-1}$ are omitted. We present two examples of such potential functions for 2-pixel cliques,

$$V_d^{c_2}(\mathbf{d}(\mathbf{x}_i), \mathbf{d}(\mathbf{x}_j)|\mathbf{l}) = ||\mathbf{d}(\mathbf{x}_i) - \mathbf{d}(\mathbf{x}_j)||^2(1 - l(\mathbf{x}_i, \mathbf{x}_j)) \tag{8.19}$$

or

$$V_d^{c_2}(\mathbf{d}(\mathbf{x}_i), \mathbf{d}(\mathbf{x}_j)|\mathbf{l}) = \left(1 - \exp(-\gamma_d||\mathbf{d}(\mathbf{x}_i) - \mathbf{d}(\mathbf{x}_j)||^2)\right)\ (1 - l(\mathbf{x}_i, \mathbf{x}_j)) \tag{8.20}$$

where γ_d is a scalar, $\mathbf{x}_i$ and $\mathbf{x}_j$ denote the elements of the two-pixel cliques c_2, and $l(\mathbf{x}_i, \mathbf{x}_j)$ denotes the line field site that is in between the pixel sites $\mathbf{x}_i$ and $\mathbf{x}_j$. As can be seen, no penalty is assigned to discontinuities in the motion vector field, if the line field site between these motion vectors is "ON." However, we must assign an appropriate penalty for turning the line field state "ON." Otherwise, the smoothness constraint can be effectively turned off by setting all of the line field sites "ON." This penalty is introduced by the motion discontinuity model discussed below.

The Occlusion Field Model

The occlusion field models the spatial distribution of occlusion labels as a discrete-valued GRF described by

$$P(\mathbf{o}|\mathbf{l}, \mathbf{g}_{k-1}) = \frac{1}{Q_o} \exp\{-U_o(\mathbf{o}|\mathbf{l}, \mathbf{g}_{k-1})/\beta_o\} \tag{8.21}$$

where

$$U_o(\mathbf{o}|\mathbf{l}, \mathbf{g}_{k-1}) = \sum_{c \in \mathcal{C}_o} V_o^c(\mathbf{o}|\mathbf{l}, \mathbf{g}_{k-1}) \tag{8.22}$$

Here $\mathcal{C}_o$ denotes the set of all cliques for the occlusion field, and $V_o^c(.)$ represents the clique potential function for $c \in \mathcal{C}_o$.

The potential functions $V_o^c(.)$ for all possible cliques are chosen to provide a penalty for turning the associated occlusion states "ON." For example, we will define the potential for singleton cliques as

$$V_o^{c_1}(o(\mathbf{x})) = o(\mathbf{x})T_o \tag{8.23}$$

which penalizes each "ON" state by an amount T_o. The quantity T_o can be viewed as a threshold on the normalized displaced frame difference,

$$\epsilon^2(\mathbf{x}) = \frac{[g_k(\mathbf{x}) - g_{k-1}(\mathbf{x} - \mathbf{d}(\mathbf{x}))]^2}{2\sigma^2} \tag{8.24}$$

That is, the occlusion state should be turned "ON" only when $\epsilon^2(\mathbf{x})$ is larger than T_o. The spatial distribution of the "ON" occlusion states and their interaction with line field states can be controlled by specifying two or more pixel cliques. Examples for the choices of occlusion field cliques can be found in Dubois and Konrad [Dub 93].

The Line Field Model

The allowed discontinuities in the motion field are represented by a line field, which is a binary GRF, modeled by the joint probability distribution

$$P(\mathbf{l}|\mathbf{g}_{k-1}) = \frac{1}{Q_l} \exp\left\{-U_l(\mathbf{l}|\mathbf{g}_{k-1})/\beta_l\right\} \tag{8.25}$$

where Q_l is the partition function, and

$$U_l(\mathbf{l}|\mathbf{g}_{k-1}) = \sum_{c \in \mathcal{C}_l} V_l^c(\mathbf{l}|\mathbf{g}_{k-1}) \tag{8.26}$$

Here $\mathcal{C}_l$ denotes the set of all cliques for the line field, and $V_l^c(.)$ represents the clique function for $c \in \mathcal{C}_l$.

Like the occlusion model, the motion discontinuity model assigns a penalty for the use of the "ON" state in the line field. The potential function $V_l^c(\mathbf{l}|\mathbf{g}_{k-1})$ may take different forms depending on the desired properties of the motion discontinuity field. Here, we will consider only singleton and four-pixel cliques; hence,

$$V_l^c(\mathbf{l}|\mathbf{g}_{k-1}) = V_l^{c_1}(\mathbf{l}|\mathbf{g}_{k-1}) + V_l^{c_4}(\mathbf{l}) \tag{8.27}$$

The singleton cliques assign a suitable penalty for the use of the "ON" state at each individual site. The motion discontinuities are, in general, correlated with intensity discontinuities; that is, every motion edge should correspond to an intensity edge, but not vice versa. Thus, a line-field site should only be turned "ON" when there is a significant gray-level gradient, resulting in the following singleton clique potential:

$$V_l^{c_1}(\mathbf{l}|\mathbf{g}_{k-1}) = \begin{cases} \frac{\alpha}{(\nabla_v \mathbf{g}_k)^2} l(\mathbf{x}_i, \mathbf{x}_j) & \text{for horizontal cliques} \\ \frac{\alpha}{(\nabla_h \mathbf{g}_k)^2} l(\mathbf{x}_i, \mathbf{x}_j) & \text{for vertical cliques} \end{cases} \tag{8.28}$$

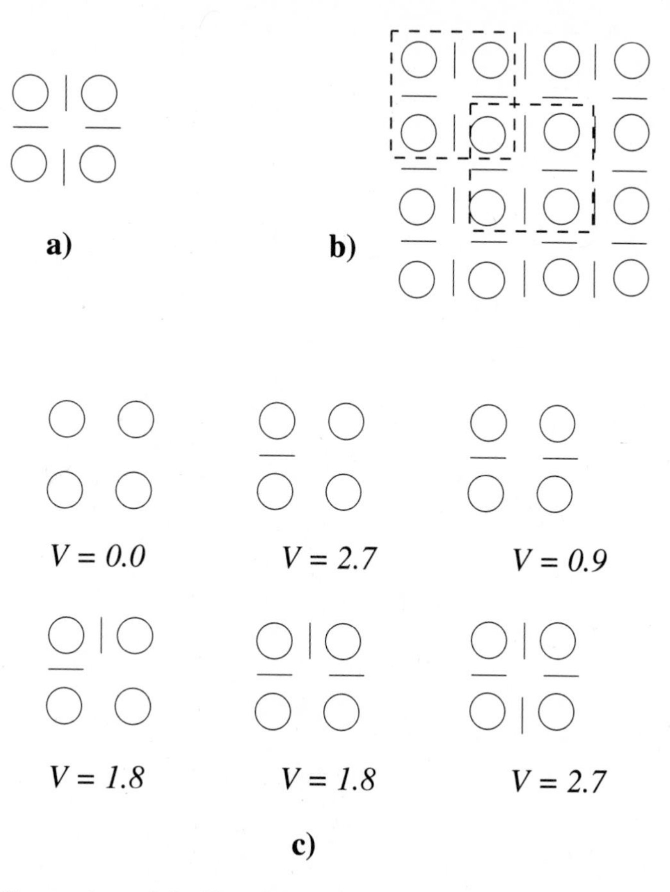

Figure 8.3: Illustration of the line field: a) four-line clique, b) all four-line cliques on a 4 × 4 image, c) potentials for four-line cliques.

where $\nabla_v \mathbf{g}_k$ and $\nabla_h \mathbf{g}_k$ denote the vertical and horizontal image gradient operators, respectively.

An example of potentials assigned to various rotation-invariant four-pixel cliques is shown in Figure 8.3 (c). More examples can be found in [Kon 92]. Next, we demonstrate the use of the line field.

Example: Demonstration of the line field

The line field model is demonstrated in Figure 8.3 using a 4×4 image. As shown in Figure 8.3 (b), a 4 × 4 image has 24 singleton line cliques and 9 distinct four-line cliques, indicating possible discontinuities between every pair of horizontal and vertical pixel sites.

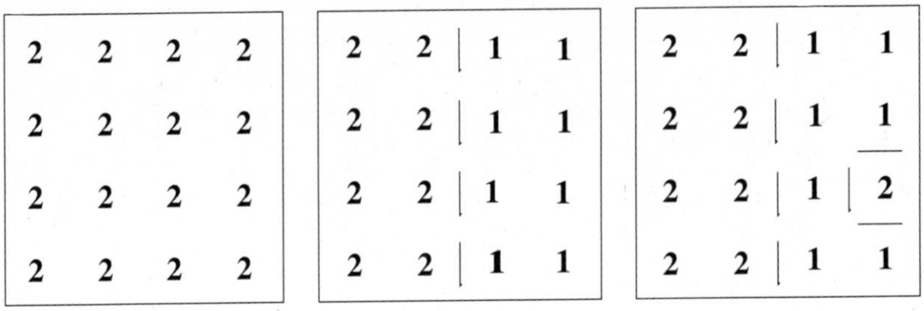

2	2	2	2
2	2	2	2
2	2	2	2
2	2	2	2

2	2	1	1
2	2	1	1
2	2	1	1
2	2	1	1

2	2	1	1
2	2	1	1
2	2	1	2
2	2	1	1

Figure 8.4: Prior probabilities with and without the line field: a) no edges, b) a vertical edge, c) a vertical edge and an isolated pixel.

The potentials shown in Figure 8.3 (c) assign *a priori* probabilities to all possible four-pixel discontinuity configurations reflecting our *a priori* expectation of their occurence. Note that these configurations are rotation-invariant; that is, the potential is equal to 2.7 whenever only one of the four line sites is ON. These potentials slightly penalize straight lines ($V = 0.9$), penalize corners ($V = 1.8$) and "T" junctions ($V = 1.8$), and heavily penalize end of a line ($V = 2.7$) and "crosses" ($V = 2.7$).

Figure 8.4 shows three pictures where there are no edges, there is a single vertical edge, and there is a vertical edge and an isolated pixel, respectively. The potential function $V_l^{c_4}(\mathbf{l})$ evaluated for each of these configurations are 0, $3 \times 0.9 = 2.7$, and $2.7 + 1.8 \times 2 = 6.3$, respectively. Recalling that the *a priori* probability is inversely proportional to the value of the potential, we observe that a smooth configuration has a higher *a priori* probability.

Konrad-Dubois Method

Given the above models, the MAP estimate of $\mathbf{d}_1$, $\mathbf{d}_2$, $\mathbf{o}$, and $\mathbf{l}$ can be expressed (neglecting the dependency of the partition functions on the unknowns, e.g., Q_d depends on the number of sites at which $\mathbf{l}$ is "ON") in terms of the potential functions as

$$\hat{\mathbf{d}}_1, \hat{\mathbf{d}}_2, \hat{\mathbf{o}}, \hat{\mathbf{l}} = \arg \min_{\mathbf{d}_1, \mathbf{d}_2, \mathbf{o}, \mathbf{l}} U(\mathbf{d}_1, \mathbf{d}_2, \mathbf{o}, \mathbf{l} | \mathbf{g}_k, \mathbf{g}_{k-1}) \tag{8.29}$$

$$= \arg \min_{\mathbf{d}_1, \mathbf{d}_2, \mathbf{o}, \mathbf{l}} \{ U_g(\mathbf{g}_k | \mathbf{d}_1, \mathbf{d}_2, \mathbf{o}, \mathbf{l}, \mathbf{g}_{k-1}) + \lambda_d U_d(\mathbf{d}_1, \mathbf{d}_2 | \mathbf{o}, \mathbf{l}, \mathbf{g}_{k-1})$$
$$+ \lambda_o U_o(\mathbf{o} | \mathbf{l}, \mathbf{g}_{k-1}) + \lambda_l U_l(\mathbf{l} | \mathbf{g}_{k-1}) \} \tag{8.30}$$

where λ_d, λ_o, and λ_l are positive constants. The minimization of (8.30) is an exceedingly difficult problem, since there are several hundreds of thousands of unknowns for a reasonable size image, and the criterion function is nonconvex. For example, for a 256 × 256 image, there are 65,536 motion vectors (131,072 components), 65,536 occlusion labels, and 131,072 line field labels for a total of 327,680 unknowns. An additional complication is that the motion vector components are continuous-valued, and the occlusion and line field labels are discrete-valued.

To address these difficulties, Dubois and Konrad [Dub 93] have proposed the following three-step iteration:

1. Given the best estimates of the auxilary fields $\hat{\mathbf{o}}$ and $\hat{\mathbf{l}}$, update the motion field $\mathbf{d}_1$ and $\mathbf{d}_2$ by minimizing

$$\min_{\mathbf{d}_1,\mathbf{d}_2} U_g(\mathbf{g}_k|\mathbf{d}_1,\mathbf{d}_2,\hat{\mathbf{o}},\mathbf{g}_{k-1}) + \lambda_d U_d(\mathbf{d}_1,\mathbf{d}_2|\hat{\mathbf{l}},\mathbf{g}_{k-1}) \tag{8.31}$$

   The minimization of (8.31) can be done by Gauss-Newton optimization, as was done in [Kon 91].

2. Given the best estimates of $\hat{\mathbf{d}}_1$, $\hat{\mathbf{d}}_2$, and $\hat{\mathbf{l}}$, update $\mathbf{o}$ by minimizing

$$\min_{\mathbf{o}} U_g(\mathbf{g}_k|\hat{\mathbf{d}}_1,\hat{\mathbf{d}}_2,\mathbf{o},\mathbf{g}_{k-1}) + \lambda_o U_o(\mathbf{o}|\hat{\mathbf{l}},\mathbf{g}_{k-1}) \tag{8.32}$$

   An exhaustive search or the ICM method can be employed to solve this step.

3. Finally, given the best estimates of $\hat{\mathbf{d}}_1$, $\hat{\mathbf{d}}_2$, and $\hat{\mathbf{o}}$, update $\mathbf{l}$ by minimizing

$$\min_{\mathbf{l}} \lambda_d U_d(\hat{\mathbf{d}}_1,\hat{\mathbf{d}}_2|\mathbf{l},\mathbf{g}_{k-1}) + \lambda_o U_o(\hat{\mathbf{o}}|\mathbf{l},\mathbf{g}_{k-1}) + \lambda_l U_l(\mathbf{l}|\mathbf{g}_{k-1}) \tag{8.33}$$

   Once all three fields are updated, the process is repeated until a suitable criterion of convergence is satisfied. This procedure has been reported to give good results.

In an earlier paper Konrad and Dubois [Kon 92] have derived a solution, using the Gibbs sampler, for the minimization of

$$U_g(\mathbf{g}_k|\mathbf{d}_1,\mathbf{d}_2,\mathbf{l},\mathbf{g}_{k-1}) + \lambda_d U_d(\mathbf{d}_1,\mathbf{d}_2|\mathbf{l},\mathbf{g}_{k-1}) + \lambda_l U_l(\mathbf{l}|\mathbf{g}_{k-1}) \tag{8.34}$$

without taking the occlusion field into account. They have shown that the Horn and Schunck iteration (5.17) constitutes a special case of this Gibbs sampler solution. Furthermore, if the motion vector field is assumed discrete-valued, then the Gibbs sampler solution generalizes the pixel-/block-matching algorithms.

While the MAP formulation proposed by Konrad and Dubois provides an elegant approach with realistic constraints for motion estimation, the implementation of this algorithm is far from easy. Alternative approaches proposed for Bayesian motion estimation include those using mean field annealing by Adelqader *et al.* [Abd 92] and Zhang *et al.* [Zha 93] to reduce the computational load of the stochastic relaxation procedure, and the multimodal motion estimation and segmentation algorithm of Heitz and Bouthemy [Hei 90]. Pixel-recursive estimators have been derived by Driessen *et al.* [Dri 92], and Kalman-type recursive estimators have been used by Chin *et al.* [Chi 93]. In the following, we discuss two other strategies to prevent blurring of motion boundaries without using line fields.

8.3.2 Estimation with Local Outlier Rejection

The local outlier rejection approach proposed by Iu [Iu 93] is an extension of the basic MAP formulation, given by (8.7), which aims at preserving the optical flow boundaries without using computationally expensive line field models. At site $\mathbf{x}$, the basic MAP method clearly favors the estimate $\hat{\mathbf{d}}(\mathbf{x})$ that is closest to all other motion estimates within the neighborhood of site $\mathbf{x}$. The local outlier rejection method is based on the observation that if the pixels within the neighborhood of site $\mathbf{x}$ exhibit two different motions, the estimate $\mathbf{d}(\mathbf{x})$ is then "pushed" towards the average of the two, resulting in blurring. To eliminate this undesirable effect, it is proposed that all the values of the clique potential function, one for each site, are ranked, and the outlier values are rejected. The outlier rejection procedure is relatively simple to incorporate into the local Gibbs potential calculation step.

To describe the approach in more detail, we revisit the basic MAP equations (8.7)-(8.9). While the likelihood model (8.8) remains unchanged, we rewrite the potential function of the prior distribution (8.9) as

$$U_{\mathbf{d}}(\mathbf{d}_1, \mathbf{d}_2 | \mathbf{g}_{k-1}) = \lambda_d \sum_{\mathbf{x}_i} \frac{1}{N_h} \sum_{\mathbf{x}_j \in \mathcal{N}_{\mathbf{x}_i}} ||\mathbf{d}(\mathbf{x}_i) - \mathbf{d}(\mathbf{x}_j)||^2 \tag{8.35}$$

where the summation $\mathbf{x}_i$ is over the whole image and the summation $\mathbf{x}_j$ runs over all the neighbors of the site $\mathbf{x}_i$, and N_h is the number of neighbors of the site $\mathbf{x}_i$. The term $||\mathbf{d}(\mathbf{x}_i) - \mathbf{d}(\mathbf{x}_j)||^2$ is the clique potential for the two-pixel clique containing sites $\mathbf{x}_i$ and $\mathbf{x}_j$. Observe that (8.35) is different than (8.9) because: i) the summation is not over all cliques, but over all pixels to simplify the outlier rejection process (it can be easily shown that in (8.35) every clique is counted twice); and ii) the potential function includes the mean clique potential, rather than the sum of all clique potentials, for each site (indeed, this corresponds to scaling of λ_d).

To incorporate outlier rejection, the potential function (8.35) is further modified as

$$U(\mathbf{d}_1, \mathbf{d}_2 | \mathbf{g}_{k-1}) = \lambda_d \sum_{\mathbf{x}_i} \frac{1}{N_h} \sum_{\mathbf{x}_j \in \mathcal{N}_{\mathbf{x}_i}} \delta_j ||\mathbf{d}(\mathbf{x}_i) - \mathbf{d}(\mathbf{x}_j)||^2 \tag{8.36}$$

where

$$\delta_j = \begin{cases} 1 & \text{if } ||\mathbf{d}(\mathbf{x}_i) - \mathbf{d}(\mathbf{x}_j)||^2 \leq T_{OR} \\ 0 & \text{otherwise} \end{cases} \tag{8.37}$$

is the indicator function of the rejected cliques, T_{OR} is a threshold, and

$$N_h = \sum_j \delta_j \tag{8.38}$$

The expression (8.36) can be used in two ways: i) given a fixed threshold T_{OR}, in which case the number of cliques N_h varies from pixel to pixel, or ii) given a fixed number of cliques, in which case the threshold T_{OR} varies from pixel to pixel.

The selection of the right threshold or number of cliques is a compromise between two conflicting requirements. A low threshold preserves the optical flow boundaries better, while a high threshold imposes a more effective smoothness constraint. For an 8-pixel neighborhood system, Iu suggests setting $N_h = 3$. For most neighborhood systems, the ranking of clique potentials and the outlier rejection procedure only account for a small computational overhead.

8.3.3 Estimation with Region Labeling

Yet another extension of the basic MAP approach (8.7) to eliminate blurring of optical flow boundaries can be formulated based on region labeling [Sti 93]. In this approach, each motion vector is assigned a label $z(\mathbf{x})$ such that the label field $\mathbf{z}$ designates regions where the motion vectors are expected to remain constant or vary slowly. Region labeling differs from the line-field approach in that here the labels belong to the motion vectors themselves as opposed to indicating the presence of discontinuities between them. Clearly, the label field is unknown, and has to be estimated along with the motion field.

The region-labeling formulation can be expressed as

$$p(\mathbf{d}_1, \mathbf{d}_2, \mathbf{z} \mid \mathbf{g}_k, \mathbf{g}_{k-1}) \propto p(\mathbf{g}_k \mid \mathbf{d}_1, \mathbf{d}_2, \mathbf{z}, \mathbf{g}_{k-1}) p(\mathbf{d}_1, \mathbf{d}_2 \mid \mathbf{z}) p(\mathbf{z}) \tag{8.39}$$

where $p(\mathbf{g}_k \mid \mathbf{d}_1, \mathbf{d}_2, \mathbf{z}, \mathbf{g}_{k-1})$ is again the likelihood model and $p(\mathbf{d}_1, \mathbf{d}_2 \mid \mathbf{z})p(\mathbf{z}) = p(\mathbf{d}_1, \mathbf{d}_2, \mathbf{z})$ is the joint prior pdf of the motion and label fields. While the likelihood model follows straightforwardly from (8.8), we will examine the prior pdf model in more detail. The prior pdf of the motion field conditioned on the label field $p(\mathbf{d}_1, \mathbf{d}_2 \mid \mathbf{z})$ is modeled by a Gibbs distribution,

$$p(\mathbf{d}_1, \mathbf{d}_2 \mid \mathbf{z}) = \frac{1}{Q} \exp\{-U(\mathbf{d}_1, \mathbf{d}_2 \mid \mathbf{z})\} \tag{8.40}$$

where

$$U(\mathbf{d}_1, \mathbf{d}_2 \mid \mathbf{z}) = \sum_{\mathbf{x}_i} \sum_{\mathbf{x}_j \in \mathcal{N}_{\mathbf{x}_i}} ||\mathbf{d}(\mathbf{x}_i) - \mathbf{d}(\mathbf{x}_j)||^2 \delta(z(\mathbf{x}_i) - z(\mathbf{x}_j)) \tag{8.41}$$

in which $\mathbf{x}_i$ ranges over all pixel sites, and $\mathbf{x}_j$ over all neighbors of $\mathbf{x}_i$. The function $\delta(z)$ is the Kronecker delta function, that is, 1 when $z = 0$ and 0 otherwise. It ensures that the local smoothness constraint is imposed only when both pixels in the clique have the same label, thus avoiding the smoothness constraint across region boundaries. The other part of this prior pdf, $p(\mathbf{z})$, enforces a connected configuration of regions (labels) over the image. A suitable prior pdf model is a discrete-valued Gibbs distribution with the potential function given by (8.11).

The MAP estimate of the triplet $\mathbf{d}_1$, $\mathbf{d}_2$, and $\mathbf{z}$ can be obtained by minimizing the potential function corresponding to the *a posteriori* pdf (8.39). An ICM optimization method has been used by Stiller [Sti 93]. While this approach provides a simpler formulation than that of Konrad-Dubois, the performance is dependent on the appropriate region assignments. Since the region labels are used only as a token to limit the motion smoothness constraint, there exists a certain degree of latitude and arbitrariness in the assignment of labels.

8.4 Examples

We have implemented four Bayesian motion estimation algorithms; namely, the basic method, the method of Konrad-Dubois (K-D), the outlier-rejection method of Iu, and the segmentation method of Stiller, on the same two frames of the Mobile and Calendar sequence that were used in the previous three chapters. The basic algorithm using a smoothness constraint evaluates (8.7) given the pdf models (8.8) and (8.9) with the potential function (8.10). In our implementation, the DFD in the exponent of (8.8) has been smoothed over the 8-neighborhood of each pixel using the motion estimate at that pixel. The initial iteration does not impose the smoothness constraint; that is, it resembles block matching. The optimization is performed by the ICM method with a temperature schedule of the form $T = 1000(1 - i/I)$, where $i = 0, \ldots$ and I stand for the iteration index and the maximum number of iterations, respectively. The algorithm usually converges after $I = 5$ iterations.

For the K-D, Iu, and Stiller algorithms, the initial motion field is set equal to the result of the basic algorithm. Our implementation of the K-D algorithm did not include an occlusion field. Hence, it is a two-step iteration, given by (8.31) and (8.33). The horizontal and vertical line fields are initialized by simple edge detection on the initial motion estimates. The method of Iu is a variation of the basic algorithm, where an outlier-rejection stage is built into the smoothness constraint. At each site, the distance between the present motion vector and those within the 8-neighborhood of that site are rank-ordered. The smallest third ($N_h = 3$) are used in imposing the smoothness constraint. Stiller's algorithm employs a segmentation field, which is initialized by a K-means segmentation of the initial motion estimates with $K = 9$. In the ICM iterations, we have used an exponential temperature schedule of the form $T = 1000(0.8)^i$, $i = 0, \ldots$.

The motion fields obtained by the K-D and the Iu estimators are shown in Figure 8.5 (a) and (b), respectively. Observe that Iu's motion field is slightly more regular. A numerical comparison of the motion fields is presented in Table 8.1.

Table 8.1: Comparison of the Bayesian methods. (Courtesy Gozde Bozdagi)

Method	PSNR (dB)	Entropy (bits)
Frame-Difference	23.45	-
Global Smoothness	33.82	3.92
Konrad-Dubois	34.10	5.02
Iu	34.54	4.99
Stiller	34.14	4.03

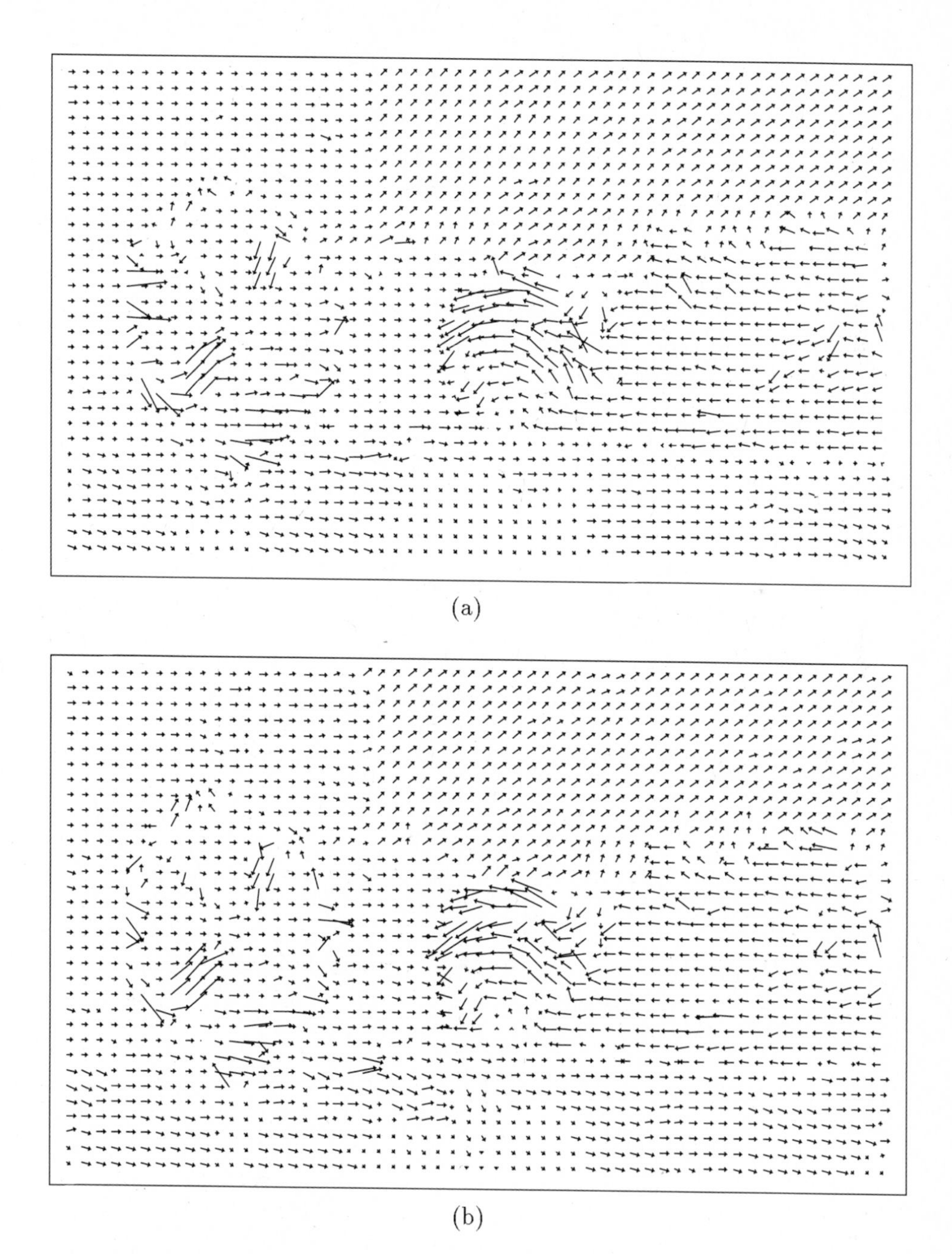

(a)

(b)

Figure 8.5: Motion field obtained by a) the Konrad-Dubois method and b) Iu's method. (Courtesy Gozde Bozdagi and Michael Chang)

8.5 Exercises

1. How do you define the joint pdf of a discrete-valued GRF?

2. The initial motion field for the MAP estimator is usually computed by a deterministic estimator such as the Horn-Schunck estimator or block matching. Suggest methods to initialize the line field and the occlusion field.

3. The MAP estimator (8.30) has been found to be highly sensitive to the values of the parameters λ_d, λ_o, and λ_l, which are free parameters. How would you select them?

4. Discuss the relationship between the MAP estimator and the Horn-Schunck algorithm. (Hint: see [Kon 92].)

5. Compare modeling motion discontinuities by line fields versus region labeling.

Bibliography

[Abd 92] I. Abdelqader, S. Rajala, W. Snyder, and G. Bilbro, "Energy minimization approach to motion estimation," *Signal Processing*, vol. 28, pp. 291–309, Sep. 1992.

[Bes 74] J. Besag, "Spatial interaction and the statistical analysis of lattice systems," *J. Royal Stat. Soc. B*, vol. 36, no. 2, pp. 192–236, 1974.

[Bil 91a] G. L. Bilbro and W. E. Snyder, "Optimization of functions with many minima," *IEEE Trans. Syst., Man and Cyber.*, vol. 21, pp. 840–849, Jul/Aug. 1991.

[Bil 91b] G. L. Bilbro, W. E. Snyder, and R. C. Mann, "Mean field approximation minimizes relative entropy," *J. Opt. Soc. Am. A*, vol. 8, pp. 290–294, Feb. 1991.

[Bil 92] G. L. Bilbro, W. E. Snyder, S. J. Garnier, and J. W. Gault, "Mean field annealing: A formalism for constructing GNC-like algorithms," *IEEE Trans. Neural Networks*, vol. 3, pp. 131–138, Jan. 1992.

[Chi 93] T. M. Chin, M. R. Luettgen, W. C. Karl, and A. S. Willsky, "An estimation theoretic prespective on image processing and the calculation of optical flow," in *Motion Analysis and Image Sequence Processing*, M. I. Sezan and R. L. Lagendijk, eds., Norwell, MA: Kluwer, 1993.

[Cho 90] P. B. Chou and C. M. Brown, "The theory and practice of Bayesian image labeling," *Int. J. Comp. Vis.*, vol. 4, pp. 185–210, 1990.

[Dep 92] R. Depommier and E. Dubois, "Motion estimation with detection of occlusion areas," *Proc. ICASSP*, pp. III.269–272, Mar. 1992.

[Dri 92] J. N. Driessen, *Motion Estimation for Digital Video*, Ph.D. Thesis, Delft University of Technology, 1992.

[Dub 93] E. Dubois and J. Konrad, "Estimation of 2-D motion fields from image sequences with application to motion-compensated processing," in *Motion Analysis and Image Sequence Processing*, M. I. Sezan and R. L. Lagendijk, eds., Norwell, MA: Kluwer, 1993.

[Gem 84] S. Geman and D. Geman, "Stochastic relaxation, Gibbs distribution, and Bayesian restoration of images," *IEEE Trans. Patt. Anal. Mach. Intel.*, vol. 6, pp. 721–741, Nov. 1984.

[Hei 90] F. Heitz and P. Bouthemy, "Motion estimation and segmentation using a global Bayesian approach," *Proc. Int. Conf. ASSP*, Albuquerque, NM, pp. 2305–2308, April 1990.

[Kir 83] S. Kirkpatrick, C. D. G. Jr., and M. P. Vecchi, "Optimization by simulated annealing," *Science*, vol. 220, pp. 671–679, May 1983.

[Kon 91] J. Konrad and E. Dubois, "Comparison of stochastic and deterministic solution methods in Bayesian estimation of 2D motion," *Image and Vis. Comput.*, vol. 9, pp. 215–228, Aug. 1991.

[Kon 92] J. Konrad and E. Dubois, "Bayesian estimation of motion vector fields," *IEEE Trans. Patt. Anal. Mach. Intel.*, vol. 14, pp. 910–927, Sep. 1992.

[Laa 87] P. J. M. Van Laarhoven and E. H. Aarts, *Simulated Annealing: Theory and Applications*, Dordrecht, Holland: Reidel, 1987.

[Met 53] N. Metropolis, A. Rosenbluth, M. Rosenbluth, H. Teller, and E. Teller, "Equation of state calculations by fast computing machines," *J. Chem. Phys.*, vol. 21, pp. 1087–1092, June 1953.

[Orl 85] H. Orland, "Mean-field theory for optimization problems," J. Physique Lett., vol. 46, 17, pp. 763–770, 1985.

[Pap 92] T. N. Pappas, "An Adaptive Clustering Algorithm for Image Segmentation," *IEEE Trans. Signal Proc.*, vol. SP-40, pp. 901–914, April 1992.

[Iu 93] S.-L. Iu, "Robust estimation of motion vector fields with discontinuity and occlusion using local outliers rejection," *SPIE*, vol. 2094, pp. 588–599, 1993.

[Sti 93] C. Stiller and B. H. Hurtgen, "Combined displacement estimation and segmentation in image sequences," *Proc. SPIE/EUROPTO Video Comm. and PACS for Medical Applications*, Berlin, Germany, vol. SPIE 1977, pp. 276–287, 1993.

[Zha 93] J. Zhang and J. Hanauer, "The mean field theory for image motion estimation," *Proc. IEEE Int. Conf. ASSP*, vol. 5, pp. 197–200, Minneapolis, MN, 1993.

Chapter 9

METHODS USING POINT CORRESPONDENCES

3-D motion estimation refers to estimating the actual motion of objects in a scene from their 2-D projections (images). Clearly, the structure (depth information) of the objects in the scene affect the projected image. Since the structure of the scene is generally unknown, 3-D motion and structure estimation need to be addressed simultaneously. Applications of 3-D motion and structure estimation include robotic vision, passive navigation, surveillance imaging, intelligent vehicle highway systems, and harbor traffic control, as well as object-based video compression. In some of these applications a camera moves with respect to a fixed environment, and in others the camera is stationary, but the objects in the environment move. In either case, we wish to determine the 3-D structure and the relative motion parameters from a time sequence of images.

Needless to say, 3-D motion and structure estimation from 2-D images is an ill-posed problem, which may not have a unique solution without some simplifying assumptions, such as rigid motion and a parametric surface. It is well-known that 3-D rigid motion can be modeled by three translation and three rotation parameters (see Chapter 2). The object surface may be approximated by a piecewise planar or quadratic model, or represented by a collection of independent 3-D points. Then, 3-D motion estimation refers to the estimation of the six rigid motion parameters, and structure estimation refers to the estimation of the parameters of the surface model, or the depth of each individual point from at least two 2-D projections.

The 3-D motion and structure estimation methods can be classified into two groups: those which require a set of feature point correspondences to be determined *a priori*, and those which do not [Agg 88]. This chapter is about the former class. Furthermore, it is assumed in this chapter that the field of view contains a single moving object. The segmentation problem in the presence of multiple moving objects will be dealt with in Chapter 11. Section 9.1 introduces parametric models for the projected displacement field that are based on the orthographic and perspective projection, respectively. 3-D motion and structure estimation using the orthographic displacement field model is covered in Section 9.2, whereas Section 9.3 deals with estimation using the perspective displacement field model.

9.1 Modeling the Projected Displacement Field

We start by deriving parametric models for the 2-D projected displacement field based on the orthographic and perspective projections of the 3-D rigid motion model, respectively.

Suppose a point $\mathbf{X} = [X_1, X_2, X_3]^T$ on a rigid object at time t moves to $\mathbf{X}' = [X_1', X_2', X_3']^T$ at time t' subject to a rotation described by the matrix $\mathbf{R}$ and a translation by the vector $\mathbf{T}$. Then, from Chapter 2, we have

$$\begin{bmatrix} X_1' \\ X_2' \\ X_3' \end{bmatrix} = \mathbf{R} \begin{bmatrix} X_1 \\ X_2 \\ X_3 \end{bmatrix} + \mathbf{T} = \begin{bmatrix} r_{11} & r_{12} & r_{13} \\ r_{21} & r_{22} & r_{23} \\ r_{31} & r_{32} & r_{33} \end{bmatrix} \begin{bmatrix} X_1 \\ X_2 \\ X_3 \end{bmatrix} + \begin{bmatrix} T_1 \\ T_2 \\ T_3 \end{bmatrix} \tag{9.1}$$

Recall that in the case of small rotation, the composite rotation matrix can be expressed, in terms of the Eulerian angles, as

$$\mathbf{R} = \begin{bmatrix} 1 & -\Delta\phi & \Delta\psi \\ \Delta\phi & 1 & -\Delta\theta \\ -\Delta\psi & \Delta\theta & 1 \end{bmatrix} \tag{9.2}$$

where $\Delta\theta$, $\Delta\psi$, and $\Delta\phi$ denote small clockwise angular displacements about the X_1, X_2, and X_3 axes, respectively.

9.1.1 Orthographic Displacement Field Model

The orthographic displacement field refers to the orthographic projection of the 3-D displacement vectors into the image plane, which is obtained by substituting X_1' and X_2' from (9.1) into

$$x_1' = X_1' \quad \text{and} \quad x_2' = X_2' \tag{9.3}$$

that define the orthographic projection as discussed in Chapter 2. Since we have $X_1 = x_1$ and $X_2 = x_2$ under the orthographic projection, the resulting model is given by

$$\begin{aligned} x_1' &= r_{11}x_1 + r_{12}x_2 + (r_{13}X_3 + T_1) \\ x_2' &= r_{21}x_1 + r_{22}x_2 + (r_{23}X_3 + T_2) \end{aligned} \tag{9.4}$$

The model (9.4) is an affine mapping of the pixel (x_1, x_2) at frame t to the pixel (x_1', x_2') at frame t' defined in terms of the six parameters r_{11}, r_{12}, $(r_{13}X_3 + T_1)$, r_{21}, r_{22}, and $(r_{23}X_3 + T_2)$. Thus, it constitutes a parametric 2-D motion field model (see Section 5.2).

It should be noted that the actual distance of the object points from the image plane is not observable in orthographic projection, since it is a parallel projection. Thus, if we express the actual depth of a point as $\tilde{X}_3 = \bar{X}_3 + X_3$, where $\bar{X}_3$ is the depth of a reference point on the object, and X_3 is the relative depth of all other object points with respect to the reference point, we can only expect to estimate X_3 associated with each image point. Further observe that in (9.4), X_3 multiplies r_{13} and r_{23}. Obviously, if we scale r_{13} and r_{23}, a new value of X_3 can be found for each pixel that would also satisfy (9.4). Hence, these variables cannot be uniquely determined for an arbitrary surface from two views, as stated in [Hua 89a].

Orthographic projection is a reasonable approximation to the actual imaging process described by the perspective projection, when the depth of object points does not vary significantly. Other approximations, such as weak perspective, paraperspective, and orthoperspective projections, also exist in the literature [Dem 92].

9.1.2 Perspective Displacement Field Model

The perspective displacement field can be derived by substituting X_1', X_2', and X_3' from (9.1) into the perspective projection model given by (from Chapter 2)

$$x_1' = f\frac{X_1'}{X_3'} \quad \text{and} \quad x_2' = f\frac{X_2'}{X_3'} \tag{9.5}$$

to obtain

$$\begin{aligned} x_1' &= f\frac{r_{11}X_1 + r_{12}X_2 + r_{13}X_3 + T_1}{r_{31}X_1 + r_{32}X_2 + r_{33}X_3 + T_3} \\ x_2' &= f\frac{r_{21}X_1 + r_{22}X_2 + r_{23}X_3 + T_2}{r_{31}X_1 + r_{32}X_2 + r_{33}X_3 + T_3} \end{aligned} \tag{9.6}$$

Letting $f = 1$, dividing both the numerator and the denominator by X_3, and expressing the object-space variables in terms of image plane coordinates using the perspective transform expression, we obtain

$$\begin{aligned} x_1' &= \frac{r_{11}x_1 + r_{12}x_2 + r_{13} + \frac{T_1}{X_3}}{r_{31}x_1 + r_{32}x_2 + r_{33} + \frac{T_3}{X_3}} \\ x_2' &= \frac{r_{21}x_1 + r_{22}x_2 + r_{23} + \frac{T_2}{X_3}}{r_{31}x_1 + r_{32}x_2 + r_{33} + \frac{T_3}{X_3}} \end{aligned} \tag{9.7}$$

The expressions (9.7) constitute a nonlinear model of the perspective projected motion field in terms of the image-plane coordinates because of the division by x_1 and x_2. Notice that this model is valid for arbitrary shaped moving surfaces in 3-D, since the depth of each point X_3 remains as a free parameter. Observe, however, that X_3 always appears in proportion to T_3 in (9.7). That is, the depth information is observable only when $T_3 \neq 0$. Furthermore, it can only be determined up to a scale factor. For example, an object at twice the distance from the image plane and moving twice as fast yields the same image under the perspective projection.

We develop 3-D motion and structure estimation algorithms based on the affine model (9.4) and the nonlinear model (9.7) in the next two sections. Because of the aforementioned limitations of these projections, we can estimate depth only up to an additive constant at best, using the orthographic displacement field, and only up to a multiplicative scale factor under the perspective displacement field model.

9.2 Methods Based on the Orthographic Model

It is well known that, from two orthographic views, we can estimate the depth of a feature point up to a scale factor α and an additive constant $\bar{X}_3$. The scale ambiguity arises because scaling X_{i3} by α, and r_{13} and r_{23} by $1/\alpha$ results in the same orthographic projection as can be seen from Equation (9.4). That is, if $\tilde{X}_{i3}$ denotes the true depth value, we expect to estimate

$$X_{i3} = \bar{X}_3 + \alpha \tilde{X}_{i3}, \quad \text{for } i = 1, \ldots, N \tag{9.8}$$

from two views. It is not possible to estimate $\bar{X}_3$ under any scheme, because this information is lost in the projection. However, the scale ambiguity may be overcome if more than two views are utilized. In his classical book Ullman [Ull 79] proved that four point correspondences over three frames, i.e. four points each traced from t_1 to t_2 and then to t_3, are sufficient to yield a unique solution to motion and structure up to a reflection. Later, Huang and Lee [Hua 89a] proposed a linear algorithm to obtain the solution in this case. Furthermore, they showed that with three point correspondences over three frames, there are 16 different solutions for motion and four for the structure plus their reflections.

In this section, we concentrate on the two-view problem, and first discuss a simple two-step iteration which was proposed by Aizawa *et al.* [Aiz 89] as part of the MBASIC algorithm for 3-D model-based compression of video. Two-step iteration is a simple and effective iterative algorithm for 3-D motion and depth estimation when the initial depth estimates are relatively accurate. However, it has been found to be particularly sensitive to random errors in the initial depth estimates. Thus, after introducing the two-step iteration algorithm, we propose an improved algorithm which yields significantly better results with only a small increase in the computational load.

9.2.1 Two-Step Iteration Method from Two Views

The two-step iteration method, proposed by Aizawa *et al.* [Aiz 89], is based on the following model of the projected motion field:

$$\begin{aligned} x_1' &= x_1 - \Delta\phi x_2 + \Delta\psi X_3 + T_1 \\ x_2' &= \Delta\phi x_1 + x_2 - \Delta\theta X_3 + T_2 \end{aligned} \tag{9.9}$$

which can be obtained by substituting the Eulerian angles definition of the matrix $\mathbf{R}$ given by (9.2) into (9.4).

In Equation (9.9), there are five unknown global motion parameters $\Delta\theta$, $\Delta\psi$, $\Delta\phi$, T_1, and T_2, and an unknown depth parameter X_3 for each given point correspondence (x_1, x_2). This is a bilinear model, since X_3 multiplies the unknown motion parameters. It is thus proposed to solve for the unknowns in two steps: First, determine the motion parameters given the depth estimates from the previous iteration, and then update the depth estimates using the new motion parameters. They are implemented as follows:

1) Given N corresponding coordinate pairs $\mathbf{x}_i = (x_{i1}, x_{i2})$ and $\mathbf{x'}_i = (x'_{i1}, x'_{i2})$ where $N \geq 3$, and the associated depth estimates X_{i3}, $i = 1, \ldots, N$, estimate the five global motion parameters.

This can be accomplished by rearranging Equation (9.9) as

$$\begin{bmatrix} x'_1 - x_1 \\ x'_2 - x_2 \end{bmatrix} = \begin{bmatrix} 0 & X_3 & -x_2 & 1 & 0 \\ -X_3 & 0 & x_1 & 0 & 1 \end{bmatrix} \begin{bmatrix} \Delta\theta \\ \Delta\psi \\ \Delta\phi \\ T_1 \\ T_2 \end{bmatrix} \tag{9.10}$$

Then, writing Equation (9.10) for N corresponding point pairs, we obtain $2N$ equations in five unknowns, which can be solved for the motion parameters using the method of least squares.

The initial estimates of the depth parameters can be obtained from an *a priori* model of the scene, and the depth estimates are not allowed to vary from these values by more than a predetermined amount, apparently because of the nonuniqueness of the solution. For example, in the case of head-and-shoulder type images, they can be obtained from a scaled wireframe model, as shown in Chapter 24.

2) Once the motion parameters are found, we can estimate the new X_{i3}, $i = 1, \ldots, N$, using

$$\begin{bmatrix} x'_1 - x_1 + \Delta\phi x_2 - T_1 \\ x'_2 - x_2 - \Delta\phi x_1 - T_2 \end{bmatrix} = \begin{bmatrix} \Delta\psi \\ -\Delta\theta \end{bmatrix} [\, X_3 \,] \tag{9.11}$$

which is again obtained by rearranging Equation (9.9). Here, we have an equation pair, for each given point correspondence in one unknown, the depth. The depth for each point correspondence can be solved in the least squares sense from the associated pair (9.11).

The procedure consists of repeating steps 1 and 2 until the estimates no longer change from iteration to iteration. Although theoretically three point correspondences are sufficient, in practice six to eight point correspondences are necessary to obtain reasonably good results due to possible fractional errors in finding the point correspondences. However, as stated even with that many points or more, the two-step iteration may converge to an erronous solution, unless we have very good initial depth estimates X_{i3}, $i = 1, \ldots, N$.

Suppose that the coordinate system is centered on the object so that $\bar{X}_3 = 0$, and we model the initial depth estimates as

$$X_{i3}^M = \beta X_{i3} + n_i \tag{9.12}$$

where β indicates a systematic error corresponding to a global underscaling or overscaling of the depth, and n_i represents the random errors, which are Gaussian distributed with zero mean. Clearly, it is not possible to estimate the scaling factor β unless the correct depth value of at least one of the N points is known. Assuming that $\beta = 1$, the performance of the MBASIC algorithm has been reported to be good when the initial depth estimates contain about 10% random error or less. However, its performance has been observed to degrade with increasing amounts of random error n_i in the initial depth estimates [Boz 94].

9.2.2 An Improved Iterative Method

In the two-step iteration there is strong correlation between the errors in the motion estimates and in the depth estimates. This can be seen from Equations (9.10) and (9.11), where the random errors in the depth estimates are fed back on the motion estimates and vice versa, repeatedly. Thus, if the initial depth estimates are not accurate enough, then the algorithm may converge to an erroneous solution.

To address this problem, we define an error criterion (9.13), and update X_{i3} in the direction of the gradient of the error with an appropriate step size, instead of computing from Equation (9.11), at each iteration. To avoid convergence to a local minimum, we also add a random perturbation to the depth estimates after each update, similar to simulated annealing. The update in the direction of the gradient increases the rate of convergence in comparison to totally random perturbations of X_{i3}. The motion parameters are still computed from Equation (9.10) at each iteration. The improved algorithm can be summarized as:

1. Initialize the depth values X_{i3} for $i = 1, \ldots, N$. Set the iteration counter $m = 0$.

2. Determine the motion parameters from (9.10) using the given depth values.

3. Compute $(\tilde{x}_{i1}^{\prime(m)}, \tilde{x}_{i2}^{\prime(m)})$, the coordinates of the matching points that are predicted by the present estimates of motion and depth parameters, using (9.9). Compute the model prediction error

$$E_m = \frac{1}{N} \sum_{i=0}^{N} e_i \tag{9.13}$$

where

$$e_i = (x_{i1}' - \tilde{x}_{i1}^{\prime(m)})^2 + (x_{i2}' - \tilde{x}_{i2}^{\prime(m)})^2 \tag{9.14}$$

Here (x_{i1}', x_{i2}') are the actual coordinates of the matching points which are known.

4. If $E_m < \epsilon$, stop the iteration. Otherwise, set $m = m + 1$, and perturb the depth parameters as

$$\hat{X}_{i3}^{(m)} \leftarrow \hat{X}_{i3}^{(m-1)} - \beta \frac{\partial e_i}{\partial X_3} + \alpha \Delta_i^{(m)} \tag{9.15}$$

   where $\Delta_i^{(m)} = N_i(0, \sigma_i^{2(m)})$ is a zero-mean Gaussian random variable with variance $\sigma_i^{2(m)} = e_i$, and α and β are constants.

5. Go to step (2).

A comparison of the performance of the improved algorithm and the two-step iteration is given in [Boz 94], where it was shown that the improved algorithm converges to the true motion and depth estimates even with 50% error in the initial depth estimates.

9.3 Methods Based on the Perspective Model

In the case of perspective projection, the equations that relate the motion and structure parameters to the image-plane coordinates (9.7) are nonlinear in the motion parameters. Early methods to estimate the motion and structure parameters from these expressions required an iterative search which might often diverge or converge to a local minimum [Roa 80, Mit 86]. Later, it was shown that with eight or more point correspondences, a two-step linear algorithm can be developed, where we first estimate some intermediate unknowns called the essential parameters [Hua 86]. The actual motion and structure parameters (for an arbitrary surface) can be subsequently derived from these essential parameters. To this effect, in the following we first present the epipolar constraint and define the essential parameters. Then, a linear least squares algorithm will be given for the estimation of the essential parameters from at least eight pixel correspondences. Finally, methods will be discussed for the estimation of the actual motion and structure parameters from the essential parameters. It should be noted that nonlinear algorithms for motion and structure estimation exist when the number of available point correspondences is five, six, or seven. One such algorithm, the homotopy method, is also briefly introduced.

9.3.1 The Epipolar Constraint and Essential Parameters

We begin by observing that the vectors $\mathbf{X}'$, $\mathbf{T}$, and $\mathbf{RX}$ are coplanar, since $\mathbf{X}' = \mathbf{RX} + \mathbf{T}$ from (9.1). Then, because $(\mathbf{T} \times \mathbf{RX})$ is orthogonal to this plane, we have

$$\mathbf{X}' \cdot (\mathbf{T} \times \mathbf{RX}) = 0 \tag{9.16}$$

where $\times$ indicates vector product, and $\cdot$ indicates dot product. It follows that

$$\mathbf{X}'^T \mathbf{E} \mathbf{X} = 0 \tag{9.17}$$

where

$$\mathbf{E} \doteq \begin{bmatrix} e_1 & e_2 & e_3 \\ e_4 & e_5 & e_6 \\ e_7 & e_8 & e_9 \end{bmatrix} = \begin{bmatrix} 0 & -T_3 & T_2 \\ T_3 & 0 & -T_1 \\ -T_2 & T_1 & 0 \end{bmatrix} \mathbf{R}$$

The elements of the matrix $\mathbf{E}$ are called the essential parameters [Hua 86]. Although there are nine essential parameters they are not all independent, because $\mathbf{E}$ is the product of a skew-symmetric matrix and a rotation matrix.

We divide both sides of (9.17) by $X_3\ X_3'$ to obtain a relationship in terms of the image plane coordinates as

$$\begin{bmatrix} x_1' & x_2' & 1 \end{bmatrix} \mathbf{E} \begin{bmatrix} x_1 \\ x_2 \\ 1 \end{bmatrix} = 0 \tag{9.18}$$

which is a linear, homogeneous equation in terms of the nine essential parameters. Equation (9.18) is known as the epipolar constraint for 3-D motion estimation, and is the basis of the linear estimation methods.

The epipolar constraint can alternatively be obtained by eliminating X_3 from the two expressions in the model (9.7) to obtain a single equation

$$\begin{aligned} (T_1 \quad - \quad & x_1' T_3) \left[x_2' (r_{31} x_1 + r_{32} x_2 + r_{33}) - (r_{21} x_1 + r_{22} x_2 + r_{23}) \right] \\ = \quad & (T_2 - x_2' T_3) \left[x_1' (r_{31} x_1 + r_{32} x_2 + r_{33}) - (r_{11} x_1 + r_{12} x_2 + r_{13}) \right] \end{aligned}$$

which can then be expressed as in (9.18).

It is well known that linear homogeneous equations have either no solution or infinitely many solutions. Thus, we set one of the essential parameters equal to one in (9.18), and solve for the remaining eight essential parameters, which is equivalent to using the essential parameter that is set equal to one as a scale factor. Recall that due to the scale ambiguity problem in the perspective projection, we can estimate the translation and depth parameters only up to a scale factor.

9.3.2 Estimation of the Essential Parameters

We first present a linear least squares method and an optimization method, which require that at least eight point correspondences be known. Then we briefly mention a nonlinear method where only five, six, or seven point correspondences may be sufficient to estimate the essential parameters.

Linear Least Squares Method

Setting $e_9 = 1$ arbitrarily, we can rearrange (9.18) as

$$\begin{bmatrix} x_1'x_1 & x_1'x_2 & x_1' & x_2'x_1 & x_2'x_2 & x_2' & x_1 & x_2 \end{bmatrix} \begin{bmatrix} e_1 \\ e_2 \\ e_3 \\ e_4 \\ e_5 \\ e_6 \\ e_7 \\ e_8 \end{bmatrix} = -1$$

which is a linear equation in the remaining eight essential parameters. Given $N \geq 8$ point correspondences, we can set up a system of $N \geq 8$ linear equations in 8 unknowns. Conditions for the coefficient matrix to have full rank were investigated by Longuet-Higgins [Lon 81]. Briefly stated, the coefficient matrix has full rank if $\mathbf{T} \neq \mathbf{0}$ and the shape of the 3-D surface satisfies certain conditions, called the surface constraint [Zhu 89]. The solution, then, gives the essential matrix $\mathbf{E}$ up to a scale factor. We note here that any of the essential parameters could have been chosen as the scale factor. In practice, it is advisable to select the e_i, $i = 1, \ldots, 9$, which would yield the coefficient matrix with the smallest condition number as the scale factor.

Optimization Method

Because it is not obvious, in general, which essential parameter should be chosen as a scale factor, an alternative is to use the norm of the solution as a scale factor. Then the estimation of the essential parameters can be posed as a constrained minimization problem [Wen 93a], such that

$$\mathbf{e} = \arg\min_{\mathbf{e}} \ || \, \mathbf{Ge} \, ||, \qquad \text{subject to} \ \ || \, \mathbf{e} \, || = 1 \tag{9.19}$$

where $|| \cdot ||$ denotes vector or matrix norm,

$$\mathbf{G} = \begin{bmatrix} x_{11}'x_{11} & x_{11}'x_{12} & x_{11}' & x_{12}'x_{11} & x_{12}'x_{12} & x_{12}' & x_{11} & x_{12} & 1 \\ \vdots & & & & \vdots & & & & \vdots \\ x_{N1}'x_{N1} & x_{N1}'x_{N2} & x_{N1}' & x_{N2}'x_{N1} & x_{N2}'x_{N2} & x_{N2}' & x_{N1} & x_{N2} & 1 \end{bmatrix} \tag{9.20}$$

is the observation matrix (derived from (9.18)), $N \geq 8$, and $\mathbf{e} = [e_1 \ e_2 \ e_3 \ e_4 \ e_5 \ e_6 \ e_7 \ e_8 \ e_9]^T$ denotes the 9×1 essential parameter vector. It is well-known that the solution of this problem is the unit eigenvector of $\mathbf{G}^T\mathbf{G}$ associated with its smallest eigenvalue.

The Homotopy Method

When fewer than eight point correspondences are known, or the rank of the coefficient matrix is less than 8, the system of linear equations is underdetermined. However, the fact that the matrix **E** can be decomposed into a skew-symmetric matrix postmultiplied by a rotation matrix (see Equation 9.21) can be formulated in the form of additional polynomial equations. In particular, the decomposability implies that one of the eigenvalues of **E** is zero, and the other two are equal. Capitalizing on this observation, Huang and Netravali [Hua 90] introduced the homotopy method to address the cases where five, six, or seven point correspondences are available. For example, with five point correspondences, they form five linear equations and three cubic equations, which are then solved by the homotopy method. It has been shown that there are at most ten solutions in this case. The interested reader is referred to [Hua 90].

9.3.3 Decomposition of the E-Matrix

Theoretically, from (9.17), the matrix **E** can be expressed as

$$\mathbf{E} = [\mathbf{e}_1 \mid \mathbf{e}_2 \mid \mathbf{e}_3] = \left[k\hat{\mathbf{T}} \times \mathbf{r}_1 \mid k\hat{\mathbf{T}} \times \mathbf{r}_2 \mid k\hat{\mathbf{T}} \times \mathbf{r}_3\right] \tag{9.21}$$

where the vectors $\mathbf{r}_i$, $i = 1, \ldots, 3$ denote the columns of the rotation matrix $\mathbf{R}$, k denotes the length of the translation vector $\mathbf{T}$, and $\hat{\mathbf{T}}$ is the unit vector along $\mathbf{T}$. In the following, we discuss methods to recover $\mathbf{R}$ and $\hat{\mathbf{T}}$ given $\mathbf{E}$ computed from noise-free and noisy point correspondence data, respectively.

Noise-Free Point Correspondences

It can easily be observed from (9.21) that each column of **E** is orthogonal to **T**. Then the unit vector along **T** can be obtained within a sign by taking the cross product of two of the three columns as

$$\hat{\mathbf{T}} = \pm \frac{\mathbf{e}_i \times \mathbf{e}_j}{||\mathbf{e}_i \times \mathbf{e}_j||} \qquad i \neq j \tag{9.22}$$

Furthermore, it can be shown that [Hua 86]

$$k^2 = \frac{1}{2}(\mathbf{e}_1 \cdot \mathbf{e}_1 + \mathbf{e}_2 \cdot \mathbf{e}_2 + \mathbf{e}_3 \cdot \mathbf{e}_3) \tag{9.23}$$

Obviously, finding k from (9.23) cannot overcome the scale ambiguity problem, since $\mathbf{e}_3$ contains an arbitrary parameter. However, it is needed to determine the correct rotation parameters.

In order to determine the correct sign of the translation vector, we utilize

$$X_3'\hat{\mathbf{T}} \times [x_1' \; x_2' \; 1]^T = X_3\hat{\mathbf{T}} \times \mathbf{R}[x_1 \; x_2 \; 1]^T$$

which is obtained by cross-multiplying both sides of (9.1) with $\hat{\mathbf{T}}$, after applying the perspective projection. Then, Zhuang [Zhu 89] shows that the vector $\hat{\mathbf{T}}$ with the correct sign satisfies

$$\sum \left[\hat{\mathbf{T}} \times [x_1' \; x_2' \; 1]^T\right]^T \mathbf{E}[x_1 \; x_2 \; 1]^T > 0 \tag{9.24}$$

where the summation is computed over all observed point correspondences.

Once the sign of the unit translation vector $\hat{\mathbf{T}}$ is determined, the correct rotation matrix can be found uniquely. To this effect, we observe that

$$\mathbf{e}_1 \times \mathbf{e}_2 = k\hat{\mathbf{T}}(k\hat{\mathbf{T}} \cdot \mathbf{r}_3) \tag{9.25}$$

$$\mathbf{e}_2 \times \mathbf{e}_3 = k\hat{\mathbf{T}}(k\hat{\mathbf{T}} \cdot \mathbf{r}_1) \tag{9.26}$$

$$\mathbf{e}_3 \times \mathbf{e}_1 = k\hat{\mathbf{T}}(k\hat{\mathbf{T}} \cdot \mathbf{r}_2) \tag{9.27}$$

The expressions (9.25)-(9.27) can be derived by employing the vector identity

$$\mathbf{A} \times (\mathbf{B} \times \mathbf{C}) = (\mathbf{A} \cdot \mathbf{C})\mathbf{B} - (\mathbf{A} \cdot \mathbf{B})\mathbf{C}$$

In particular,

$$\begin{aligned} \mathbf{e}_1 \times \mathbf{e}_2 &= \mathbf{e}_1 \times (k\hat{\mathbf{T}} \times \mathbf{r}_2) \\ &= [(k\hat{\mathbf{T}} \times \mathbf{r}_1) \cdot \mathbf{r}_2]k\hat{\mathbf{T}} - [(k\hat{\mathbf{T}} \times \mathbf{r}_1) \cdot k\hat{\mathbf{T}}]\mathbf{r}_2 \end{aligned}$$

Using the properties of the cross product, the first term simplifies as

$$[(\mathbf{r}_1 \times \mathbf{r}_2) \cdot k\hat{\mathbf{T}}]k\hat{\mathbf{T}} = (\mathbf{r}_3 \cdot k\hat{\mathbf{T}})k\hat{\mathbf{T}}$$

since $\mathbf{r}_1$, $\mathbf{r}_2$, and $\mathbf{r}_3$ are mutually orthogonal and have unity length (recall that they are the column vectors of a rotation matrix). The second term is zero since $(k\hat{\mathbf{T}} \times \mathbf{r}_1)$ is orthogonal to $k\hat{\mathbf{T}}$, yielding the relation (9.25).

Next, observe that we can express the vector $\mathbf{r}_1$ as

$$\mathbf{r}_1 = (\hat{\mathbf{T}} \cdot \mathbf{r}_1)\hat{\mathbf{T}} + (\hat{\mathbf{T}} \times \mathbf{r}_1) \times \hat{\mathbf{T}} \tag{9.28}$$

Here, the first term denotes the orthogonal projection of $\mathbf{r}_1$ onto $\hat{\mathbf{T}}$, and the second term gives the orthogonal complement of the projection, since

$$(\hat{\mathbf{T}} \times \mathbf{r}_1) \times \hat{\mathbf{T}} = ||\mathbf{r}_1|| \sin\beta = \sin\beta$$

where β denotes the angle between $\mathbf{r}_1$ and $\hat{\mathbf{T}}$, as depicted in Figure 9.1.

It follows from (9.28) that if we know the cross product and the dot product of an unknown vector $\mathbf{r}_1$ with a known vector $\hat{\mathbf{T}}$, we can determine the unknown through the relation (9.28). Evaluating the dot product of $\hat{\mathbf{T}}$ with both sides of (9.26) yields

$$\hat{\mathbf{T}} \cdot \mathbf{r}_1 = \frac{1}{k^2}\hat{\mathbf{T}} \cdot (\mathbf{e}_2 \times \mathbf{e}_3)$$

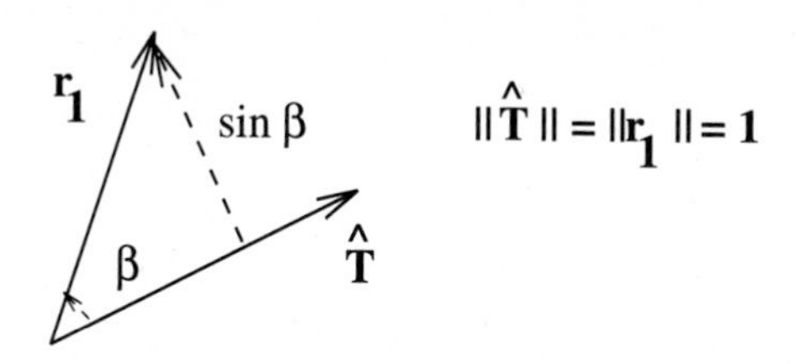

Figure 9.1: Orthogonal projection of $\mathbf{r}_1$ onto $\hat{\mathbf{T}}$.

Recall that we already have

$$\hat{\mathbf{T}} \times \mathbf{r}_1 = \frac{1}{k}\,\mathbf{e}_1$$

from the definition of the E-matrix (9.21). Substituting these dot and cross product expressions in (9.28), we find

$$\mathbf{r}_1 = \left[\frac{1}{k^2}\hat{\mathbf{T}} \cdot (\mathbf{e}_2 \times \mathbf{e}_3)\right]\hat{\mathbf{T}} + \frac{1}{k}(\mathbf{e}_1 \times \hat{\mathbf{T}}) \tag{9.29}$$

The other column vectors of the rotation matrix $\mathbf{R}$ are given similarly as

$$\mathbf{r}_2 = \left[\frac{1}{k^2}\hat{\mathbf{T}} \cdot (\mathbf{e}_3 \times \mathbf{e}_1)\right]\hat{\mathbf{T}} + \frac{1}{k}(\mathbf{e}_2 \times \hat{\mathbf{T}}) \tag{9.30}$$

$$\mathbf{r}_3 = \left[\frac{1}{k^2}\hat{\mathbf{T}} \cdot (\mathbf{e}_1 \times \mathbf{e}_2)\right]\hat{\mathbf{T}} + \frac{1}{k}(\mathbf{e}_3 \times \hat{\mathbf{T}}) \tag{9.31}$$

Noisy Point Correspondences

When feature point correspondences are contaminated by noise, we may obtain different estimates of the unit translation vector by using different combinations of the column vectors e_i, $i = 1, 2, 3$ in (9.22), and the estimate of the rotation matrix obtained from (9.29)-(9.31) may no longer satisfy the properties of a rotation matrix. To address these problems, let

$$\mathbf{W} = [\mathbf{w}_1\ \mathbf{w}_2\ \mathbf{w}_3] \tag{9.32}$$

denote an estimate of the rotation matrix where $\mathbf{w}_1$, $\mathbf{w}_2$, and $\mathbf{w}_3$ are obtained from (9.29)-(9.31). It has been shown that the estimates of $\hat{\mathbf{T}}$ and $\mathbf{R}$, in the case of noisy point correspondences, are given by [Wen 93a]

$$\hat{\mathbf{T}} = \arg\min_{\hat{\mathbf{T}}} \|\ \mathbf{E}^T\hat{\mathbf{T}}\ \|, \qquad \text{subject to } \|\ \hat{\mathbf{T}}\ \| = 1 \tag{9.33}$$

and

$$\hat{\mathbf{R}} = \arg\min_{\mathbf{R}} \| \mathbf{R} - \mathbf{W} \|, \quad \text{subject to } \mathbf{R} \text{ is a rotation matrix} \qquad (9.34)$$

respectively.

The solution $\hat{\mathbf{T}}$ of (9.33) is the unit eigenvector of $\mathbf{E}\mathbf{E}^T$ associated with its smallest eigenvalue. Note that the correct sign of $\hat{\mathbf{T}}$ must satisfy (9.24). Let the solution of (9.34) be expressed in terms of the quaternion representation (2.10). Then, $\mathbf{q} = [q_0\ q_1\ q_2\ q_3]^T$ is the unit eigenvector of the 4×4 matrix

$$\mathbf{B} = \sum_{i=1}^{3} \mathbf{B}_i^T \mathbf{B}_i$$

where

$$\mathbf{B}_i = \begin{bmatrix} 0 & (\mathbf{I}_i - \mathbf{w}_i)^T \\ \mathbf{w}_i - \mathbf{I}_i & [\mathbf{w}_i - \mathbf{I}_i]_\times \end{bmatrix}$$

such that $\mathbf{I}_i$, $i = 1, 2, 3$ denote the columns of a 3×3 identity matrix, and

$$[(x_1\ x_2\ x_3)^T]_\times = \begin{bmatrix} 0 & -x_3 & x_2 \\ x_3 & 0 & -x_1 \\ -x_2 & x_1 & 0 \end{bmatrix}$$

The reader is referred to [Wen 93a] for a derivation of this result.

9.3.4 Algorithm

The method can then be summarized as:

1) Given eight or more point correspondences, estimate $\mathbf{E}$ up to a scale factor using either the least squares or the optimization method.

2) Compute $\mathbf{T}$ (up to a scale factor) using (9.33).

3) Find $\mathbf{W}$ from (9.29)-(9.31) and $\mathbf{R}$ from (9.34).

Given the rotation matrix $\mathbf{R}$, the axis of rotation in terms of its directional cosines (n_1, n_2, n_3) and the incremental angle of rotation $\Delta\alpha$ about this axis can then be determined, if desired, from

$$trace\{\mathbf{R}\} = 1 + 2\cos\Delta\alpha \qquad (9.35)$$

and

$$\mathbf{R} - \mathbf{R}^T = 2\sin\Delta\alpha \begin{pmatrix} 0 & -n_3 & n_2 \\ n_3 & 0 & -n_1 \\ -n_2 & n_1 & 0 \end{pmatrix} \qquad (9.36)$$

4) Once the motion parameters are estimated, solve for the depth parameter X_3 in the least squares sense (up to a scale factor) from

$$\begin{bmatrix} T_1 - x_1' T_3 \\ T_2 - x_2' T_3 \end{bmatrix} = \begin{bmatrix} x_1'(r_{31}x_1 + r_{32}x_2 + r_{33}) - (r_{11}x_1 + r_{12}x_2 + r_{13}) \\ x_2'(r_{31}x_1 + r_{32}x_2 + r_{33}) - (r_{21}x_1 + r_{22}x_2 + r_{23}) \end{bmatrix} (X_3) \quad (9.37)$$

We note that the translation vector $\mathbf{T}$ and the depth parameters can only be determined up to a scale factor due to the scale ambiguity inherent in the perspective projection.

Alternative algorithms for the estimation of the 3-D motion and structure parameters from feature correspondences exist in the literature. The two-step linear algorithm presented in this section is favored because it provides an algebraic solution which is simple to compute. However, it is known to be sensitive to noise in the pixel correspondences [Wen 89, Phi 91, Wen 92]. This sensitivity may be attributed to the fact that the epipolar constraint constrains only one of the two components of the image plane position of a 3-D point [Wen 93]. It is well known that the matrix $\mathbf{E}$ possesses nice properties, such as that two of its eigenvalues must be equal and the third has to be zero [Hua 89b], which the linear method is unable to use. To this effect, Weng *et al.* [Wen 93] proposed maximum likelihood and minimum variance estimation algorithms which use properties of the matrix $\mathbf{E}$ as constraints in the estimation of the essential parameters to improve the accuracy of the solution in the presence of noise.

9.4 The Case of 3-D Planar Surfaces

Planar surfaces are an important special case because most real-world surfaces can be approximated as planar at least on a local basis. This fact leads to representation of arbitrary surfaces by 3-D mesh models composed of planar patches, such as the wireframe model widely used in computer graphics. The main reason for treating planar surfaces as a special case is that they do not satisfy the surface assumption required in the general case provided above. Fortunately, it is possible to derive simple linear algorithms for the case of planar surfaces, as described in this section.

9.4.1 The Pure Parameters

We start by deriving a simplified model for the case of planar surfaces. Let the 3-D points that we observe all lie on a plane described by

$$aX_1 + bX_2 + cX_3 = 1 \quad (9.38)$$

where $[\, a \; b \; c \,]^T$ denotes the normal vector of this plane. Then, the 3-D displacement model (9.1) can be rewritten as

$$\begin{bmatrix} X_1' \\ X_2' \\ X_3' \end{bmatrix} = \mathbf{R} \begin{bmatrix} X_1 \\ X_2 \\ X_3 \end{bmatrix} + \mathbf{T} [\, a \; b \; c \,] \begin{bmatrix} X_1 \\ X_2 \\ X_3 \end{bmatrix}$$

or

$$\begin{bmatrix} X_1' \\ X_2' \\ X_3' \end{bmatrix} = \begin{bmatrix} a_1 & a_2 & a_3 \\ a_4 & a_5 & a_6 \\ a_7 & a_8 & a_9 \end{bmatrix} \begin{bmatrix} X_1 \\ X_2 \\ X_3 \end{bmatrix} = \mathbf{A} \begin{bmatrix} X_1 \\ X_2 \\ X_3 \end{bmatrix} \tag{9.39}$$

where

$$A = R + T \begin{bmatrix} a & b & c \end{bmatrix}$$

Next, we map the 3-D displacements onto the 2-D image plane using the perspective transformation, and normalize $a_9 = 1$ due to the well-known scale ambiguity, to obtain the image plane mapping from t to t' given by

$$\begin{aligned} x_1' &= \frac{a_1 x_1 + a_2 x_2 + a_3}{a_7 x_1 + a_8 x_2 + 1} \\ x_2' &= \frac{a_4 x_1 + a_5 x_2 + a_6}{a_7 x_1 + a_8 x_2 + 1} \end{aligned} \tag{9.40}$$

The constants $a_1, \ldots, a_8$ are generally known as the pure parameters [Tsa 81]. Next we present a linear algorithm for the estimation of the pure parameters.

9.4.2 Estimation of the Pure Parameters

By cross-multiplying each equation, we can rearrange (9.40) for each given point correspondence, as follows:

$$\begin{bmatrix} x_1 & x_2 & 1 & 0 & 0 & 0 & -x_1 x_1' & -x_2 x_1' \\ 0 & 0 & 0 & x_1 & x_2 & 1 & -x_1 x_2' & -x_2 x_2' \end{bmatrix} \begin{bmatrix} a_1 \\ a_2 \\ a_3 \\ a_4 \\ a_5 \\ a_6 \\ a_7 \\ a_8 \end{bmatrix} = \begin{bmatrix} x_1' \\ x_2' \end{bmatrix} \tag{9.41}$$

Therefore, given at least four point correspondences, we can set up eight or more linear equations to solve for the eight pure parameters. It has been shown that the rank of the coefficient matrix is 8 if and only if no three of the four observed points are collinear in three dimensions [Hua 86].

9.4.3 Estimation of the Motion and Structure Parameters

Once the matrix $\mathbf{A}$ has been determined, the motion parameters $\mathbf{R}$ and $\mathbf{T}$, and the structure parameters, i.e., the normal vector of the plane, can be estimated by means of a singular value decomposition (SVD) of the matrix $\mathbf{A}$ as described in [Tsa 82]. We start by expressing the matrix $\mathbf{A}$ as

$$\mathbf{A} = \mathbf{U} \begin{bmatrix} \lambda_1 & 0 & 0 \\ 0 & \lambda_2 & 0 \\ 0 & 0 & \lambda_3 \end{bmatrix} \mathbf{V}^T \tag{9.42}$$

where $\mathbf{U} = [\mathbf{u}_1 \mid \mathbf{u}_2 \mid \mathbf{u}_3]$ and $\mathbf{V} = [\mathbf{v}_1 \mid \mathbf{v}_2 \mid \mathbf{v}_3]$ are 3×3 orthogonal matrices and $\lambda_1 \geq \lambda_2 \geq \lambda_3 \geq 0$ are the singular values of $\mathbf{A}$. There are three possibilities depending on the singular values:

Case 1) The singular values are distinct with $\lambda_1 > \lambda_2 > \lambda_3$, which indicates that the motion can be decomposed into a rotation about an axis through the origin, followed by a translation in a direction other than the direction of the normal vector of the plane. Then two solutions for the motion parameters exist, which can be expressed as

$$\begin{aligned}
\mathbf{R} &= \mathbf{U} \begin{bmatrix} \alpha & 0 & \beta \\ 0 & 1 & 0 \\ -s\beta & 0 & s\alpha \end{bmatrix} \mathbf{V}^T \\
\mathbf{T} &= k\left(-\beta \mathbf{u}_1 + (\frac{\lambda_3}{\lambda_2} - s\alpha)\mathbf{u}_3\right) \\
[\, a\ b\ c\,]^T &= \frac{1}{k}(\delta \mathbf{v}_1 + \mathbf{v}_3)
\end{aligned}$$

where

$$\begin{aligned}
\delta &= \pm\left(\frac{\lambda_1^2 - \lambda_2^2}{\lambda_2^2 - \lambda_3^2}\right)^{\frac{1}{2}} \\
\alpha &= \frac{\lambda_1 + s\lambda_3\delta^2}{\lambda_2(1+\delta^2)} \\
\beta &= \frac{1}{\delta}\left(\alpha - \frac{\lambda_1}{\lambda_2}\right) \\
s &= \det(\mathbf{U})\det(\mathbf{V})
\end{aligned}$$

and k is an arbitrary scale factor (positive or negative). The sign ambiguity may be resolved by requiring $1/X_3 > 0$ for all points.

Case 2) If the multiplicity of the singular values is two, e.g., $\lambda_1 = \lambda_2 \neq \lambda_3$, then a unique solution for the motion parameters exist, which is given by

$$\begin{aligned}
\mathbf{R} &= \frac{1}{\lambda_1}\mathbf{A} - \left(\frac{\lambda_3}{\lambda_1} - s\right)\mathbf{u}_3\mathbf{v}_3^T \\
\mathbf{T} &= k\left(\frac{\lambda_3}{\lambda_1} - s\right) \\
[\, a\ b\ c\,]^T &= \frac{1}{k}\mathbf{v}_3 \\
s &= \det(\mathbf{U})\det(\mathbf{V})
\end{aligned}$$

and k is an arbitrary scale factor. In this case, the motion can be described by a rotation about an axis through the origin followed by a translation along the normal vector of the plane.

Case 3) If the multiplicity of the singular values is three, i.e., $\lambda_1 = \lambda_2 = \lambda_3$, then the motion is a pure rotation around an axis through the origin, and $\mathbf{R}$ is uniquely determined by

$$\mathbf{R} = (\frac{1}{\lambda_1})\mathbf{A}$$

However, it is not possible to determine $\mathbf{T}$ and $[\ a\ b\ c\ ]^T$.

As an alternative method, Tsai and Huang [Tsa 81] obtained a sixth-order polynomial equation to solve for one of the unknowns, and then solved for the remaining unknowns. However, the SVD method provides a closed-form solution.

9.5 Examples

This section provides an experimental evaluation of the performances of some leading 3-D motion and structure estimation algorithms in order to provide the reader with a better understanding of their relative strengths and weaknesses. Results are presented for numerical simulations as well as with two frames of the sequence "Miss America." In each case, we have evaluated methods based on both the orthographic projection (two-step iteration and the improved method) and the perspective projection (E-matrix and A-matrix methods).

9.5.1 Numerical Simulations

Simulations have been performed to address the following questions: 1) Under what conditions can we successfully use the methods based on orthographic projection? 2) Which method, among the four that we compare, performs the best? 3) How sensitive are these methods to errors in the point correspondences? To this effect, we have generated two 3-D data sets, each containing 30 points. The first set consists of 3-D points that are randomly selected within a 100 cm × 100 cm × 0.6 cm rectangular prism whose center is 51.7 cm away from the center of projection; that is, we assume X_1 and X_2 are uniformly distributed in the interval [-50,50] cm, and X_3 is uniformly distributed within [51.4,52] cm. The second set consists of points such that X_1 and X_2 are uniformly distributed in the interval [-25,25] cm, and X_3 is uniformly distributed within [70,100] cm. The image-plane points are computed via the perspective projection, $x_i = fX_i/X_3$, $i = 1, 2$, with $f = 50$ mm. In order to find the matching points in the next frame, the 3-D data set is rotated and translated by the "true" motion parameters, and the corresponding image-plane points are recomputed. Clearly, the orthographic projection model provides a good approximation for the first data set, where the range of X_3 ($\Delta X_3 = 0.6$ cm) is small compared to the average value of X_3 ($\bar{X}_3 = 51.7$ cm), since $x_i = fX_i/51.7$, $i = 1, 2$,

Table 9.1: Comparison of the motion parameters. All angles are in radians, and the translations are in pixels. (Courtesy Gozde Bozdagi)

True	10% Error		30% Error		50% Error	
	Two-Step	Improved	Two-Step	Improved	Two-Step	Improved
$\Delta\theta = 0.007$	0.00689	0.00690	0.00626	0.00641	0.00543	0.00591
$\Delta\psi = 0.010$	0.00981	0.00983	0.00898	0.00905	0.00774	0.00803
$\Delta\phi = 0.025$	0.02506	0.02500	0.02515	0.02500	0.02517	0.02501
$T_1 = 0.100$	0.09691	0.09718	0.08181	0.08929	0.06172	0.07156
$T_2 = 0.180$	0.18216	0.18212	0.19240	0.19209	0.20660	0.17011

more-or-less describes all image-plane points. On the contrary, for the second set $\Delta X_3 = 30$ cm is not small in comparison to $\bar{X}_3 = 85$ cm; hence, the orthographic projection would not be a good approximation.

We have tested the performance of the methods based on the orthographic projection, the two-step iteration, and the improved algorithm on the first data set. The true motion parameters that are used in the simulations are listed in Table 9.1. Recall that both the two-step and the improved iterative algorithms require an initial estimate of the depth values for each point pair. In order to test the sensitivity of these algorithms to random errors in the initial depth estimates, $\pm 10\%$, $\pm 30\%$, or $\pm 50\%$ error has been added to each depth parameter X_{i3}. The sign of the error ($+$ or $-$) was chosen randomly for each point. The parameter values $\alpha = 0.95$ and $\beta = 0.3$ have been used to obtain the reported results. In the case of the improved algorithm, we iterate until E_m given by (9.13) is less than an acceptable level. In order to minimize the effect of random choices, the results are repeated three times using different seed values. The average results are reported.

Table 9.1 provides a comparison of the motion parameter estimates obtained by the two-step algorithm and the improved algorithm at the conclusion of the iterations. In order to compare the results of depth estimation, we define the following error measure:

$$Error = \sqrt{\frac{1}{N} \sum_{i=1}^{N} \frac{(X_{i3} - \hat{X}_{i3})^2}{(X_{i3})^2}} \tag{9.43}$$

where N is the number of matching points, and X_{i3} and $\hat{X}_{i3}$ are the "true" and estimated depth parameters, respectively. Figures 9.2 (a) and (b) show the error measure plotted versus the iteration number for the cases of 30% and 50% initial error, respectively. Note that the scale of the vertical axis is not the same in both plots. Although the results are depicted for 500 iterations, convergence has resulted in about 100 iterations in almost all cases.

It can be seen from Table 9.1 that the error in the initial depth estimates directly

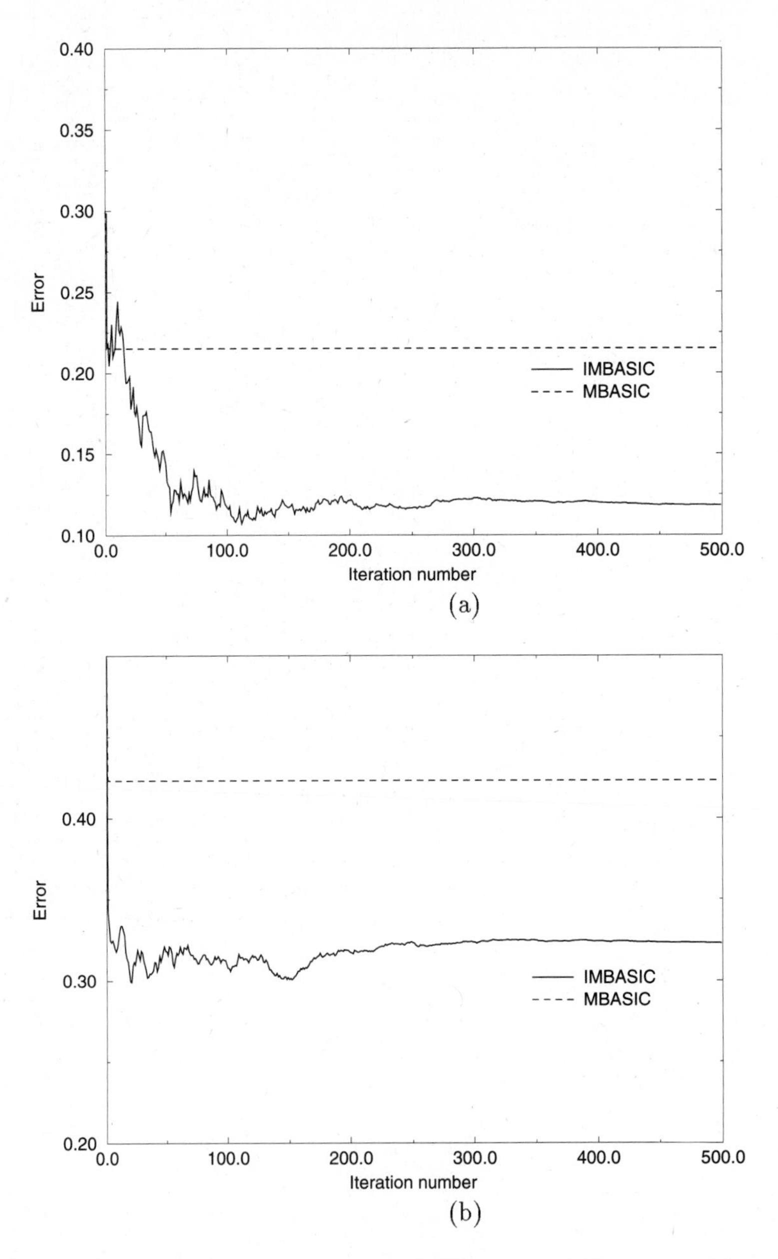

Figure 9.2: Comparison of depth parameters for a) 30% and b) 50% error. (Courtesy Gozde Bozdagi)

affects the accuracy of $\Delta\theta$ and $\Delta\psi$, which are multiplied by X_3 in both equations. Thus, in the two-step algorithm, the error in $\Delta\theta$ and $\Delta\psi$ estimates increases as we increase the error in the initial depth estimates. Furthermore, the error in the depth estimates (at the convergence) increases with increasing error in the initial depth parameters (see Figure 9.2). However, in the improved algorithm, as can be seen from Figure 9.2, the depth estimates converge closer to the correct parameters even in the case of 50% error in the initial depth estimates. For example, in the case of 50% error in the initial depth estimates, the improved method results in about 10% error after 500 iterations, whereas the two-step algorithm results in 45% error, demonstrating the robustness of the improved method to errors in the initial depth estimates.

The maximum difference in the image plane coordinates due to orthographic versus perspective projection of the 3-D object points is related to the ratio of the width of the object to the average distance of the points from the center of projection, which is $\Delta X_3/2\bar{X}_3 \approx 1\%$ for the first data set used in the above experiments. However, this ratio is approximately 18% for the second set, on which our experiments did not yield successful results using either the two-step or the improved algorithm. To provide the reader with a feeling of when we can use methods based on the orthographic projection successfully, we have shown the overall error in the motion and depth estimates given by

$$\begin{aligned} Error &= (\Delta\phi - \hat{\Delta\phi})^2 + (\Delta\psi - \hat{\Delta\psi}/\alpha)^2(\Delta\theta - \hat{\Delta\theta}/\alpha)^2 + (T_1 - \hat{T}_1)^2 \\ &\quad + (T_2 - \hat{T}_2)^2 + 1/N\sum_{i=1}^{N}(X_{i3} - \alpha\hat{X}_{i3})^2 \end{aligned}$$

where α is a scale factor, as a function of the ratio $\Delta X_3/2\bar{X}_3$ in Figure 9.3. The results indicate that the orthographic approximation yields acceptable results for $\Delta X_3/2\bar{X}_3 < 5\%$, while methods based on the perspective projection are needed otherwise.

We have tested two methods based on the perspective projection, namely the E-matrix and the A-matrix methods, on the second data set. In order to test the sensitivity of both methods to random errors in the point correspondences, $P\%$ error is added to the coordinates of the matching points according to

$$x_i' = x_i + (x_i' - x_i)(1 \pm \frac{P}{100}), \quad i = 1, 2$$

Table 9.2 shows the results of estimation with the true point correspondences as well as with 3% and 10% error in the point correspondences. Recall that the A-matrix method assumes all selected points lie on a planar surface, which is not the case with our data set. As a result, when the amount of noise is small, the E-matrix method outperforms the A-matrix method. Interestingly, the A-matrix method seems to be more robust in the presence of moderate noise in the coordinates of matching points, which may be due to use of a surface model.

Table 9.2: Comparison of the motion parameters. All angles are in radians, and the translations are in pixels. (Courtesy Yucel Altunbasak)

True	E-Matrix No Error	E-Matrix 3% Error	E-Matrix 10% Error	A-Matrix No Error	A-Matrix 3% Error	A-Matrix 10% Error
$\Delta\theta$ =0.007	0.00716	0.00703	0.00938	0.00635	0.00587	0.00470
$\Delta\psi = 0.010$	0.00991	0.00920	0.00823	0.01467	0.01450	0.01409
$\Delta\phi = 0.025$	0.02500	0.02440	0.02600	0.02521	0.02513	0.02501
$T_2/T_1 = 1.80$	1.80014	1.89782	2.15015	1.05759	1.12336	1.28980
$T_3/T_1 = 0.48$	0.47971	0.50731	-0.98211	0.25627	0.29159	0.37145
Depth error	0.00298	0.02284	0.15549	0.09689	0.09744	0.09974
Match error	1.73 E-6	2.20 E-4	6.57 E-4	1.15 E-3	1.20 E-3	1.61 E-3

The "depth error" refers to the root mean square (RMS) error in the X_3 coordinates of the 30 points selected. We have also calculated the "matching error" which

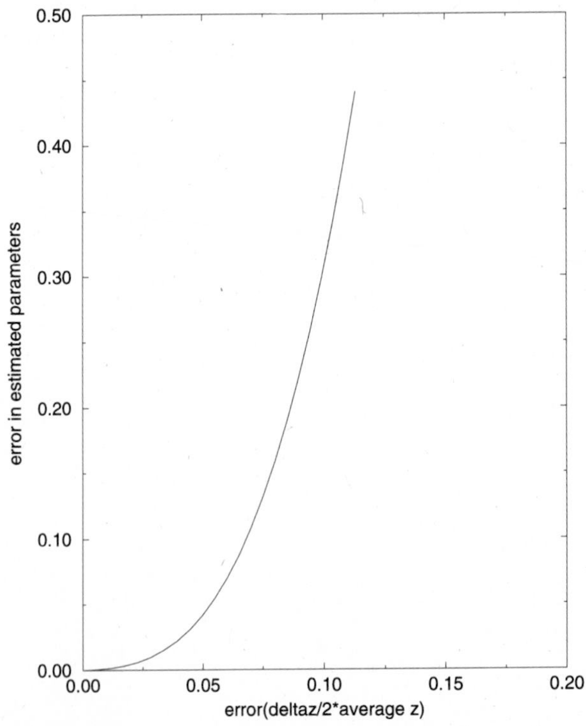

Figure 9.3: The performance of methods based on the orthographic projection. (Courtesy Gozde Bozdagi)

is the RMS difference between the coordinates of the matching points calculated with the true parameter values and those predicted by the estimated 3-D motion and structure parameters. The matching error with no motion compensation has been found as $1.97\ E-3$.

9.5.2 Experiments with Two Frames of Miss America

We next tested all four methods on two frames of the "Miss America" sequence, which are shown in Figure 9.4 (a) and (b), respectively. We note that the ratio of the maximum depth on a front view of a typical face to the average distance of the face from the camera falls within the range for which the orthographic projection is a reasonable approximation. Marked in Figure 9.4 (a) are the coordinates of 20 points (x_1, x_2) which are selected as features. The corresponding feature points (x_1', x_2') are likewise marked in Figure 9.4 (b). The coordinates of the matching points have been found by hierarchical block matching to the nearest integer. This truncation corresponds to adding approximately 15% noise to the matching point coordinates.

Table 9.3: Comparison of methods on two frames of the Miss America sequence.

Method	Coordinate RMSE
No Motion	4.4888
Two-Step	1.7178
Improved	0.5731
E-matrix	1.2993
A-matrix	0.7582

Since we have no means of knowing the actual 3-D motion parameters between the two frames or the actual depth of the selected feature points, we have calculated the RMS difference between the coordinates of the matching points determined by hierarchical block matching and those predicted by estimated 3-D motion and the structure parameters in order to evaluate the performance of the four methods.

Inspection of the RMSE values in Table 9.3 indicates that the improved iterative algorithm based on the orthographic projection performed best on the head-and-shoulder-type sequence. The reader is reminded that this sequence can be well approximated by the orthographic projection. The A-matrix method has been found to be the second best. We note that in the case of the A-matrix method, the image-plane points were compensated by the "a-parameters" without actually decomposing the A-matrix to determine the rotation and translation parameters. In the case of the E-matrix method, the E-matrix needs to be decomposed to find the motion parameters in order to be able compensate the image-plane points. We have observed that in the presence of errors in the coordinates of the matching points, the decomposition step tends to increase the coordinate RMSE.

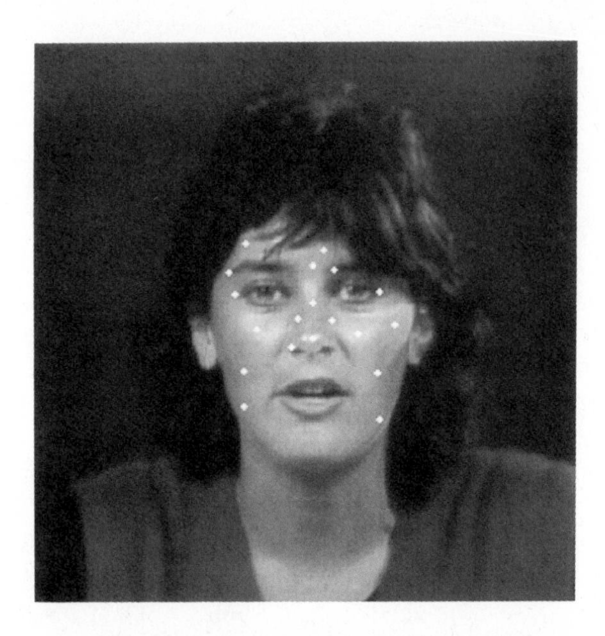

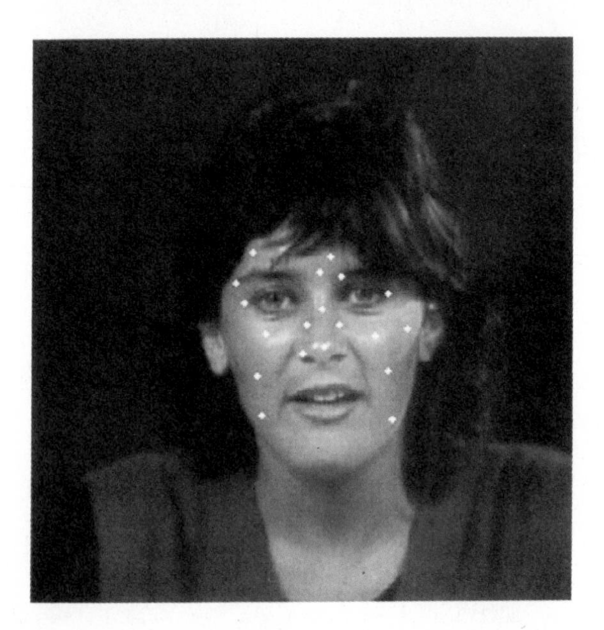

Figure 9.4: a) The first and b) the third frames of Miss America with matching points marked by white circles. (Courtesy Yucel Altunbasak)

9.6 Exercises

1. Is it possible to estimate $\Delta\theta$, $\Delta\psi$, and X_3 uniquely from (9.9)? Discuss all ambiguities that arise from the orthographic projection.

2. Observe from (9.7) that no depth information can be estimated when there is no translation, i.e., $\mathbf{T} = \mathbf{0}$. What is the minimum number of point correspondences in this case to determine the rotation matrix $\mathbf{R}$ uniquely?

3. Discuss the sensitivity of the proposed methods to noise in the point correspondences. Is the sensitivity related to the size of the object within the field of view? Explain.

Bibliography

[Agg 88] J. K. Aggarwal and N. Nandhakumar, "On the computation of motion from sequences of images," *Proc. IEEE*, vol. 76, pp. 917–935, Aug. 1988.

[Aiz 89] K. Aizawa, H. Harashima, and T. Saito, "Model-based analysis-synthesis image coding (MBASIC) system for a person's face," *Signal Processing: Image Communication*, no. 1, pp. 139–152, Oct. 1989.

[Boz 94] G. Bozdagi, A. M. Tekalp, and L. Onural, "An improvement to MBASIC algorithm for 3-D motion and depth estimation," *IEEE Trans. on Image Processing* (special issue), vol. 3, pp. 711–716, June 1994.

[Dem 92] D. DeMenthon and L. S. Davis, "Exact and approximate solutions of the perspective-three-point problem," *IEEE Trans. Patt. Anal. Mach. Intel.*, vol. 14, pp. 1100–1104, Nov. 1992.

[Hua 86] T. S. Huang, "Determining three-dimensional motion and structure from two perspective views," Chp. 14 in *Handbook of Patt. Recog. and Image Proc.*, Academic Press, 1986.

[Hua 89a] T. S. Huang and C. H. Lee, "Motion and structure from orthographic projections," *IEEE Trans. Patt. Anal. Mach. Intel.*, vol. 11, pp. 536–540, May 1989.

[Hua 89b] T. S. Huang and O. D. Faugeras, "Some properties of the E-matrix in two-view motion estimation," *IEEE Trans. Patt. Anal. Mach. Intel.*, vol. 11, No. 12, pp. 1310–1312, Dec. 1989.

[Hua 90] T. S. Huang and A. N. Netravali, "3D motion estimation," in *Machine Vision for Three-Dimensional Scenes*, Academic Press, 1990.

[Hua 94] T. S. Huang and A. N. Netravali, "Motion and structure from feature correspondences: A review," *Proc. IEEE*, vol. 82, pp. 252–269, Feb. 1994.

[Lon 81] H. C. Longuet-Higgins, "A computer program for reconstructing a scene from two projections," *Nature*, vol. 392, pp. 133–135, 1981.

[Mit 86] A. Mitiche and J. K. Aggarwal, "A computational analysis of time-varying images," in *Handbook of Pattern Recognition and Image Processing*, T. Y. Young and K. S. Fu, eds., New York, NY: Academic Press, 1986.

[Phi 91] J. Philip, "Estimation of three-dimensional motion of rigid objects from noisy observations," *IEEE Trans. Patt. Anal. Mach. Intel.*, vol. 13, no. 1, pp. 61–66, Jan. 1991.

[Roa 80] J. W. Roach and J. K. Aggarwal, "Determining the movement of objects from a sequence of images," *IEEE Trans. Patt. Anal. Mach. Intel.*, vol. 6, pp. 554–562, 1980.

[Tsa 81] R. Y. Tsai and T. S. Huang, "Estimating three-dimensional motion parameters of a rigid planar patch," *IEEE Trans. Acoust. Speech Sign. Proc.*, vol. 29, no. 6, pp. 1147–1152, Dec. 1981.

[Tsa 82] R. Y. Tsai, T. S. Huang, and W. L. Zhu, "Estimating 3-D motion parameters of a rigid planar patch II: Singular value decomposition," *IEEE Trans. Acoust. Speech Sign. Proc.*, vol. 30, no. 4, pp. 525–534, 1982; and vol. ASSP-31, no. 2, p. 514, 1983.

[Ull 79] S. Ullman, *The Interpretation of Visual Motion*, Cambridge, MA: MIT Press, 1979.

[Wen 89] J. Weng, T. S. Huang, and N. Ahuja, "Motion and structure from two perspective views: Algorithms, error analysis, and error estimation," *IEEE Trans. Patt. Anal. Mach. Intel.*, vol. 11, no. 5, pp. 451–476, May 1989.

[Wen 91] J. Weng, N. Ahuja, and T. S. Huang, "Motion and structure from point correspondences with error estimation: Planar surfaces," *IEEE Trans. Sign. Proc.*, vol. 39, no. 12, pp. 2691–2717, Dec. 1991.

[Wen 92] J. Weng, N. Ahuja, and T. S. Huang, "Motion and structure from line correspondences: Closed-form solution, uniqueness, and optimization," *IEEE Trans. Patt. Anal. Mach. Intel.*, vol. 14, no. 3, pp. 318–336, Mar. 1992.

[Wen 93] J. Weng, N. Ahuja, and T. S. Huang, "Optimal motion and structure estimation," *IEEE Trans. Patt. Anal. Mach. Intel.*, vol. 15, no. 9, pp. 864–884, Sep. 1993.

[Wen 93a] J. Weng, T. S. Huang, and N. Ahuja, *Motion and Structure from Image Sequences*, Springer-Verlag (Series in Information Sciences), 1993.

[Zhu 89] X. Zhuang, "A simplification to linear two-view motion algorithms," *Comp. Vis. Graph. Image Proc.*, vol. 46, pp. 175–178, 1989.

Chapter 10

OPTICAL FLOW AND DIRECT METHODS

This chapter discusses 3-D motion and structure estimation from two orthographic or perspective views based on an estimate of the optical flow field (optical flow methods) or on spatio-temporal image intensity gradients using the optical flow constraint (direct methods). The main differences between the methods in this chapter and the previous one are: optical flow methods utilize a projected velocity model as opposed to a projected displacement model (see Section 10.1.3 for a comparison of the two models); and optical flow methods require a dense flow field estimate or estimation of the image intensity gradient everywhere, rather than selecting a set of distinct feature points and matching them in two views. Here, we assume that the optical flow field is generated by a single object subject to 3-D rigid motion. The case of multiple moving objects, hence motion segmentation, will be dealt with in the next chapter.

Section 10.1 introduces 2-D velocity field models under orthographic and perspective projections. Estimation of the 3-D structure of the scene from the optical flow field in the case of pure 3-D translational motion is discussed in Section 10.2. Motion and structure estimation from the optical flow field in the case of general 3-D rigid motion using algebraic methods and optimization methods are the subjects of Section 10.3 and 10.4, respectively. Finally, in Section 10.5, we cover direct methods which do not require estimation of the optical flow field or any feature correspondence, but utilize spatio-temporal image intensity gradients directly.

10.1 Modeling the Projected Velocity Field

In this section, we present models for the 2-D (projected) velocity field starting with the 3-D velocity expression. Recall from Chapter 2 that, for the case of small angular rotation, the 3-D velocity vector $(\dot{X}_1, \dot{X}_2, \dot{X}_3)$ of a rigid object is given by

$$\begin{bmatrix} \dot{X}_1 \\ \dot{X}_2 \\ \dot{X}_3 \end{bmatrix} = \begin{bmatrix} 0 & -\Omega_3 & \Omega_2 \\ \Omega_3 & 0 & -\Omega_1 \\ -\Omega_2 & \Omega_1 & 0 \end{bmatrix} \begin{bmatrix} X_1 \\ X_2 \\ X_3 \end{bmatrix} + \begin{bmatrix} V_1 \\ V_2 \\ V_3 \end{bmatrix}$$

which can be expressed in vector notation as a cross product

$$\dot{\mathbf{X}} = \boldsymbol{\Omega} \times \mathbf{X} + \mathbf{V} \tag{10.1}$$

where $\boldsymbol{\Omega} = [\Omega_1 \ \Omega_2 \ \Omega_3]^T$ is the angular velocity vector and $\mathbf{V} = [V_1 \ V_2 \ V_3]^T$ represents the translational velocity vector. In scalar form, we have

$$\begin{aligned} \dot{X}_1 &= \Omega_2 X_3 - \Omega_3 X_2 + V_1 \\ \dot{X}_2 &= \Omega_3 X_1 - \Omega_1 X_3 + V_2 \\ \dot{X}_3 &= \Omega_1 X_2 - \Omega_2 X_1 + V_3 \end{aligned} \tag{10.2}$$

We will present two models for the 2-D velocity field, based on the orthographic and perspective projections of the model (10.2), respectively, in the following.

10.1.1 Orthographic Velocity Field Model

The orthographic projection of the 3-D velocity field onto the image plane can be computed from

$$\begin{aligned} v_1 = \dot{x}_1 &= \dot{X}_1 \\ v_2 = \dot{x}_2 &= \dot{X}_2 \end{aligned}$$

which results in

$$\begin{aligned} v_1 &= V_1 + \Omega_2 X_3 - \Omega_3 x_2 \\ v_2 &= V_2 + \Omega_3 x_1 - \Omega_1 X_3 \end{aligned} \tag{10.3}$$

in terms of the image plane coordinates. The orthographic model can be thought of as an approximation of the perspective model as the distance of the object from the image plane gets larger and the field of view becomes narrower.

10.1.2 Perspective Velocity Field Model

In order to obtain the perspective projection of the 3-D velocity field, we first apply the chain rule of differentiation to the perspective projection expression as

$$\begin{aligned} v_1 = \dot{x}_1 = f\frac{X_3\dot{X}_1 - X_1\dot{X}_3}{X_3^2} &= f\frac{\dot{X}_1}{X_3} - x_1\frac{\dot{X}_3}{X_3} \\ v_2 = \dot{x}_2 = f\frac{X_3\dot{X}_2 - X_2\dot{X}_3}{X_3^2} &= f\frac{\dot{X}_2}{X_3} - x_2\frac{\dot{X}_3}{X_3} \end{aligned} \tag{10.4}$$

Now, substituting the 3-D velocity model (10.2) in (10.4), and rewriting the resulting expressions in terms of the image plane coordinates, we have

$$\begin{aligned} v_1 &= f\left(\frac{V_1}{X_3} + \Omega_2\right) - \frac{V_3}{X_3}x_1 - \Omega_3 x_2 - \frac{\Omega_1}{f}x_1 x_2 + \frac{\Omega_2}{f}x_1^2 \\ v_2 &= f\left(\frac{V_2}{X_3} - \Omega_1\right) + \Omega_3 x_1 - \frac{V_3}{X_3}x_2 + \frac{\Omega_2}{f}x_1 x_2 - \frac{\Omega_1}{f}x_2^2 \end{aligned} \quad (10.5)$$

When the projection is normalized, $f = 1$, the model (10.5) can be rearranged as

$$\begin{aligned} v_1 &= \frac{-V_1 + x_1 V_3}{X_3} + \Omega_1 x_1 x_2 - \Omega_2(1 + x_1^2) + \Omega_3 x_2 \\ v_2 &= \frac{-V_2 + x_2 V_3}{X_3} + \Omega_1(1 + x_2^2) - \Omega_2 x_1 x_2 - \Omega_3 x_1 \end{aligned} \quad (10.6)$$

Note that the perspective velocities in the image plane depend on the X_3 coordinate of the object. The dependency of the model on X_3 can be relaxed by assuming a parametric surface model, at least on a local basis (see Sections 10.3.1 and 10.3.2).

For arbitrary surfaces, the depth X_3 of each scene point can be eliminated from the model (10.6) by solving both expressions for X_3 and equating them. This results in a single nonlinear expression, which relates each measured flow vector to the 3-D motion and structure parameters, given by

$$\begin{aligned} -v_2 e_1 + v_1 e_2 - x_1(\Omega_1 + \Omega_3 e_1) - x_2(\Omega_2 + \Omega_3 e_2) - x_1 x_2(\Omega_2 e_1 + \Omega_1 e_2) \\ +(x_1^2 + x_2^2)\Omega_3 + (1 + x_2^2)\Omega_1 e_1 + (1 + x_1^2)\Omega_2 e_2 = v_1 x_2 - v_2 x_1 \end{aligned} \quad (10.7)$$

where $e_1 = V_1/V_3$ and $e_2 = V_2/V_3$ denote the focus of expansion (see Section 10.2). Recall from the discussion of perspective projection that we can find the translational velocities and the depth up to a scaling constant, which is set equal to V_3 above.

10.1.3 Perspective Velocity vs. Displacement Models

Because optical flow estimation from two views indeed corresponds to approximate displacement estimation using spatio-temporal image intensity gradients (see Section 7.1), it is of interest to investigate the relation between the perspective velocity (10.5) and displacement (9.7) models for finite Δt. To this effect, let

$$v_1 = \dot{x}_1 \approx \frac{x_1' - x_1}{\Delta t} \quad (10.8)$$

Substituting (9.7) into (10.8), we can write

$$\frac{x_1' - x_1}{\Delta t} = \frac{x_1 + \Delta x_1}{\Delta t(1 + \Delta X_3)} - \frac{x_1}{\Delta t}$$

where

$$\Delta x_1 = -\Delta\phi x_2 + \Delta\psi + \frac{T_1}{X_3} \tag{10.9}$$

$$\Delta X_3 = -\Delta\psi x_1 + \Delta\theta x_2 + \frac{T_3}{X_3} \tag{10.10}$$

Now, invoking the approximation $1/(1+x) \approx 1 - x$ for $x << 1$, we have

$$\begin{aligned}\frac{x_1 + \Delta x_1}{1 + \Delta X_3} &\approx (x_1 + \Delta x_1)(1 - \Delta X_3) \\ &\approx x_1 + \Delta x_1 - x_1 \Delta X_3 \end{aligned} \tag{10.11}$$

where we assume $\Delta X_3 << 1$, and $\Delta x_1 \Delta X_3$ is negligible. Substituting (10.11) into (10.8), we obtain

$$v_1 = \frac{\Delta x_1}{\Delta t} - x_1 \frac{\Delta X_3}{\Delta t} \tag{10.12}$$

which yields (10.6) after replacing Δx_1 and ΔX_3 with (10.9) and (10.10), respectively. This leads us to two conclusions: the relations (10.5) and (9.7) are equivalent in the limit as Δt goes to zero; and second, for Δt finite, the two relations are equivalent only when the approximation (10.11) is valid.

10.2 Focus of Expansion

Estimation of the structure of a 3-D scene from a set of images is a fundamental problem in computer vision. There exist various means for structure estimation, such as structure from stereo, structure from motion, and structure from shading. In the previous chapter, we were able to estimate the depth (from motion) at only selected feature points. Recovery of the structure of the moving surface then requires 3-D surface interpolation. In this section, we estimate the structure of a moving surface by analyzing a dense optical flow field in the special case of 3-D translational motion. Structure from optical flow in the more general case, with rotational motion, is treated in the subsequent sections.

For the case of pure 3-D translation, the optical flow vectors (in the image plane) all appear to either emanate from a single point, called the focus of expansion (FOE), or converge to a single point, called the focus of contraction. Here, we consider only the case of expansion. The FOE can be defined as the intersection of the 3-D vector representing the instantaneous direction of translation and the image plane. More specifically, if an object is in pure translation, the 3-D coordinates of a point on the object at time t are

$$\begin{bmatrix} X_1(t) \\ X_2(t) \\ X_3(t) \end{bmatrix} = \begin{bmatrix} X_1(0) + V_1 t \\ X_2(0) + V_2 t \\ X_3(0) + V_3 t \end{bmatrix}$$

where $[X_1(0), X_2(0), X_3(0)]^T$ denote the coordinates of the point at time $t = 0$. Under perspective projection, the image of this point is given by

$$\begin{bmatrix} x_1(t) \\ x_2(t) \end{bmatrix} = \begin{bmatrix} \frac{X_1(0)+V_1 t}{X_3(0)+V_3 t} \\ \frac{X_2(0)+V_2 t}{X_3(0)+V_3 t} \end{bmatrix}$$

Then it appears that all points on the object emanate from a fixed point $\mathbf{e}$ in the image plane, called the FOE, given by

$$\mathbf{e} \doteq lim_{t \to -\infty} \begin{bmatrix} x_1(t) \\ x_2(t) \end{bmatrix} = \begin{bmatrix} V_1/V_3 \\ V_2/V_3 \end{bmatrix} \qquad (10.13)$$

Observe that the vector $[V_1/V_3 \;\; V_2/V_3 \;\; 1]^T$ indicates the instantaneous direction of the translation. Several approaches exist to calculate the FOE from two or more frames, either by first estimating the optical flow vectors [Law 83] or by means of direct search [Law 83, Jai 83].

Given two frames at times t and t', we can determine the relative depths of image points in the 3-D scene as follows:

1. Estimate the optical flow vectors from time t to t', and the location of FOE at t.

2. The depth $X_{3,i}(t')$ of a single image point $\mathbf{x}_i = [x_{1,i}, x_{2,i}]^T$ at time t' can be determined in terms of ΔX_3, called time-until-contact by Lee [Lee 76], as

$$\frac{X_{3,i}(t')}{\Delta X_3} = \frac{d_i}{\Delta d_i} \qquad (10.14)$$

where d_i is the distance of the image point from the FOE at time t, and Δd_i is the displacement of this image point from t to t'. Note that (10.14) is derived by using the similar triangles in the definition of the perspective transform, and ΔX_3 corresponds to the displacement of the 3-D point from t to t' along the X_3 axis, which cannot be determined.

3. The relative depths of two image points $\mathbf{x}_i$ and $\mathbf{x}_j$ at time t' is then given by the ratio

$$\frac{X_{3,i}(t')}{X_{3,j}(t')} = \frac{d_i}{d_j} \frac{\Delta d_i}{\Delta d_j} \qquad (10.15)$$

canceling ΔX_3.

10.3 Algebraic Methods Using Optical Flow

Optical flow methods consist of two steps: estimation of the optical flow field, and recovery of the 3-D motion parameters and the scene structure by analyzing the estimated optical flow. Methods to estimate the optical flow were discussed in Chapter 5. Various approaches exist to perform the second step which differ in the assumptions they make about the structure of the optical flow field and the criteria they employ. We can broadly classify these techniques in two groups: algebraic methods, which seek for a closed-form solution; and optimization methods, which can be characterized as iterative refinement methods towards optimization of a criterion function.

This section is devoted to algebraic methods. After a brief discussion about the uniqueness of the solution, we first assume a planar surface model to eliminate X_3 from (10.2). In this case, the orthographic and perspective projection of the 3-D velocity field results in an affine flow model and a quadratic flow model, respectively [Ana 93]. A least squares method can then be employed to estimate the affine and quadratic flow parameters. We then present two linear algebraic methods to estimate 3-D motion and structure from arbitrary flow fields.

10.3.1 Uniqueness of the Solution

Because 3-D motion from optical flow requires the solution of a nonlinear equation (10.7), there may, in general, be multiple solutions which yield the same observed optical flow. Observe that (10.7) has five unknowns, Ω_1, Ω_2, Ω_3, e_1 and e_2. However, in the presence of noise-free optical flow data, it has been shown, using algebraic geometry and homotopy continuation, that [Hol 93]:

- there are at most 10 solutions with five optical flow vectors,
- optical flow at six or more points almost always determines 3-D motion uniquely,
- if the motion is purely rotational, then it is uniquely determined by two optical flow values, and
- in the case of 3-D planar surfaces, optical flow at four points almost always gives two solutions.

With these uniqueness results in mind, we now present a number of algorithms to determine 3-D motion and structure from optical flow.

10.3.2 Affine Flow

A planar surface undergoing rigid motion yields an affine flow field under the orthographic projection. This can easily be seen by approximating the local surface structure with the equation of a plane

$$X_3 = z_0 + z_1 X_1 + z_2 X_2 \tag{10.16}$$

Substituting (10.16) into the orthographic velocity expressions (10.3), we obtain the six-parameter affine flow model

$$\begin{aligned} v_1 &= a_1 + a_2 x_1 + a_3 x_2 \\ v_2 &= a_4 + a_5 x_1 + a_6 x_2 \end{aligned} \tag{10.17}$$

where

$$a_1 = V_1 + z_0\Omega_2, \quad a_2 = z_1\Omega_2, \quad a_3 = z_2\Omega_2 - \Omega_3$$
$$a_4 = V_2 - z_0\Omega_1, \quad a_5 = \Omega_3 - z_1\Omega_1, \quad a_6 = -z_2\Omega_1$$

Observe that, if we know the optical flow at three or more points, we can set up six or more equations in six unknowns to solve for $(a_1, \ldots, a_6)$. However, because of the orthographic projection, it is not possible to determine all eight motion and structure parameters from $(a_1, \ldots, a_6)$ uniquely. For example, z_0 is not observable under the orthographic projection.

10.3.3 Quadratic Flow

The quadratic flow is the most fundamental flow field model, because it is an exact model for the case of planar surfaces under perspective projection; otherwise, it is locally valid under a first-order Taylor series expansion of the surface [Wax 87]. In the case of a planar surface, we observe from (10.16) that

$$\frac{1}{X_3} = \frac{1}{z_0} - \frac{z_1}{z_0}x_1 - \frac{z_2}{z_0}x_2 \tag{10.18}$$

Substituting the expression (10.18) into the perspective velocity field model (10.5), we obtain the eight-parameter quadratic flow field

$$\begin{aligned} v_1 &= a_1 + a_2x_1 + a_3x_2 + a_7x_1^2 + a_8x_1x_2 \\ v_2 &= a_4 + a_5x_1 + a_6x_2 + a_7x_1x_2 + a_8x_2^2 \end{aligned} \tag{10.19}$$

where

$$a_1 = f\left(\frac{V_1}{z_0} + \Omega_2\right), \quad a_2 = -\left(f\frac{V_1z_1}{z_0} + \frac{V_3}{z_0}\right), \quad a_3 = -\left(f\frac{V_1z_2}{z_0} + \Omega_3\right)$$
$$a_4 = f\left(\frac{V_2}{z_0} - \Omega_1\right), \quad a_5 = -\left(f\frac{V_2z_1}{z_0} - \Omega_3\right), \quad a_6 = -\left(f\frac{V_2z_2}{z_0} + \frac{V_3}{z_0}\right)$$
$$a_7 = \left(\frac{V_3z_1}{z_0} + \frac{\Omega_2}{f}\right), \quad a_8 = \left(\frac{V_3z_2}{z_0} - \frac{\Omega_1}{f}\right)$$

If we know the optical flow on at least four points that lie on a planar patch, we can set up eight or more equations in eight unknowns to solve for $(a_1, \ldots, a_8)$. Note that z_0, which is the distance between the surface point and the camera, always appears as a scale factor. Other approaches have also been reported to solve for the model parameters [Die 91, Ana 93].

In order to recover the 3-D motion and structure parameters using the second-order flow model, Waxman *et al.* [Wax 87] defined 12 kinematic deformation parameters that are related to the first- and second-order partial derivatives of the optical flow. Although closed-form solutions can be obtained, partials of the flow estimates are generally sensitive to noise.

10.3.4 Arbitrary Flow

When the depth of each object point varies arbitrarily, the relationship between the measured flow vectors and the 3-D motion and structure parameters is given by (10.6) or (10.7). An important observation about the expressions (10.6) is that they are bilinear; that is, they are linear in $(V_1, V_2, V_3, \Omega_1, \Omega_2, \Omega_3)$ for a given value of X_3. We discuss two methods, for the solution of (10.6) and (10.7), respectively, in the following. The first method obtains a set of linear equations in terms of 8 intermediate unknowns, three of which are redundant, whereas the second method proposes a two-step least squares solution for the motion parameters.

A Closed-Form Solution

Following Zhuang *et al.* [Zhu 88] and Heikkonen [Hei 93], the expression (10.7) can be expressed in vector-matrix form as

$$[-v_2 \; v_1 \; -x_1 \; -x_2 \; -x_1x_2 \; (x_1^2+x_2^2) \; (1+x_2^2) \; (1+x_1^2)]\mathbf{H} = v_1x_2 - v_2x_1 \quad (10.20)$$

where

$$\begin{aligned}\mathbf{H} &\doteq [\, h_1 \; h_2 \; h_3 \; h_4 \; h_5 \; h_6 \; h_7 \; h_8 \,] \\ &= [e_1 \; e_2 \; \Omega_1 + \Omega_3 e_1 \; \Omega_2 + \Omega_3 e_2 \; \Omega_2 e_1 + \Omega_1 e_2 \; \Omega_3 \; \Omega_1 e_1 \; \Omega_2 e_2]^T\end{aligned}$$

Given the optical flow vectors at a minimum of eight image points, we can set up eight linear equations to solve for $\mathbf{H}$. Usually we need more than eight image points to alleviate the effects of errors in the optical flow estimates, in which case the intermediate unknowns h_i can be solved in the least squares sense. The parameters h_i can be estimated uniquely as long as the rank of the coefficient matrix is 8, which is almost always the case provided that not all the selected points are coplanar. It can be easily seen from (10.20) that when all points are coplanar, we have a quadratic flow field, and the columns of the coefficient matrix are dependent on each other.

The five motion parameters, Ω_1, Ω_2, Ω_3, e_1, and e_2, can subsequently be recovered from h_i, $i = 1, \ldots, 8$. Next, the depth X_3 of each point can be estimated from the model (10.6). We note here, however, that the estimation of the motion parameters from h_i is not unique in the presence of errors in the optical flow estimates. This is due to the fact that only five of the eight h_i are independent. For example, motion parameters estimated from h_1, h_2, h_6, h_7, and h_8 do not necessarily satisfy h_3, h_4, and h_5. Estimation of the motion parameters to satisfy all h_i in the least squares sense can be considered, though it is not trivial.

Heeger and Jepson Method

Alternatively, Heeger and Jepson [Hee 92] develop a two-step method where no redundant variables are introduced. They express the arbitrary flow equations (10.6) in vector-matrix form as

$$\begin{bmatrix} v_1(x_1, x_2) \\ v_2(x_1, x_2) \end{bmatrix} = p(x_1, x_2)\mathbf{A}(x_1, x_2)\mathbf{V} + \mathbf{B}(x_1, x_2)\mathbf{\Omega} \tag{10.21}$$

where

$$\begin{aligned} \mathbf{A}(x_1, x_2) &= \begin{bmatrix} f & 0 & -x_1 \\ 0 & f & -x_2 \end{bmatrix}, \quad \mathbf{B}(x_1, x_2) = \begin{bmatrix} -x_1 x_2/f & f + x_1^2/f & -x_2 \\ -(f + x_2^2/f) & x_1 x_2/f & x_1 \end{bmatrix} \\ \mathbf{V} &= [\, V_1 \; V_2 \; V_3 \,]^T, \quad \mathbf{\Omega} = [\, \Omega_1 \; \Omega_2 \; \Omega_3 \,]^T, \text{ and } p(x_1, x_2) = \frac{1}{X_3} \end{aligned}$$

Consider stacking (10.21) for flow vectors at N different spatial locations, (x_{11}, x_{12}), ..., (x_{N1}, x_{N2}), to obtain

$$\begin{aligned} \mathbf{v} &= \mathbf{A}(\mathbf{V})\mathbf{p} + \mathbf{B}\mathbf{\Omega} \\ &= \mathbf{C}(\mathbf{V})\mathbf{q} \end{aligned} \tag{10.22}$$

where

$$\mathbf{A}(\mathbf{V}) = \begin{bmatrix} \mathbf{A}(x_{11}, x_{12})\mathbf{V} & \cdots & 0 \\ \vdots & \vdots & \vdots \\ 0 & \cdots & \mathbf{A}(x_{N1}, x_{N2})\mathbf{V} \end{bmatrix}, \quad \mathbf{B} = \begin{bmatrix} \mathbf{B}(x_{11}, x_{12}) \\ \vdots \\ \mathbf{B}(x_{N1}, x_{N2}) \end{bmatrix}$$

$$\mathbf{C}(\mathbf{V}) = \begin{bmatrix} | & | \\ \mathbf{A}(\mathbf{V}) & \mathbf{B} \\ | & | \end{bmatrix}, \quad \mathbf{q} = \begin{bmatrix} \mathbf{p} \\ \mathbf{\Omega} \end{bmatrix}$$

In order to find the least squares estimates of the motion and structure parameters $\mathbf{V}$ and $\mathbf{q}$, we minimize the error in (10.22) given by

$$E(\mathbf{V}, \mathbf{q}) = ||\mathbf{v} - \mathbf{C}(\mathbf{V})\mathbf{q}||^2 \tag{10.23}$$

Evidently, to minimize (10.23) with respect to $\mathbf{V}$ and $\mathbf{q}$ we need to compute the partial derivatives of (10.23) with respect to each variable and set them equal to zero, and solve the resulting equations simultaneously. Each equation would give a surface in a multidimensional space, and the solution lies at the intersection of all of these surfaces.

Alternatively, we can compute the partial of (10.23) with respect to $\mathbf{q}$ only. Setting it equal to zero yields

$$\hat{\mathbf{q}} = \left[\mathbf{C}(\mathbf{V})^T \mathbf{C}(\mathbf{V})\right]^{-1} \mathbf{C}(\mathbf{V})^T \mathbf{v} \tag{10.24}$$

which is a surface in terms of $\mathbf{V}$. Obviously, Equation (10.24) cannot be directly used to find an estimate of $\mathbf{q}$, since $\mathbf{V}$ is also unknown. However, because the actual solution lies on this surface, it can be substituted back into (10.23) to express the criterion function (10.23) in terms of $\mathbf{V}$ only. Now that we have a nonlinear criterion function in terms of $\mathbf{V}$ only, the value of $\mathbf{V}$ that minimizes (10.23) can be determined by means of a search procedure. Straightforward implementation of this strategy suffers from a heavy computational burden, since Equation (10.24) needs to be solved for every perturbation of $\mathbf{V}$ during the search process. Observe that the dimensions of the matrix in (10.24) to be inverted is related to the number of optical flow vectors used.

Fortunately, Heeger and Jepson have proposed an efficient method to search for the minimum of this index as a function of only $\mathbf{V}$, without evaluating (10.24) at each step, using some well-known results from linear algebra. Furthermore, since we can estimate $\mathbf{V}$ only up to a scale factor, they have restricted the search space, without loss of generality, to the unit sphere, which can be described by two angles in the spherical coordinates. Once the best estimate of $\mathbf{V}$ is determined, the least squares estimate of $\mathbf{q}$ can be evaluated from (10.24). Experimental results suggest that this method is quite robust to optical flow estimation errors.

10.4 Optimization Methods Using Optical Flow

Optimization methods are based on perturbing the 3-D motion and structure parameters independently until the projected velocity field is consistent with the observed optical flow field. These techniques are most successful for tracking small changes in the motion and structure parameters from frame to frame in a long sequence of frames, starting with reasonably good initial estimates, usually obtained by other means. Some sort of smoothness constraint on the 3-D motion and structure parameters is usually required to prevent the optimization algorithm from diverging or converging to a local minimum of the cost function.

Morikawa and Harashima [Mor 91] have proposed an optimization method using the orthographic velocity field (v_1, v_2) given by

$$\begin{aligned} v_1 = \dot{x}_1 = V_1 + \Omega_2 X_3 - \Omega_3 x_2 \\ v_2 = \dot{x}_2 = V_2 + \Omega_3 x_1 - \Omega_1 X_3 \end{aligned} \tag{10.25}$$

Note that the motion parameters are global parameters assuming there is a single rigid object in motion. However, the depth parameters vary spatially. Given initial estimates of the 3-D motion and depth parameters, we can update them incrementally using

$$\begin{aligned} \Omega_1(k) &= \Omega_1(k-1) + \Delta\Omega_1 \\ \Omega_2(k) &= \Omega_2(k-1) + \Delta\Omega_2 \\ \Omega_3(k) &= \Omega_3(k-1) + \Delta\Omega_3 \\ V_1(k) &= V_1(k-1) + \Delta V_1 \\ V_2(k) &= V_2(k-1) + \Delta V_2 \\ X_3(x_1, x_2)(k) &= X_3(x_1, x_2)(k-1) + \Delta X_3(x_1, x_2) \end{aligned}$$

where the problem reduces to finding the incremental parameters from frame to frame. The smoothness of motion constraint means that the update terms (the incremental parameters) should be small. Consequently, we introduce a measure of the smoothness in terms of the functional

$$||P||^2 = \frac{\alpha}{N}||\mathbf{\Delta\Omega}||^2 + \frac{\beta}{N}||\mathbf{\Delta V}||^2 + \gamma \sum_{i=1}^{N} (\Delta X_{i3})^2$$

where α, β, and γ are scale parameters, $||.||$ is the L_2 norm, and N is the number of points considered.

Then the overall cost functional measures the sum of how well the projected velocity field conforms with the observed flow field, and how smooth the variations in the 3-D motion and structure parameters are from frame to frame, as follows

$$E(\mathbf{\Delta\Omega}, \mathbf{\Delta T}, \Delta X_3) = \sum_i \left[(v_{i1} - \hat{u}_{i1})^2 + (v_{i2} - \hat{u}_{i2})^2 \right] + ||P||^2 \tag{10.26}$$

where

$$\hat{u}_{i1} = (V_1 + \Delta V_1) + (\Omega_2 + \Delta\Omega_2)(X_{i3} + (\Delta X)_{i3}) - (\Omega_3 + \Delta\Omega_3)x_2$$
$$\hat{u}_{i2} = (V_2 + \Delta V_2) + (\Omega_3 + \Delta\Omega_3)x_1 - (\Omega_1 + \Delta\Omega_1)(X_{i3} + (\Delta X)_{i3})$$

The optimization can be performed by gradient-based methods or simulated annealing procedures. We note that because this procedure uses the orthographic velocity field, the depth parameters can be found up to an additive constant and a scale parameter, while two of the rotation angles can be determined up to the reciprocal of this scaling constant. Other successful methods also exist for estimating motion and structure under the orthographic model [Kan 86].

10.5 Direct Methods

Direct methods utilize only the spatio-temporal image intensity gradients to estimate the 3-D motion and structure parameters. In this section, we present two examples of direct methods, one as an extension of optical-flow-based methods, and another starting from the projected motion field model and the optical flow equation. For other examples of direct methods, the reader is referred to the literature [Hor 88].

10.5.1 Extension of Optical Flow-Based Methods

Almost all optical-flow-based estimation methods can be extended as direct methods by replacing the optical flow vectors with their estimates given in terms of spatio-temporal image intensity gradients as derived in Chapter 5. Recall that an estimate of the flow field was given by (5.12)

$$\begin{bmatrix} v_1 \\ v_2 \end{bmatrix} = \begin{bmatrix} \sum \frac{\partial s_c(\mathbf{x})}{\partial x_1} \frac{\partial s_c(\mathbf{x})}{\partial x_1} & \sum \frac{\partial s_c(\mathbf{x})}{\partial x_1} \frac{\partial s_c(\mathbf{x})}{\partial x_2} \\ \sum \frac{\partial s_c(\mathbf{x})}{\partial x_1} \frac{\partial s_c(\mathbf{x})}{\partial x_2} & \sum \frac{\partial s_c(\mathbf{x})}{\partial x_2} \frac{\partial s_c(\mathbf{x})}{\partial x_2} \end{bmatrix}^{-1} \begin{bmatrix} -\sum \frac{\partial s_c(\mathbf{x})}{\partial x_1} \frac{\partial s_c(\mathbf{x})}{\partial t} \\ -\sum \frac{\partial s_c(\mathbf{x})}{\partial x_2} \frac{\partial s_c(\mathbf{x})}{\partial t} \end{bmatrix} \quad (10.27)$$

Substituting this expression in any optical-flow-based method for the optical flow vectors, we can obtain an estimate of the 3-D motion and structure parameters directly in terms of the spatio-temporal image intensity gradients. However, the reader is alerted that (5.12) may not provide the best possible estimates for the optical flow field. An alternative approach is presented in the following starting with the projected motion and optical flow equations.

10.5.2 Tsai-Huang Method

Tsai and Huang [Tsa 81] considered the case of planar surfaces, where we can model the displacement of pixels from frame t to t' by (9.40)

$$x_1' = T_1(a_1, \ldots, a_8) = \frac{a_1 x_1 + a_2 x_2 + a_3}{a_7 x_1 + a_8 x_2 + 1}$$

$$x_2' = T_2(a_1, \ldots, a_8) = \frac{a_4 x_1 + a_5 x_2 + a_6}{a_7 x_1 + a_8 x_2 + 1} \quad (10.28)$$

where the pure parameters $a_1, \ldots, a_8$ are usually represented in vector notation by $\mathbf{a}$. Note that $\mathbf{a} = \mathbf{e}$, where $\mathbf{e} = (1,0,0,0,1,0,0,0)^T$ corresponds to no motion, i.e., $x_1 = T_1(\mathbf{e})$, $x_2 = T_2(\mathbf{e})$.

In order to obtain a linear estimation algorithm, the mapping (10.28) will be linearized by means of a first-order Taylor series expansion about $\mathbf{a} = \mathbf{e}$, assuming small motion, as

$$T_1(\mathbf{a}) - T_1(\mathbf{e}) = \Delta x_1 = \sum_{i=1}^{8} \frac{\partial T_1(\mathbf{a})}{\partial a_i}\Big|_{\mathbf{a}=\mathbf{e}} (a_i - e_i)$$

$$T_2(\mathbf{a}) - T_2(\mathbf{e}) = \Delta x_2 = \sum_{i=1}^{8} \frac{\partial T_2(\mathbf{a})}{\partial a_i}\Big|_{\mathbf{a}=\mathbf{e}} (a_i - e_i) \quad (10.29)$$

where $\mathbf{a} = (a_1, \ldots, a_8)$, $\Delta x_1 = x_1' - x_1$ and $\Delta x_2 = x_2' - x_2$.

Now, referring to the optical flow equation (5.5) and approximating the partials by finite differences, we obtain the discrete expression

$$\frac{s_k(x_1,x_2) - s_{k+1}(x_1,x_2)}{\Delta t} = \frac{\partial s_{k+1}(x_1,x_2)}{\partial x_1}\frac{\Delta x_1}{\Delta t} + \frac{\partial s_{k+1}(x_1,x_2)}{\partial x_2}\frac{\Delta x_2}{\Delta t} \quad (10.30)$$

where Δt is the time interval between the frames k and $k+1$ at times t and t', respectively. After cancelling Δt, we obtain

$$s_k(x_1,x_2) = s_{k+1}(x_1,x_2) + \frac{\partial s_{k+1}(x_1,x_2)}{\partial x_1}\Delta x_1 + \frac{\partial s_{k+1}(x_1,x_2)}{\partial x_2}\Delta x_2 \qquad (10.31)$$

Observe that (10.31) may also be interpreted as the linearization of image intensity function in the vicinity of (x_1, x_2) in frame $k+1$, as

$$s_{k+1}(x_1',x_2') - s_{k+1}(x_1,x_2) = \frac{\partial s_{k+1}(x_1,x_2)}{\partial x_1}\Delta x_1 + \frac{\partial s_{k+1}(x_1,x_2)}{\partial x_2}\Delta x_2$$

since we have

$$s_{k+1}(x_1',x_2') = s_k(x_1,x_2)$$

Substituting the linearized expressions for Δx_1 and Δx_2 from (10.29) into (10.31), we can write the frame difference between the frames k and $k+1$ as

$$\begin{aligned} FD(x_1,x_2) &= s_k(x_1,x_2) - s_{k+1}(x_1,x_2) \\ &= \frac{\partial s_{k+1}(x_1,x_2)}{\partial x_1}\left[\sum_{i=1}^{8}\frac{\partial T_1(\mathbf{a})}{\partial a_i}(a_i - e_i)\right] \\ &\quad + \frac{\partial s_{k+1}(x_1,x_2)}{\partial x_2}\left[\sum_{i=1}^{8}\frac{\partial T_2(\mathbf{a})}{\partial a_i}(a_i - e_i)\right] \end{aligned} \qquad (10.32)$$

In order to express (10.32) in vector matrix form, we define

$$\mathbf{H} = [H_1\ H_2\ H_3\ H_4\ H_5\ H_6\ H_7\ H_8]^T \qquad (10.33)$$

where

$$H_1 = \frac{T_1(\mathbf{a})}{\partial a_1}\frac{\partial s_{k+1}}{\partial x_1} = x_1\frac{\partial s_{k+1}}{\partial x_1}, \qquad H_2 = \frac{T_1(\mathbf{a})}{\partial a_2}\frac{\partial s_{k+1}}{\partial x_1} = x_2\frac{\partial s_{k+1}}{\partial x_1}$$

$$H_3 = \frac{T_1(\mathbf{a})}{\partial a_3}\frac{\partial s_{k+1}}{\partial x_1} = \frac{\partial s_{k+1}}{\partial x_1}, \qquad H_4 = \frac{T_2(\mathbf{a})}{\partial a_4}\frac{\partial s_{k+1}}{\partial x_2} = x_1\frac{\partial s_{k+1}}{\partial x_2}$$

$$H_5 = \frac{T_2(\mathbf{a})}{\partial a_5}\frac{\partial s_{k+1}}{\partial x_2} = x_2\frac{\partial s_{k+1}}{\partial x_2}, \qquad H_6 = \frac{T_2(\mathbf{a})}{\partial a_6}\frac{\partial s_{k+1}}{\partial x_2} = \frac{\partial s_{k+1}}{\partial x_2}$$

$$H_7 = \frac{T_1(\mathbf{a})}{\partial a_7}\frac{\partial s_{k+1}}{\partial x_1} + \frac{T_2(\mathbf{a})}{\partial a_7}\frac{\partial s_{k+1}}{\partial x_2} = -x_1^2\frac{\partial s_{k+1}}{\partial x_1} - x_1x_2\frac{\partial s_{k+1}}{\partial x_2}$$

$$H_8 = \frac{T_1(\mathbf{a})}{\partial a_8}\frac{\partial s_{k+1}}{\partial x_1} + \frac{T_2(\mathbf{a})}{\partial a_8}\frac{\partial s_{k+1}}{\partial x_2} = -x_1x_2\frac{\partial s_{k+1}}{\partial x_1} - x_2^2\frac{\partial s_{k+1}}{\partial x_2}$$

Then

$$FD(x_1, x_2) = \mathbf{H} \cdot (\mathbf{a} - \mathbf{e}) \tag{10.34}$$

Note that this relation holds for every image point that originates from a single planar object.

To summarize, the following algorithm is proposed for the estimation of pure parameters without the need to identify any point correspondences:

1. Select at least eight points (x_1, x_2) in the kth frame that are on the same plane.
2. Compute $FD(x_1, x_2)$ between the frames k and $k+1$ at these points.
3. Estimate the image gray level gradients $\frac{\partial s_{k+1}(x_1,x_2)}{\partial x_1}$ and $\frac{\partial s_{k+1}(x_1,x_2)}{\partial x_2}$ at these eight points in the frame $k+1$.
4. Compute the vector $\mathbf{H}$.
5. Solve for $\mathbf{\Delta a} = \mathbf{a} - \mathbf{e}$ in the least squares sense.

Once the pure parameters have been estimated, the corresponding 3-D motion and structure parameters can be determined by a singular value decomposition of the matrix $\mathbf{A}$ as described in Section 9.4. We note here that the estimator proposed by Netravali and Salz [Net 85] yields the same results in the case of a planar object.

10.6 Examples

Optical-flow-based and direct methods do not require establishing feature point correspondences, but they rely on estimated optical flow field and spatio-temporal image intensity gradients, respectively. Recall that the estimation of the optical flow and image intensity gradients are themselves ill-posed problems, which are highly sensitive to observation noise.

We compare the performances of four algorithms, three optical flow-based methods and one direct method. They are: i) Zhuang's closed-form solution and ii) the Heeger-Jepson method, which are optical-flow-based methods using a perspective model; iii) the iterative method of Morikawa, which is based on the orthographic flow model; and iv) the direct method of Tsai and Huang, using the perspective projection and a planar surface model. It is also of interest to compare these methods with the feature-based techniques discussed in Chapter 9 to determine which class of techniques is more robust in the presence of errors in the point correspondences, optical flow field, and image intensity values, respectively.

The methods will be assessed on the basis of the accuracy of the resulting 3-D motion and depth estimates, and the goodness of the motion compensation that can be achieved by synthesizing the second frame from the first, given the 3-D motion and depth estimates. We start with some numerical simulations to provide the

Table 10.1: Comparison of the motion parameters. All angles are in radians, and the translations are in pixels. (Courtesy Yucel Altunbasak)

True	Zhuang		Heeger-Jepson		
	No Error	3% Error	No Error	3% Error	10% Error
$\Omega_1 = 0.007$	0.0070	0.0133	0.0064	0.0094	0.0036
$\Omega_2 = 0.010$	0.0099	0.0058	0.0105	0.0066	0.0008
$\Omega_3 = 0.025$	0.0250	0.0232	0.0249	0.0251	0.0249
$V_2/V_1 = 1.80$	1.7999	2.9571	1.7872	1.9439	2.7814
$V_3/V_1 = 0.48$	0.4799	10.121	0.4693	0.5822	0.8343
Depth Error	1.73 E-5	0.824	1.41 E-3	0.0181	0.0539
Match Error	2.17 E-7	0.0152	2.71 E-5	5.99 E-4	0.0018

reader with a comparison of the true and estimated 3-D motion and depth estimates obtained by these methods. The accuracy of the resulting motion compensation will be demonstrated on both simulated and real video sequences.

10.6.1 Numerical Simulations

We have simulated an optical flow field using the perspective flow model (10.5) and a set of 3-D motion and depth parameters, which are given in Table 10.1 under the "True" column. The flow field has been simulated for 30 points such that X_1 and X_2 are uniformly distributed in the interval [-25,25] cm, and X_3 is uniformly distributed within [70,100] cm. The focal length parameter has been set to $f = 50$ mm. Indeed, the simulation parameters are identical to those used in Chapter 9 to facilitate comparison of the results with those of the methods using point correspondences. In order to test the sensitivity of the methods to the errors in optical flow estimation, 3% and 10% errors have been added to the synthesized flow vectors, respectively.

Table 10.1 provides a comparison of the results obtained by the methods of Zhuang and Heeger-Jepson (H-J). In the table, "Depth Error" corresponds to the normalized RMS error in the depth estimates given by

$$\text{Depth Error} = \sqrt{\frac{1}{N}\sum_{i=1}^{N}\frac{(X_3^{(i)} - \hat{X}_3^{(i)})^2}{(X_3^{(i)})^2}} \tag{10.35}$$

where $N = 30$, and the "Match Error" measures the RMS deviation of the flow field $\hat{\mathbf{v}}$ generated by the estimated 3-D motion and depth parameters from the input flow field $\mathbf{v}$, given by

$$\text{Match Error} = \frac{1}{N}\sqrt{\sum_{i=1}^{N}(v_1 - \hat{v}_1)^2 + (v_2 - \hat{v}_2)^2}$$

It can easily be seen that the method of Zhuang is sensitive to noise. This is mainly because it introduces three redundant intermediate unknowns, which makes the linear formulation (10.20) possible. Observe that only five of the eight intermediate unknowns are sufficient to solve for the motion parameters, and different combinations of the five intermediate unknowns may yield different motion estimates in the presence of noise. Although the intermediate unknowns are solved in the least squares sense, the final motion estimates are not really least squares estimates. On the other hand, the H-J method always finds the motion and depth parameters that minimize the least squares error criterion, and hence is more robust to errors in the optical flow estimates. We remark that the accuracy of the estimates in the H-J method depends on the step size (angular increments on the unit sphere) used in the search procedure. Two-degree increments were used to obtain the results given in Table 10.1.

Because the method of Morikawa is based on the orthographic flow model, its performance depends on how much the given flow vectors deviate from the orthographic model (10.3). Recall from Chapter 9 that this deviation is related to the ratio of the width of the object to the average distance of the points from the center of projection, given by $R = \Delta X_3 / 2\bar{X}_3$. For the data set used in the above simulation, we have $R \approx 18\%$, and Morikawa's method did not give successful results. Table 10.2 provides a comparison of the performance of Morikawa's method on a perfectly orthographic flow field simulated from (10.3) and a perspective flow field with $R \approx 1\%$, where the true parameter values are identical to those of the first simulation set in Chapter 9. Another issue of practical importance in Morikawa's method is how to initialize the unknown parameters. In the above simulations, all parameters have been initialized at $\pm 5\%$ of their true values. Note that, in general, the cost function has multiple minima, and the results very much depend on the initial point. It is thus recommended that Morikawa's method be used for frame-to-frame tracking of small motion parameters, where the initial estimates for the first frame are obtained through another method, such as the improved two-step algorithm in Chapter 9.

Table 10.2: Comparison of the motion parameters. All angles are in radians, and the translations are in pixels. (Courtesy Gozde Bozdagi)

True	Orthographic	Perspective, $R \approx 1\%$
$\Omega_1 = 0.007$	0.00713	0.00677
$\Omega_2 = 0.010$	0.01013	0.00977
$\Omega_3 = 0.025$	0.02513	0.02477
$V_1 = 0.100$	0.09987	0.10023
$V_2 = 0.180$	0.17987	0.18023
Depth Error	0.00011	0.00023
Match Error	0.00322	0.00489

To demonstrate the performance of the Tsai-Huang method, we have mapped the first frame of "Miss America" onto the plane given by $n_1X_1 + n_2X_2 + n_3X_3 = 1$. The resulting planar object is subjected to 3-D rigid motion, and an image of the new object is computed by means of the prespective projection with $f = 50$ mm. In this simulation, we have a planar object, no local motion, and no uncovered background. The only sources of error are gradient estimation and linearized model assumptions. Two sets of parameters for the 3-D motion and the planar surface used in the simulation are tabulated in Table 10.3. For the first simulation, the plane is approximately 70 cm away from the image plane. The size of the object is assumed to be 50 cm × 50 cm, and 1 pel = 0.1395 mm. A requirement for the success of the direct methods is undoubtedly the accuracy of the spatio-temporal gradient estimates. To ensure reasonably accurate gradient estimates, pixels where the Laplacian of the image intensity is above a threshold have been selected for 3-D motion estimation. This is because in flat regions, gradient values are in the same order of magnitude as the noise. The gradient estimation is performed by 9th order spatio-temporal polynomial fitting.

Table 10.3 shows the results of 3-D motion and structure estimation using the Tsai-Huang method for two different motions. Clearly, the results are better in the

Table 10.3: Comparison of the motion parameters. All angles are in radians, and the translations are in pixels. (Courtesy Yucel Altunbasak)

Motion 1: True							
Ω_1	Ω_2	Ω_3	V_2/V_1	V_3/V_1	n_1	n_2	n_3
0.0070	0.0100	0.0250	1.8000	0.4800	0	0	0.01429
a_1	a_2	a_3	a_4	a_5	a_6	a_7	a_8
0.9964	-0.0249	6.0981	0.0249	0.9964	2.5109	-2.7 E-5	1.9 E-5
Estimate							
Ω_1	Ω_2	Ω_3	V_2/V_1	V_3/V_1	n_1	n_2	n_3
0.0355	0.1034	-0.0086	-0.3775	-0.1599	-15.871	17.8351	115.99
a_1	a_2	a_3	a_4	a_5	a_6	a_7	a_8
0.9983	-0.0065	1.4815	-0.0099	0.9952	0.7108	-0.0003	0.0001
Motion 2: True							
Ω_1	Ω_2	Ω_3	V_2/V_1	V_3/V_1	n_1	n_2	n_3
0.004	0.005	0.006	-1.25	0.625	0	0	0.01
a_1	a_2	a_3	a_4	a_5	a_6	a_7	a_8
0.9975	-0.0059	0.0049	0.0059	0.9975	0.0049	-0.0009	0.0008
Estimate							
Ω_1	Ω_2	Ω_3	V_2/V_1	V_3/V_1	n_1	n_2	n_3
0.0019	0.0053	0.0055	-0.6552	0.6842	-0.3325	0.0542	-1.4870
a_1	a_2	a_3	a_4	d_5	a_6	a_7	a_8
0.9972	-0.0054	0.0049	0.0062	0.9980	0.0049	-0.0009	0.0004

case of Motion 2. To see how well we can compensate for the motion using the estimated a-parameters, we synthesized the second frame from the first for both motions. The resulting mean square displaced frame differences (MS-DFD) were compared with plain frame differences (MS-FD) (i.e., no motion compensation). In the first case, the MS-DFD was 16.02 compared to MS-FD equal to 19.84. However, in the second case, the MS-DFD has dropped to 0.24 in comparison to MS-FD equal to 4.63. Inspection of the results indicates that the Tsai-Huang method performs better when both the frame difference and the a-parameter values are relatively small. In fact, this is expected since the method relies on the linearization of both the frame difference (the discrete optical flow equation) and the pixel-to-pixel mapping in terms of the a-parameters.

10.6.2 Experiments with Two Frames of Miss America

These methods have also been tested on the same two frames of "Miss America" that were used in Chapter 9, shown in Figure 9.4 (a) and (b). The optical flow field between these two frames has been estimated by 3-level hierarchical block matching. All algorithms have been applied to those pixels, where the Laplacian of the image intensity is above a predetermined threshold (T=50) to work with reliable optical

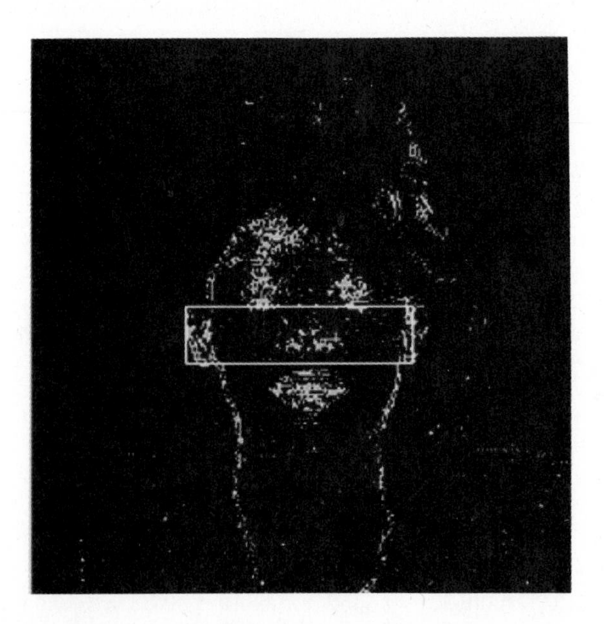

Figure 10.1: The map of pixels selected for 3-D motion estimation marked on the first frame of Miss America. (Courtesy Yucel Altunbasak)

flow estimates. In order to exclude points which may exhibit nonrigid motion, only pixels within the white box depicted in Figure 10.1 have been used. Although we expect to have reliable optical flow and/or spatio-temporal gradient estimates at these pixels, some of these pixels may still fall into regions of local motion or uncovered background, violating the assumption of rigid motion. However, because we work with at least 500 pixels, it is expected that the effect of such pixels will be negligible. The results are compared on the basis of how well we can synthesize the estimated optical flow field at those pixels shown in Figure 10.1 through the estimated 3-D motion and structure parameters, and in some cases how well we can synthesize the next frame from the first and estimated motion parameters.

We have initialized Morikawa's algorithm with the 3-D motion estimates obtained from the improved algorithm discussed in Chapter 9. The initial depth estimates at all selected points have been computed using the optical flow and the initial motion parameter values. After 100 iterations, the criterion function has dropped from 1.94 to 1.04. The value of the criterion function with zero motion estimation was computed as 6.79. Clearly, Morikawa's method is suitable to estimate small changes in the motion and depth parameters, provided a good set of initial estimates are available.

Next, the method of Heeger and Jepson (H-J) was used with the same two frames, where the search for the best $\mathbf{V}$ was performed with angular increments of 1 degree on the unit sphere. The RMSE difference between the optical flow synthesized using the parameters estimated by the H-J method and the one estimated by hierarchical block matching was found to be 0.51, which ranks among the best results. Finally, we have applied the direct method of Tsai-Huang on the same set of pixels in the two frames, and compare the results with those of the A-matrix method where each optical flow vector is used to determine a point correspondence. Table 10.4 tabulates the respective velocity and intensity synthesis errors. The results indicate that the A-matrix method is more successful in compensating for the motion.

Table 10.4: Comparison of sythesis errors. (Courtesy Yucel Altunbasak)

Method	No Compensation	Tsai-Huang	A-matrix
Velocity Sythesis Error	6.96	5.52	2.32
Intensity Sythnesis Error	20.59	17.63	8.10

10.7 Exercises

1. How do you compare 3-D motion estimation from point correspondences versus that from optical flow? List the assumptions made in each case. Suppose we have 3-D motion with constant acceleration and precession; which method would you employ?

2. Which one can be estimated more accurately, token matching or spatio-temporal image intensity gradients?

3. Suppose we assume that the change in X_3 (ΔX_3) in the time interval between the two views can be neglected. Can you propose a linear algorithm, similar to the one in Section 9.3, to estimate the 3-D motion and structure parameters from optical flow?

4. Show that (10.31) may also be interpreted as the linearization of the image intensity function in the vicinity of (x_1, x_2) in frame $k+1$.

Bibliography

[Adi 85] G. Adiv, "Determining three-dimensional motion and structure from optical flow generated by several moving objects," *IEEE Trans. Patt. Anal. Mach. Intel.*, vol. 7, pp. 384–401, July 1985.

[Agg 88] J. K. Aggarwal and N. Nandhakumar, "On the computation of motion from sequences of images," *Proc. IEEE*, vol. 76, pp. 917–935, Aug. 1988.

[Ana 93] P. Anandan, J. R. Bergen, K. J. Hanna, and R. Hingorani, "Hierarchical model-based motion estimation," in *Motion Analysis and Image Sequence Processing*, M. I. Sezan and R. L. Lagendijk, eds., Kluwer, 1993.

[Cam 92] M. Campani and A. Verri, "Motion analysis from first-order properties of optical flow," *CVGIP: Image Understanding*, vol. 56, no. 1, pp. 90–107, Jul. 1992.

[Die 91] N. Diehl, "Object-oriented motion estimation and segmentation in image sequences," *Signal Processing: Image Comm.*, vol. 3, pp. 23–56, 1991.

[Hee 92] D. J. Heeger and A. Jepson, "Subspace methods for recovering rigid motion I: Algorithm and implementation," *Int. J. Comp. Vis.*, vol. 7, pp. 95–117, 1992.

[Hei 93] J. Heikkonen, "Recovering 3-D motion from optical flow field," *Image Processing: Theory and Applications*, Vernazza, Venetsanopoulos, and Braccini, eds., Elsevier, 1993.

[Hol 93] R. J. Holt and A. N. Netravali, "Motion from optic flow: Multiplicity of solutions," *J. Vis. Comm. Image Rep.*, vol. 4, no. 1, pp. 14–24, 1993.

[Hor 88] B. K. P. Horn and E. J. Weldon Jr., "Direct methods for recovering motion," *Int. J. Comp. Vision*, vol. 2, pp. 51–76, 1988.

[Jai 83] R. Jain, "Direct computation of the focus of expansion," *IEEE Trans. Patt. Anal. Mach. Intel.*, vol. 5, pp. 58–63, Jan. 1983.

[Jer 91] C. P. Jerian and R. Jain, "Structure from motion - A critical analysis of methods," *IEEE System, Man and Cyber.*, vol. 21, pp. 572–588, 1991.

[Kan 86] K. Kanatani, "Structure and motion from optical flow under orthographic projection," *Comp. Vision Graph. Image Proc.*, vol. 35, pp. 181-199, 1986.

[Law 83] D. T. Lawton, "Processing translational motion sequences," *Comp. Vis. Grap. Image Proc.*, vol. 22, pp. 116–144, 1983.

[Lee 76] D. N. Lee, "A theory of visual control of braking based on information about time to collision," *Perception*, vol. 5, pp. 437–459, 1976.

[Mor 91] H. Morikawa and H. Harashima, "3D structure extraction coding of image sequences," *J. Visual Comm. and Image Rep.*, vol. 2, no. 4, pp. 332–344, Dec. 1991.

[Net 85] A. N. Netravali and J. Salz, "Algorithms for estimation of three-dimensional motion," *AT&T Tech. J.*, vol. 64, no. 2, pp. 335–346, Feb. 1985.

[Tsa 81] R. Y. Tsai and T. S. Huang, "Estimating three-dimensional motion parameters of a rigid planar patch," *IEEE Trans. Acoust. Speech Sign. Proc.*, vol. 29, no. 6, pp. 1147–1152, Dec. 1981.

[Ver 89] A. Verri and T. Poggio, "Motion field and optical flow: Qualitative properties," *IEEE Trans. Patt. Anal. Mach. Intel.*, vol. PAMI-11, no. 5, pp. 490–498, May 1989.

[Wax 87] A. M. Waxman, B. Kamgar-Parsi, and M. Subbarao, "Closed-form solutions to image flow equations for 3-D structure and motion," *Int. J. Comp. Vision*, vol. 1, pp. 239–258, 1987.

[Zhu 88] X. Zhuang, T. S. Huang, N. Ahuja, and R. M. Haralick, "A simplified linear optic-flow motion algorithm," *Comp. Vis. Graph. and Image Proc.*, vol. 42, pp. 334–344, 1988.

Chapter 11

MOTION SEGMENTATION

Most real image sequences contain multiple moving objects or multiple motions. For example, the Calendar and Train sequence depicted in Chapter 5 exhibits five different motions, shown in Figure 11.1. Optical flow fields derived from multiple motions usually display discontinuities (motion edges). Motion segmentation refers to labeling pixels that are associated with each independently moving 3-D object in a sequence featuring multiple motions. A closely related problem is optical flow segmentation, which refers to grouping together those optical flow vectors that are associated with the same 3-D motion and/or structure. These two problems are identical when we have a dense optical flow field with an optical flow vector for every pixel.

It should not come as a surprise that motion-based segmentation is an integral part of many image sequence analysis problems, including: i) improved optical flow estimation, ii) 3-D motion and structure estimation in the presence of multiple moving objects, and iii) higher-level description of the temporal variations and/or the content of video imagery. In the first case, the segmentation labels help to identify optical flow boundaries and occlusion regions where the smoothness constraint should be turned off. The approach presented in Section 8.3.3 constitutes an example of this strategy. Segmentation is required in the second case, because a distinct parameter set is needed to model the flow vectors associated with each independently moving 3-D object. Recall that in Chapters 9 and 10, we have assumed that all feature points or flow vectors belong to a single rigid object. Finally, in the third case, segmentation information may be considered as a high-level (object-level) description of the frame-to-frame motion information as opposed to the low-level (pixel-level) motion information provided by the individual flow vectors.

As with any segmentation problem, proper feature selection facilitates effective motion segmentation. In general, application of standard image segmentation

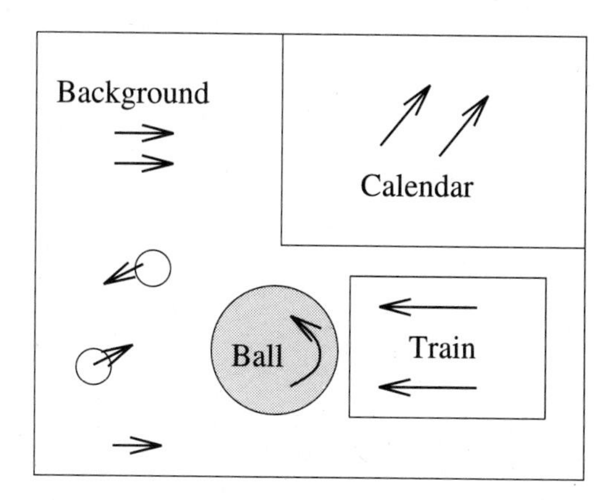

Figure 11.1: Example of optical flow generated by multiple moving objects.

methods (see Appendix B for a brief overview) directly to optical flow data may not yield meaningful results, since an object moving in 3-D usually generates a spatially varying optical flow field. For example, in the case of a single rotating object, there is no flow at the center of the rotation, and the magnitude of the flow vectors grows as we move away from the center of rotation. Therefore, in this chapter, a parametric model-based approach has been adopted for motion-based video segmentation where the model parameters constitute the features. Examples of parametric mappings that can be used with direct methods include the 8-parameter mapping (10.28), and affine mapping (9.4). Recall, from Chapter 9, that the mapping parameters depend on: i) the 3-D motion parameters, the rotation matrix $\mathbf{R}$ and the translation vector $\mathbf{T}$, and ii) the model of the object surface, such as the orientation of the plane in the case of a piecewise planar model. Since each independently moving object and/or different surface structure will best fit a different parametric mapping, parameters of a suitably selected mapping will be used as features to distinguish between different 3-D motions and surface structures.

Direct methods, which utilize spatio-temporal image gradients, are presented in Section 11.1. These techniques may be considered as extensions of the direct methods discussed in Chapter 10 to the case of multiple motion. In Section 11.2, a two-step procedure is followed, where first the optical flow field is estimated using one of the techniques covered in Chapters 5 through 8. A suitable parametric motion model has subsequently been used for optical flow segmentation using clustering or maximum *a posteriori* (MAP) estimation. The accuracy of segmentation results clearly depends on the accuracy of the estimated optical flow field. As mentioned earlier, optical flow estimates are usually not reliable around moving object boundaries due to occlusion and use of smoothness constraints. Thus, optical flow estimation and segmentation are mutually interrelated, and should be addressed simultaneously for best results. We present methods for simultaneous optical flow estimation and segmentation using parametric flow models in Section 11.3.

11.1 Direct Methods

In this section, we consider direct methods for segmentation of images into independently moving regions based on spatio-temporal image intensity and gradient information. This is in contrast to first estimating the optical flow field between two frames and then segmenting the image based on the estimated optical flow field. We start with a simple thresholding method that segments images into "changed" and "unchanged regions." Methods using parametric displacement field models will be discussed next.

11.1.1 Thresholding for Change Detection

Thresholding is often used to segment a video frame into "changed" versus "unchanged" regions with respect to the previous frame. The unchanged regions denote the stationary background, while the changed regions denote the moving and occlusion areas.

We define the frame difference $FD_{k,k-1}(x_1, x_2)$ between the frames k and $k-1$ as

$$FD_{k,k-1}(x_1, x_2) = s(x_1, x_2, k) - s(x_1, x_2, k-1) \tag{11.1}$$

which is the pixel-by-pixel difference between the two frames. Assuming that the illumination remains more or less constant from frame to frame, the pixel locations where $FD_{k,k-1}(x_1, x_2)$ differ from zero indicate "changed" regions. However, the frame difference hardly ever becomes exactly zero, because of the presence of observation noise.

In order to distinguish the nonzero differences that are due to noise from those that are due to a scene change, segmentation can be achieved by thresholding the difference image as

$$z_{k,k-1}(x_1, x_2) = \begin{cases} 1 & \text{if } |FD_{k,k-1}(x_1, x_2)| > T \\ 0 & \text{otherwise} \end{cases} \tag{11.2}$$

where T is an appropriate threshold. The value of the threshold T can be chosen according to one of the threshold determination algorithms described in Appendix B. Here, $z_{k,k-1}(x_1, x_2)$ is called a segmentation label field, which is equal to "1" for changed regions and "0" otherwise. In practice, thresholding may still yield isolated 1s in the segmentation mask $z_{k,k-1}(x_1, x_2)$, which can be eliminated by postprocessing; for example, forming 4- or 8-connected regions, and discarding any region(s) with less than a predetermined number of entries.

Example: Change Detection

The changed region for the same two frames of the Miss America that were used in Chapters 9 and 10, computed as described above, is depicted in Figure 11.2. The threshold was set at 7, and no post-filtering was performed.

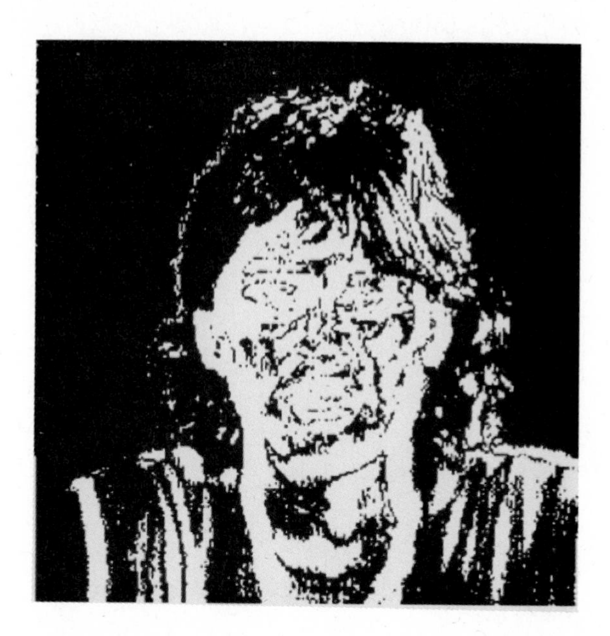

Figure 11.2: Changed region between the first and third frames of the Miss America sequence. (Courtesy Gozde Bozdagi)

Another approach to eliminating isolated 1's is to consider accumulative differences which add memory to the motion detection process. Let $s(x_1, x_2, k)$, $s(x_1, x_2, k-1)$, ..., $s(x_1, x_2, k-N)$ be a sequence of N frames, and let $s(x_1, x_2, k)$ be the reference frame. An accumulative difference image is formed by comparing this reference image with every subsequent image in the sequence. A counter for each pixel location in the accumulative image is incremented every time the difference between the reference image and the next image in the sequence at that pixel location is bigger than the threshold. Thus, pixels with higher counter values are more likely to correspond to actual moving regions.

11.1.2 An Algorithm Using Mapping Parameters

The method presented here, based on the works of Hotter and Thoma [Hoe 88] and Diehl [Die 91], can be considered as a hierarchically structured top-down approach. It starts by fitting a parametric model, in the least squares sense, to the entire changed region from one frame to the next, and then breaks this region into successively smaller regions depending on how well a single model fits each region or subregion. This is in contrast to the clustering and MAP approaches, to be discussed in the next section, which start with many small subregions and group them

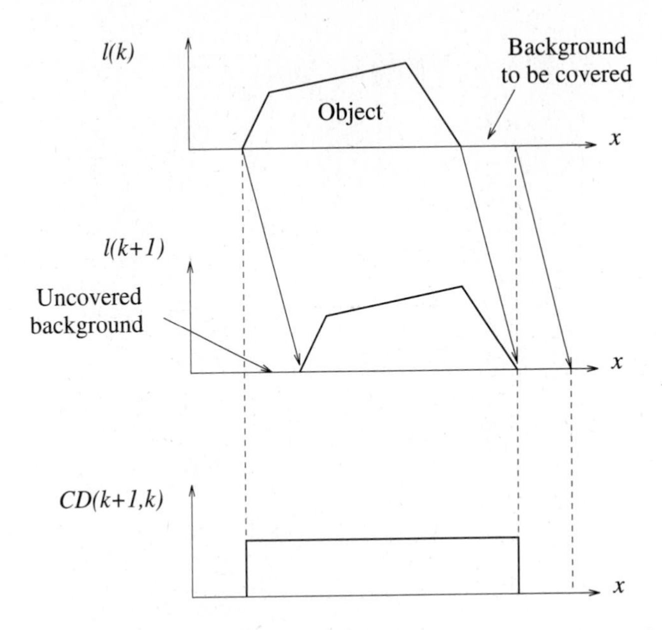

Figure 11.3: Detection of uncovered background [Hoe 88].

together according to some merging criterion to form segments. The hierarchically structured approach can be summarized through the following steps:

1) In the first step, a change detector, described above, initializes the segmentation mask separating the changed and unchanged regions from frame k to $k+1$. Median filtering or morphological filtering can be employed to eliminate small regions in the change detection mask. Each spatially connected changed region is interpreted as a different object.

2) For each object a different parametric model is estimated. The methods proposed by Hotter and Thoma [Hoe 88] and Diehl [Die 91] differ in the parametric models employed and in the estimation of the model parameters. Estimation of the parameters for each region is discussed in the next subsection.

3) The changed region(s) found in step 1 is (are) divided into moving region(s) and the uncovered background using the mapping parameters computed in step 2. This is accomplished as follows: All pixels in frame $k+1$ that are in the changed region are traced backwards, with the inverse of the motion vector computed from the mapping parameters found in step 2. If the inverse of the motion vector points to a pixel in frame k that is within the changed region, then the pixel in frame $k+1$ is classified as a moving pixel; otherwise, it is assigned to the uncovered background. Figure 11.3 illustrates this process, where CD refers to the change detection mask between the lines $l(k)$ and $l(k+1)$.

Next, the validity of the model parameters for those pixels within the moving region is verified by evaluating the displaced frame difference. The regions where the respective parameter vector is not valid are marked as independent objects for the second hierarchical level. The procedure iterates between steps 2 and 3 until the parameter vectors for each region are consistent with the region.

11.1.3 Estimation of Model Parameters

Let the kth frame of the observed sequence be expressed as

$$g_k(\mathbf{x}) = s_k(\mathbf{x}) + n_k(\mathbf{x}) \tag{11.3}$$

where $n_k(\mathbf{x})$ denotes the observation noise. Assuming no occlusion effects, we have $s_{k+1}(\mathbf{x}) = s_k(\mathbf{x}')$, where

$$\mathbf{x}' = h(\mathbf{x}, \theta) \tag{11.4}$$

is a tranformation of pixels from frame k to $k+1$, with the parameter vector θ. Then

$$g_{k+1}(\mathbf{x}) = s_k(\mathbf{x}') + n_{k+1}(\mathbf{x}) \tag{11.5}$$

The transformation $h(\mathbf{x}, \theta)$ must be unique and invertible. Some examples for the transformation are as follows:

i) Assuming a planar surface and using the perspective projection, we obtain the eight-parameter mapping

$$\begin{aligned} x_1' &= \frac{a_1x_1 + a_2x_2 + a_3}{a_7x_1 + a_8x_2 + 1} \\ x_2' &= \frac{a_4x_1 + a_5x_2 + a_6}{a_7x_1 + a_8x_2 + 1} \end{aligned} \tag{11.6}$$

where $\theta = (a_1, a_2, a_3, a_4, a_5, a_6, a_7, a_8)^T$ is the vector of so-called pure parameters.

ii) Alternatively, assuming a planar surface and using the orthographic projection, we have the affine transform

$$\begin{aligned} x_1' &= c_1x_1 + c_2x_2 + c_3 \\ x_2' &= c_4x_1 + c_5x_2 + c_6 \end{aligned} \tag{11.7}$$

where $\theta = (c_1, c_2, c_3, c_4, c_5, c_6)^T$ is the vector of mapping parameters.

iii) Finally, let's assume a quadratic surface, given by

$$X_3 = a_{11}X_1^2 + a_{12}X_1X_2 + a_{22}X_2^2 + a_{13}X_1 + a_{23}X_2 + a_{33} \tag{11.8}$$

Substituting (11.8) into the 3-D motion model, using the orthographic projection equations, and grouping the terms with the same exponent, we arrive at the quadratic transform

$$\begin{aligned} x_1' &= a_1x_1^2 + a_2x_2^2 + a_3x_1x_2 + a_4x_1 + a_5x_2 + a_6 \\ x_2' &= b_1x_1^2 + b_2x_2^2 + b_3x_1x_2 + b_4x_1 + b_5x_2 + b_6 \end{aligned} \tag{11.9}$$

which has 12 parameters.

Note that we may not always be able to determine the actual 3-D motion and surface structure parameters from the mapping parameters. However, for image coding applications this does not pose a serious problem, since we are mainly interested in predicting the next frame from the current frame. Furthermore, the approach presented here is not capable of handling occlusion effects.

The method of Hotter and Thoma uses the eight-parameter model (11.6). They estimated the model parameters for each region using a direct method which is similar to the method of Tsai and Huang [Tsa 81] that was presented in Chapter 10. On the other hand, Diehl [Die 91] suggests using the quadratic transform, because it provides a good approximation to many real-life images. He proposed estimating the mapping parameters to minimize the error function

$$J(\hat{\theta}) = \frac{1}{2} E \left\{ (\ g_{k+1}(\mathbf{x}) - g_k(\mathbf{x}', \hat{\theta})\ )^2 \right\}$$

where $E\{.\}$ is the expectation operator, and $g_k(\mathbf{x}', \hat{\theta})$ denotes the prediction of frame $k+1$ from frame k using the mapping $h(\mathbf{x}, \theta)$. A gradient-based minimization algorithm, the modified Newton's method, is used to find the best parameter vector $\hat{\theta}$. The contents of the images $g_k(\mathbf{x})$ and $g_{k+1}(\mathbf{x})$ must be sufficiently similar in order for the error function to have a unique minimum.

11.2 Optical Flow Segmentation

In this section, we treat segmentation of a given flow field using parameters of a flow field model as features. We assume that there are K independently moving objects, and each flow vector corresponds to the projection of a 3-D rigid motion of a single opaque object. Then each distinct motion can be accurately described by a set of mapping parameters. Most common examples of parametric models, such as eight-parameter mapping (10.19) and affine mapping (10.17), implicitly assume a 3-D planar surface in motion. Approximating the surface of a real object by a union of a small number of planar patches, the optical flow generated by a real object can be modeled by a piecewise quadratic flow field, where the parameters $a_1, \ldots, a_8$ in (10.19) vary in a piecewise fashion. It follows that flow vectors corresponding to the same surface and 3-D motion would have the same set of mapping parameters, and optical flow segmentation can be achieved by assigning the flow vectors with the same mapping parameters into the same class.

The underlying principle of parametric, model-based segmentation methods can be summarized as follows: Suppose we have K sets of parameter vectors, where each set defines a correspondence or a flow vector at each pixel. Flow vectors defined by the mapping parameters are called model-based or synthesized flow vectors. Thus, we have K synthesized flow vectors at each pixel. The segmentation procedure then assigns the label of the synthesized vector which is closest to the estimated flow vector at each site. However, there is a small problem with this simple scheme: both the number of classes, K, and the mapping parameters for each class are not known *a priori*. Assuming a particular value for K, the mapping parameters for each class could be computed in the least squares sense provided that the estimated optical flow vectors associated with the respective classes are known. That is, we need to know the mapping parameters to find the segmentation labels, and the segmentation labels are needed to find the mapping parameters. This suggests an iterative procedure, similar to the K-means clustering algorithm where both the segmentation labels and the class means are unknown (see Appendix B). The modified Hough transform approach of Adiv [Adi 85], the modified K-means approach of Wang and Adelson [Wan 94], and the MAP method of Murray and Buxton [Mur 87], which are described in the following, all follow variations of this strategy.

11.2.1 Modified Hough Transform Method

The Hough transform is a well-known clustering technique where the data samples "vote" for the most representative feature values in a quantized feature space. In a straightforward application of the Hough transform method to optical flow segmentation using the six-parameter affine flow model (10.17), the six-dimensional feature space $a_1, \ldots, a_6$ would be quantized to certain parameter states after the minimal and maximal values for each parameter are determined. Then, each flow vector $\mathbf{v}(\mathbf{x}) = [v_1(\mathbf{x})\ v_2(\mathbf{x})]^T$ votes for a set of quantized parameters which minimizes

$$\eta^2(\mathbf{x}) \doteq \eta_1^2(\mathbf{x}) + \eta_2^2(\mathbf{x}) \tag{11.10}$$

where $\eta_1(\mathbf{x}) = v_1(\mathbf{x}) - a_1 - a_2x_1 - a_3x_2$ and $\eta_2(\mathbf{x}) = v_2(\mathbf{x}) - a_4 - a_5x_1 - a_6x_2$. The parameter sets that receive at least a predetermined amount of votes are likely to represent candidate motions. The number of classes K and the corresponding parameter sets to be used in labeling individual flow vectors are hence determined. The drawback of this scheme is the significant amount of computation involved.

In order to keep the computational burden at a reasonable level, Adiv proposed a two-stage algorithm that involves a modified Hough transform procedure. In the first stage of his algorithm, connected sets of flow vectors are grouped together to form components which are consistent with a single parameter set. Several simplifications were proposed to ease the computational load, including: i) decomposition of the parameters space into two disjoint subsets $\{a_1, a_2, a_3\} \times \{a_4, a_5, a_6\}$ to perform two 3-D Hough transforms, ii) a multiresolution Hough transform, where at each resolution level the parameter space is quantized around the estimates obtained at the previous level, and iii) a multipass Hough technique, where the flow vectors which are most consistent with the candidate parameters are grouped first. In the second stage, those components formed in the first stage which are consistent with the same quadratic flow model (10.19) in the least squares sense are merged together to form segments. Several merging criteria have been proposed. In the third and final stage, ungrouped flow vectors are assimilated into one of their neighboring segments.

In summary, the modified Hough transform approach is based on first clustering the flow vectors into small groups, each of which is consistent with the flow generated by a moving planar facet. The fusion of these small groups into segments is then performed based on some ad-hoc merging criteria.

11.2.2 Segmentation for Layered Video Representation

A similar optical flow segmentation method using the K-means clustering technique has been employed in the so-called layered video representation strategy. Instead of trying to capture the motion of multiple overlapping objects in a single motion field, Wang and Adelson [Wan 94] proposed a layered video representation in which the image sequence is decomposed into layers by means of an optical-flow-based image segmentation, and ordered in depth along with associated maps defining their motions, opacities, and intensities. In the layered representation, the segmentation labels denote the layer in which a particular pixel resides.

The segmentation method is based on the affine motion model (10.17) and clustering in a six-dimensional parameter space. The image is initially divided into small blocks. Given an optical flow field (e.g., computed by using (5.12)), a set of affine parameters are estimated for each block. To determine the reliability of the parameter estimates, the sum of squared distances between the synthesized and estimated flow vectors is computed as

$$\bar{\eta}^2 = \sum_{\mathbf{x} \in \mathcal{B}} ||\mathbf{v}(\mathbf{x}) - \tilde{\mathbf{v}}(\mathbf{x})||^2 \qquad (11.11)$$

where $\mathcal{B}$ refers to a block of pixels. Obviously, if the flow within the block complies with an affine model, the residual will be small. On the other hand, if the block falls on the boundary between two distinct motions, the residual will be large. The motion parameters for blocks with acceptably small residuals are selected as the candidate layer models. To determine the appropriate number K of layers, the motion parameters of the candidate layers are clustered in the six-dimensional parameter space. The initial set of affine model parameters are set equal to the mean of the K clusters. Then the segmentation label of each pixel site is selected as the index of the parameter set that yields the closest optical flow vector at that site. After all sites are labeled, the affine parameters of each layer are recalculated based on the new segmentation labels. This procedure is repeated until the segmentation labels no longer change or a fixed number of iterations is reached.

It should be noted that if the smallest difference between an observed vector and its parameter-based estimate exceeds a threshold, then the site is not labeled in the above iterative procedure, and the observed flow vector is ignored in the parameter estimation that follows. After the iterative procedure converges to reliable affine models, all sites without labels are assigned one according to the motion compensation criterion, which assigns the label of the parameter vector that gives the best motion compensation at that site. This feature ensures more robust parameter estimation by eliminating the outlier vectors. A possible limitation of this segmentation method is that it lacks constraints to enforce spatial and temporal continuity of the segmentation labels. Thus, rather ad-hoc steps are needed to eliminate small, isolated regions in the segmentation label field. The Bayesian segmentation strategy promises to impose continuity constraints in an optimization framework.

11.2.3 Bayesian Segmentation

The Bayesian method searches for the maximum of the *a posteriori* probability of the segmentation labels given the optical flow data, which is a measure of how well the current segmentation explains the observed optical flow data and how well it conforms to our prior expectations. Murray and Buxton [Mur 87] first proposed a MAP segmentation method where the optical flow data was modeled by a piecewise quadratic flow field, and the segmentation field modeled by a Gibbs distribution. The search for the labels that maximize the *a posteriori* probability was performed by simulated annealing. Here we briefly present their approach.

Problem Formulation

Let $\mathbf{v}_1$, $\mathbf{v}_2$, and $\mathbf{z}$ denote the lexicographic ordering of the components of the flow vector $\mathbf{v}(\mathbf{x}) = [v_1(\mathbf{x})\ v_2(\mathbf{x})]^T$ and the segmentation labels $z(\mathbf{x})$ at each pixel. The *a posteriori* probability density function (pdf) $p(\mathbf{z}|\mathbf{v}_1, \mathbf{v}_2)$ of the segmentation label field $\mathbf{z}$ given the optical flow data $\mathbf{v}_1$ and $\mathbf{v}_2$ can be expressed, using the Bayes theorem, as

$$p(\mathbf{z}|\mathbf{v}_1, \mathbf{v}_2) = \frac{p(\mathbf{v}_1, \mathbf{v}_2|\mathbf{z})p(\mathbf{z})}{p(\mathbf{v}_1, \mathbf{v}_2)} \tag{11.12}$$

where $p(\mathbf{v}_1, \mathbf{v}_2|\mathbf{z})$ is the conditional pdf of the optical flow data given the segmentation $\mathbf{z}$, and $p(\mathbf{z})$ is the *a priori* pdf of the segmentation. Observe that, i) $\mathbf{z}$ is a discrete-valued random vector with a finite sample space Ω, and ii) $p(\mathbf{v}_1, \mathbf{v}_2)$ is constant with respect to the segmentation labels, and hence can be ignored for the purposes of segmentation. The MAP estimate, then, maximizes the numerator of (11.12) over all possible realizations of the segmentation field $\mathbf{z} = \omega$, $\omega \in \Omega$.

The conditional probability $p(\mathbf{v}_1, \mathbf{v}_2|\mathbf{z})$ is a measure of how well the piecewise quadratic flow model (10.19), where the model parameters $a_1, \ldots, a_8$ depend on the segmentation label $\mathbf{z}$, fits the estimated optical flow field $\mathbf{v}_1$ and $\mathbf{v}_2$. Assuming that the mismatch between the observed flow $\mathbf{v}(\mathbf{x})$ and the synthesized flow,

$$\begin{aligned} \tilde{v}_1(\mathbf{x}) &= a_1 x_1 + a_2 x_2 - a_3 + a_7 x_1^2 + a_8 x_1 x_2 \\ \tilde{v}_2(\mathbf{x}) &= a_4 x_1 + a_5 x_2 - a_6 + a_7 x_1 x_2 + a_8 x_2^2 \end{aligned} \tag{11.13}$$

is modeled by white, Gaussian noise with zero mean and variance σ^2, the conditional pdf of the optical flow field given the segmentation labels can be expressed as

$$p(\mathbf{v}_1, \mathbf{v}_2|\mathbf{z}) = \frac{1}{(2\pi\sigma^2)^{M/2}} \exp\left\{ - \sum_{i=1}^{M} \eta^2(\mathbf{x}_i)/2\sigma^2 \right\} \tag{11.14}$$

where M is the number of flow vectors available at the sites $\mathbf{x}_i$, and

$$\eta^2(\mathbf{x}_i) = (v_1(\mathbf{x}_i) - \tilde{v}_1(\mathbf{x}_i))^2 + (v_2(\mathbf{x}_i) - \tilde{v}_2(\mathbf{x}_i))^2 \tag{11.15}$$

is the norm-squared deviation of the actual flow vectors from what is predicted by the quadratic flow model. Assuming that the quadratic flow model is more or less accurate, this deviation is due to segmentation errors and the observation noise.

The prior pdf is modeled by a Gibbs distribution which effectively introduces local constraints on the interpretation (segmentation). It is given by

$$p(\mathbf{z}) = \frac{1}{Q} \sum_{\omega \in \Omega} \exp\{-U(\mathbf{z})\}\, \delta(\mathbf{z} - \omega) \tag{11.16}$$

where Ω denotes the discrete sample space of $\mathbf{z}$, Q is the partition function

$$Q = \sum_{\omega \in \Omega} \exp\{-U(\omega)\} \tag{11.17}$$

and $U(\omega)$ is the potential function which can be expressed as a sum of local clique potentials $V_C(z(\mathbf{x}_i), z(\mathbf{x}_j))$. The prior constraints on the structure of the segmentation labels can be specified in terms of local clique potential functions as discussed in Chapter 8. For example, a local smoothness constraint on the segmentation labels can be imposed by choosing $V_C(z(\mathbf{x}_i), z(\mathbf{x}_j))$ as in Equation (8.11). Temporal continuity of the labels can similarly be modeled [Mur 87].

Substituting (11.14) and (11.16) into the criterion (11.12) and taking the logarithm of the resulting expression, maximization of the *a posteriori* probability distribution can be performed by minimizing the cost function

$$E = \frac{1}{2\sigma^2} \sum_{i=1}^{M} \eta^2(\mathbf{x}_i) + U(\omega) \tag{11.18}$$

The first term describes how well the predicted data fit the actual optical flow measurements (in fact, optical flow is estimated from the image sequence at hand), and the second term measures how much the segmentation conforms to our prior expectations.

The Algorithm

Because the model parameters corresponding to each label are not known *a priori*, the MAP segmentation alternates between estimation of the model parameters and assignment of the segmentation labels to optimize the cost function (11.18) based on a simulated annealing (SA) procedure. Given the flow field $\mathbf{v}$ and the number of independent motions K, MAP segmentation via the Metropolis algorithm can be summarized as follows:

1. Start with an initial labeling $\mathbf{z}$ of the optical flow vectors. Calculate the mapping parameters $\mathbf{a} = [a_1 \ \ldots \ a_8]^T$ for each region using least squares fitting as in Section 10.3.2. Set the initial temperature for SA.

2. Scan the pixel sites according to a predefined convention. At each site $\mathbf{x}_i$:

   (a) Perturb the label $z_i = z(\mathbf{x}_i)$ randomly.

   (b) Decide whether to accept or reject this perturbation, as described in Section 8.1.1, based on the change ΔE in the cost function (11.18),

$$\Delta E = \frac{1}{2\sigma^2}\Delta\eta^2(\mathbf{x}_i) + \sum_{\mathbf{x}_j \in \mathcal{N}_{\mathbf{x}_i}} \Delta V_C(z(\mathbf{x}_i), z(\mathbf{x}_j)) \qquad (11.19)$$

   where $\mathcal{N}_{\mathbf{x}_i}$ denotes a neighborhood of the site $\mathbf{x}_i$ and $V_C(z(\mathbf{x}_i), z(\mathbf{x}_j))$ is given by Equation (8.11). The first term indicates whether or not the perturbed label is more consistent with the given flow field determined by the residual (11.15), and the second term reflects whether or not it is in agreement with the prior segmentation field model.

3. After all pixel sites are visited once, re-estimate the mapping parameters for each region in the least squares sense based on the new segmentation label configuration.

4. Exit if a stopping criterion is satisfied. Otherwise, lower the temperature according to the temperature schedule, and go to step 2.

The following observations about the MAP segmentation algorithm are in order:
i) The procedure proposed by Murray-Buxton suggests performing step 3, the model parameter update, after each and every perturbation. Because such a procedure will be computationally more demanding, the parameter updates are performed only after all sites have been visited once.
ii) This algorithm can be applied to any parametric model relating to optical flow, although the original formulation has been developed on the basis of vernier velocities [Mur 87] and the associated eight-parameter model.
iii) The actual 3-D motion and depth parameters are not needed for segmentation purposes. If desired, they can be recovered from the parameter vector $\mathbf{a}$ for each segmented flow region at convergence.

We conclude this section by noting that all methods discussed so far are limited by the accuracy of the available optical flow estimates. Next, we introduce a novel framework, in which optical flow estimation and segmentation interact in a mutually beneficial manner.

11.3 Simultaneous Estimation and Segmentation

By now, it should be clear that the success of optical flow segmentation is closely related to the accuracy of the estimated optical flow field, and vice versa. It follows that optical flow estimation and segmentation have to be addressed simultaneously for best results. Here, we present a simultaneous Bayesian approach based on a representation of the motion field as the sum of a parametric field and a residual field. The interdependence of optical flow and segmentation fields are expressed in terms of a Gibbs distribution within the MAP framework. The resulting optimization problem, to find estimates of a dense set of motion vectors, a set of segmentation labels, and a set of mapping parameters, is solved using the highest confidence first (HCF) and iterated conditional mode (ICM) algorithms. It will be seen that several existing motion estimation and segmentation algorithms can be formulated as degenerate cases of the algorithm presented here.

11.3.1 Motion Field Model

Suppose that there are K independently moving, opaque objects in a scene, where the 2-D motion induced by each object can be approximated by a parametric model, such as (11.13) or a 6-parameter affine model. Then, the optical flow field $\mathbf{v}(\mathbf{x})$ can be represented as the sum of a parametric flow field $\tilde{\mathbf{v}}(\mathbf{x})$ and a nonparametric residual field $\mathbf{v}_r(\mathbf{x})$, which accounts for local motion and other modeling errors [Hsu 94]; that is,

$$\mathbf{v}(\mathbf{x}) = \tilde{\mathbf{v}}(\mathbf{x}) + \mathbf{v}_r(\mathbf{x}) \tag{11.20}$$

The parametric component of the motion field clearly depends on the segmentation label $z(\mathbf{x})$, which takes on the values $1, \cdots, K$.

11.3.2 Problem Formulation

The simultaneous MAP framework aims at maximizing the a posteriori pdf

$$p(\mathbf{v}_1, \mathbf{v}_2, \mathbf{z} \mid \mathbf{g}_k, \mathbf{g}_{k+1}) = \frac{p(\mathbf{g}_{k+1} \mid \mathbf{g}_k, \mathbf{v}_1, \mathbf{v}_2, \mathbf{z}) p(\mathbf{v}_1, \mathbf{v}_2 \mid \mathbf{z}, \mathbf{g}_k) p(\mathbf{z} \mid \mathbf{g}_k)}{p(\mathbf{g}_{k+1} \mid \mathbf{g}_k)} \tag{11.21}$$

with respect to the optical flow $\mathbf{v}_1$, $\mathbf{v}_2$ and the segmentation labels $\mathbf{z}$. Through careful modeling of these pdfs, we can express an interrelated set of constraints that help to improve the estimates.

The first conditional pdf $p(\mathbf{g}_{k+1} \mid \mathbf{g}_k, \mathbf{v}_1, \mathbf{v}_2, \mathbf{z})$ provides a measure of how well the present displacement and segmentation estimates conform with the observed frame $k+1$ given frame k. It is modeled by a Gibbs distribution as

$$p(\mathbf{g}_{k+1} \mid \mathbf{g}_k, \mathbf{v}_1, \mathbf{v}_2, \mathbf{z}) = \frac{1}{Q_1} \exp\{-U_1(\mathbf{g}_{k+1} \mid \mathbf{g}_k, \mathbf{v}_1, \mathbf{v}_2, \mathbf{z})\} \tag{11.22}$$

where Q_1 is the partition function (a constant), and

$$U_1(\mathbf{g}_{k+1} \mid \mathbf{g}_k, \mathbf{v}_1, \mathbf{v}_2, \mathbf{z}) = \sum_{\mathbf{x}} [g_k(\mathbf{x}) - g_{k+1}(\mathbf{x} + \mathbf{v}(\mathbf{x})\Delta t)]^2 \tag{11.23}$$

is called the Gibbs potential. Here, the Gibbs potential corresponds to the norm-square of the displaced frame difference (DFD) between the frames $\mathbf{g}_k$ and $\mathbf{g}_{k+1}$. Thus, maximization of (11.22) imposes the constraint that $\mathbf{v}(\mathbf{x})$ minimizes the DFD.

The second term in the numerator in (11.21) is the conditional pdf of the displacement field given the motion segmentation and the search image. It is modeled by a Gibbs distribution

$$p(\mathbf{v}_1, \mathbf{v}_2 \mid \mathbf{z}, \mathbf{g}_k) = p(\mathbf{v}_1, \mathbf{v}_2 \mid \mathbf{z}) = \frac{1}{Q_2} \exp\left\{-U_2(\mathbf{v}_1, \mathbf{v}_2 \mid \mathbf{z})\right\} \tag{11.24}$$

where Q_2 is a constant, and

$$\begin{aligned} U_2(\mathbf{v}_1, \mathbf{v}_2 \mid \mathbf{z}) &= \alpha \sum_{\mathbf{x}} ||\mathbf{v}(\mathbf{x}) - \tilde{\mathbf{v}}(\mathbf{x})||^2 \\ &+ \beta \sum_{\mathbf{x}_i} \sum_{\mathbf{x}_j \in \mathcal{N}_{\mathbf{x}_i}} ||\mathbf{v}(\mathbf{x}_i) - \mathbf{v}(\mathbf{x}_j)||^2 \, \delta(z(\mathbf{x}_i) - z(\mathbf{x}_j)) \end{aligned} \tag{11.25}$$

is the corresponding Gibbs potential, $||\cdot||$ denotes the Euclidian distance, and $\mathcal{N}_{\mathbf{x}}$ is the set of neighbors of site $\mathbf{x}$. The first term in (11.25) enforces a minimum norm estimate of the residual motion field $\mathbf{v}_r(\mathbf{x})$; that is, it aims to minimize the deviation of the motion field $\mathbf{v}(\mathbf{x})$ from the parametric motion field $\tilde{\mathbf{v}}(\mathbf{x})$ while minimizing the DFD. Note that the parametric motion field $\tilde{\mathbf{v}}(\mathbf{x})$ is calculated from the set of model parameters $\mathbf{a}_i$, $i = 1, \ldots, K$, which in turn is a function of $\mathbf{v}(\mathbf{x})$ and $z(\mathbf{x})$. The second term in (11.25) imposes a piecewise local smoothness constraint on the optical flow estimates without introducing any extra variables such as line fields. Observe that this term is active only for those pixels in the neighborhood $\mathcal{N}_{\mathbf{x}}$ which share the same segmentation label with the site $\mathbf{x}$. Thus, spatial smoothness is enforced only on the flow vectors generated by a single object. The parameters α and β allow for relative scaling of the two terms.

The third term in (11.21) models the *a priori* probability of the segmentation field given by

$$p(\mathbf{z} \mid \mathbf{g}_k) = p(\mathbf{z}) = \frac{1}{Q_3} \sum_{\omega \in \Omega} \exp\{-U_3(\mathbf{z})\} \delta(\mathbf{z} - \omega) \tag{11.26}$$

where Ω denotes the sample space of the discrete-valued random vector $\mathbf{z}$, Q_3 is given by Equation (11.17),

$$U_3(\mathbf{z}) = \sum_{\mathbf{x}_i} \sum_{\mathbf{x}_j \in \mathcal{N}_{\mathbf{x}_i}} V_C(z(\mathbf{x}_i), z(\mathbf{x}_j)) \tag{11.27}$$

$\mathcal{N}_{\mathbf{x}_i}$ denotes the neighborhood system for the label field, and

$$V_C(z(\mathbf{x}_i), z(\mathbf{x}_j)) = \begin{cases} -\gamma & \text{if } z(\mathbf{x}_i) = z(\mathbf{x}_j) \\ +\gamma & \text{otherwise} \end{cases} \tag{11.28}$$

The dependence of the labels on the image intensity is usually neglected, although region boundaries generally coincide with intensity edges.

11.3.3 The Algorithm

Maximizing the a posteriori pdf (11.21) is equivalent to minimizing the cost function,

$$E = U_1(\mathbf{g}_{k+1} \mid \mathbf{g}_k, \mathbf{v}_1, \mathbf{v}_2, \mathbf{z}) + U_2(\mathbf{v}_1, \mathbf{v}_2 \mid \mathbf{z}) + U_3(\mathbf{z}) \tag{11.29}$$

that is composed of the potential functions in Equations (11.22), (11.24), and (11.26). Direct minimization of (11.29) with respect to all unknowns is an exceedingly difficult problem, because the motion and segmentation fields constitute a large set of unknowns. To this effect, we perform the minimization of (11.29) through the following two-steps iterations [Cha 94]:

1. Given the best available estimates of the parameters $\mathbf{a}_i$, $i = 1, \ldots, K$, and $\mathbf{z}$, update the optical flow field $\mathbf{v}_1$, $\mathbf{v}_2$. This step involves the minimization of a modified cost function

$$E_1 = \sum_{\mathbf{x}} [g_k(\mathbf{x}) - g_{k+1}(\mathbf{x} + \mathbf{v}(\mathbf{x})\Delta t)]^2 + \alpha \sum_{\mathbf{x}} ||\mathbf{v}(\mathbf{x}) - \tilde{\mathbf{v}}(\mathbf{x})||^2$$
$$+\beta \sum_{\mathbf{x}_i} \sum_{\mathbf{x}_j \in \mathcal{N}_{\mathbf{x}_i}} ||\mathbf{v}(\mathbf{x}_i) - \mathbf{v}(\mathbf{x}_j)||^2 \, \delta(z(\mathbf{x}_i) - z(\mathbf{x}_j)) \tag{11.30}$$

which is composed of all the terms in (11.29) that contain $\mathbf{v}(\mathbf{x})$. While the first term indicates how well $\mathbf{v}(\mathbf{x})$ explains our observations, the second and third terms impose prior constraints on the motion estimates that they should conform with the parametric flow model, and that they should vary smoothly within each region. To minimize this energy function, we employ the HCF method recently proposed by Chou and Brown [Cho 90]. HCF is a deterministic method designed to efficiently handle the optimization of multivariable problems with neighborhood interactions.

2. Update the segmentation field $\mathbf{z}$, assuming that the optical flow field $\mathbf{v}(\mathbf{x})$ is known. This step involves the minimization of all the terms in (11.29) which contain $\mathbf{z}$ as well as $\tilde{\mathbf{v}}(\mathbf{x})$, given by

$$E_2 = \alpha \sum_{\mathbf{x}} ||\mathbf{v}(\mathbf{x}) - \tilde{\mathbf{v}}(\mathbf{x})||^2 + \sum_{\mathbf{x}_i} \sum_{\mathbf{x}_j \in \mathcal{N}_{\mathbf{x}_i}} V_C(z(\mathbf{x}_i), z(\mathbf{x}_j)) \tag{11.31}$$

The first term in (11.31) quantifies the consistency of $\tilde{\mathbf{v}}(\mathbf{x})$ and $\mathbf{v}(\mathbf{x})$. The second term is related to the *a priori* probability of the present configuration of the segmentation labels. We use an ICM procedure to optimize E_2 [Cha 93]. The mapping parameters $\mathbf{a}_i$ are updated by least squares estimation within each region.

An initial estimate of the optical flow field can be found using the Bayesian approach with a global smoothness constraint. Given this estimate, the segmentation labels can be initialized by a procedure similar to Wang and Adelson's [Wan 94]. The determination of the free parameters α, β, and γ is a design problem. One strategy is to choose them to provide a dynamic range correction so that each term

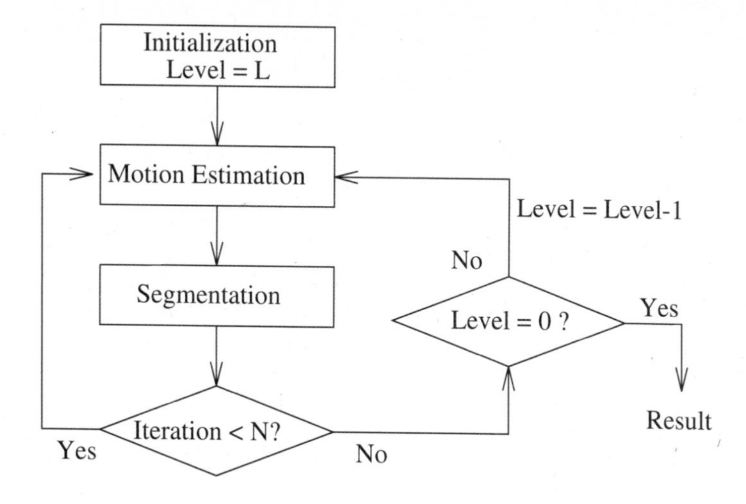

Figure 11.4: The block diagram of the simultaneous MAP algorithm.

in the cost function (11.29) has equal emphasis. However, because the optimization is implemented in two steps, the ratio α/γ also becomes of consequence. We recommend to select $1 \leq \alpha/\gamma \leq 5$, depending on how well the motion field can be represented by a piecewise-parametric model and whether we have a sufficient number of classes.

A hierarchical implementation of this algorithm is also possible by forming successive low-pass filtered versions of the images $\mathbf{g}_k$ and $\mathbf{g}_{k+1}$. Thus, the quantities $\mathbf{v}_1$, $\mathbf{v}_2$, and $\mathbf{z}$ can be estimated at different resolutions. The results of each hierarchy are used to initialize the next lower level. A block diagram of the hierarchical algorithm is depicted in Figure 11.4. Note that the Gibbsian model for the segmentation labels has been extended to include neighbors in scale by Kato *et al.* [Kat 93].

11.3.4 Relationship to Other Algorithms

It is important to recognize that this simultaneous estimation and segmentation framework not only enables 3-D motion and structure estimation in the presence of multiple moving objects, but also provides improved optical flow estimates. Several existing motion analysis algorithms can be formulated as special cases of this framework. If we retain only the first and the third terms in (11.29), and assume that all sites possess the same segmentation label, then we have Bayesian motion estimation with a global smoothness constraint. The motion estimation algorithm proposed by Iu [Iu 93] utilizes the same two terms, but replaces the $\delta(\cdot)$ function by a local outlier rejection function (Section 8.3.2).

The motion estimation and region labeling algorithm proposed by Stiller [Sti 94] (Section 8.3.3) involves all terms in (11.29), except the first term in (11.25). Furthermore, the segmentation labels in Stiller's algorithm are used merely as tokens to allow for a piecewise smoothness constraint on the flow field, and do not attempt to enforce consistency of the flow vectors with a parametric component. We also note that the motion estimation algorithms of Konrad and Dubois [Kon 92, Dub 93] and Heitz and Bouthemy [Hei 93] which use line fields are fundamentally different in that they model discontinuities in the motion field, rather than modeling regions that correspond to different physical motions (Section 8.3.1).

On the other hand, the motion segmentation algorithm of Murray and Buxton [Mur 87] (Section 11.2.2) employs only the second term in (11.25) and third term in (11.29) to model the conditional and prior pdf, respectively. Wang and Adelson [Wan 94] relies on the first term in (11.25) to compute the motion segmentation (Section 11.2.3). However, they also take the DFD of the parametric motion vectors into consideration when the closest match between the estimated and parametric motion vectors, represented by the second term, exceeds a threshold.

11.4 Examples

Examples are provided for optical flow segmentation using the Wang-Adelson (W-A) and Murray-Buxton (M-B) methods, as well as for simultaneous motion estimation and segmentation using the proposed MAP algorithm with the same two frames of the Mobile and Calendar sequence shown in Figure 5.9 (a) and (b). This is a challenging sequence, since there are several objects with distinct motions as depicted in Figure 11.1.

The W-A and M-B algorithms use the optical flow estimated by the Horn-Schunck algorithm, shown in Figure 5.11 (b), as input. We have set the number of regions to four. In order to find the representative affine motion parameters in the W-A algorithm, 8×8 and 16×16 seed blocks have been selected, and the affine parameters estimated for each of these blocks are clustered using the K-means algorithm. It is important that the size of the blocks be large enough to identify rotational motion. The result of the W-A segmentation is depicted in Figure 11.5 (a). We have initialized the M-B algorithm with the result of the W-A algorithm, and set the initial temperature equal to 1. The resulting segmentation after 50 iterations is shown in Figure 11.5 (b). Observe that the M-B algorithm eliminates small isolated regions on the calendar, and grows the region representing the rotating ball in the right direction by virtue of the probabilistic smoothness constraint that it employs.

Next, we initialized the simultaneous MAP estimation and segmentation method, also by the result of the W-A algorithm. We have set $\alpha = \beta = 10$ and $\alpha/\gamma = 5$, since the motion field can be well-represented by a piecewise parametric model. The estimated optical flow and segmentation label fields are shown in Figure 11.6 (a) and (b), respectively. Note that the depicted motion field corresponds to the lower right portion of the segmentation field. The results show some improvement over the M-B method.

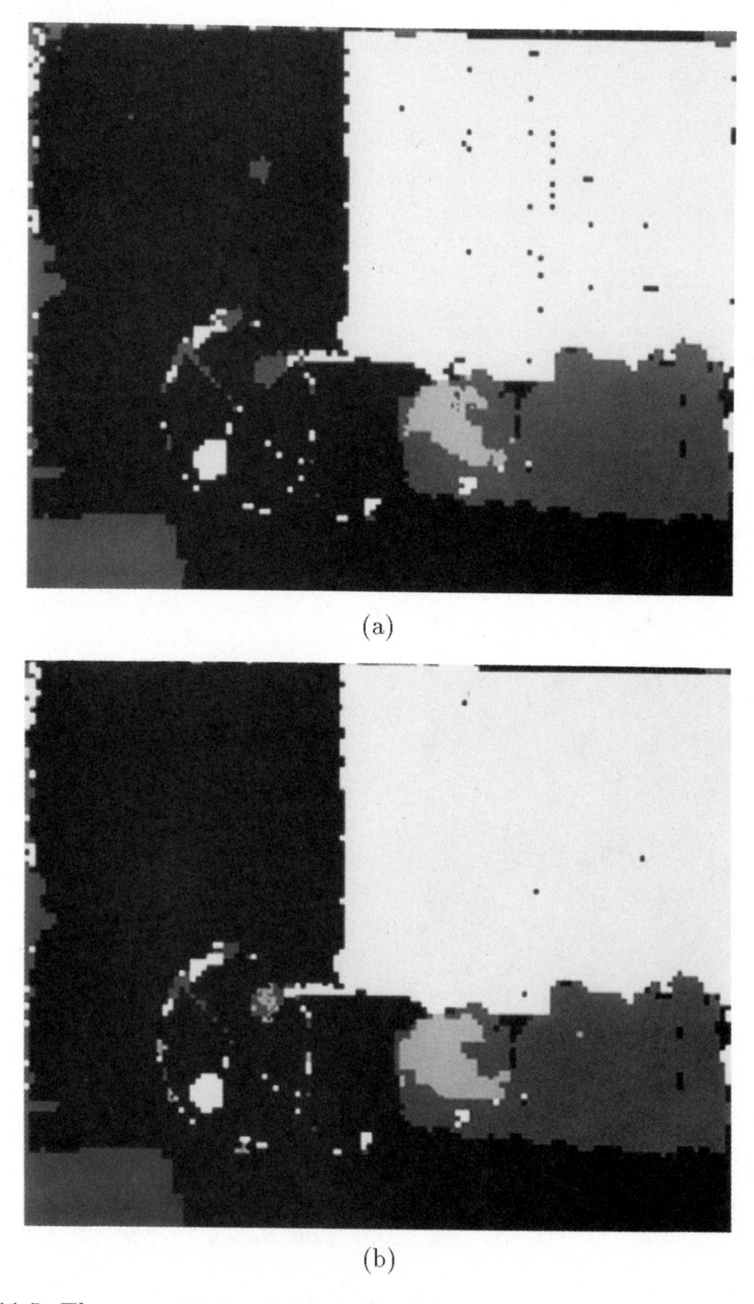

Figure 11.5: The segmentation field obtained by the a) Wang-Adelson method and b) Murray-Buxton method. (Courtesy Michael Chang)

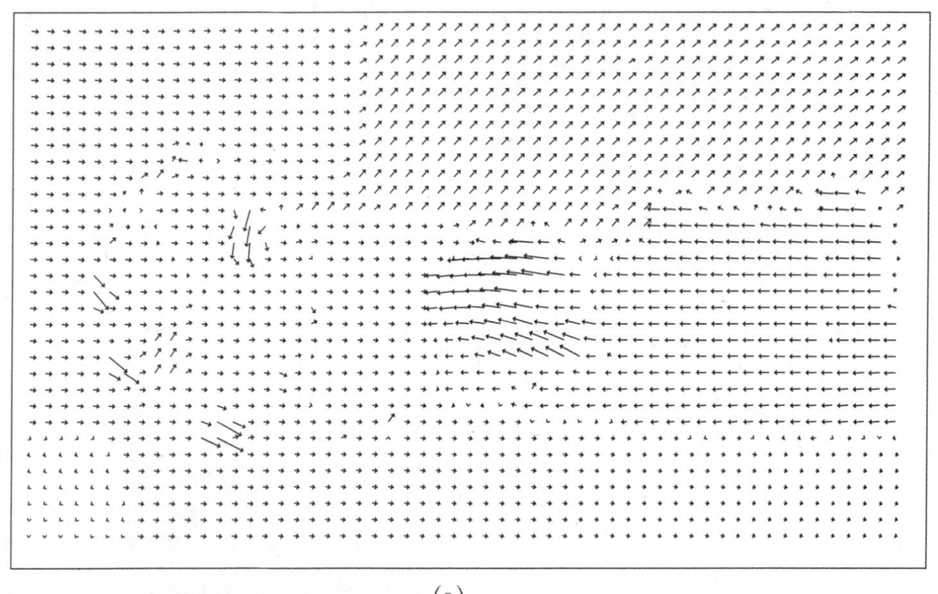

(a)

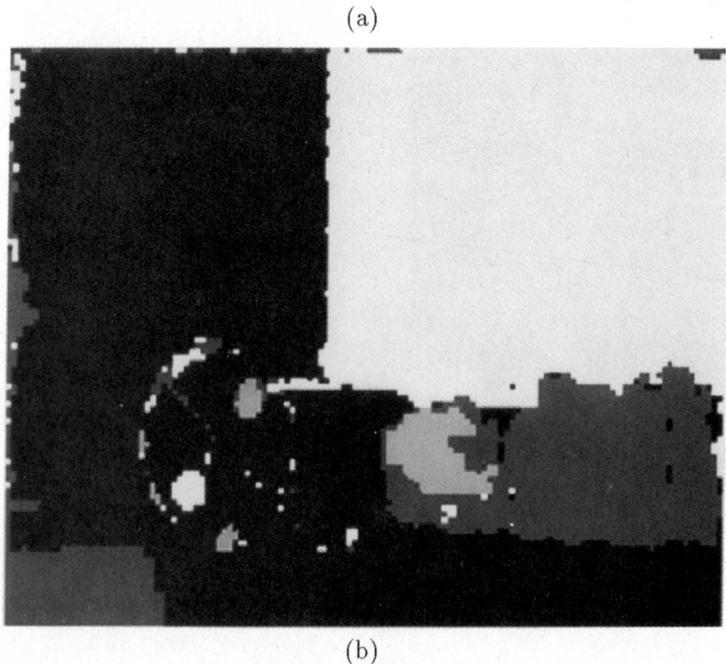

(b)

Figure 11.6: a) The optical flow field and b) segmentation field estimated by the simultaneous MAP method after 100 iterations. (Courtesy Michael Chang)

11.5 Exercises

1. Show that the MAP segmentation reduces to the K-means algorithm if we assume the conditional pdf is Gaussian and no *a priori* information is available.
2. How would you apply the optimum threshold selection method discussed in Appendix B.1.1 to change detection?
3. How would you modify (11.15) considering that (v_1, v_2) is the measured normal flow rather than the projected flow?
4. Do you prefer to model the flow discontinuities through a segmentation field or through line fields? Why?
5. Verify the relationships claimed in Section 11.3.3.
6. Discuss how to choose the scale factors α, λ, γ, and ψ in (11.30) and (11.31).

Bibliography

[Adi 85] G. Adiv, "Determining three-dimensional motion and structure from optical flow generated by several moving objects," *IEEE Trans. Pattern Anal. Mach. Intell.*, vol. 7, pp. 384–401, 1985.

[Ber 92] J. R. Bergen, P. J. Burt, R. Hingorani, and S. Peleg, "A three-frame algorithm for estimating two-component image motion," *IEEE Trans. Patt. Anal. Mach. Intel.*, vol. 14, pp. 886–896, Sep. 1992.

[Cha 93] M. M. Chang, A. M. Tekalp, and M. I. Sezan, "Motion field segmentation using an adaptive MAP criterion," *Proc. Int. Conf. ASSP*, Minneapolis, MN, April 1993.

[Cha 94] M. M. Chang, M. I. Sezan, and A. M. Tekalp, "An algorithm for simultaneous motion estimation and scene segmentation," *Proc. Int. Conf. ASSP*, Adelaide, Australia, April 1994.

[Cho 90] P. B. Chou and C. M. Brown, "The theory and practice of Bayesian image labeling," *Int. J. Comp. Vision*, vol. 4, pp. 185–210, 1990.

[Die 91] N. Diehl, "Object-oriented motion estimation and segmentation in image sequences," *Signal Processing: Image Comm.*, vol. 3, pp. 23–56, 1991.

[Dub 93] E. Dubois and J. Konrad, "Estimation of 2-D motion fields from image sequences with application to motion-compensated processing," in *Motion Analysis and Image Sequence Processing*, M. I. Sezan and R. L. Lagendijk, eds., Norwell, MA: Kluwer, 1993.

[Hoe 88] M. Hoetter and R. Thoma, "Image segmentation based on object oriented mapping parameter estimation," *Signal Proc.*, vol. 15, pp. 315–334, 1988.

[Hsu 94] S. Hsu, P. Anandan, and S. Peleg, "Accurate computation of optical flow by using layered motion representations," *Proc. Int. Conf. Patt. Recog.*, Jerusalem, Israel, pp. 743–746, Oct. 1994.

[Iu 93] S.-L. Iu, "Robust estimation of motion vector fields with discontinuity and occlusion using local outliers rejection," *SPIE*, vol. 2094, pp. 588–599, 1993.

[Kat 93] Z. Kato, M. Berthod, and J. Zerubia, "Parallel image classification using multiscale Markov random fields," *Proc. IEEE Int. Conf. ASSP*, Minneapolis, MN, pp. V137–140, April 1993.

[Mur 87] D. W. Murray and B. F. Buxton, "Scene segmentation from visual motion using global optimization," *IEEE Trans. Patt. Anal. Mach. Intel.*, vol. 9, no. 2, pp. 220–228, Mar. 1987.

[Sti 93] C. Stiller, "A statistical image model for motion estimation," *Proc. Int. Conf. ASSP*, Minneapolis, MN, pp. V193–196, April 1993.

[Sti 94] C. Stiller, "Object-oriented video coding employing dense motion fields," *Proc. Int. Conf. ASSP*, Adelaide, Australia, April 1994.

[Tho 80] W. B. Thompson, "Combining motion and contrast for segmentation," *IEEE Trans. Pattern Anal. Mach. Intel.*, vol. 2, pp. 543–549, 1980.

[Wan 94] J. Y. A. Wang and E. Adelson, "Representing moving images with layers," *IEEE Trans. on Image Proc.*, vol. 3, pp. 625–638, Sep. 1994.

[Wu 93] S. F. Wu and J. Kittler, "A gradient-based method for general motion estimation and segmentation," *J. Vis. Comm. Image Rep.*, vol. 4, no. 1, pp. 25–38, Mar. 1993.

Chapter 12

STEREO AND MOTION TRACKING

Because 3-D motion and structure estimation from two monocular views has been found to be highly noise-sensitive, this chapter presents two new approaches, stereo imaging and motion tracking, that offer more robust estimates. They are discussed in Sections 12.1 and 12.2, respectively. In stereo imaging, we acquire a pair of right and left images, at each instant. More robust motion and structure estimation from two stereo pairs is possible, because structure parameters can be estimated using stereo triangulation, hence decoupling 3-D motion and structure estimation. Moreover, the mutual relationship between stereo disparity and 3-D motion parameters can be utilized for stereo-motion fusion, thereby improving the accuracy of both stereo disparity and 3-D motion estimation. In motion tracking, it is assumed that a long sequence of monocular or stereo video is available. In this case, more robust estimation is achieved by assuming a temporal dynamics; that is, a model describing the temporal evolution of the motion. Motion between any pair of frames is expected to obey this model. Then batch or recursive filtering techniques can be employed to track the 3-D motion parameters in time. At present, stereo-motion fusion and motion tracking using Kalman filtering are active research topics of significant interest.

12.1 Motion and Structure from Stereo

It is well known that 3-D motion and structure estimation from monocular video is an ill-posed problem, especially when the object is relatively far away from the camera. Several researchers have employed stereo sequences for robust 3-D motion and structure estimation [Dho 89, Hua 94]. Stereo imaging also alleviates the scale ambiguity (between depth and translation) inherent in monocular imaging, since a properly registered stereo pair contains information about the scene structure.

In the following, we first briefly present the principles of still-frame stereo imaging. Subsequently, motion and structure estimation from a sequence of stereo pairs will be discussed.

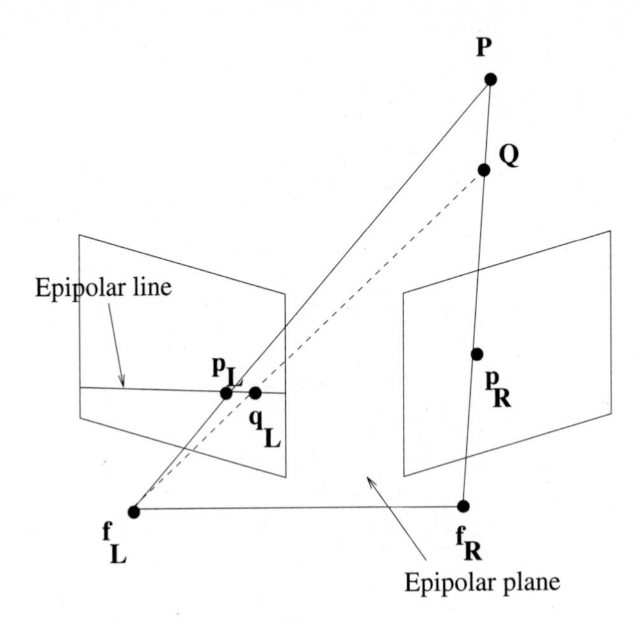

Figure 12.1: The epipolar plane and lines.

12.1.1 Still-Frame Stereo Imaging

Let $\mathbf{p}_L = (x_{1L}, x_{2L})$ and $\mathbf{p}_R = (x_{1R}, x_{2R})$ denote the perspective projections of a point $\mathbf{P} = (X_1, X_2, X_3)$ onto the left and right image planes, respectively. In general, the point $\mathbf{P}$ and the focal points of the left and right cameras, $\mathbf{f}_L$ and $\mathbf{f}_R$, define the so-called epipolar plane, as depicted in Figure 12.1. The intersection of the epipolar plane with the left and right image planes is called the epipolar lines. It follows that the perspective projection of a point anywhere on the epipolar plane falls on the epipolar lines.

A cross-section of the imaging geometry, as seen in the $X_1 - X_3$ plane, is depicted in Figure 12.2. Here, $\mathbf{C}_W$, $\mathbf{C}_R$, and $\mathbf{C}_L$ denote the world, right camera, and left camera coordinate systems, respectively, and the focal lengths of both cameras are assumed to be equal, $f_R = f_L = f$. Let $\mathbf{P}_R = (X_{1R}, X_{2R}, X_{3R})$ and $\mathbf{P}_L = (X_{1L}, X_{2L}, X_{3L})$ denote the representation of the point $\mathbf{P} = (X_1, X_2, X_3)$ in $\mathbf{C}_R$ and $\mathbf{C}_L$, respectively. Then the coordinates of $\mathbf{P}$ in the right and left coordinate systems are given by

$$\mathbf{P}_R = \mathbf{R}_R \mathbf{P} + \mathbf{T}_R \tag{12.1}$$

$$\mathbf{P}_L = \mathbf{R}_L \mathbf{P} + \mathbf{T}_L \tag{12.2}$$

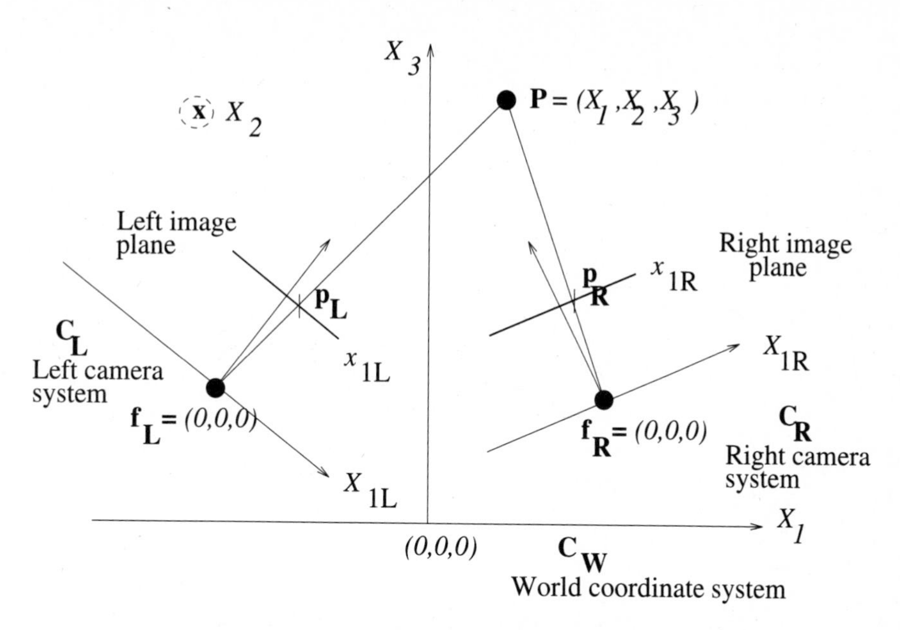

Figure 12.2: Stereo image formation.

where $\mathbf{R}_R$ and $\mathbf{T}_R$, and $\mathbf{R}_L$ and $\mathbf{T}_L$ are the extrinsic parameters of the camera model indicating relative positions of $\mathbf{C}_R$ and $\mathbf{C}_L$ with respect to $\mathbf{C}_W$, respectively. Combining equations (12.1) and (12.2), we have

$$\begin{aligned} \mathbf{P}_L &= \mathbf{R}_L \mathbf{R}_R^{-1} \mathbf{P}_R - \mathbf{R}_L \mathbf{R}_R^{-1} \mathbf{T}_R + \mathbf{T}_L \\ &\doteq \mathbf{M} \mathbf{P}_R + \mathbf{B} \end{aligned} \tag{12.3}$$

where $\mathbf{M}$ and $\mathbf{B}$ are known as the relative configuration parameters [Wen 92]. Then, based on similar triangles, the perspective projection of the point $\mathbf{P}$ into the left and right image planes can be expressed as

$$\begin{aligned} x_{1L} &= f \frac{X_{1L}}{X_{3L}}, \qquad & x_{2L} &= f \frac{X_{2L}}{X_{3L}} \\ x_{1R} &= f \frac{X_{1R}}{X_{3R}}, \qquad & x_{2R} &= f \frac{X_{2R}}{X_{3R}} \end{aligned} \tag{12.4}$$

respectively. Substituting (12.4) into (12.3), we have

$$\frac{X_{3L}}{f} \begin{bmatrix} x_{1L} \\ x_{2L} \\ f \end{bmatrix} = \frac{X_{3R}}{f} \mathbf{M} \begin{bmatrix} x_{1R} \\ x_{2R} \\ f \end{bmatrix} + \mathbf{B} \tag{12.5}$$

The structure from stereo problem refers to estimating the coordinates (X_1, X_2, X_3) of a 3-D point $\mathbf{P}$, given the corresponding points (x_{1L}, x_{2L}) and

(x_{1R}, x_{2R}) in the left and right image planes, and the extrinsic camera calibration parameters. We first solve for X_{3L} and X_{3R} from (12.5). It can be shown that only two of the three equations in (12.5) are linearly independent in the event of noise-free point correspondences. In practice, (12.5) should be solved in the least squares sense to find X_{3L} and X_{3R}. Next, the 3-D coordinates of the point $\mathbf{P}$ in the left and right camera coordinate systems can be computed from (12.4). Finally, a least squares estimate of $\mathbf{P}$ in the world coordinate system can be obtained from (12.1) and (12.2). A closed-form expression for $\mathbf{P}$ in the world coordinate system can be written in the special case when the left and right camera coordinates are parallel and aligned with the $X_1 - X_2$ plane of the world coordinates. Then

$$X_1 = \frac{b\,(x_{1L} + x_{1R})}{x_{1L} - x_{1R}}, \qquad X_2 = \frac{2\,b\,x_{2L}}{x_{1L} - x_{1R}}, \qquad X_3 = \frac{2\,f\,b}{x_{1L} - x_{1R}} \tag{12.6}$$

where b refers to half of the distance between the two cameras (assuming that they are symmetrically placed on both sides of the origin).

Finding corresponding pairs of image points is known as the stereo matching problem [Mar 78, Dho 89, Bar 93]. We define

$$\mathbf{w}(\mathbf{x}_R) \doteq (w_1(\mathbf{x}_R), w_2(\mathbf{x}_R)) = (x_{1R} - x_{1L}, x_{2R} - x_{2L}) \tag{12.7}$$

as the stereo disparity (taking the right image as the reference). Observe that the matching problem always involves a 1-D search along one of the epipolar lines. Suppose we start with the point $\mathbf{p}_R$. The object point $\mathbf{P}$ must lie along the line combining $\mathbf{p}_R$ and $\mathbf{f}_R$. Note that the loci of the projection of all points $\mathbf{Q}$ along this line onto the left image plane define the epipolar line for the left image plane (see Figure 12.1). Thus, it suffices to search for the matching point $\mathbf{p}_L$ along the left epipolar line.

12.1.2 3-D Feature Matching for Motion Estimation

The simplest method for motion and structure estimation from stereo would be to decouple the structure estimation and motion estimation steps by first estimating the depth at selected image points from the respective stereo pairs at times t and t' using 2-D feature matching between the left and right pairs, and temporal matching in one of the left or right channels, independently. Three-D rigid motion parameters can then be estimated by 3-D feature matching between the frames t and t' [Aru 87, Hua 89c]. However, this scheme neglects the obvious relationship between the estimated disparity fields at times t and t'. Alternatively, more sophisticated stereo-motion fusion schemes, which aim to enforce the mutual consistency of the resulting motion and disparity values, have also been proposed [Wax 86]. Both approaches are introduced in the following.

There are, in general, two approaches for the estimation of the 3-D motion parameters $\mathbf{R}$ and $\mathbf{T}$, given two stereo pairs at times t and t': 3-D to 3-D feature matching and 3-D to 2-D feature matching. If we start with an arbitrary point (x_{1R}, x_{2R}) in the right image at time t, we can first find the matching point (x_{1L}, x_{2L}) in the left image by still-frame stereo matching, which determines a 3-D feature point $\mathbf{P} = (X_1, X_2, X_3)$ in the world coordinate system at time t, as discussed in the previous section. Next, we locate the corresponding 2-D point (x'_{1R}, x'_{2R}) at time t' that matches (x_{1R}, x_{2R}) by 2-D motion estimation. At this moment, we can estimate the 3-D motion parameters using a 3-D to 2-D point-matching algorithm based on a set of matching (x'_{1R}, x'_{2R}) and $\mathbf{P}$. Alternatively, we can determine (x'_{1L}, x'_{2L}) at time t' again by still-frame stereo matching or by 2-D motion estimation using the left images at time t and t', and then use a 3-D to 3-D point-matching algorithm based on a set of matching $\mathbf{P}'$ and $\mathbf{P}$.

3-D to 3-D Methods

Given N 3-D point correspondences $(\mathbf{P}_i,\ \mathbf{P}'_i)$ (expressed in the world coordinate system) at two different times, obtained by still-frame stereo or other range-finding techniques, which lie on the same rigid object; the rotation matrix $\mathbf{R}$ (with respect to the world coordinate system) and the translation vector $\mathbf{T}$ can be found from

$$\mathbf{P}'_i = \mathbf{R}\mathbf{P}_i + \mathbf{T}, \quad i = 1, \ldots, N \tag{12.8}$$

It is well known that, in general, three noncollinear point correspondences are necessary and sufficient to determine $\mathbf{R}$ and $\mathbf{T}$ uniquely [Hua 94]. In practice, point correspondences are subject to error, and therefore, one prefers to work with more than the minimum number of point correspondences. In this case, $\mathbf{R}$ and $\mathbf{T}$ can be found by minimizing

$$\sum_{i=1}^{N} ||\mathbf{P}'_i - (\mathbf{R}\mathbf{P}_i + \mathbf{T})||^2 \tag{12.9}$$

subject to the constraints that $\mathbf{R}$ is a valid rotation matrix. Robust estimation procedures can be employed to eliminate outliers from the set of feature correspondences to improve results. Observe that if the small angle assumption is applicable, the problem reduces to solving a set of linear equations.

Establishing temporal relationships between 3-D lines and points from a sequence of depth maps computed independently from successive still-frame pairs to estimate 3-D motion was also proposed [Kim 87].

3-D to 2-D Methods

From (9.1) and (12.1), we can express the equation of 3-D motion with respect to the right camera coordinate system as

$$\mathbf{X}'_R = \grave{\mathbf{R}}_R \mathbf{X}_R + \grave{\mathbf{T}}_R \tag{12.10}$$

where

$$\begin{aligned} \grave{\mathbf{R}}_R &= \mathbf{R}_R \mathbf{R} \mathbf{R}_R^{-1} \\ \grave{\mathbf{T}}_R &= \mathbf{R}_R \mathbf{T} + \mathbf{T}_R - \grave{\mathbf{R}}_R \mathbf{T}_R \end{aligned}$$

denote the rotation matrix and the translation vector, respectively, with respect to the right camera coordinate system. Then, substituting (12.10) into (12.4), we have a computed correspondence in the right image given by

$$\tilde{x}'_{1R} = f\frac{X'_{1R}}{X'_{3R}} \quad \text{and} \quad \tilde{x}'_{2R} = f\frac{X'_{2R}}{X'_{3R}} \tag{12.11}$$

Given N 3-D to 2-D point correspondence measurements, $(\mathbf{P}_{iR}, \mathbf{p}'_{iR})$, where $\mathbf{P}_R = (X_{1R}, X_{2R}, X_{3R})$ refers to a 3-D point with respect to the right camera coordinate system at time t, and $\mathbf{p}'_R = (x'_{1R}, x'_{2R})$ denotes the corresponding right image point at time t'; we can estimate $\grave{\mathbf{R}}_R$ and $\grave{\mathbf{T}}_R$ by minimizing

$$\sum_{i=1}^{N} ||\mathbf{p}'_{iR} - \tilde{\mathbf{p}}'_{iR}||^2 \tag{12.12}$$

where $\tilde{\mathbf{p}}'_{iR}$ refers to the corresponding image point computed by (12.11) as a function of $\grave{\mathbf{R}}_R$ and $\grave{\mathbf{T}}_R$. The minimization of (12.12) can be posed as a nonlinear least squares problem in the six unknown 3-D motion parameters. Once again, a linear formulation is possible if the small angle of rotation approximation is applicable. Assuming that the extrinsic parameters of the camera system $\mathbf{R}_R$ and $\mathbf{T}_R$ are known, the rotation and translation parameters $\mathbf{R}$ and $\mathbf{T}$ in the world coordinates can be easily recovered from $\grave{\mathbf{R}}_R$ and $\grave{\mathbf{T}}_R$ based on (12.10).

The difficulty with both 3-D to 3-D and 2-D to 3-D approaches is in the process of establishing the matching 3-D and/or 2-D features. Stereo disparity estimation and estimation of feature correspondences in the temporal direction are, individually, both ill-posed problems, complicated by the aperture and occlusion problems; hence, the motivation to treat these problems simultaneously.

12.1.3 Stereo-Motion Fusion

Theoretically, in a time-varying stereo pair, the stereo matching need be performed only for the initial pair of frames. For subsequent pairs, stereo correspondences can be predicted by means of the left and right optical flow vectors. Stereo matching would only be required for those features that newly enter the field of view. However, because both optical flow estimation and disparity estimation are individually ill-posed problems, several studies have recently been devoted to fusion of stereo disparity and 3-D motion estimation in a mutually beneficial way. Stereo-motion fusion methods impose the constraint that the loop formed by the disparity and motion correspondence estimates, in Figure 12.3, must be a closed loop. Richards [Ric 85] and Waxman *et al.* [Wax 86] derived an analytical expression for the temporal rate of change of the disparity to disparity ratio as a function of the 3-D motion and structure parameters. They then proposed a simultaneous stereo and motion estimation method based on this relation. Aloimonos *et al.* [Alo 90] proposed an algebraic solution which requires no point-to-point correspondence estimation for the

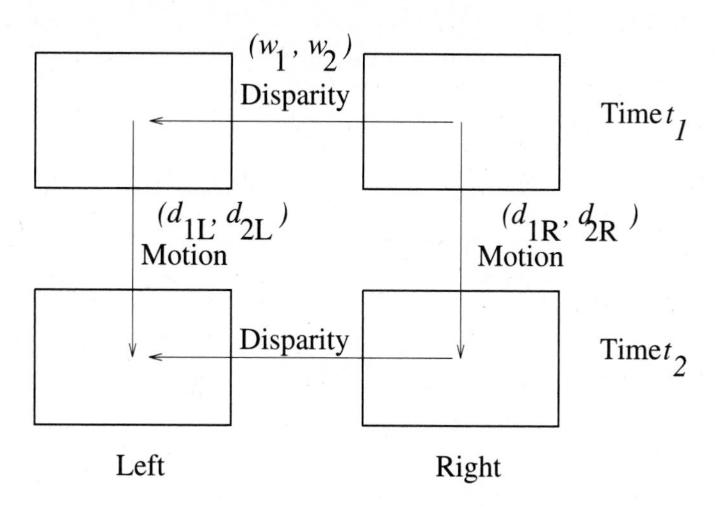

Figure 12.3: Stereo-motion fusion.

special case of 3-D planar objects. Methods to integrate stereo matching and optical flow estimation using multiresolution edge matching and dynamic programming [Liu 93] and constrained optimization [Tam 91] have also been proposed.

In the following, we first develop a maximum *a posteriori* probability (MAP) estimation framework (see Chapter 8) for stereo-motion fusion, based on dense displacement and disparity field models, in the case of a single rigid object in motion. Note that stereo-motion fusion with a set of isolated feature points (instead of dense fields) can be posed as a maximum-likelihood problem, which is a special case of this MAP framework. The MAP framework is extended to the case of multiple moving objects in the next section.

Dense Fields: Let $\mathbf{d}_R(\mathbf{x}_R) = (d_{1R}(\mathbf{x}_R), d_{2R}(\mathbf{x}_R))$ and $\mathbf{d}_L(\mathbf{x}_L) = (d_{1L}(\mathbf{x}_L), d_{2L}(\mathbf{x}_L))$ denote the 2-D displacement fields computed at each pixel $\mathbf{x}_R = (x_{1R}, x_{2R})$ and $\mathbf{x}_L = (x_{1L}, x_{2L})$ of the right and left images, respectively, and $\mathbf{w}(\mathbf{x}_R)$ denote the disparity field at each pixel $\mathbf{x}_R$ of the right image. Let $\mathbf{d}_{1R}$, $\mathbf{d}_{2R}$, $\mathbf{d}_{1L}$, $\mathbf{d}_{2L}$, $\mathbf{w}_1$, and $\mathbf{w}_2$ denote vectors formed by lexicographic ordering of the scalars $d_{1R}(\mathbf{x}_R)$, $d_{2R}(\mathbf{x}_R)$, $d_{1L}(\mathbf{x}_L)$, $d_{2L}(\mathbf{x}_L)$, $w_1(\mathbf{x}_R)$, and $w_2(\mathbf{x}_R)$ at all pixels, respectively. We wish to estimate $\mathbf{d}_{1R}$, $\mathbf{d}_{2R}$, $\mathbf{d}_{1L}$, $\mathbf{d}_{2L}$, $\mathbf{w}_1$, and $\mathbf{w}_2$, given $\mathbf{I}_L$, $\mathbf{I}_R$, $\mathbf{I}'_L$, $\mathbf{I}'_R$, the left and right images at times t and t', respectively, in order to maximize the joint *a posteriori* probability density function (pdf)

$$
\begin{aligned}
p(\mathbf{d}_{1R}, \mathbf{d}_{2R}, \mathbf{d}_{1L}, \mathbf{d}_{2L}, \mathbf{w}_1, \mathbf{w}_2 | \mathbf{I}'_L, \mathbf{I}'_R, \mathbf{I}_L, \mathbf{I}_R) \propto \\
p(\mathbf{I}'_L, \mathbf{I}'_R, \mathbf{I}_L | \mathbf{d}_{1R}, \mathbf{d}_{2R}, \mathbf{d}_{1L}, \mathbf{d}_{2L}, \mathbf{w}_1, \mathbf{w}_2, \mathbf{I}_R) \\
p(\mathbf{d}_{1R}, \mathbf{d}_{2R}, \mathbf{d}_{1L}, \mathbf{d}_{2L} | \mathbf{w}_1, \mathbf{w}_2, \mathbf{I}_R) p(\mathbf{w}_1, \mathbf{w}_2 | \mathbf{I}_R)
\end{aligned}
\quad (12.13)
$$

where $p(\mathbf{I}'_L, \mathbf{I}'_R, \mathbf{I}_L | \mathbf{d}_{1R}, \mathbf{d}_{2R}, \mathbf{d}_{1L}, \mathbf{d}_{2L}, \mathbf{w}_1, \mathbf{w}_2, \mathbf{I}_R)$ provides a measure of how well the present displacement and disparity estimates conform with the observed frames, $p(\mathbf{d}_{1R}, \mathbf{d}_{2R}, \mathbf{d}_{1L}, \mathbf{d}_{2L} | \mathbf{w}_1, \mathbf{w}_2, \mathbf{I}_R)$ quantifies the consistency of the 2-D motion field with the 3-D motion and structure parameters and imposes a local smoothness constraint on it, and $p(\mathbf{w}_1, \mathbf{w}_2 | \mathbf{I}_R)$ enforces a local smoothness constraint on the disparity field. Given the MAP estimates of $\mathbf{d}_{1R}$, $\mathbf{d}_{2R}$, $\mathbf{d}_{1L}$, $\mathbf{d}_{2L}$, $\mathbf{w}_1$, and $\mathbf{w}_2$, the 3-D rotation matrix $\mathbf{R}$, the 3-D translation vector $\mathbf{T}$, and the scene depth X_3 at each pixel can be estimated by one of the 3-D to 3-D or 3-D to 2-D matching techniques discussed in Section 12.1.2.

The three pdfs on the right hand side of (12.13) are assumed to be Gibbsian with the potential functions $U_1(\cdot)$, $U_2(\cdot)$, and $U_3(\cdot)$, respectively, that are given by

$$U_1(\mathbf{I}'_L, \mathbf{I}'_R, \mathbf{I}_L | \mathbf{d}_{1R}, \mathbf{d}_{2R}, \mathbf{d}_{1L}, \mathbf{d}_{2L}, \mathbf{w}_1, \mathbf{w}_2, \mathbf{I}_R) = \tag{12.14}$$
$$\sum_{\mathbf{x}_R \in I_R} \left[(I_L(\mathbf{x}_R + \mathbf{w}(\mathbf{x}_R)) - I_R(\mathbf{x}_R))^2 + (I'_R(\mathbf{x}_R + \mathbf{d}_R(\mathbf{x}_R)) - I_R(\mathbf{x}_R))^2 \right.$$
$$\left. + (I'_L(\mathbf{x}_L + \mathbf{d}_L(\mathbf{x}_L)) - I_L(\mathbf{x}_L))^2 \right]$$

$$U_2(\mathbf{d}_{1R}, \mathbf{d}_{2R}, \mathbf{d}_{1L}, \mathbf{d}_{2L} | \mathbf{w}_1, \mathbf{w}_2, \mathbf{I}_R) = \tag{12.15}$$
$$\sum_{\mathbf{x}_R \in I_R} \left(||\mathbf{d}_R(\mathbf{x}_R) - \tilde{\mathbf{d}}_R(\mathbf{x}_R)||^2 + ||\mathbf{d}_L(\mathbf{x}_L) - \tilde{\mathbf{d}}_L(\mathbf{x}_L)||^2 \right)$$
$$+ \alpha \sum_{\mathbf{x}_{Ri}} \left(\sum_{\mathbf{x}_{Rj} \in \mathcal{N}_{\mathbf{x}_{Ri}}} ||\mathbf{d}_R(\mathbf{x}_{Ri}) - \mathbf{d}_R(\mathbf{x}_{Rj})||^2 + \sum_{\mathbf{x}_{Lj} \in \mathcal{N}_{\mathbf{x}_{Li}}} ||\mathbf{d}_L(\mathbf{x}_{Li}) - \mathbf{d}_L(\mathbf{x}_{Lj})||^2 \right)$$

where $\mathbf{x}_L = \mathbf{x}_R + \mathbf{w}(\mathbf{x}_R)$ is the corresponding point on the left image,

$$\tilde{\mathbf{d}}_R(\mathbf{x}_R) \doteq \tilde{\mathbf{x}}'_R - \mathbf{x}_R \qquad \text{and} \qquad \tilde{\mathbf{d}}_L(\mathbf{x}_L) \doteq \tilde{\mathbf{x}}'_L - \mathbf{x}_L \tag{12.16}$$

denote the projected 3-D displacement field onto the right and left image planes, respectively, in which $\tilde{\mathbf{x}}'_R$ and $\tilde{\mathbf{x}}'_L$ are computed as functions of the 3-D motion and disparity parameters (step 1 of the algorithm below), α is a constant, $\mathcal{N}_{\mathbf{x}_{Ri}}$ and $\mathcal{N}_{\mathbf{x}_{Li}}$ denote neighborhoods of the sites $\mathbf{x}_{Ri}$ and $\mathbf{x}_{Li} = \mathbf{x}_{Ri} + \mathbf{w}(\mathbf{x}_{Ri})$, respectively, and

$$U_3(\mathbf{w}_1, \mathbf{w}_2 | \mathbf{I}_R) = \sum_{\mathbf{x}_{Ri}} \sum_{\mathbf{x}_{Rj} \in \mathcal{N}_{\mathbf{x}_{Ri}}} ||\mathbf{w}(\mathbf{x}_{Ri}) - \mathbf{w}(\mathbf{x}_{Rj})||^2 \tag{12.17}$$

The maximization of (12.13) can, then, be performed by the following two-step iteration process:

1. Given the present estimates of $\mathbf{d}_{1R}$, $\mathbf{d}_{2R}$, $\mathbf{d}_{1L}$, $\mathbf{d}_{2L}$, $\mathbf{w}_1$, and $\mathbf{w}_2$:

   i) reconstruct a set of 3-D points (X_1, X_2, X_3) as described in Section 12.1.1,

   ii) estimate $\mathbf{R}$ and $\mathbf{T}$ by means of 3-D to 2-D point matching or 3-D to 3-D point matching, and

   iii) calculate the projected displacement fields $\tilde{\mathbf{d}}_R(\mathbf{x}_R)$ and $\tilde{\mathbf{d}}_L(\mathbf{x}_L)$ from (12.1), (12.2), and (12.4).

2. Perturb the displacement $\mathbf{d}_{1R}$, $\mathbf{d}_{2R}$, $\mathbf{d}_{1L}$, $\mathbf{d}_{2L}$ and the disparity $\mathbf{w}_1$, $\mathbf{w}_2$ fields in order to minimize

$$E = U_1(\mathbf{I}'_L, \mathbf{I}'_R, \mathbf{I}_L | \mathbf{d}_{1R}, \mathbf{d}_{2R}, \mathbf{d}_{1L}, \mathbf{d}_{2L}, \mathbf{w}_1, \mathbf{w}_2, \mathbf{I}_R) + \gamma_1 U_2(\mathbf{d}_{1R}, \mathbf{d}_{2R}, \mathbf{d}_{1L}, \mathbf{d}_{2L} | \mathbf{w}_1, \mathbf{w}_2, \mathbf{I}_R) + \gamma_2 U_3(\mathbf{w}_1, \mathbf{w}_2 | \mathbf{I}_R)$$

where γ_1 and γ_2 are some constants.

The algorithm is initialized with motion and disparity estimates obtained by existing decoupled methods. A few iterations using the ICM algorithm are usually sufficient. Observe that $U_1(\cdot)$ requires that the 2-D displacement and disparity estimates be consistent with all four frames, while the first term of $U_2(\cdot)$ enforces consistency of the 2-D and 3-D motion estimates. The second term of $U_2(\cdot)$ and $U_3(\cdot)$ impose smoothness constraints on the motion and disparity fields, respectively.

Isolated Feature Points: Because stereo-motion fusion with dense motion and disparity fields may be computationally demanding, the proposed MAP formulation can be simplified into a maximum-likelihood (ML) estimation problem by using selected feature points. This is achieved by turning off the smoothness constraints imposed by the potential (12.17) and the second term of (12.15), which does not apply in the case of isolated feature points [Alt 95]. Results using 14 feature points are shown in Section 12.3. The methods cited here are mere examples of many possible approaches. Much work remains for future research in stereo-motion fusion.

12.1.4 Extension to Multiple Motion

The simultaneous MAP estimation and segmentation framework presented in Section 11.3 can be easily extended to stereo-motion fusion in the presence of multiple moving objects. Here, pairs of disparity $\mathbf{w}(\mathbf{x}_R)$ and displacement vectors $\mathbf{d}_R(\mathbf{x}_R)$ are segmented into K regions, where within each region the displacement vectors are expected to be consistent with a single set of 3-D motion parameters, and the disparity is allowed to vary smoothly. The regions are identified by the label field $\mathbf{z}$.

We seek the disparity, motion, and segmentation field configuration that would maximize the *a posteriori* probability density function

$$p(\mathbf{d}_{1R}, \mathbf{d}_{2R}, \mathbf{d}_{1L}, \mathbf{d}_{2L}, \mathbf{w}_1, \mathbf{w}_2, \mathbf{z} | \mathbf{I}'_L, \mathbf{I}'_R, \mathbf{I}_L, \mathbf{I}_R,) \propto p(\mathbf{I}'_L, \mathbf{I}'_R, \mathbf{I}_L | \mathbf{d}_{1R}, \mathbf{d}_{2R}, \mathbf{d}_{1L}, \mathbf{d}_{2L}, \mathbf{w}_1, \mathbf{w}_2, \mathbf{z}, \mathbf{I}_R) \; p(\mathbf{d}_{1R}, \mathbf{d}_{2R}, \mathbf{d}_{1L}, \mathbf{d}_{2L} | \mathbf{w}_1, \mathbf{w}_2, \mathbf{z}, \mathbf{I}_R) \; p(\mathbf{w}_1, \mathbf{w}_2 | \mathbf{z}, \mathbf{I}_R) \; p(\mathbf{z} | \mathbf{I}_R) \quad (12.18)$$

given $\mathbf{I}_L$, $\mathbf{I}_R$, $\mathbf{I}'_L$, $\mathbf{I}'_R$. We assume that all pdfs on the right-hand side of (12.18) are Gibbsian with the potential functions

$$U_1(\mathbf{I}'_L, \mathbf{I}'_R, \mathbf{I}_L | \mathbf{d}_{1R}, \mathbf{d}_{2R}, \mathbf{d}_{1L}, \mathbf{d}_{2L}, \mathbf{w}_1, \mathbf{w}_2, \mathbf{z}, \mathbf{I}_R) = \sum_{\mathbf{x}_R \in \mathbf{I}_R} \Big[(I_L(\mathbf{x}_R + \mathbf{w}(\mathbf{x}_R)) - I_R(\mathbf{x}_R))^2 + (I'_R(\mathbf{x}_R + \mathbf{d}_R(\mathbf{x}_R)) - I_R(\mathbf{x}_R))^2 + (I'_L(\mathbf{x}_L + \mathbf{d}_L(\mathbf{x}_L)) - I_L(\mathbf{x}_L))^2 \Big] \quad (12.19)$$

$$U_2(\mathbf{d}_{1R}, \mathbf{d}_{2R}, \mathbf{d}_{1L}, \mathbf{d}_{2L}|\mathbf{w}_1, \mathbf{w}_2, \mathbf{z}) = \sum_{\mathbf{x}_R \in \mathbf{I}_R} \left(||\mathbf{d}_R(\mathbf{x}_R) - \tilde{\mathbf{d}}_R(\mathbf{x}_R)||^2 + ||\mathbf{d}_L(\mathbf{x}_L) - \tilde{\mathbf{d}}_L(\mathbf{x}_L)||^2 \right)$$
$$+ \sum_{\mathbf{x}_{Ri}} \left(\sum_{\mathbf{x}_{Rj} \in \mathcal{N}_{\mathbf{x}_{Ri}}} ||\mathbf{d}_R(\mathbf{x}_{Ri}) - \mathbf{d}_R(\mathbf{x}_{Rj})||^2 \right.$$
$$\left. + \sum_{\mathbf{x}_{Lj} \in \mathcal{N}_{\mathbf{x}_{Li}}} ||\mathbf{d}_L(\mathbf{x}_{Li}) - \mathbf{d}_L(\mathbf{x}_{Lj})||^2 \right) \delta(z(\mathbf{x}_{Ri}) - z(\mathbf{x}_{Rj})) \quad (12.20)$$

where $\tilde{\mathbf{d}}_R$ is the projected 3-D motion given by (12.16), and $\mathbf{x}_L$, $\mathcal{N}_{\mathbf{x}_{Ri}}$, and $\mathcal{N}_{\mathbf{x}_{Li}}$ are as defined in the previous section,

$$U_3(\mathbf{w}_1, \mathbf{w}_2|\mathbf{z}) = \sum_{\mathbf{x}_{Ri}} \sum_{\mathbf{x}_{Rj} \in \mathcal{N}_{\mathbf{x}_{Ri}}} ||\mathbf{w}(\mathbf{x}_{Ri}) - \mathbf{w}(\mathbf{x}_{Rj})||^2 \, \delta(z(\mathbf{x}_{Ri}) - z(\mathbf{x}_{Rj}))$$

imposes a piecewise smoothness constraint on the disparity field, and

$$U_4(\mathbf{z}) = \sum_{\mathbf{x}_{Ri}} \sum_{\mathbf{x}_{Rj} \in \mathcal{N}_{\mathbf{x}_{Ri}}} V_C(z(\mathbf{x}_{Ri}), z(\mathbf{x}_{Rj})) \quad (12.21)$$

where $V_C(z(\mathbf{x}_{Ri}), z(\mathbf{x}_{Rj}))$ is defined by (11.28). Observe that, in the case of multiple motion, the smoothness constraints are turned on for motion and disparity vectors which possess the same segmentation label.

The maximization of (12.18) can be performed in an iterative manner. Each iteration consists of the following steps:

1. Given the initial estimates of the optical flow, $\mathbf{d}_{1R}$, $\mathbf{d}_{2R}$, $\mathbf{d}_{1L}$, $\mathbf{d}_{2L}$ the disparity $\mathbf{w}_1$, $\mathbf{w}_2$, and the segmentation $\mathbf{z}$, estimate the 3-D motion parameters $\mathbf{R}$ and $\mathbf{T}$ for each segment. (Similar to step 1 of the algorithm in the previous subsection.)

2. Update the optical flow $\mathbf{d}_{1R}$, $\mathbf{d}_{2R}$, $\mathbf{d}_{1L}$, and $\mathbf{d}_{2L}$ assuming that the disparity field and the segmentation labels are given by minimizing

$$E_1 = U_1(\mathbf{I}'_L, \mathbf{I}'_R, \mathbf{I}_L|\mathbf{d}_{1R}, \mathbf{d}_{2R}, \mathbf{d}_{1L}, \mathbf{d}_{2L}, \mathbf{w}_1, \mathbf{w}_2, \mathbf{z}, \mathbf{I}_R) + U_2(\mathbf{d}_{1R}, \mathbf{d}_{2R}, \mathbf{d}_{1L}, \mathbf{d}_{2L}|\mathbf{w}_1, \mathbf{w}_2, \mathbf{z})$$

3. Update the disparity field, $\mathbf{w}_1$, $\mathbf{w}_2$ assuming that the motion field and the segmentation labels are given. This step involves minimization of

$$E_2 = U_1(\mathbf{I}'_L, \mathbf{I}'_R, \mathbf{I}_L|\mathbf{d}_{1R}, \mathbf{d}_{2R}, \mathbf{d}_{1L}, \mathbf{d}_{2L}, \mathbf{w}_1, \mathbf{w}_2, \mathbf{z}, \mathbf{I}_R) + U_3(\mathbf{w}_1, \mathbf{w}_2|\mathbf{z})$$

4. Update the segmentation labels $\mathbf{z}$, assuming that disparity field and motion estimates are given. This step involves the minimization of all terms that contain $\mathbf{z}$, given by

$$E_3 = U_1(\mathbf{I}'_L, \mathbf{I}'_R, \mathbf{I}_L|\mathbf{d}_{1R}, \mathbf{d}_{2R}, \mathbf{d}_{1L}, \mathbf{d}_{2L}, \mathbf{w}_1, \mathbf{w}_2, \mathbf{z}, \mathbf{I}_R) + U_2(\mathbf{d}_{1R}, \mathbf{d}_{2R}, \mathbf{d}_{1L}, \mathbf{d}_{2L}|\mathbf{w}_1, \mathbf{w}_2, \mathbf{z}) + U_3(\mathbf{w}_1, \mathbf{w}_2|\mathbf{z}) + U_4(\mathbf{z})$$

The initial estimates of the optical flow, disparity, and segmentation fields can be obtained from existing still-image disparity estimation and monocular motion estimation and segmentation (see Chapter 11) algorithms, respectively.

12.2 Motion Tracking

Physical objects generally exhibit temporally smooth motion. In motion estimation from two views, whether monocular or stereo, it is not possible to make use of this important cue to resolve ambiguities and/or obtain more robust estimates. For example, acceleration information and the actual location of the center of rotation cannot be determined from two views. To this effect, here we present some recent results on motion estimation (tracking) from long sequences of monocular and stereo video based on an assumed kinematic model of the motion. We first discuss basic principles of motion tracking. Subsequently, some examples of 2-D and 3-D motion models and algorithms to track the parameters of these models are presented.

12.2.1 Basic Principles

Fundamental to motion tracking are a set of feature (token) matches or optical flow estimates over several frames, and a dynamic model describing the evolution of the motion in time. The feature matches or optical flow estimates, which serve as observations for the tracking algorithm, are either assumed available or estimated from pairs of frames by other means. The tracking algorithm in essence determines the parameters of the assumed motion model that best fit (usually in the least squares or minimum mean square error sense) the entire set of observations. In the following, we provide an overview of the main components of a tracking system.

Motion Model

Motion models vary in complexity, ranging from constant velocity models to more sophisticated local constant angular momentum models [Wen 87] depending upon the application. The performance of a tracking algorithm is strongly dependent on the accuracy of the dynamical model it employs. We may classify temporal motion models as 2-D motion models, to represent image-plane trajectory of 3-D points, and 3-D motion models, to represent the kinematics of physical motion.

- *2-D Trajectory Models:* Temporal trajectory of pixels in the image plane can be approximated by affine, perspective, or polynomial spatial transformations (see Sections 6.1.2 and 9.1), where the parameters a_i, $i = 1, \ldots, 6$ or 8, become functions of time (or the frame index k). Such trajectory models can be employed to track the motion of individual tokens or group of pixels (regions) (see Section 12.2.2 for examples). The temporal evolution of the transformation parameters can be modeled by either relating them to some 2-D rotation, translation, and/or dilation dynamics, or by a low-order Taylor series expansion of an unknown dynamics. Examples of the former approach include assuming a purely translational constant acceleration trajectory (see the token-tracking example in Section 12.2.2), and assuming a spatially local simplified affine model given by

$$
\begin{aligned}
x_1(k+1) &= x_1(k)\cos\alpha(k) - x_2(k)\sin\alpha(k) + t_1(k) \\
x_2(k+1) &= x_1(k)\sin\alpha(k) + x_2(k)\cos\alpha(k) + t_2(k)
\end{aligned}
$$

with some 2-D rotation (denoted by the angle $\alpha(k)$) and translation (represented by $t_1(k)$ and $t_2(k)$) dynamics. An example of the latter approach (see region tracking) is presented in Section 12.2.2 based on the work of Meyer *et al.* [Mey 94].

- *3-D Rigid Motion Models:* There are some important differences between 3-D rigid motion modeling for two-view problems and for tracking problems. The models used in two-view problems, discussed in Chapters 2, 9, and 10, do not include acceleration and precession, since they cannot be estimated from two views. Another difference is in choosing the center of rotation. There are, in general, two alternatives:

i) Rotation is defined with respect to a fixed world coordinate system, where it is assumed that the center of rotation coincides with the origin of the world coordinate system. This approach, which has been adopted in Chapter 2, is generally unsuitable for tracking problems. To see why, suppose we wish to track a rolling wheel. The rotation parameters computed with respect to the origin of the world coordinates are different for each frame (due to the translation of the center of the wheel), although the wheel rotates with a constant angular velocity about its center [Sha 90]. Thus, it is unnecessarily difficult to model the kinematics of the rotation in this case.

ii) Rotation is defined about an axis passing through the actual center of rotation. This approach almost always leads to simpler kinematic models for tracking applications. The coordinates of the center of rotation, which are initially unknown and translate in time, can only be determined after solving the equations of motion. It has been shown that the center of rotation can be uniquely determined if there exists precessional motion [You 90]. In the case of rotation with constant angular velocity, only an axis of rotation can be determined. It follows that one can estimate at best the axis of rotation in two-view problems.

2-D and 3-D motion models can each be further classified as rigid versus deformable motion models. The use of active contour models [Ley 93], such as snakes, and deformable templates [Ker 94] for tracking 2-D deformable motion, and superquadrics [Met 93] for tracking 3-D deformable motion have recently been proposed. Tracking of 2-D and 3-D deformable motion are active research topics of current interest.

Observation Model

All tracking algorithms require the knowledge of a set of 2-D or 3-D feature correspondences or optical flow data as observations. In practice, these feature correspondences or optical flow data are not directly available, and need to be estimated from the observed spatio-temporal image intensity. Feature matching and optical flow estimation are ill-posed problems themselves (see Chapter 5), because of ambiguities such as multiple matches (similar features) within the search area or no matches (occlusion). Several approaches have been proposed for estimating feature correspondences, including multifeature image matching [Wen 93] and probabilistic data association techniques [Cox 93]. These techniques involve a search within a finite window centered about the location predicted by the tracking algorithm. It has been shown that the nearest match (to the center of the window) may not always give the correct correspondence. To this effect, probabilistic criteria have been proposed to determine the most likely correspondence within the search window [Cox 93]. In optical-flow-based tracking, a multiresolution iterative refinement algorithm has been proposed to determine the observations [Mey 94]. In stereo imagery, stereo-motion fusion using dynamic programing [Liu 93] or constrained optimization [Tam 91] have been proposed for more robust correspondence estimation.

Batch vs. Recursive Estimation

Once a dynamical model and a number of feature correspondences over multiple frames have been determined, the best motion parameters consistent with the model and the observations can be computed using either batch or recursive estimation techniques. Batch estimators, such as the nonlinear least squares estimator, process the entire data record at once after all data have been collected. On the other hand, recursive estimators, such as Kalman filters or extended Kalman filters (see Appendix C), process each observation as it becomes available to update the motion parameters. It can be easily shown that both batch and recursive estimators are mathematically equivalent when the observation model is linear in the unknowns (state variables). Relative advantages and disadvantages of batch versus recursive estimators are:

1) Batch estimators tend to be numerically more robust than recursive estimators. Furthermore, when the state-transition or the observation equation is nonlinear, the performance of batch methods is generally superior to that of recursive methods (e.g., extended Kalman filtering).
2) Batch methods would require processing of the entire data record every time a new observation becomes available. Hence, recursive methods are computationally more attractive when estimates are needed in real-time (or "almost" real-time).

To combine the benefits of both methods, that is, computational efficiency and robustness of the results, hybrid methods called recursive-batch estimators have been proposed [Wen 93].

12.2.2 2-D Motion Tracking

Two-dimensional motion tracking refers to the ability to follow a set of tokens or regions over multiple frames based on a polynomial approximation of the image-plane trajectory of individual points or the temporal dynamics of a collection of flow vectors. Two examples are presented in the following.

Token Tracking

Several possibilities exist for 2-D tokens, such as points, line segments, and corners. Here, an example is provided, where 2-D lines, given by

$$x_2 = m\, x_1 + b \quad \text{or} \quad x_1 = m\, x_2 + b$$

where m is the slope and b is the intercept, are chosen as tokens. Each such line segment can be represented by a 4-D feature vector $\mathbf{p} = [\mathbf{p}_1\ \mathbf{p}_2]^T$ consisting of the two end points, $\mathbf{p}_1$ and $\mathbf{p}_2$.

Let the 2-D trajectory of each of the endpoints be approximated by a second-order polynomial model, given by

$$\begin{aligned} \mathbf{x}(k) &= \mathbf{x}(k-1) + \mathbf{v}(k-1)\Delta t + \frac{1}{2}\mathbf{a}(k-1)(\Delta t)^2 \\ \mathbf{v}(k) &= \mathbf{a}(k-1)\Delta t \qquad \text{and} \qquad \mathbf{a}(k) = \mathbf{a}(k-1) \end{aligned} \tag{12.22}$$

where $\mathbf{x}(k)$, $\mathbf{v}(k)$, and $\mathbf{a}(k)$ denote the position, velocity, and acceleration of the pixel at time k, respectively. Note that (12.22) models an image plane motion with constant acceleration.

Assuming that the tracking will be performed by a Kalman filter, we define the 12-dimensional state of the line segment as

$$\mathbf{z}(k) = \begin{bmatrix} \mathbf{p}(k) \\ \dot{\mathbf{p}}(k) \\ \ddot{\mathbf{p}}(k) \end{bmatrix} \tag{12.23}$$

where $\dot{\mathbf{p}}(k)$ and $\ddot{\mathbf{p}}(k)$ denote the velocity and the acceleration of the coordinates, respectively. Then the state propagation and observation equations can be expressed as

$$\mathbf{z}(k) = \mathbf{\Phi}(k, k-1)\mathbf{z}(k-1) + \mathbf{w}(k), \quad k = 1, \ldots, N \tag{12.24}$$

where

$$\mathbf{\Phi}(k, k-1) = \begin{bmatrix} \mathbf{I}_4 & \mathbf{I}_4\Delta t & \frac{1}{2}\mathbf{I}_4(\Delta t)^2 \\ \mathbf{0}_4 & \mathbf{I}_4 & \mathbf{I}_4\Delta t \\ \mathbf{0}_4 & \mathbf{0}_4 & \mathbf{I}_4 \end{bmatrix} \tag{12.25}$$

is the state-transition matrix, $\mathbf{I}_4$ and $\mathbf{0}_4$ are 4×4 identity and zero matrices, respectively, $\mathbf{w}(k)$ is a zero-mean, white random sequence, with the covariance matrix $\mathbf{Q}(k)$ representing the state model error, and

$$\mathbf{y}(k) = \mathbf{p}(k) + \mathbf{v}(k), \quad k = 1, \ldots, N \tag{12.26}$$

respectively. Note that the observation equation (12.26) simply states that the noisy coordinates of the end points of the line segment at each frame can be observed. It is assumed that the observations can be estimated from pairs of frames using some token-matching algorithm. The application of Kalman filtering follows straightforwardly using (C.5)-(C.9) given the state (12.24) and the observation (12.26) models.

Region Tracking

Tracking regions rather than discrete tokens may yield more robust results and provide new means to detect occlusion. Meyer *et al.* [Mey 94] recently proposed a region-tracking algorithm that is based on motion segmentation and region boundary propagation using affine modeling of a dense flow field within each region. Their method employs two Kalman filters, a motion filter that tracks the affine flow model parameters and a geometric filter that tracks boundaries of the regions. A summary of the state-space formulation for both Kalman filters is provided in the following.

1. *Motion Filter:* Let the affine flow field within each region be expressed as

$$\begin{bmatrix} v_1(k) \\ v_2(k) \end{bmatrix} = \begin{bmatrix} \dot{x}_1(k) \\ \dot{x}_2(k) \end{bmatrix} = \mathbf{A}(k) \begin{bmatrix} x_1(k) \\ x_2(k) \end{bmatrix} + \mathbf{b}(k) \tag{12.27}$$

where the matrix $\mathbf{A}$ and the vector $\mathbf{b}$ are composed of the six affine parameters $a_i, \ i = 1, \ldots 6$.

The temporal dynamics of the affine parameters, $a_i, \ i = 1, \ldots 6$, have been modeled by a second-order Taylor series expansion of unknown temporal dynamics, resulting in the state-space model

$$\begin{bmatrix} a_i(k) \\ \dot{a}_i(k) \end{bmatrix} = \begin{bmatrix} 1 & \Delta t \\ 0 & 1 \end{bmatrix} \begin{bmatrix} a_i(k-1) \\ \dot{a}_i(k-1) \end{bmatrix} + \begin{bmatrix} \epsilon_{i1}(k) \\ \epsilon_{i2}(k) \end{bmatrix}, \quad i = 1, \ldots, 6 \tag{12.28}$$

where $[\epsilon_{i1}(k) \ \epsilon_{i2}(k)]^T, \ i = 1, \ldots, 6$ are uncorrelated, identically distributed sequences of zero-mean Gaussian random vectors with the covariance matrix

$$\mathbf{P} = \sigma_P^2 \begin{bmatrix} \frac{\Delta t^3}{3} & \frac{\Delta t^2}{2} \\ \frac{\Delta t^2}{2} & \Delta t \end{bmatrix}$$

Observe that the state-space model (12.28) models the dynamics of each of the affine parameters separately, assuming that the temporal evolution of the six parameters are independent. Then, each parameter can be tracked by a different Kalman filter with a 2-D state vector.

The measurements for the motion filters, $\tilde{a}_i, \ i = 1, \ldots 6$ are modeled as noisy observations of the true parameters, given by

$$\tilde{a}_i(k) = a_i(k) + \beta_i(k) \tag{12.29}$$

where $\beta_i(k)$ is a sequence of zero-mean, white Gaussian random variables with the variance σ_β^2. As mentioned before, the measurements $\tilde{a}_i,\ i = 1, \ldots 6$ are estimated from the observed image sequence, two views at a time. Combining (12.27) with the well-known optical flow equation (5.5), a multiresolution estimation approach has been proposed by Meyer *et al.* [Mey 94] using least-squares fitting and iterative refinement within each region.

2. *Geometric Filter:* The geometric filter is concerned with tracking each region process that is identified by a motion segmentation algorithm [Bou 93]. The convex hull (the smallest convex polygon covering the region) of a polygonal approximation of each region is taken as the region descriptor. The region descriptor vector consists of $2N$ components, consisting of the (x_1, x_2) coordinates of the N vertices of the convex hull.

   A state-space model for tracking the evolution of the region descriptor vector over multiple frames can be written as

$$\begin{pmatrix} x_{11}(k) \\ x_{12}(k) \\ \vdots \\ x_{N1}(k) \\ x_{N2}(k) \end{pmatrix} = \begin{bmatrix} \mathbf{\Phi}(k,k-1) & 0 & \cdots & 0 \\ \vdots & & & \vdots \\ 0 & \cdots & 0 & \mathbf{\Phi}(k,k-1) \end{bmatrix} \begin{pmatrix} x_{11}(k-1) \\ x_{12}(k-1) \\ \vdots \\ x_{N1}(k-1) \\ x_{N2}(k-1) \end{pmatrix} + \begin{bmatrix} \mathbf{I}_2 \\ \vdots \\ \mathbf{I}_2 \end{bmatrix} \mathbf{u}(k) + \mathbf{w}(k) \tag{12.30}$$

   where

$$\mathbf{\Phi}(k,k-1) = \mathbf{I}_2 + \Delta t \mathbf{A}(k)$$

   which can be seen by considering a first-order Taylor expansion of $[x_1(k)\ x_2(k)]^T$ given by

$$\begin{bmatrix} x_1(k) \\ x_2(k) \end{bmatrix} = \begin{bmatrix} x_1(k-1) \\ x_2(k-1) \end{bmatrix} + \Delta t \begin{bmatrix} \dot{x}_1(k) \\ \dot{x}_2(k) \end{bmatrix}$$

   about the time $k-1$, and substituting the flow field model (12.27) for $[\dot{x}_1(k)\ \dot{x}_2(k)]^T$,

$$\mathbf{u}(k) = \Delta t\ \mathbf{b}(k)$$

   is a deterministic input vector which follows from the above derivation, and $\mathbf{w}(k)$ is a white-noise sequence with the covariance matrix $\mathbf{Q}(k)$.

The measurement equation can be expressed as

$$\begin{bmatrix} \tilde{x}_{11}(k) \\ \tilde{x}_{12}(k) \\ \vdots \\ \tilde{x}_{N1}(k) \\ \tilde{x}_{N2}(k) \end{bmatrix} = \begin{bmatrix} x_{11}(k) \\ x_{12}(k) \\ \vdots \\ x_{N1}(k) \\ x_{N2}(k) \end{bmatrix} + \mathbf{n}(k) \tag{12.31}$$

where $\mathbf{n}(k)$ is a sequence of zero-mean, white Gaussian random vectors. An observation of the region descriptor vector can be estimated by two-view optical flow analysis as described in [Bou 93].

All motion-tracking algorithms should contain provisions for detecting occlusion and de-occlusion, as moving objects may exit the field of view, or new objects may appear or reappear in the picture. An occlusion and de-occlusion detection algorithm based on the divergence of the motion field has been proposed in [Mey 94]. The main idea of the detector is to monitor the difference in the area of the regions from frame to frame, taking into account any global zooming. The descriptor is updated after each frame using occlusion/de-occlusion information. The region tracking algorithm presented in this section has recently been extended to include active contour models for representation of complex primitives with deformable B-splines [Bas 94].

12.2.3 3-D Rigid Motion Tracking

Three-dimensional motion tracking refers to the ability to predict and monitor 3-D motion and structure of moving objects in a scene from long sequences of monocular or stereo video. In what follows, we present examples of 3-D rigid motion tracking from monocular and stereo video based on the works of Broida and Chellappa [Bro 91] and Young and Chellappa [You 90].

From Monocular Video

Suppose we wish to track the 3-D motion of M feature points on a rigid object. We define two coordinate systems, depicted in Figure 12.4: $\mathbf{C}_o$, the object coordinate system whose origin $\mathbf{X}_o = [X_{o1}(k)\; X_{o2}(k)\; X_{o3}(k)]^T$, with respect to the camera and world coordinate systems, coincides with the center of rotation (which is unknown); and $\mathbf{C}_s$, the structure coordinate system, whose origin is located at a known point on the object. It is assumed that $\mathbf{C}_s$ and $\mathbf{C}_o$ are related by an unknown translation $\mathbf{T}_s$. Let $\tilde{\mathbf{X}}_i = [\tilde{X}_{i1}\; \tilde{X}_{i2}\; \tilde{X}_{i3}]^T$ denote the known coordinates of a feature point i with respect to the structure coordinate system. Then the coordinates of the feature point with respect to the world and camera coordinate systems are given by

$$\mathbf{X}_i(k) = \mathbf{X}_o(k) + \mathbf{R}(k)(\tilde{\mathbf{X}}_i - \mathbf{T}_s), \quad k = 0, \ldots, N \tag{12.32}$$

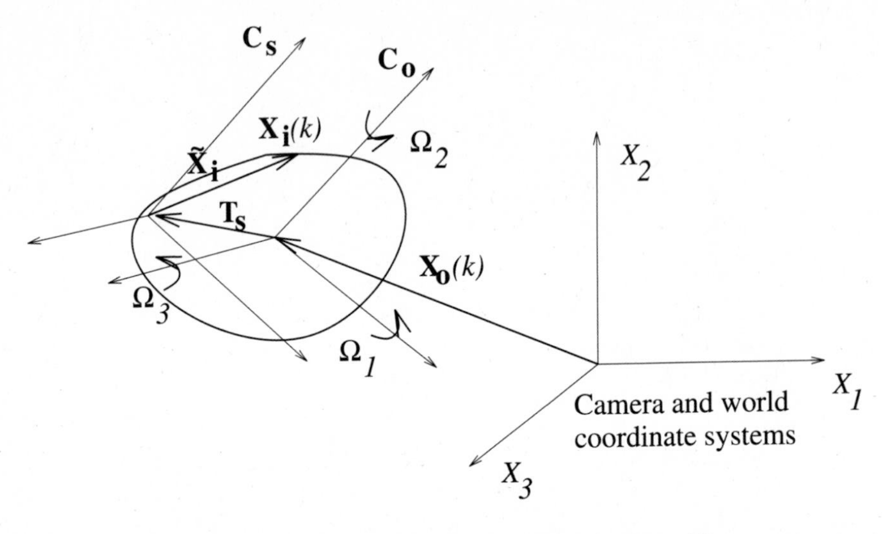

Figure 12.4: Motion tracking with monocular video.

where N is the number of frames, $\mathbf{R}(k)$ is the rotation matrix defined with respect to the center of rotation (object reference frame), $\mathbf{R}(0)$ is taken as the 3×3 identity matrix, the origin of the object coordinate system translates in time with respect to the camera and world coordinate systems, and the object is assumed rigid; that is, the coordinates of the feature points $\tilde{\mathbf{X}}_i$ with respect to the object coordinate system remains fixed in time.

The translational component of the motion will be represented by a constant acceleration model given by

$$\mathbf{X}_o(k) = \mathbf{X}_o(k-1) + \mathbf{V}_o(k-1)\Delta t + \frac{1}{2}\mathbf{A}_o(k-1)(\Delta t)^2 \quad (12.33)$$

$$\mathbf{V}_o(k) = \mathbf{V}_o(k-1) + \mathbf{A}_o(k-1)\Delta t \quad (12.34)$$

$$\mathbf{A}_o(k) = \mathbf{A}_o(k-1) \quad (12.35)$$

where $\mathbf{V}_o(k)$ and $\mathbf{A}_o(k)$ denote 3-D translational velocity and acceleration vectors.

The rotational component of the motion can be represented by a constant precession (rate of change of angular velocity) model, where $\mathbf{\Omega}(k) = [\Omega_1(k)\ \Omega_2(k)\ \Omega_3(k)]^T$ and $\mathbf{P} = [P_1\ P_2\ P_3]^T$ denote the angular velocity and precession vectors, respectively. Assuming that the rotation matrix $\mathbf{R}$ is expressed in terms of the unit quaternion $\mathbf{q}(k) = [q_0(k)\ q_1(k)\ q_2(k)\ q_3(k)]^T$, given by (2.10), the evolution of the rotation matrix in time can be expressed by that of the unit quaternion $\mathbf{q}(k)$. It has been shown that the temporal dynamics of the unit quaternion, in this case, can be expressed in closed form as [You 90]

$$\mathbf{q}(k) = \Phi[\mathbf{\Omega}(k-1), \mathbf{P}; \Delta t]\mathbf{q}(k-1), \quad k = 1, \ldots, N \quad (12.36)$$

where $\mathbf{q}(0) = [0\ 0\ 0\ 1]^T$,

$$\Phi[\mathbf{\Omega}(k-1), \mathbf{P}; \Delta t] = \Lambda[\mathbf{P}; \Delta t]\Lambda[\mathbf{\Omega}(k-1) - \mathbf{P}; \Delta t] \tag{12.37}$$

$$\Lambda[\mathbf{G}; \Delta t] = \begin{cases} \mathbf{I}_4 \cos\frac{|\mathbf{G}|\Delta t}{2} + \frac{2}{|\mathbf{G}|}\mathbf{\Gamma}[\mathbf{G}] \sin\frac{|\mathbf{G}|\Delta t}{2} & \text{if } \mathbf{G} \neq \mathbf{0} \\ \mathbf{I}_4 & \text{if } \mathbf{G} = \mathbf{0} \end{cases} \tag{12.38}$$

and

$$\mathbf{\Gamma}[\mathbf{G}] = \frac{1}{2} \begin{bmatrix} 0 & -G_3 & G_2 & -G_1 \\ G_3 & 0 & -G_1 & -G_2 \\ -G_2 & G_1 & 0 & -G_3 \\ G_1 & G_2 & G_3 & 0 \end{bmatrix} \tag{12.39}$$

In Equation (12.38), $\mathbf{G}$ stands for a 3×1 vector, which takes the values of $\mathbf{P}$ or $\mathbf{\Omega}(k-1) - \mathbf{P}$. Observe that this model is also valid for the special case $\mathbf{P} = \mathbf{0}$.

A multiframe batch algorithm, proposed by Broida and Chellappa [Bro 91], when both translational and rotational motion are with constant velocity is briefly summarized below. Assuming that all M feature points have the same 3-D motion parameters, the unknown parameters are

$$\mathbf{u} = \begin{bmatrix} \frac{X_{o1}}{X_{o3}}(0) \\ \frac{X_{o2}}{X_{o3}}(0) \\ \frac{V_1}{X_{o3}}(0) \\ \frac{V_2}{X_{o3}}(0) \\ \frac{V_3}{X_{o3}}(0) \\ \Omega_1(0) \\ \Omega_2(0) \\ \Omega_3(0) \\ \frac{X_{11}}{X_{o3}}(0) \\ \frac{X_{12}}{X_{o3}}(0) \\ \frac{X_{21}}{X_{o3}}(0) \\ \frac{X_{22}}{X_{o3}}(0) \\ \frac{X_{23}}{X_{o3}}(0) \\ \vdots \\ \frac{X_{M1}}{X_{o3}}(0) \\ \frac{X_{M2}}{X_{o3}}(0) \\ \frac{X_{M3}}{X_{o3}}(0) \end{bmatrix} \tag{12.40}$$

Observe that all position and translational motion parameters are scaled by X_{o3}, which is equivalent to setting $X_{o3} = 1$ to account for the scale ambiguity inherent in monocular imaging. Furthermore, the X_3 component of the first feature point is left as a free variable, because the origin of the object-coordinate frame (center of rotation) cannot be uniquely determined (only the axis of rotation can be computed) in the absence of precessional motion.

Suppose we have point correspondences measured for these M points over N frames, given by

$$x_{i1}(k) = \frac{X_{i1}(k)}{X_{i3}(k)} + n_{i1}(k) \tag{12.41}$$

$$x_{i2}(k) = \frac{X_{i2}(k)}{X_{i3}(k)} + n_{i2}(k) \tag{12.42}$$

where $X_{i1}(k)$, $X_{i2}(k)$, and $X_{i3}(k)$ can be computed in terms of the unknown parameters $\mathbf{u}$ using (12.32), (12.35), (12.36), and (2.10). Assuming the noise terms $n_{i1}(k)$ and $n_{i2}(k)$ are uncorrelated, identically distributed zero-mean Gaussian random variables, the maximum-likelihood estimate of the parameter vector $\mathbf{u}$ can be computed by minimizing the summed square residuals

$$E(\mathbf{u}) = \sum_{k=1}^{N}\sum_{i=1}^{M}[x_{i1}(k) - \frac{X_{i1}(k)}{X_{i3}(k)}]^2 + [x_{i2}(k) - \frac{X_{i2}(k)}{X_{i3}(k)}]^2 \tag{12.43}$$

The minimization can be performed by the conjugate-gradient descent or Levenberg-Marquardt methods. Alternatively, defining the value of the parameter vector $\mathbf{u}$ at each time k, we can define a recursive estimator. However, because of the nonlinearity of the rotational motion model and the observation equation in the unknowns, this would result in an extended Kalman filter.

From Stereo Video

In stereo imaging, the camera and the world coordinate systems no longer coincide, as depicted in Figure 12.5. Hence, it is assumed that feature correspondences both between the right and left images and in the temporal direction are available at each time k as observations. They can be obtained either separately or simultaneously using stereo-motion fusion techniques with two stereo-pairs [Tam 91, Liu 93].

Given these feature correspondence data and the kinematics of the 3-D motion modeled by (12.32), (12.35), (12.36), and (2.10), batch or recursive estimation algorithms can be derived using either 2-D or 3-D measurement equations [You 90]. In the case of 2-D measurements, the corresponding points in the right and left images are separately expressed in terms of the state vector at each time and included in the measurement vector. In the case of 3-D measurements, stereo triangularization is employed at each time k to reconstruct a time sequence of 3-D feature matches, which are then expressed in terms of the state vector. A comparison of both recursive and batch estimators has been reported in [Wu 94]. It has been concluded that, in the case of recursive estimation, experiments with 2-D measurements have reached steady-state faster. More recently, simultaneous stereo-motion fusion and 3-D motion tracking have been proposed [Alt 95].

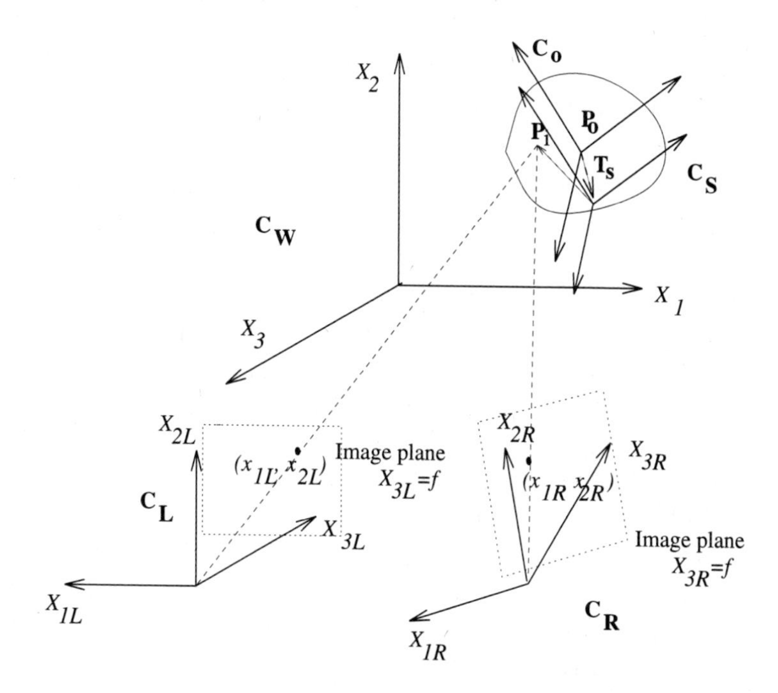

Figure 12.5: Motion tracking with stereo video.

12.3 Examples

We have experimented with some of the algorithms presented in this section using a stereo image sequence, known as a static indoor scene, produced at the Computer Vision Laboratory of Ecole Polytechnique de Montreal. The sequence consists of 10 frame pairs, where each left and right image is 512 pixels × 480 lines with 8 bits/pixel. The calibration parameters of the two cameras were known [Wen 92]. The first and second frame pairs of the sequence are shown in Figure 12.6.

We have interactively marked 14 feature points on the right image of the first frame (at time t_1), which are depicted by white circles in Figure 12.6 (b). The corresponding points on the other three images are computed by the maximum likelihood stereo-motion fusion algorithm described in Section 12.1.3. The initial estimates of the displacement $\mathbf{d}_{1R}$, $\mathbf{d}_{2R}$, $\mathbf{d}_{1L}$, $\mathbf{d}_{2L}$ and the disparity $\mathbf{w}_1$, $\mathbf{w}_2$ are computed by three-level hierarchical block matching. Next, we have tracked the motion of one of these feature points over the 10 frames using a batch algorithm. Observe that the selected feature point leaves the field of view after nine frames on the right image and eight frames on the left, as shown in Figure 12.7. We note that tracking using a temporal trajectory not only improves 3-D motion estimation, but also disparity and hence depth estimation.

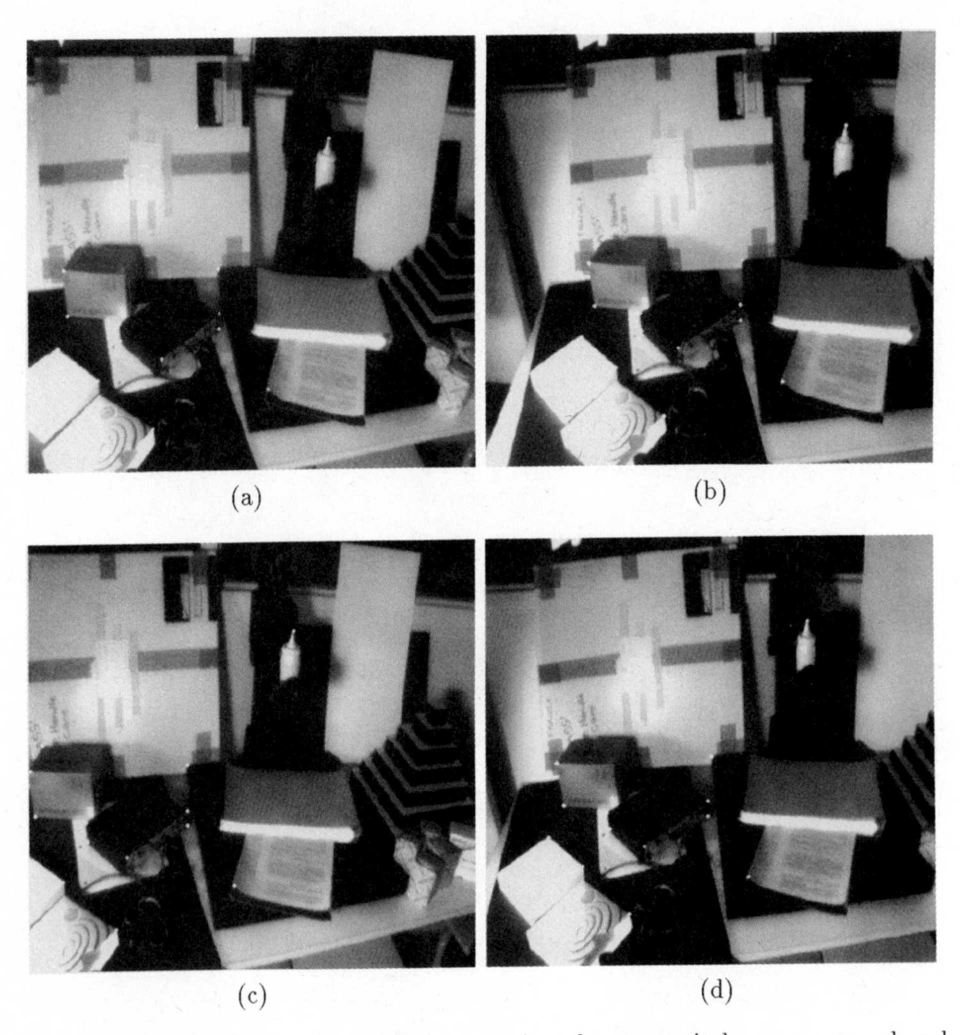

(a) (b)

(c) (d)

Figure 12.6: The first and second frame pairs of a stereo indoor scene produced at the Computer Vision Laboratory of Ecole Polytechnique de Montreal: a) left image at time t_1, b) right image at time t_1, c) left image at time t_2, and d) right image at time t_2. Fourteen feature points (marked by white circles) have been selected interactively on the right image at time t_1. Point correspondences on the other images are found by the maximum likelihood stereo-motion fusion algorithm. (Courtesy Yucel Altunbasak)

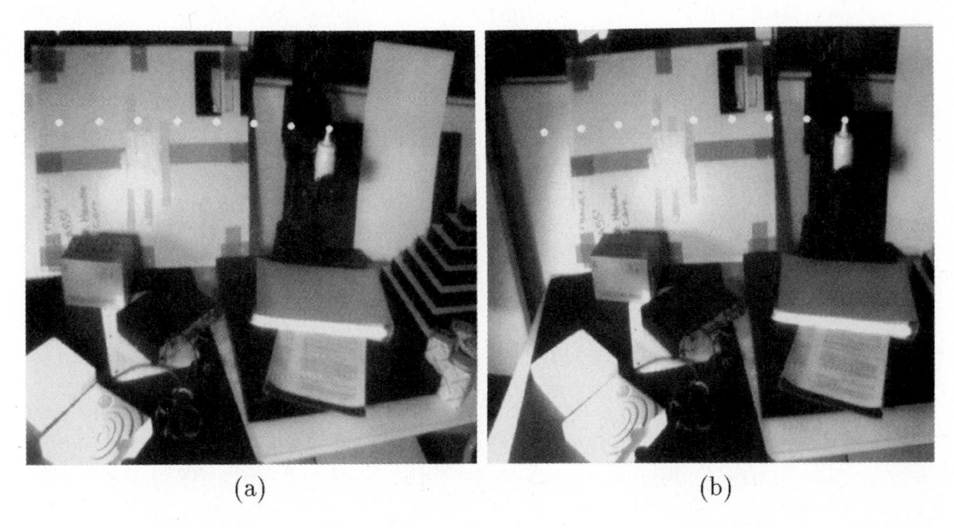

(a) (b)

Figure 12.7: Temporal trajectory of a single point marked on the a) left and b) right images of the first frame. The white marks indicate the position of the feature point in the successive frames. (Courtesy Yucel Altunbasak)

12.4 Exercises

1. Show that only two of the three equations in (12.5) are linearly independent if we have exact left-right correspondences.

2. Derive (12.6) given that the left and right camera coordinates are parallel and aligned with the $X_1 - X_2$ plane of the world coordinate system.

3. Derive (12.10).

4. Write the Kalman filter equations for the 2-D token-matching problem discussed in Section 12.2.2 given the state-transition model (12.24) and the observation model (12.26).

5. Write the equations of a recursive estimator for the 3-D motion-tracking problem discussed in Section 12.2.3. Discuss the relationship between this recursive estimator and the batch solution that was provided.

Bibliography

[Alo 90] Y. Aloimonos and J.-Y. Herve, "Correspondenceless stereo and motion: Planar surfaces," *IEEE Trans. Patt. Anal. Mach. Intel.*, vol. 12, pp. 504–510, May 1990.

[Alt 95] Y. Altunbasak, A. M. Tekalp, and G. Bozdagi, "Simultaneous stereo-motion fusion and 3-D motion tracking," *Proc. Int. Conf. Acoust. Speech and Sign. Proc.*, Detroit, MI, May 1995.

[Aru 87] K. S. Arun, T. S. Huang, and S. D. Blostein, "Least squares fitting of two 3-D point sets," *IEEE Trans. Patt. Anal. Mach. Intel.*, vol. 9, pp. 698–700, Sep. 1987.

[Bar 93] S. T. Barnard, "Stereo matching," in *Markov Random Fields*, R. Chellappa and A. Jain, eds., Academic Press, 1993.

[Bas 94] B. Bascle, P. Bouthemy, R. Deriche, and F. Meyer, "Tracking complex primitives in an image sequence," *Int. Conf. Patt. Recog.*, Jerusalem, Israel, pp. 426–431, Oct. 1994.

[Bou 93] P. Bouthemy and E. Francois, "Motion segmentation and qualitative dynamic scene analysis from an image sequence," *Int. J. Comp. Vis.*, vol. 10:2, pp. 157–182, 1993.

[Bro 91] T. J. Broida and R. Chellappa, "Estimating the kinematics and structure of a rigid object from a sequence of monocular images," *IEEE Trans. Patt. Anal. Mach. Intel*, vol. 13, pp. 497–513, June 1991.

[Che 94] R. Chellappa, T.-H. Wu, P. Burlina, and Q. Zheng, "Visual motion analysis," in *Control and Dynamic Systems*, C. T. Leondes, ed., vol. 67, pp. 199–261, 1994.

[Cox 93] I. J. Cox, "A review of statistical data association techniques for motion correspondence," *Int. J. Comp. Vis.*, vol. 10:1, pp. 53–66, 1993.

[Dho 89] U. R. Dhond, and J. K. Aggarwal, "Structure from stereo - A review," *IEEE Trans. Syst., Man, and Cyber.*, vol 19, no 6, Nov. 1989.

[Fau 93] O. Faugeras, *Three-Dimensional Computer Vision*, MIT Press, 1993.

[Fuj 93] K. Fujimura, N. Yokoya, and K. Yamamoto, "Motion tracking of deformable objects by active contour models using multiscale dynamic programming," *J. Vis. Comm. and Image Rep.*, vol. 4, no. 4, pp. 382–391, Dec. 1993.

[Hua 89c] T. S. Huang, "Motion estimation from stereo sequences," in *Machine Vision for Inspection and Measurement*, H. Freeman, ed., pp. 127–135, New York, NY: Academic Press, 1989.

[Hua 94] T. S. Huang and A. Netravali, "Motion and structure from feature correspondences: A review," *Proc. IEEE*, vol. 82, pp. 252–269, Feb. 1994.

[Ker 94] C. Kervrann and F. Heitz, "Robust tracking of stochastic deformable models in long image sequences," *Proc. IEEE Int. Conf. Image Proc.*, Austin, TX, vol. III, pp. 88–92, Nov. 1994.

[Kim 87] Y. C. Kim and J. K. Aggarwal, "Determining object motion in a sequence of stereo images," *IEEE J. Robotics and Auto.*, vol. 3, pp. 599–614, Dec. 1987.

[Ley 93] F. Leymarie and M. D. Levine, "Tracking deformable objects in the plane using an active contour model," *IEEE Trans. Patt. Anal. Mach. Intel*, vol. 15, pp. 617–634, June 1993.

[Liu 93] J. Liu and R. Skerjanc, "Stereo and motion correspondence in a sequence of stereo images," *Signal Proc.: Image Comm.*, vol. 5, pp. 305–318, 1993.

[Mar 78] D. Marr, G. Palm, and T. Poggio, "Analysis of a cooperative stereo algorithm," *Biol. Cybern.*, vol. 28, pp. 223–229, 1978.

[Mat 89] L. Matthies, R. Szeliski and T. Kanade, "Kalman-filter based algorithms for estimating depth from image sequences," *Int. J. Comp. Vision*, vol. 3, pp. 181–208, 1989.

[Met 93] D. Metaxas and D. Terzopoulos, "Shape and nonrigid motion estimation through physics-based synthesis," *IEEE Trans. Patt. Anal. Mach. Intel*, vol. 15, pp. 580–591, June 1993.

[Mey 94] F. G. Meyer and P. Bouthemy, "Region-based tracking using affine motion models in long image sequences," *CVGIP: Image Understanding*, vol. 60, no. 2, pp. 119–140, Sep. 1994.

[Ric 85] W. Richards, "Structure from stereo and motion," *J. Opt. Soc. Am. A*, vol. 2, pp. 343–349, Feb. 1985.

[Sha 90] H. Shariat and K. Price, "Motion estimation with more than two frames," *IEEE Trans. Patt. Anal. Mach. Intel*, vol. 12, pp. 417–434, May 1990.

[Spe 91] M. Spetsakis and J. Aloimonos, "A multi-frame approach to visual motion perception," *Int. J. Comp. Vision*, vol. 6, no. 3, pp. 245–255, 1991.

[Tam 91] A. Tamtaoui and C. Labit, "Constrained disparity and motion estimators for 3DTV image sequence coding," *Sign. Proc.: Image Comm.*, vol. 4, pp. 45–54, 1991.

[Wax 86] A. M. Waxman and J. H. Duncan, "Binocular image flows: Steps towards stereo-motion fusion," *IEEE Trans. Patt. Anal. Mach. Intel.*, vol. 8, pp. 715–729, Nov. 1986.

[Wen 87] J. Weng, T. S. Huang, and N. Ahuja, "3-D motion estimation, understanding, and prediction from noisy image sequences," *IEEE Trans. Patt. Anal. Mach. Intel*, vol. 9, pp. 370–389, May 1987.

[Wen 92] J. Weng, P. Cohen and M. Herniou, "Camera calibration with distortion models and accuracy evaluation," *IEEE Trans. Patt. Anal. Mach. Intel.*, vol. 14, no. 10, pp. 965–980, Oct. 1992.

[Wen 93] J. Weng, T. S. Huang, and N. Ahuja, *Motion and Structure from Image Sequences*, Springer-Verlag (Series in Information Sciences), 1993.

[Wu 88] J. J. Wu, R. E. Rink, T. M. Caelli, and V. G. Gourishankar, "Recovery of the 3-D location and motion of a rigid object through camera image (An extended Kalman filter approach)," *Int. J. Comp. Vision*, vol. 3, pp. 373–394, 1988.

[Wu 94] T.-H. Wu and R. Chellappa, "Stereoscopic recovery of egomotion and structure: Models, uniqueness and experimental results," *Proc. Int. Conf. Patt. Recog.*, Jerusalem, Israel, vol. I, pp. 645–648, Oct. 1994.

[You 90] G.-S. J. Young and R. Chellappa, "3-D motion estimation using a sequence of noisy stereo images: Models, estimation, and uniqueness results," *IEEE Trans. Patt. Anal. Mach. Intel.*, vol. 12, pp. 735–759, Aug. 1990.

[Zha 92] Z. Zhang and O. D. Faugeras, "Estimation of displacements from two 3-D frames obtained from stereo," *IEEE Trans. Patt. Anal. Mach. Intel.*, vol. 14, pp. 1141–1156, Dec. 1992.

[Zhe 92] Z. Zhang and O. Faugeras, "Three-dimensional motion computation and object segmentation in a long sequence of stereo frames," *Int. J. Comp. Vision*, vol. 7, no. 3, pp. 211–241, 1992.

Chapter 13

MOTION COMPENSATED FILTERING

This chapter is devoted to the theory of motion-compensated filtering, which forms the basis of several popular methods for noise filtering, restoration, standards conversion, and high-resolution reconstruction, presented in the next three chapters. Motion compensation is also an essential part of interframe compression methods, which are discussed in Part 6 [Gir 93]. In essence, this chapter incorporates specific motion models into the theory of sampling and reconstruction that was presented in Chapters 3 and 4 for video with arbitrary temporal variations. These results are presented in a separate chapter, because an in-depth understanding of motion-compensated filtering is fundamental for a complete appreciation of how and why spatio-temporal filtering is superior to intraframe filtering.

In Section 13.1, we derive the spatio-temporal frequency spectrum of video, assuming that the intensity variations in the temporal direction follow simplified motion models. It is shown that motion models introduce constraints on the support of the spatio-temporal frequency spectrum of video. The implications of the resulting spatio-temporal spectral characterizations for sub-Nyquist sampling of video are investigated in Section 13.2. The frequency responses of spatio-temporal filters that operate along motion trajectories have been derived for the special cases of global constant velocity and accelerated motions in Section 13.3. Section 13.4 discusses the application of motion-compensated filtering in noise reduction and reconstruction of a continuous video from digital data possibly sampled at sub-Nyquist rates. Specific details of designing motion-compensated noise reduction and standards conversion filters will be addressed in Chapters 14 and 16, respectively.

13.1 Spatio-Temporal Fourier Spectrum

The aim of this section is to derive the spatio-temporal frequency spectrum of video under the assumption that the temporal variations in the image intensity pattern can be represented by simplified motion models. The fundamental principle of motion compensation is that the intensity of a pixel remains unchanged along a well-defined motion trajectory. In the following, we first define a motion trajectory and then derive the frequency spectrum of video in the case of two specific motion models.

Motion in the image plane refers to the 2-D displacement or velocity of the projection of scene points, under either orthographic or perspective projection. Each image plane point follows a curve in the (x_1, x_2, t) space, which is called a " motion trajectory." The motion trajectory can be formally defined as a vector function $\mathbf{c}(t; x_1, x_2, t_0)$ which specifies the horizontal and vertical coordinates at time t of a reference point (x_1, x_2) at time t_0 [Dub 92]. The concept of motion trajectory is illustrated in Figure 13.1, where $\mathbf{c}(t'; x_1, x_2, t_0) = (x_1', x_2')$.

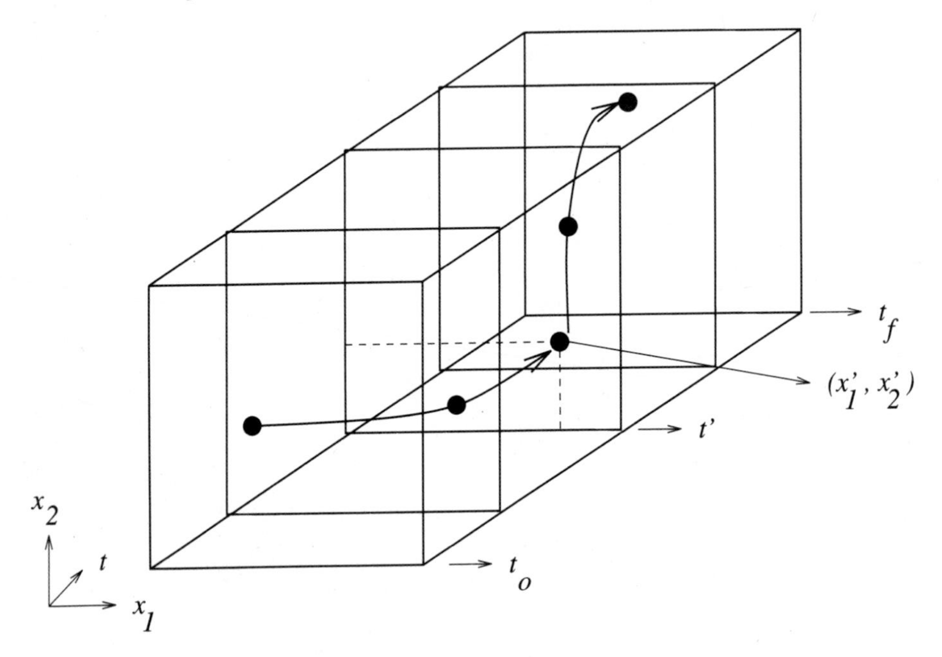

Figure 13.1: Motion trajectory.

Given the motion trajectory $c(t; x_1, x_2, t_0)$, the velocity at time t' at a point (x_1', x_2') along the trajectory can be defined by

$$v(x_1', x_2', t') = \frac{dc}{dt}(t; x_1, x_2, t_0)\,|_{t=t'} \tag{13.1}$$

In the following, we model the motion trajectory with closed-form expressions in two cases: global motion with constant velocity, and global motion with acceleration. These expressions are then incorporated into the spatio-temporal frequency domain analysis of video.

13.1.1 Global Motion with Constant Velocity

A simplified model of the projected motion is that of global motion with constant velocity (v_1, v_2) in the image plane. Then the frame-to-frame intensity variations can be modeled as

$$\begin{aligned} s_c(x_1, x_2, t) &= s_c(x_1 - v_1 t, x_2 - v_2 t, 0) \\ &\doteq s_0(x_1 - v_1 t, x_2 - v_2 t) \end{aligned} \tag{13.2}$$

where the reference frame is chosen as $t_0 = 0$, and $s_0(x_1, x_2)$ denotes the 2-D intensity distribution within the reference frame. This model is commonly used in many practical applications, including international video compression standards such as H.261 and MPEG 1-2, and many motion-compensated standards conversion methods, because it leads to analytically tractable algorithms, as shown in the following.

In order to derive the spatio-temporal spectrum of such video, we first define the Fourier transform of an arbitrary spatio-temporal function as

$$S_c(F_1, F_2, F_t) = \int\int\int s_c(x_1, x_2, t)\ e^{-j2\pi(F_1 x_1 + F_2 x_2 + F_t t)} dx_1 dx_2 dt \tag{13.3}$$

and the inverse Fourier transform relationship is given by

$$s_c(x_1, x_2, t) = \int\int\int S_c(F_1, F_2, F_t)\ e^{j2\pi(F_1 x_1 + F_2 x_2 + F_t t)} dF_1 dF_2 dF_t \tag{13.4}$$

The support of $S_c(F_1, F_2, F_t)$ generally occupies the entire (F_1, F_2, F_t) space for arbitrary $s_c(x_1, x_2, t)$.

Next, we substitute (13.2) into (13.3) to obtain

$$S_c(F_1, F_2, F_t) = \int_{-\infty}^{\infty} s_0(x_1 - v_1 t, x_2 - v_2 t)\ e^{-j2\pi(F_1 x_1 + F_2 x_2 + F_t t)} dx_1 dx_2 dt \tag{13.5}$$

Making the change of variables $x_i' = x_i - v_i t, \quad$ for $i = 1, 2$, we get

$$\begin{aligned} S_c(F_1, F_2, F_t) &= \int_{-\infty}^{\infty}\int_{-\infty}^{\infty} s_0(x_1, x_2)\ e^{-j2\pi(F_1 x_1 + F_2 x_2)}\ dx_1 dx_2 \\ &\quad \cdot \int_{-\infty}^{\infty} e^{-j2\pi(F_1 v_1 + F_2 v_2 + F_t)t}\ dt \end{aligned} \tag{13.6}$$

which can be simplified as

$$S_c(F_1, F_2, F_t) = S_0(F_1, F_2) \cdot \delta(F_1 v_1 + F_2 v_2 + F_t) \tag{13.7}$$

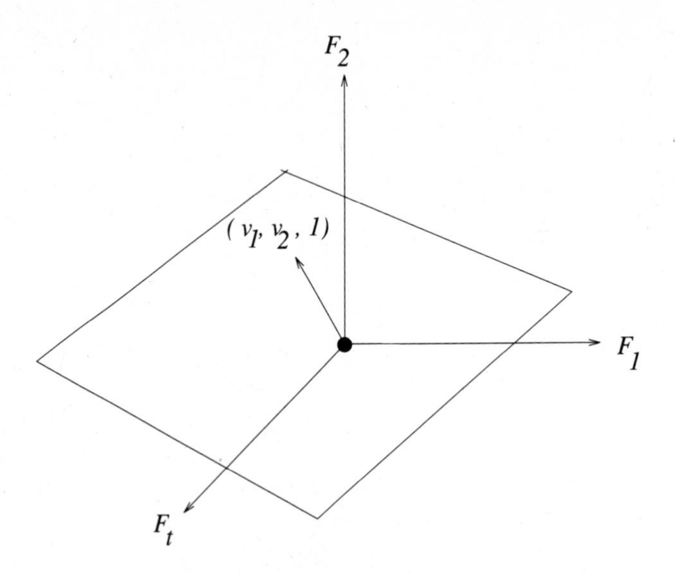

Figure 13.2: The spectral support of video with global, uniform velocity motion.

where $S_0(F_1, F_2)$ is the 2-D Fourier transform of $s_0(x_1, x_2)$ and $\delta(\cdot)$ is the 1-D Dirac delta function.

Defining $\mathbf{F} \doteq [F_1\ F_2\ F_t]^T$ and $\mathbf{v} \doteq [v_1\ v_2\ 1]^T$, the delta function in (13.7) is nonzero only when the argument $\mathbf{F}^T\mathbf{v} = 0$. The delta function thus confines the support of $S_c(F_1, F_2, F_t)$ to a plane in R^3 given by

$$F_1 v_1 + F_2 v_2 + F_t = 0 \tag{13.8}$$

which passes through the origin, and is orthogonal to the vector $\mathbf{v}$. The spectral support of video with global, uniform velocity motion is depicted in Figure 13.2.

This result is intuitively clear. Since the reference frame is sufficient to determine all future frames by the nature of the motion model, we expect that the Fourier transform of the reference frame would be sufficient to represent the entire spatio-temporal frequency spectrum of the video. The extent of the support of the continuous video spectrum $S_c(F_1, F_2, F_t)$ on the plane $\mathbf{F}^T\mathbf{v} = 0$ is determined by the support of $S_0(F_1, F_2)$. We assume that $s_0(x_1, x_2)$ is bandlimited, i.e., $S_0(F_1, F_2) = 0$ for $|F_1| > B_1$ and $|F_2| > B_2$, then clearly, $s_c(x_1, x_2, t)$ is bandlimited in the temporal variable; that is, $S_c(F_1, F_2, F_t) = 0$ for $|F_t| > B_t$, where

$$B_t = B_1 v_1 + B_2 v_2 \tag{13.9}$$

For simplicity of notation and the illustrations, in the following we base our discussions on the projection of the support of $S_c(F_1, F_2, F_t)$ into the (F_1, F_t) plane, which is defined by $F_1 v_1 + F_t = 0$ and depicted in Figure 13.3.

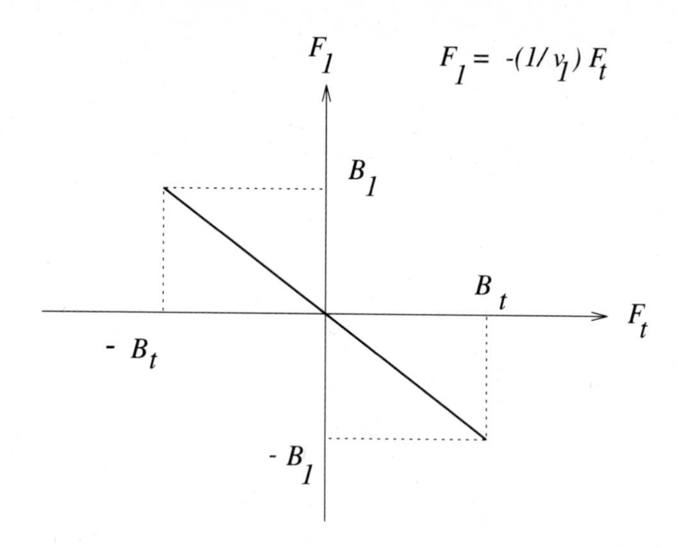

Figure 13.3: Projection of the support into the (F_1, F_t) plane.

13.1.2 Global Motion with Acceleration

In the case of motion with constant acceleration, the trajectory function $\mathbf{c}(t; x_1, x_2, t_0)$ may be approximated by a second-order polynomial. Considering global (space-invariant) motion, we then have

$$\begin{aligned} s_c(x_1, x_2, t) &= s_c(x_1 - v_1 t - \frac{1}{2}a_1 t^2, x_2 - v_2 t - \frac{1}{2}a_2 t^2, 0) \\ &= s_0(x_1 - v_1 t - \frac{1}{2}a_1 t^2, x_2 - v_2 t - \frac{1}{2}a_2 t^2) \qquad (13.10) \end{aligned}$$

where v_1, v_2, a_1, and a_2 denote the components of instantaneous velocity and acceleration, respectively.

The spatio-temporal spectrum of video containing accelerated motion has been shown to have infinite support in the temporal dimension. This is because the velocity increases without a bound as time goes to infinity. Short-time spectral analysis techniques can be used for spectral characterization of video with accelerated motion by applying a window to the video signal in the temporal dimension (see Exercise 1). Patti *et al.* [Pat 94] show that the support of the short-time spectrum (STS) of the video is concentrated along a line whose orientation changes by the instantaneous velocity vector. The interested reader is referred to [Pat 94] for further details. A similar analysis is possible for the case of spatially varying motion fields.

13.2 Sub-Nyquist Spatio-Temporal Sampling

The goal of this section is to extend the sampling theory presented in Chapter 3 (for arbitrary spatio-temporal signals) to video with global, constant-velocity motion. We assume that most video sources can be represented by such a simplified model at least on a block-by-block basis over a short period of time. We will see that the nature of the spatio-temporal frequency support of video with global, uniform (constant) velocity motion makes it possible to sample it at sub-Nyquist rates without introducing aliasing effects. That is, in most cases a full-resolution video can be recovered from samples at a sub-Nyquist rate, without any aliasing artifacts, by motion-compensated reconstruction/interpolation filtering.

13.2.1 Sampling in the Temporal Direction Only

For simplicity, we first study sampling in the temporal direction only, with the spatial variables taken as continuous variables. According to the Nyquist sampling theorem, the temporal sampling frequency should be greater than $2B_t$, where B_t denotes the temporal bandwidth. However, in the case of global constant-velocity motion, the spectral replications due to temporal sampling can come arbitrarily close to each other without overlapping, as depicted in Figure 13.4. Observe that this is true even if B_t is equal to infinity.

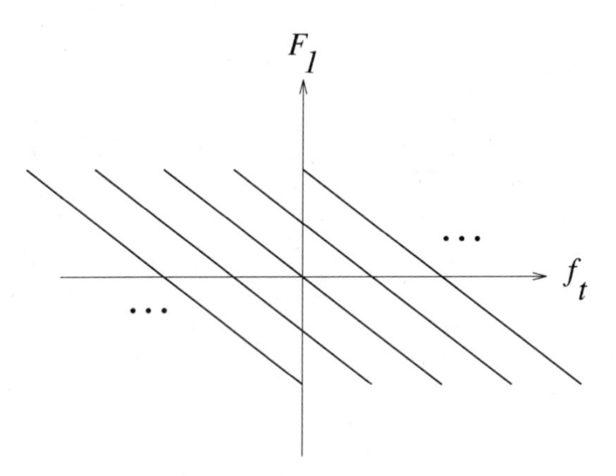

Figure 13.4: Spectral replications due to temporal sampling only.

The above frequency-domain result indeed states that just a single reference frame, $s_c(x_1, x_2, 0)$, is sufficient to represent the entire video $s_c(x_1, x_2, t)$ without loss of information, when x_1 and x_2 are continuous variables, provided that the global velocity vector (v_1, v_2) is known. Thus, the sampling interval in the temporal direction can be chosen to be arbitrarily large, since any desired frame can be generated by shifting the reference frame appropriately. Of course, hidden here is the assumption that the images are infinitely wide in the spatial coordinates, so that there are no boundary effects.

13.2.2 Sampling on a Spatio-Temporal Lattice

Recall that if we sample $s_c(x_1, x_2, t)$ on a spatio-temporal lattice Λ with the sampling matrix $\mathbf{V}$, the Fourier transform of the sampled video $S(f_1, f_2, f_t)$ is periodic with the periodicity matrix $\mathbf{U}$ defining the reciprocal lattice Λ^*. That is, we have

$$S(f_1, f_2, f_t) = \frac{1}{|detV|} \sum_{\mathbf{k}} S_c(\mathbf{F} + \mathbf{Uk}) \tag{13.11}$$

where $\mathbf{F} = (F_1, F_2, F_t)$ and $\mathbf{k} = (k_1, k_2, k_t)$. It follows from (13.7) and (13.11) that we can sample video with global, constant-velocity motion at sub-Nyquist rates in both the temporal and spatial dimensions without introducing aliasing. This can be achieved by choosing an appropriate sampling lattice such that the resulting spectral replications do not overlap for a given velocity.

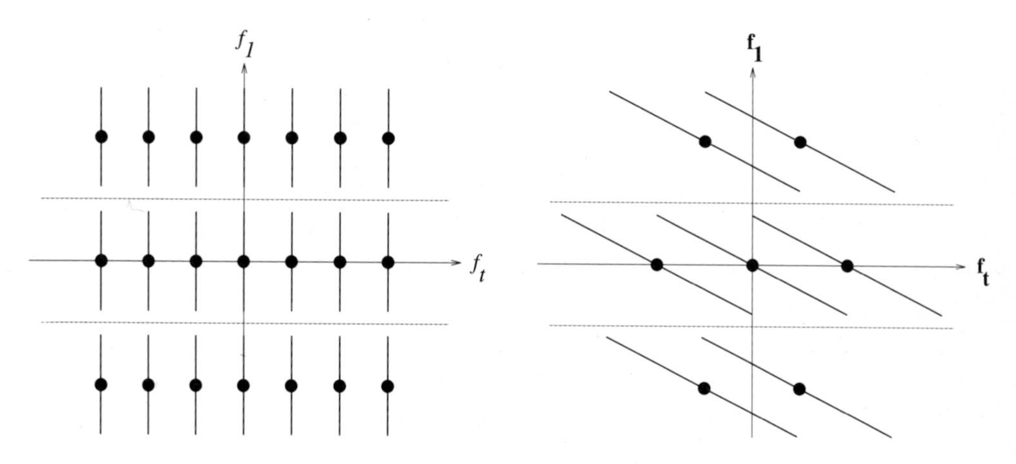

Figure 13.5: a) Progressive sampling, $v = 0$; b) interlaced sampling, $v \neq 0$.

Let's first treat spatio-temporal sampling where the spatial sampling frequencies are above the Nyquist rate, i.e., above $2B_1$ and $2B_2$, respectively. The sampled video spectrum in this case is depicted in Figure 13.5, where the spectral replications are contained within nonoverlapping strips, shown by horizontal dotted lines, in the spatial frequency direction. Inspection of Figure 13.5 indicates that the temporal sampling rate in this case is arbitrary. This is intuitively clear, because perfect reconstruction of $s_c(x_1, x_2, 0)$ is possible from $s(m, n, 0)$ as long as the spatial sampling frequencies are above the Nyquist rate. Then, the case where the spatial sampling frequencies are above the Nyquist rate reduces to that of temporal sampling only.

In the more general case, where both the spatial and temporal sampling rates are sub-Nyquist, the spectral replications will be interleaved with each other in both dimensions. Figure 13.6 shows an example of such a case, where the spectral replications do not overlap with each other; thus, reconstruction or up-conversion without aliasing is possible by means of motion-compensated filtering, as will be described subsequently. In general, reconstruction or up-conversion from spatio-temporally sub-Nyquist sampled video without aliasing is possible for all velocities except the critical velocities.

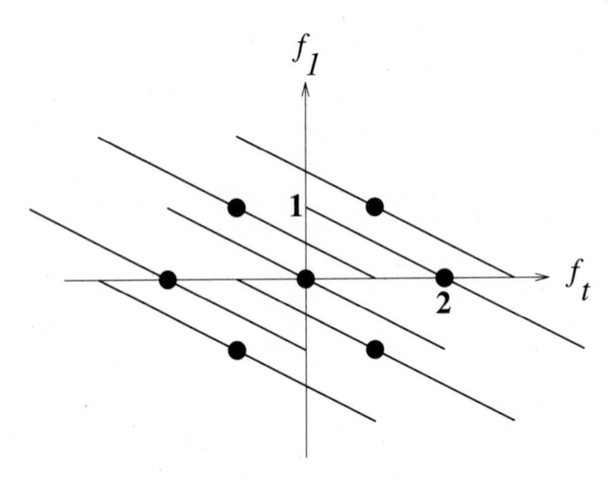

Figure 13.6: Spectral replications due to spatio-temporal sampling.

13.2.3 Critical Velocities

Aliasing results if the spectral replicas overlap with each other, which occurs when a special relationship exists between the components of the velocity vector, the sampling lattice coordinates, and the spatial/temporal bandwidth of the video. (Recall that, under our model, the temporal bandwidth is a function of the spatial bandwidth and velocity.) With reference to Figure 13.6, the spatial bandwidth and velocity specify the extent and orientation of the spatio-temporal frequency support, respectively, while the sampling lattice determines the centers of the replicas. For a fixed spatial bandwidth and sampling lattice, the velocities that would result in overlapping of the replicas are called the "critical velocities."

The critical velocities are demonstrated by means of two examples. Figure 13.7 (a) shows the case of sub-Nyquist sampling on an orthogonal grid when there is no motion. The support of the continuous video is depicted on the left side, whereas that of the sampled video is shown on the right. In this example, the sampling rate is half the Nyquist rate; hence, the supports fully overlap. Figure 13.7 (b) demonstrates sub-Nyquist sampling on a hexagonal (interlaced) grid.

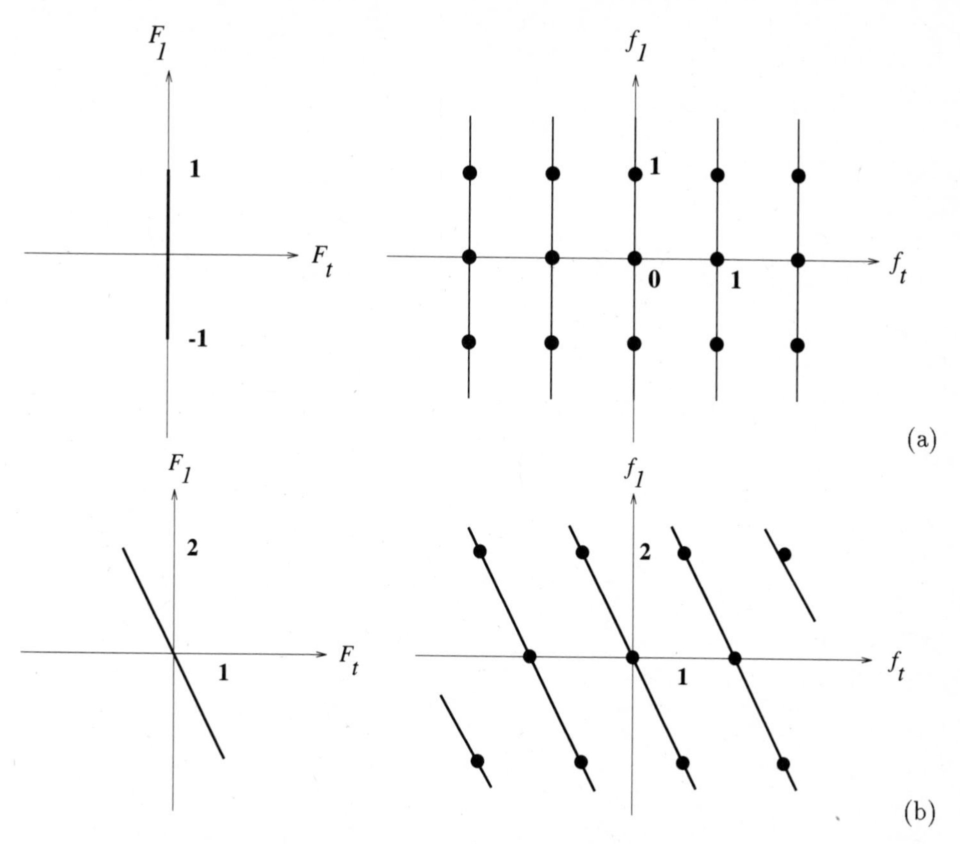

Figure 13.7: Spectral replications a) with progressive sampling, $v = 0$, and b) with interlaced sampling on a hexagonal lattice.

Here, the velocity is such that the spatio-temporal frequency support is perfectly aligned to pass through the centers of the neighboring replicas; hence, the supports again overlap.

The critical velocities can equivalently be described in the spatio-temporal domain. It is straightforward to see that if the continuous video is not spatially bandlimited, a critical velocity arises should the motion trajectory pass through a single existing spatio-temporal sample. If the continuous video is spatially bandlimited, but subsampled by a factor of $N \times N$, then a critical velocity is encountered if the motion trajectory passes through an existing sample in any $N^2 - 1$ consecutive frames. This interpretation is illustrated by means of examples. The case of a single spatial variable and orthogonal sampling is depicted in Figure 13.8 (a) for several velocities. First, observe that if there is no motion, i.e., $v = 0$, no "new" information is present in the spatio-temporal signal beyond that is present at $t = 0$; hence, $v = 0$ is a critical velocity in the case of orthogonal sub-Nyquist sampling. Furthermore,

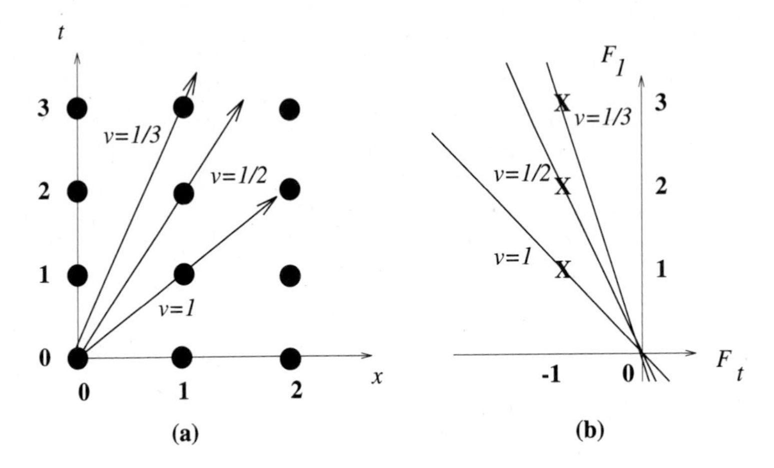

Figure 13.8: Critical velocities in one spatial dimension.

$v = 1$ and all integer velocities are also critical velocities, since the motion trajectory would pass through an existing sample in all frames. However, $v = 1/2$ and $v = 1/3$ are critical velocities only for sub-Nyquist spatial sampling by factors greater than 2 and 3, respectively; since the corresponding motion trajectories pass through existing samples every second and third frames (see Figure 13.8 (a)). Observe that the frequency support of a signal subsampled by 2 and 3 does not overlap along the lines depicted in Figure 13.8 (b) for $v = 1/2$ and $v = 1/3$, respectively. The full-resolution video signal can then be recovered by simply copying samples along the motion trajectory back onto a frame of interest (assuming impulse sampling and zero aperture-time).

An example for the case of two spatial dimensions is depicted in Figure 13.9, where the filled-in circles denote the original samples in frame k. Assuming that the original frames are spatially sub-Nyquist sampled by a factor of 2 by means of impulse sampling, the full-resolution video can be reconstructed by copying pixels along the motion trajectory within the next three frames, $k+1$, $k+2$, and $k+3$, back onto frame k, provided that the motion trajectories do not pass through existing samples in these three frames. Should the motion trajectory pass through an existing sample in these three frames, we have a critical velocity, and the full-resolution video cannot be reconstructed. This would correspond to the situation where the planar replicas in the 3-D spatio-temporal frequency domain overlap with each other.

13.3 Filtering Along Motion Trajectories

Motion-compensated filtering refers to filtering along the motion trajectories at each pixel of each frame. Motion-compensated filtering can easily be shown to be the optimum filtering strategy for both noise reduction and standards conversion in the case of linear, shift-invariant filtering of video with global, constant-velocity motion. This is because the support of the spatio-temporal frequency response of a properly motion-compensated linear, shift-invariant filter matches the support of the spatio-temporal frequency spectrum of video with global, constant-velocity motion. We further show here, using a short-time Fourier analysis, that this is also the case with global accelerated motion.

13.3.1 Arbitrary Motion Trajectories

We first define motion-compensated filtering along an arbitrary motion trajectory. Note that in the most general case, a different motion trajectory $\mathbf{c}(\tau; x_1, x_2, t)$ may be defined for each pixel (x_1, x_2) of the present frame at t, and τ ranges over the frames within the temporal support of the filter. We define the output of the filter at (x_1, x_2, t) as

$$y(x_1, x_2, t) = \mathcal{F} s_1(\tau; \mathbf{c}(\tau; x_1, x_2, t)) \tag{13.12}$$

where $s_1(\tau; \mathbf{c}(\tau; x_1, x_2, t)) = s_c(\mathbf{c}(\tau; x_1, x_2, t), \tau)$ is a 1-D signal representing the input image along the motion trajectory passing through (x_1, x_2, t), and $\mathcal{F}$ is a 1-D filter operating along the motion trajectory. The filter $\mathcal{F}$ may be a linear or nonlinear.

The resulting motion-compensated filter is shift-invariant only if all motion trajectories at each pixel of each frame are parallel to each other. This is the case when we have global, constant velocity (in time) motion. Motion-compensated filtering

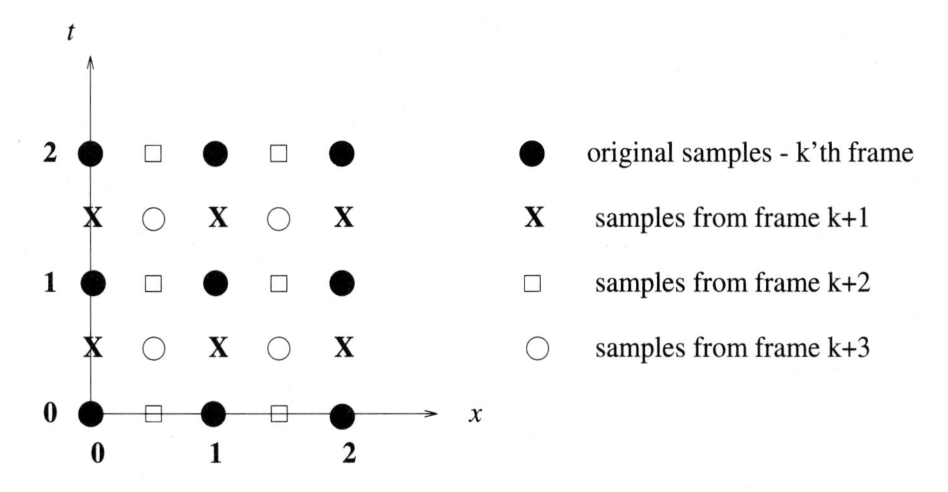

Figure 13.9: Critical velocities in two spatial dimensions.

in the case of global but accelerated (in time) motion, or in the case of any space-varying motion will result in a shift-varying filter. Obviously, a spatio-temporal frequency domain analysis of motion-compensated filtering can only be performed in the linear, shift-invariant case. This is discussed next.

13.3.2 Global Motion with Constant Velocity

A linear, shift-invariant filtering operation along a constant-velocity motion trajectory can be expressed as

$$\begin{aligned} y(x_1, x_2, t) \\ &= \int\int\int h_1(\tau)\delta(z_1 - v_1\tau, z_2 - v_2\tau)s_c(x_1 - z_1, x_2 - z_2, t-\tau)dz_1 dz_2 d\tau \\ &= \int h_1(\tau)s_c(x_1 - v_1\tau, x_2 - v_2\tau, t-\tau)d\tau \\ &= \int h_1(\tau)s_1(t-\tau; x_1, x_2, t)d\tau \end{aligned} \tag{13.13}$$

where $h_1(\tau)$ is the impulse response of the 1-D filter applied along the motion trajectory, and v_1 and v_2 are the components of the motion vector. The impulse response of this spatio-temporal filter can be expressed as

$$h(x_1, x_2, t) = h_1(t)\delta(x_1 - v_1 t, x_2 - v_2 t) \tag{13.14}$$

The frequency response of the motion-compensated filter can be found by taking the 3-D Fourier transform of (13.14)

$$\begin{aligned} H(F_1, F_2, F_t) \\ &= \int\int\int h_1(t)\delta(x_1 - v_1 t, x_2 - v_2 t)e^{-j2\pi(F_1x_1+F_2x_2+f_t t)}dF_1 dF_2 dt \\ &= \int h_1(t)e^{-j2\pi(F_1v_1+F_2v_2+f_t)t}dt \\ &= H_1(F_1v_1 + F_2v_2 + f_t) \end{aligned} \tag{13.15}$$

The shaded areas in Figure 13.10 show the support of the frequency response of the motion-compensated filter projected into the (F_1, F_t) plane for the cases when $v_1 = 0$, and when v_1 matches the input video signal. The case when $v_1 = 0$ corresponds to pure temporal filtering with no motion compensation. Inspection of the figure shows that proper motion compensation is achieved when v_1 matches the velocity of the input video signal.

13.3.3 Accelerated Motion

Analogous to the case of global constant-velocity motion, it is desirable that the support of the motion-compensated filter matches that of the signal. Since it was

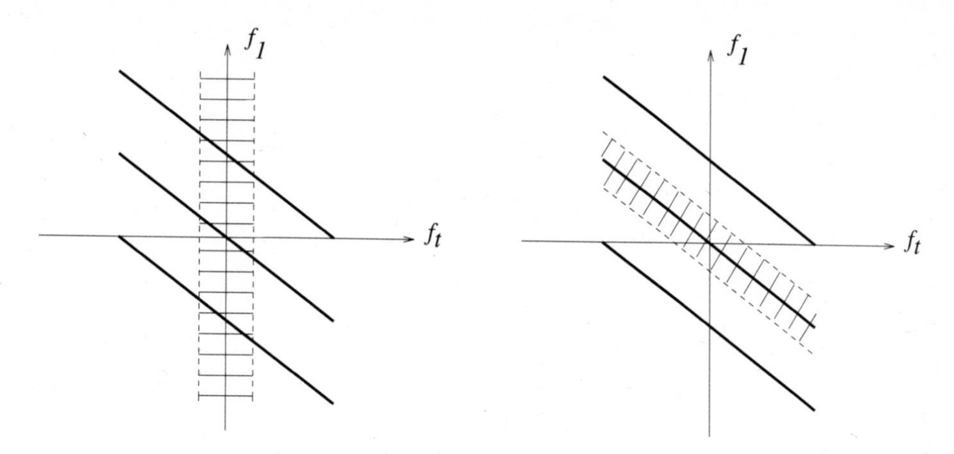

Figure 13.10: a) No motion-compensation, $v_1 = 0$; b) motion-compensated filter.

shown in [Pat 94] that in the case of accelerated motion the support of the STS (the support of the FT of the windowed signal) varies with t_0, we would expect that the motion-compensated filter in this case has a time varying frequency response, i.e., it is a linear time-varying (LTV) filter.

If we let $h_1(t)$ be a 1-D filter operating along an accelerated motion trajectory at time t_0, then the time-varying impulse response $h_{t_o}(x, t)$ of the motion-compensated spatio-temporal filter can be expressed as

$$\begin{aligned} h_{t_o}(x_1, t) &= h_1(t)\delta(x_1 - (v_1 + a_1 t_o)t - \frac{a_1}{2}t^2) \\ &= h_1(t)\delta(x_1 - v_1' t - \frac{a_1}{2}t^2) \end{aligned} \quad (13.16)$$

The support of this filter coincides with the motion trajectory for all t_o. The time-varying frequency response of this filter can be written as

$$\begin{aligned} H_{t_o}(f_1, f_t) &= \int\int h_1(t)\delta(x_1 - v_1' t - \frac{a_1}{2}t^2) e^{-j2\pi f_1 x_1} e^{-j2\pi f_t t} dx_1 dt \\ &= \int h_1(t) e^{-j2\pi f_1 (v_1' t + \frac{a_1}{2}t^2)} e^{-j2\pi f_t t} dt \end{aligned} \quad (13.17)$$

It was shown in [Pat 94] that the support of this time-varying frequency response matches that of the input video with global accelerated motion for the proper choice of the motion trajectory parameters v_1 and a_1.

The facts that the LTV filter operating along the accelerated motion trajectory has STS magnitude support characteristics matching those of the signal, and that it has zero phase in the passband, make this filter an ideal choice for the sampled video reconstruction problem in the presence of global accelerated motion. The function

$h_1(t)$, which is the filter weighting function on the motion trajectory, should be designed by specifying a prototype filter response $H_1(f_t)$ that is an ideal LPF. The cutoff frequency of $H_1(f_t)$ is chosen to avoid aliasing, and will depend on the sampling lattice used and the spectral support of the signal $S_0(f_1, f_t)$. An example of this filter design for a particular standards conversion problem is demonstrated in [Pat 94].

Note that the cases of spatially varying constant-velocity motion and spatially varying accelerated motion can similarly be analyzed within the framework of local Fourier transform and short-time local Fourier transform, respectively.

13.4 Applications

In this section, we motivate the need for motion-compensated filtering by introducing two of its major applications, noise filtering and reconstruction/interpolation of sub-Nyquist sampled video in the case of global, constant-velocity motion.

13.4.1 Motion-Compensated Noise Filtering

An observed video $g(n_1, n_2, k)$ is commonly modeled as contaminated by zero-mean, additive spatio-temporal noise, given by

$$g(n_1, n_2, k) = s(n_1, n_2, k) + v(n_1, n_2, k) \tag{13.18}$$

where $s(n_1, n_2, k)$ and $v(n_1, n_2, k)$ denote the ideal video and noise at frame k, respectively. Assuming the noise $v(n_1, n_2, k)$ is spatio-temporally white, its spectrum is flat over the cube, $-\pi \leq f_1, f_2, f_t < \pi$ in the normalized spatio-temporal frequency domain.

Recall from Section 13.1 that in the case of global, constant-velocity motion, the spatio-temporal spectral support of digital video is limited to a plane

$$f_1 v_1 + f_2 v_2 + f_t = 0 \tag{13.19}$$

Hence, in theory, a properly motion-compensated filter that is, a filter whose spatio-temporal frequency support matches this plane can effectively suppress all of the noise energy outside the plane without any spatio-temporal blurring. Equivalently, in the spatio-temporal domain, the effect of zero-mean white noise can be perfectly eliminated provided that we filter along the correct motion trajectory, that is, the motion trajectory contains different noisy realizations of the same "ideal" gray level. We elaborate on various methods for spatio-temporal noise filtering, including the motion-compensated methods, in Chapter 14.

13.4.2 Motion-Compensated Reconstruction Filtering

Here, we extend the theory of ideal reconstruction filtering presented in Chapter 3 to ideal motion-compensated reconstruction of sub-Nyquist sampled video under

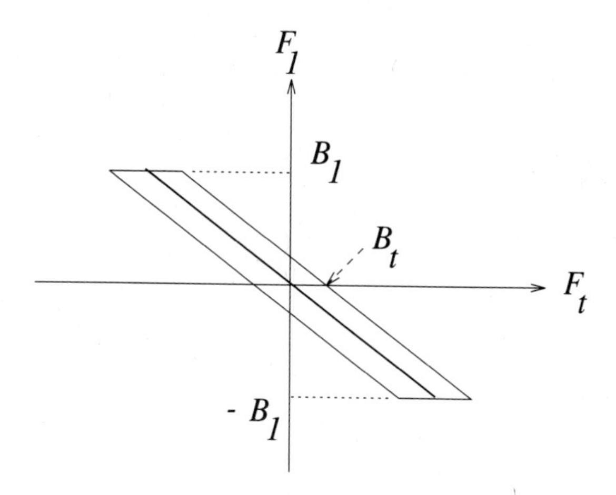

Figure 13.11: The support of a motion-compensated ideal low-pass filter.

certain assumptions. Because the sampling structure conversion problem can be posed as a reconstruction problem followed by resampling on the desired sampling structure, this theory also applies to motion-compensated standards conversion, which will be discussed in Chapter 16.

Recall that an ideal reconstruction filter to reconstruct a continuous video from spatio-temporal samples is an ideal spatio-temporal low-pass filter whose passband is limited to the unit cell of the reciprocal sampling lattice. In analogy, we can define a motion-compensated ideal low-pass filter as

$$H(F_1, F_2, F_t) = \begin{cases} 1 & F_1 < |B_1|, \quad F_2 < |B_2|, \text{ and} \\ & -v_1 F_1 - v_2 F_2 - B_t < F_t < -v_1 F_1 - v_2 F_2 + B_t \\ 0 & \text{otherwise} \end{cases} \tag{13.20}$$

where B_1, B_2 and B_t denote the cutoff frequencies in the respective directions. The support of the motion-compensated ideal low-pass filter is depicted in Figure 13.11.

Observe that the support of the motion-compensated ideal low-pass filter can be matched to that of video with global, constant-velocity motion by choosing the appropriate velocity v_1. Then it can be used to reconstruct a continuous video, without aliasing, from sub-Nyquist sampled video with global, constant-velocity motion. The spatio-temporal spectrum of such sub-Nyquist sample video is shown in Figure 13.6. Because the passband of the filter can be oriented in the direction of the motion, all other replicas will be effectively eliminated, except for the case of critical velocities. Of course, reconstruction at full resolution is only possible assuming impulse sampling with no prior anti-alias filtering.

It is interesting to note that although each frame is spatially sub-Nyquist sampled, and thus would contain spatial aliasing artifacts if individually reconstructed, a number of frames collectively may contain sufficient information to reconstruct

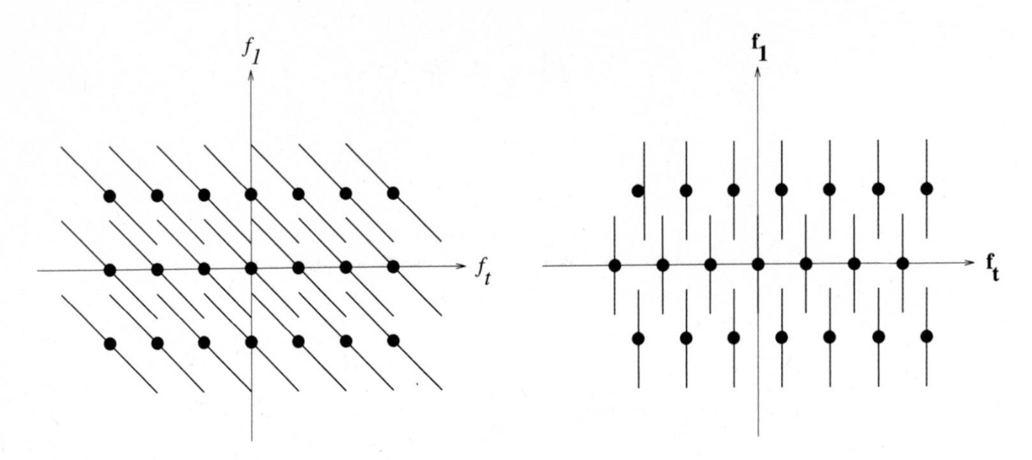

Figure 13.12: (a) Progressive sampling, $v \neq 0$; b) interlaced sampling, $v = 0$.

$s_c(x_1, x_2, t)$. The "new" information may be provided by

- the subpixel valued motion vectors,
- the sampling lattice, or
- a combination of the two.

Two examples in which "new" information is present in consecutive frames are shown in Figure 13.12. Note that in the first example the new information is provided by the interframe motion, whereas in the second the new information is provided by the sampling lattice. In theory, perfect reconstruction is possible from multiple frames even if they are not spatially bandlimited. Of course, this would require an infinite number of sampled frames. Various algorithms for standards conversion and high-resolution reconstruction will be discussed in Chapters 16 and 17.

13.5 Exercises

1. This problem extends the spectral analysis of motion compensation to video with global, accelerated motion. Let the short-time spectrum (STS) of video be defined by

$$\begin{aligned} S_{t_0}^{(h)}(F_1, F_2, F_t) &= \int\int\int s_c(x_1, x_2, t) h(t - t_0) \\ &\qquad \exp\{-j2\pi(F_1 x_1 + F_2 x_2 + F_t t)\}\, dx_1 dx_2 dt \end{aligned}$$

where $h(t)$ denotes the temporal analysis window. Show that the STS of a video signal modeled by (13.10) can be expressed as

$$\begin{aligned} S_{t_0}^{(h)}(F_1, F_2, F_t) &= S_0(F_1, F_2) \\ &\quad [H(F_t + F_1 v_1 + F_2 v_2) \exp\{-j2\pi(F_t + F_1 v_1 + F_2 v_2) t_0\}] *_{F_t} Q(F_1, F_2, F_t) \end{aligned}$$

where

$$Q(F_1, F_2, F_t) = \mathcal{F}_t \left\{ \exp\{-j2\pi(F_1 \frac{a_1 t^2}{2} + F_2 \frac{a_2 t^2}{2})\} \right\}$$

$*_{F_t}$ denotes 1-D convolution over the variable F_t, and $\mathcal{F}_t$ denotes 1-D Fourier transform with respect to t.

Show that this expression reduces to (13.7) when $a_1 = a_2 = 0$, and $h(t) = 1$ for all t.

2. Consider the case of a space-varying 2-D motion, which can be modeled locally by a piecewise quadratic flow model. Extend the STS analysis framework developed in Exercise 1 to this case by employing a spatio-temporal analysis window.

3. Can you conclude that the best motion-compensated filtering strategy is filtering along the motion trajectory, even in the case of an arbitrary motion trajectory?

4. Suppose we have a video with global, constant-velocity motion, modeled by (13.2). Show that the motion-compensated ideal low-pass filter (13.20), tuned to the nominal velocity (V_1, V_2), passes the video if the actual velocity (v_1, v_2) satisfies

$$B_1|v_1 - V_1| + B_2|v_2 - V_2| < B_t$$

Plot the velocity passband (the range of velocities for which the filter allows video to pass with no attenuation) in the velocity plane (v_1, v_2).

5. Comment on the tradeoffs between the width of the passband versus the number of framestores required to implement a motion-compensated low-pass filter, and the ability to reject close replicas versus motion estimation errors.

Bibliography

[Dub 92] E. Dubois, "Motion-compensated filtering of time-varying images," *Multidim. Syst. Sign. Proc.*, vol. 3, pp. 211–239, 1992.

[Gir 93] B. Girod, "Motion compensation: Visual aspects, accuracy, and fundamental limits," in *Motion Analysis and Image Sequence Processing*, M. I. Sezan and R. L. Lagendijk, eds., Norwell, MA: Kluwer, 1993.

[Pat 94] A. J. Patti, M. I. Sezan, and A. M. Tekalp, "Digital video standards conversion in the presence of accelerated motion," *Signal Proc.: Image Communication*, vol. 6, pp. 213–227, June 1994.

Chapter 14

NOISE FILTERING

Image recording systems are not perfect. As a result, all images are contaminated by noise to various levels, which may or may not be visible. Sources of noise include electronic noise, photon noise, film-grain noise, and quantization noise. In video scanned from motion picture film, streaks due to possible scratches on film are generally modeled by impulsive noise. Speckle noise is common in radar image sequences and biomedical cine-ultrasound sequences. The level of noise in an image is generally specified in terms of a signal-to-noise ratio (SNR), which may be defined in a number of ways. If the SNR is below a certain level, the noise becomes visible as a pattern of graininess, resulting in a degradation of the image quality. Even if the noise may not be perceived at full-speed video due to the temporal masking effect of the eye, it often leads to unacceptable "freeze-frames." In addition to degradation of visual quality, the noise pattern may mask important image detail and increase the entropy of the image, hindering effective compression.

The SNR of video can be enhanced by spatio-temporal filtering, which is based on some sort of modeling of the image and noise characteristics. It should be rather obvious that exact separation of variations in the recorded image intensity which are due to noise from genuine image detail is impossible. To this effect, both the image and noise will be characterized by statistical models. In general, the noise can be modeled as additive or multiplicative, signal-dependent or signal-independent, and white or colored. For example, photon noise and film-grain are signal-dependent, whereas CCD sensor noise and quantization noise are usually modeled as white, Gaussian distributed, and signal-independent. Ghosts in TV images can also be modeled as signal-dependent noise. In this chapter, we will assume a simple additive noise model given by

$$g(n_1, n_2, k) = s(n_1, n_2, k) + v(n_1, n_2, k) \tag{14.1}$$

where $s(n_1, n_2, k)$ and $v(n_1, n_2, k)$ denote the ideal video and noise at frame k, respectively.

Noise filters can be classified as spatial (intraframe) filters and spatio-temporal (interframe) filters. Spatial filters, covered in Section 14.1, may be used to filter each frame independently. However, in intraframe filtering, there is a tradeoff between noise reduction and spatial blurring of the image detail. Spatio-temporal filters are 3-D filters with two spatial coordinates and one temporal coordinate [Sam 85, Alp 90, Arc 91, Kat 91]. They utilize not only the spatial correlations, but also the temporal correlations between the frames. We classify spatio-temporal noise filters as motion-adaptive and motion-compensated filters, which are treated in Sections 14.2 and 14.3, respectively. Motion-adaptive filtering utilizes a motion-detection scheme, but does not require explicit estimation of the interframe motion vectors. This is distinct from motion-compensated filtering schemes, which operate along motion trajectories, thus requiring exact knowledge of the trajectories at each pixel.

14.1 Intraframe Filtering

Spatial filtering for image enhancement and restoration has received significant attention in the image processing literature over the last three decades. Image enhancement generally refers to eliminating the noise as well as sharpening the image detail. In this section, we provide an overview of noise filtering techniques. Many other still-frame image enhancement techniques, including contrast adjustment by histogram equalization and unsharp masking, will not be treated in this book. The reader is referred to [Lim 90] for a detailed presentation of these techniques and edge detection methods. The discussion of intraframe image restoration methods for deblurring of blurred images is deferred until Chapter 15. We can classify intraframe noise filters into three categories: (i) linear, shift-invariant (LSI) filters, such as the weighted averaging filters and linear minimum mean square error (LMMSE) filters, also known as Wiener filters, (ii) nonlinear filters, such as median filters and other order statistics filters, and (iii) adaptive filters, such as directional smoothing filters and local (space-varying) LMMSE filters.

LSI noise filters can be designed and analyzed using frequency domain concepts, and are easier to implement. However, they present a tradeoff between noise reduction and sharpness of the image detail. This is because any LSI noise reduction filter is essentially a low-pass filter which eliminates high-frequency noise. Images usually feature a low-pass Fourier spectra, and the Fourier spectrum of the noise is flat, assuming white noise. As illustrated in Figure 14.1, an LSI noise reduction filter attenuates those frequencies where the noise power exceeds the signal power. As a result, the high-frequency content of the image is also attenuated, causing blurring. In the following, we study the optimal LSI filtering, and two approaches for adaptive filtering where the filter characteristics vary with the spatial coordinates to preserve edges and other image detail.

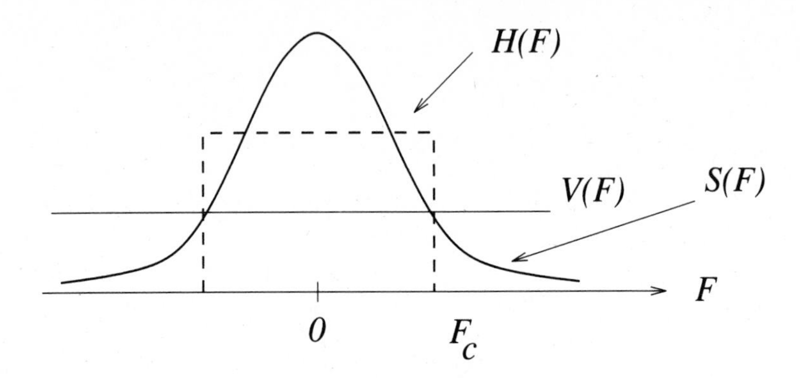

Figure 14.1: Tradeoff between noise reduction and blurring.

14.1.1 LMMSE Filtering

The LMMSE filter gives the minimum mean square error estimate of the ideal image among all linear filters. That is, it is the optimal linear filter in the minimum mean square error sense. Loosely speaking, it determines the best cutoff frequency for low-pass filtering based on the statistics of the ideal image and the noise. In the following, we present the derivation of both infinite impulse response (IIR) and finite impulse response (FIR) LLMSE filters, using the principle of orthogonality. To this effect, we assume that the image and noise are wide-sense stationary; that is, their means are constant and their correlation functions are shift-invariant. The mean of the image and noise are taken to be zero without loss of generality, since any nonzero mean can be removed prior to filtering.

The Unrealizable Wiener Filter

The input-output relationship for the IIR LMMSE filter can be expressed in the form of a convolution, given by

$$\hat{s}(n_1, n_2) = \sum_{i_1=-\infty}^{\infty} \sum_{i_2=-\infty}^{\infty} h(i_1, i_2) g(n_1 - i_1, n_2 - i_2) \tag{14.2}$$

where $g(n_1, n_2)$ is the noisy image, $\hat{s}(n_1, n_2)$ denotes the LMMSE estimate of the ideal image $s(n_1, n_2)$, and $h(n_1, n_2)$ is the impulse response of the filter.

The principle of orthogonality states that the estimation error, $s(n_1, n_2) - \hat{s}(n_1, n_2)$, at each pixel should be orthogonal to every sample of the observed image, which can be expressed as

$$< [s(n_1, n_2) - \hat{s}(n_1, n_2)], g(k_1, k_2) > =$$
$$E\{[s(n_1, n_2) - \hat{s}(n_1, n_2)] g(k_1, k_2)\} = 0, \quad \forall \ (n_1, n_2) \text{ and } (k_1, k_2) \tag{14.3}$$

Here the inner product $< \cdot, \cdot >$ is defined in terms of the expectation operator $E\{\cdot\}$. Thus, orthogonality is the same as uncorrelatedness.

Substituting (14.2) into (14.3) yields

$$E\{\sum_{i_1}\sum_{i_2} h(i_1, i_2) g(n_1 - i_1, n_2 - i_2) g(k_1, k_2)\} = E\{s(n_1, n_2) g(k_1, k_2)\}, \quad \forall\ (n_1, n_2) \text{ and } (k_1, k_2) \tag{14.4}$$

Simplifying this expression, we obtain

$$\sum_{i_1}\sum_{i_2} h(i_1, i_2) R_{gg}(n_1 - i_1 - k_1, n_2 - i_2 - k_2) = R_{sg}(n_1 - k_1, n_2 - k_2), \quad \forall\ (n_1, n_2) \text{ and } (k_1, k_2) \tag{14.5}$$

where $R_{gg}(\cdot)$ is the autocorrelation function of the observations, and $R_{sg}(\cdot)$ is the cross-correlation between the ideal image and the observed image. The double summation can be written as a 2-D convolution as

$$h(n_1, n_2) ** R_{gg}(n_1, n_2) = R_{sg}(n_1, n_2) \tag{14.6}$$

which is called the discrete Wiener-Hopf equation. The expression (14.6) defines the impulse response of the noncausal, IIR Wiener filter, also known as the unrealizable Wiener filter. This filter is unrealizable because an infinite time delay is required to compute an output sample.

We can obtain the frequency response of the unrealizable Wiener filter by taking the 2-D Fourier transform of both sides of (14.6)

$$H(f_1, f_2) = \frac{P_{sg}(f_1, f_2)}{P_{gg}(f_1, f_2)} \tag{14.7}$$

where $P_{gg}(\cdot)$ and $P_{sg}(\cdot)$ denote the auto and cross power spectra, respectively. The derivation, up to this point, is quite general, in the sense that it only uses the linear filter constraint (14.2), and the principle of orthogonality (14.3), and is independent of the particular estimation problem. However, the derivation of the expressions $P_{gg}(\cdot)$ and $P_{sg}(\cdot)$ requires an observation model, which is problem-dependent.

For the noise-filtering problem, the auto and cross power spectra in (14.7) can be derived from the observation model (14.1) as

$$\begin{aligned} R_{sg}(n_1, n_2) &= E\{s(i_1, i_2) g(i_1 - n_1, i_2 - n_2)\} \\ &= E\{s(i_1, i_2) s(i_1 - n_1, i_2 - n_2)\} + \\ &\qquad E\{s(i_1, i_2) v(i_1 - n_1, i_2 - n_2)\} \\ &= R_{ss}(n_1, n_2) \end{aligned} \tag{14.8}$$

and

$$\begin{aligned} R_{gg}(n_1, n_2) &= E\{(s(i_1, i_2) + v(i_1, i_2)) \\ &\qquad (s(i_1 - n_1, i_2 - n_2) + v(i_1 - n_1, i_2 - n_2))\} \\ &= R_{ss}(n_1, n_2) + R_{vv}(n_1, n_2) \end{aligned} \tag{14.9}$$

where we assumed that the image and noise are uncorrelated. The auto and cross power spectra can be obtained by taking the Fourier transform of the respective correlation functions, and the filter frequency response becomes

$$H(f_1, f_2) = \frac{P_{ss}(f_1, f_2)}{P_{ss}(f_1, f_2) + P_{vv}(f_1, f_2)} \tag{14.10}$$

We observe that the unrealizable Wiener noise filter is a low-pass-type filter, since the image power spectrum diminishes at high frequencies, which implies that the filter frequency response goes to zero at those frequencies, whereas at low frequencies the noise power is negligible compared with the image power so that the frequency response of the filter is approximately equal to one.

A realizable approximation to the filter (14.10) can be obtained by a technique known as *frequency-sampling design*, where the filter frequency response $H(f_1, f_2)$ is sampled in the frequency domain using $N_1 \times N_2$ samples. The samples can be efficiently computed using a $N_1 \times N_2$ fast Fourier transform (FFT). The frequency sampling design is equivalent to approximating the IIR filter with the impulse response $h(n_1, n_2)$ with an $N_1 \times N_2$ FIR filter with the impulse response $\tilde{h}(n_1, n_2)$, given by

$$\tilde{h}(n_1, n_2) = \sum_{r_1=-\infty}^{\infty} \sum_{r_2=-\infty}^{\infty} h(n_1 - r_1 N_1, n_2 - r_2 N_2) \tag{14.11}$$

Observe that the frequency-sampling design method may suffer from aliasing effects. In practice, this effect is negligible, provided that N_1 and N_2 are reasonably large, for example $N_1 = N_2 = 256$.

FIR LMMSE Filter

Alternatively, we can pose the problem to design the optimal linear, shift-invariant FIR filter. Assuming that the observed image and the estimate are $N \times N$ arrays, the FIR LMMSE filter can be expressed in vector-matrix form as

$$\hat{\mathbf{s}} = \mathbf{H}\mathbf{g} \tag{14.12}$$

where $\hat{\mathbf{s}}$ and $\mathbf{g}$ are $N^2 \times 1$ vectors formed by lexicographic ordering of the estimated and observed image pixels, and $\mathbf{H}$ is an $N^2 \times N^2$ matrix operator formed by the coefficients of the FIR filter impulse response.

The principle of orthogonality can be stated in vector-matrix form as

$$E\{(\mathbf{s} - \hat{\mathbf{s}})\mathbf{g}^T\} = \mathbf{0} \quad \text{(zero matrix)} \tag{14.13}$$

which states that every element of $(\mathbf{s} - \hat{\mathbf{s}})$ is uncorrelated with every element of $\mathbf{g}$.

Susbtituting (14.12) into (14.13), we obtain

$$E\{(\mathbf{s} - \mathbf{H}\mathbf{g})\mathbf{g}^T\} = \mathbf{0} \tag{14.14}$$

which can be simplified as

$$E\{\mathbf{s}\mathbf{g}^T\} = \mathbf{H}E\{\mathbf{g}\mathbf{g}^T\} \tag{14.15}$$

Then the FIR LMMSE filter operator $\mathbf{H}$ can be obtained as

$$\mathbf{H} = \mathbf{R}_{sg}\mathbf{R}_{gg}^{-1} \tag{14.16}$$

where $\mathbf{R}_{gg}$ is the auto-correlation matrix of the observed image, and $\mathbf{R}_{sg}$ is the cross-correlation matrix of the ideal and observed images.

Given the observation model, we can easily show, as in the derivation of the IIR filter, that $\mathbf{R}_{sg} = \mathbf{R}_{ss}$, and $\mathbf{R}_{gg} = \mathbf{R}_{ss} + \mathbf{R}_{vv}$, where $\mathbf{R}_{ss}$ and $\mathbf{R}_{vv}$ are the auto-correlation matrices of the ideal image and the observation noise, respectively. Then the filter operator becomes

$$\mathbf{H} = \mathbf{R}_{ss}\left[\mathbf{R}_{ss} + \mathbf{R}_{vv}\right]^{-1} \tag{14.17}$$

Observe that the implementation of the filter (14.17) requires inversion of a $N^2 \times N^2$ matrix. For a typical image, e.g., $N = 256$, this is a formidable task.

However, assuming that the image and the noise are wide-sense stationary, that is, they have constant mean vectors (taken as zero without loss of generality) and spatially invariant correlation matrices, the matrices $\mathbf{R}_{ss}$ and $\mathbf{R}_{vv}$ are block-Toeplitz. It is common to approximate block-Toeplitz matrices by block-circulant ones, which can be diagonalized through the 2-D DFT operation [Gon 92]. The reader can readily show that the resulting frequency-domain FIR filter expression is identical to that obtained by sampling the frequencies (f_1, f_2) in (14.10).

14.1.2 Adaptive (Local) LMMSE Filtering

Linear shift-invariant filters limit our ability to separate genuine image variations from noise, because they are based on wide-sense stationary (homogeneous) image models. As a compromise between the complexity of the implementation of (14.17) and desire to preserve important image detail, Kuan *et al.* [Kua 85] proposed a simple space-varying image model, where the local image characteristics were captured in a space-varying mean, and the residual after removing the local mean was modeled by a white Gaussian process. In this section, a spatially adaptive LMMSE estimator that is based on this model is presented. The resulting filter is easy to implement, yet avoids excessive blurring in the vicinity of edges and other image detail.

We define the ideal residual image as

$$r_s(n_1, n_2) = s(n_1, n_2) - \mu_s(n_1, n_2) \tag{14.18}$$

where $\mu_s(n_1, n_2)$ is a spatially varying mean function. The residual image is modeled by a white process; that is, its correlation matrix, given by

$$R_{rr}(n_1, n_2) = \sigma_s^2(n_1, n_2)\delta(n_1, n_2) \tag{14.19}$$

is diagonal. Note that the variance, $\sigma_s^2(n_1, n_2)$ of the residual is allowed to change from pixel to pixel.

It follows that the residual of the observed image, defined as

$$r_g(n_1, n_2) = g(n_1, n_2) - \mu_g(n_1, n_2) \quad (14.20)$$

where $\mu_g(n_1, n_2)$ is the local mean of the observed image, is also zero-mean and white, because the noise is assumed to be zero-mean and white. Furthermore, from (14.1),

$$\mu_g(n_1, n_2) = \mu_s(n_1, n_2) \quad (14.21)$$

since the noise is zero-mean.

Applying the filter (14.17) to the residual observation vector, $\mathbf{r}_g = \mathbf{g} - \mu_g$ which is a zero-mean, wide-sense stationary image, we have

$$\begin{aligned} \hat{\mathbf{r}}_s - \mu_s &= \mathbf{H}\mathbf{r}_g \\ &= \mathbf{R}_{rr}\left[\mathbf{R}_{rr} + \mathbf{R}_{vv}\right]^{-1}\left[\mathbf{g} - \mu_g\right] \end{aligned} \quad (14.22)$$

The matrix $\mathbf{R}_{rr}$ is diagonal because $r_s(n_1, n_2)$ is white. Then the above vector-matrix expression simplifies to the scalar form

$$\hat{s}(n_1, n_2) = \mu_g(n_1, n_2) + \frac{\sigma_s^2(n_1, n_2)}{\sigma_s^2(n_1, n_2) + \sigma_v^2}[g(n_1, n_2) - \mu_g(n_1, n_2)] \quad (14.23)$$

which has a predictor-corrector structure.

Note that the adaptive LMMSE filter (14.23) requires the estimation of the mean $\mu_g(n_1, n_2)$ and the variance $\sigma_s^2(n_1, n_2)$ at each pixel. We estimate the local sample mean and sample variance over an $M \times M$ window, $\mathcal{W}$, as

$$\mu_s = \mu_g = \sum_{(n_1, n_2) \in \mathcal{W}} g(n_1, n_2) \quad (14.24)$$

and

$$\sigma_g^2 = \sum_{(n_1, n_2) \in \mathcal{W}} [g(n_1, n_2) - \mu_g(n_1, n_2)]^2 \quad (14.25)$$

We have

$$\sigma_s^2 = \max\{\sigma_g^2 - \sigma_v^2, 0\} \quad (14.26)$$

so that σ_s^2 is always nonnegative. It is assumed that the variance of the noise σ_v^2 is either known or can be estimated from a uniform image region.

It is interesting to note that when σ_s^2 is small, the second term in (14.23) is negligible, and the adaptive LMMSE filter approaches a direct averaging filter. On the other hand, when σ_s^2 is large compared with σ_v^2, the filter is turned off. Because large σ_s^2 usually indicates presence of edges, the adaptive LMMSE filter preserves edges by effectively turning the filter off across edges. Consequently, some noise is left in the vicinity of edges which may be visually disturbing.

14.1.3 Directional Filtering

An alternative approach for edge-preserving filtering is the directional filtering approach, where we filter along the edges, but not across them. The directional filtering approach may in fact be superior to adaptive LMMSE filtering, since noise around the edges can effectively be eliminated by filtering along the edges, as opposed to turning the filter off in the neighborhood of edges.

In directional filtering, possible edge orientations are generally quantized into four, 0 deg, 45 deg, 90 deg, and 135 deg; and five FIR filter kernels, one for each orientation and one for nonedge regions, are defined. The supports of the edge-oriented FIR filters are depicted in Figure 14.2. In general, there exist two approaches for directional filtering: i) select the most uniform support out of the five at each pixel according to a criterion of uniformity or by edge-detection [Dav 75], and ii) apply an edge-adaptive filter within each kernel at each pixel, and cascade the results. They are discussed in more detail in the following.

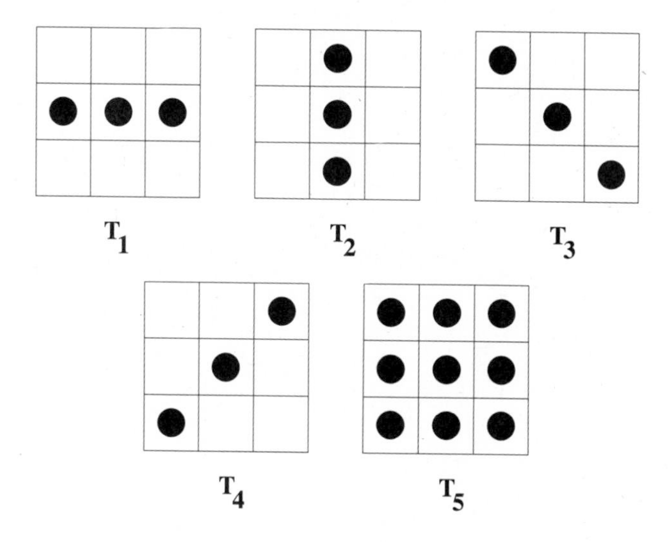

Figure 14.2: Directional filtering kernels.

i) Method I: The variance of pixels within each support can be used as a selection criterion. If the variance of the pixels within the nonedge kernel is more than a predetermined threshold, we decide that an edge is present at that pixel. Then the edge-kernel with the lowest variance indicates the most likely edge orientation at that pixel. Filtering can be performed by averaging the pixels in the direction of the smallest variance. Filtering along the edge avoids spatial blurring to a large extent.

ii) Method II: Alternatively, we can use a spatially adaptive filter, such as the local LMMSE filter, within each of the five supports at each pixel. Recall that the local LMMSE filter is effectively off within those supports with a high variance, and approaches direct averaging as the signal variance goes to zero. Thus, cascading the five filters as [Cha 85]

$$T = T_1 T_2 T_3 T_4 T_5 \tag{14.27}$$

where T_i is the local LMMSE filter applied over the respective kernel, we expect that effective filtering is performed only over those kernels with a small variance. This method avoids the support selection process by using adaptive filtering within each support.

Directional filtering, using either method, usually offers satisfactory noise reduction around edges, since at least one of the filters should be active at every pixel. In contrast, the local LMMSE filter performs very little filtering in windows with an edge, leaving a significant amount of noise around edges.

14.1.4 Median and Weighted Median Filtering

Median filtering is a nonlinear operation that is implicitly edge-adaptive. The output $\hat{s}(n_1, n_2)$ of the median filter is given by the median of the pixels within the support of the filter, that is,

$$\hat{s}(n_1, n_2) \quad = \quad \text{Med}\left\{g(i_1, i_2)\right\}, \qquad \text{for} \quad (i_1, i_2) \in \mathcal{B}_{(n_1, n_2)} \tag{14.28}$$

where $\mathcal{B}_{(n_1,n_2)}$ is the filter support centered about (n_1, n_2), and "Med" denotes the median operation. The median filter is edge-preserving because the median operation easily rejects outliers, avoiding blurring across edges [Arc 86]. Fast algorithms for one- and two-dimensional (separable) median filtering exist for real-time implementations [Ata 80].

In median filtering each sample in the filter support is given an equal emphasis. The weighted median filter is an extension of the median filter where each sample (i_1, i_2) is assigned a weight w_{i_1,i_2}. The weighting is achieved by replicating the (i_1, i_2)th sample w_{i_1,i_2} times. Then, the output of the weighted median filter is given by

$$\hat{s}(n_1, n_2) \quad = \quad \text{Med}\left\{w_{i_1,i_2} \diamond g(i_1, i_2)\right\}, \qquad \text{for} \quad (i_1, i_2) \in \mathcal{B}_{(n_1, n_2)} \tag{14.29}$$

where $\diamond$ is the replication operator. The properties of the filter vary depending on how the weights are assigned. The reader is referred to [Arc 91] and the references therein for other generalizations of median filtering, including order statistic and multistage order statistic filters.

14.2 Motion-Adaptive Filtering

In video processing, interframe noise filtering may provide several advantages over intraframe filtering. A special class of interframe filters is temporal filters, which employ 1-D filtering in the temporal direction [McM 79, Den 80, Hua 81, Dub 84, Mar 85, Sez 91, Rei 91, Boy 92]. In principle, temporal filters can avoid spatial and temporal blurring. In this section, we discuss motion-adaptive noise filters where there is no explicit motion estimation. The filters that are presented here are applied over a fixed spatio-temporal support at each pixel. We start with direct filtering where there is no adaptivity at all, or the adaptivity is implicit in the filter design. Next, we discuss filter structures, where some coefficients vary as a function of a so-called "motion-detection" signal.

14.2.1 Direct Filtering

The simplest form of direct filtering is frame averaging, where we average pixels occupying the same spatial coordinates in consecutive frames. Direct temporal averaging is well-suited to stationary parts of the image, because averaging multiple observations of essentially the same pixel in different frames eliminates noise while resulting in no loss of spatial image resolution. It is well known that direct averaging in this case corresponds to maximum likelihood estimation under the assumption of white, Gaussian noise, and reduces the variance of the noise by a factor of N, where N is the number of samples. It follows that in purely temporal filtering a large number of frames may be needed for effective noise reduction, which requires a large number of frame stores. Spatio-temporal filtering provides a compromise between the number of frame stores needed for effective noise reduction and the amount of spatial blurring introduced.

Although direct temporal averaging serves well for stationary image regions, it may lead to smearing and chrominance separation in the moving areas. Direct temporal averaging causes temporal blurring in the same way that direct spatial averaging leads to spatial blurring, that is, when there are sudden temporal variations. These degradations may be avoided if the filter makes use of interframe motion information. A fundamental question is how to distinguish the temporal variations due to motion from those due to noise. Motion-adaptive filtering is the temporal counterpart of edge-preserving spatial filtering in that frame-to-frame motion gives rise to temporal edges. It follows that spatio-temporal noise filters that adapt to motion can be obtained by using structures similar to those of the edge-preserving filters. Examples of such filters include directional filters and order statistic filters, including median, weighted median, and multistage median filters [Arc 91].

For instance, Martinez and Lim [Mar 85] proposed a cascade of five one-dimensional finite impulse response (FIR) linear minimum mean square error (LMMSE) estimators over a set of five hypothesized motion trajectories at each pixel. These trajectories correspond to no motion, motion in the $+x_1$ direction, motion in the $-x_1$ direction, motion in the $+x_2$ direction, and motion in the $-x_2$ direction. Due to the adaptive nature of the LMMSE estimator, filtering is effective only along hypothesized trajectories that are close to actual ones. This approach has been reported to be successful in cases where one of the hypothesized motion trajectories is close to the actual one.

14.2.2 Motion-Detection Based Filtering

In this approach, the selected filter structure has parameters which can be tuned according to a motion detection signal, such as the frame difference. Both FIR and IIR filter structures can be employed in motion adaptive filtering. A simple example of a motion-adaptive FIR filter may be given by

$$\hat{s}(n_1, n_2, k) = (1-\gamma)\, g(n_1, n_2, k) + \gamma\, g(n_1, n_2, k-1) \tag{14.30}$$

and that of an IIR filter by

$$\hat{s}(n_1, n_2, k) = (1-\gamma)\, g(n_1, n_2, k) + \gamma\, \hat{s}(n_1, n_2, k-1) \tag{14.31}$$

where

$$\gamma \doteq \max\{0, \frac{1}{2} - \alpha |g(n_1, n_2, k) - g(n_1, n_2, k-1)|\}$$

is the motion detection signal, and α is a scaling constant. Observe that these filters tend to turn off filtering when a large motion is detected in an attempt to prevent artifacts. The FIR structure has limited noise reduction ability, especially when used as a purely temporal filter with a small number of frames, because the reduction in the noise variance is proportional to the number of samples in the filter support. IIR filters are more effective in noise reduction, but they generally cause Fourier phase distortions.

Several implementations of noise filters based on the above structures have been proposed in the literature, which generally differ in the way they compute the motion-detection signal. Use of motion-adaptive IIR filters in practical systems have been demonstrated by McMann *et al.* [McM 79] and Dennis [Den 80].

14.3 Motion-Compensated Filtering

The motion-compensated approach is based on the assumption that the variation of the pixel gray levels over any motion trajectory $\mathbf{c}(\tau; x_1, x_2, t)$ is due mainly to noise. Recall from the definition of the motion trajectory in Chapter 13 that $\mathbf{c}(\tau; x_1, x_2, t)$ is a continuous-valued vector function returning the coordinates of a point at time τ that corresponds to the point (x_1, x_2) at time t. Thus, noise in both the stationary and moving areas of the image can effectively be reduced by low-pass filtering over the respective motion trajectory at each pixel. Motion-compensated filters differ according to (i) the motion estimation method, (ii) the support of the filter, (e.g., temporal vs. spatio-temporal) and (iii) the filter structure (e.g., FIR vs. IIR, adaptive vs. nonadaptive).

The concept and estimation of a motion trajectory in spatio-temporally sampled image sequences are illustrated in Figure 14.3. Suppose we wish to filter the kth frame of an image sequence using N frames $k-M, \ldots, k-1, k, k+1, \ldots, k+M$, where $N = 2M+1$. The first step is to estimate the discrete motion trajectory

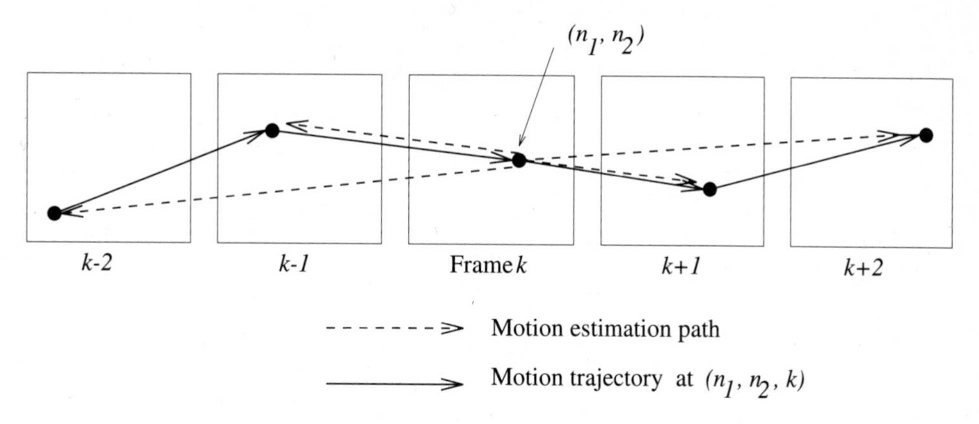

Figure 14.3: Estimation of the motion trajectory (N=5).

$\mathbf{c}(l; n_1, n_2, k), l = k-M, \ldots, k-1, k, k+1, \ldots, k+M$, at each pixel (n_1, n_2) of the kth frame. The function $\mathbf{c}(l; n_1, n_2, k)$ is a continuous-valued vector function returning the (x_1, x_2) coordinates of the point in frame l which corresponds to the pixel (n_1, n_2) of the kth frame. The discrete motion trajectory is depicted by the solid line in Figure 14.3 for the case of $N = 5$ frames. In estimating the trajectory, the displacement vectors are usually estimated in reference to the frame k as indicated by the dotted lines. The trajectory in general passes through subpixel locations, where the intensities can be determined via spatial or spatio-temporal interpolation. The support $\mathcal{S}_{n_1,n_2,k}$ of a motion-compensated spatio-temporal filter is defined as the union of predetermined spatial neighborhoods (e.g., 3×3 regions) centered about the pixel (subpixel) locations along the motion trajectory. In temporal filtering, the filter support $\mathcal{S}_{n_1,n_2,k}$ coincides with the motion trajectory $\mathbf{c}(l; n_1, n_2, k)$. Clearly, the effectiveness of motion-compensated spatio-temporal filtering is strongly related to the accuracy of the motion estimates.

Various filtering techniques, ranging from averaging to more sophisticated adaptive filtering, can be employed given the motion-compensated filter support. In the ideal case, where the motion estimation is perfect, direct averaging of image intensities along a motion trajectory provides effective noise reduction [Hua 81, Mar 85]. In practice, motion estimation is hardly ever perfect due to noise and sudden scene changes, as well as changing camera views. As a result, image intensities over an estimated motion trajectory may not necessarily correspond to the same image structure, and direct temporal averaging may yield artifacts. In the case of spatio-temporal filtering, there may also be spatial variations within the support; hence the need for adaptive filter structures over the motion-compensated filter support. Here, we consider two adaptive filter structures: the adaptive LMMSE filter, which is aimed at reducing the amount of filtering whenever a nonuniformity is detected within the motion-compensated support, and the adaptive weighted averaging (AWA) filter, which is aimed at weighting down the effect of the outliers that create the nonuniformity and achieving effective filtering by concentrating on the remaining similar image intensities.

14.3.1 Spatio-Temporal Adaptive LMMSE Filtering

The motion-compensated adaptive LMMSE filter [Sam 85, Sez 91] is an extension of the edge-preserving spatial filter proposed by Lee [Lee 80] and Kuan *et al.* [Kua 85] to the spatio-temporal domain, where the local spatial statistics are replaced by their spatio-temporal counterparts. Then, the estimate of the pixel value at (n_1, n_2, k) is given by

$$\hat{s}(n_1, n_2, k) = \frac{\sigma_s^2(n_1, n_2, k)}{\sigma_s^2(n_1, n_2, k) + \sigma_v^2(n_1, n_2, k)} [g(n_1, n_2, k) - \mu_g(n_1, n_2, k)] + \mu_s(n_1, n_2, k) \quad (14.32)$$

where $\mu_{.}(n_1, n_2, k)$ and $\sigma_{.}^2(n_1, n_2, k)$ denote the ensemble mean and variance of the corresponding signal, respectively. Depending on whether we use spatio-temporal or temporal statistics, this filter will be referred to as the LMMSE-ST or the LMMSE-T filter, respectively.

We assume that the noise can be modeled as

$$v(n_1, n_2, k) = s^\alpha(n_1, n_2, k)\, u(n_1, n_2, k)$$

where $u(n_1, n_2, k)$ is a wide sense stationary process with zero mean, which is independent of the signal, and α is a real number. The film grain noise is commonly modeled with $\alpha = 1/3$ [Hua 66]. Observe that the signal-independent noise is a special case of this model with $\alpha = 0$. Under this model, we have $\mu_s(n_1, n_2, k) = \mu_g(n_1, n_2, k)$, and $\sigma_s^2(n_1, n_2, k) = \sigma_g^2(n_1, n_2, k) - \sigma_v^2(n_1, n_2, k)$, where $\sigma_v^2(n_1, n_2, k)$ denotes the spatio-temporal variance of the noise process.

In practice, the ensemble mean $\mu_g(n_1, n_2, k)$ and variance $\sigma_g^2(n_1, n_2, k)$ are replaced with the sample mean $\hat{\mu}_g(n_1, n_2, k)$ and variance $\hat{\sigma}_g^2(n_1, n_2, k)$, which are computed within the support $\mathcal{S}_{n_1, n_2, k}$ as

$$\hat{\mu}_g(n_1, n_2, k) = \frac{1}{L} \sum_{(i_1, i_2; \ell) \in \mathcal{S}_{n_1, n_2, k}} g(i_1, i_2; \ell) \quad (14.33)$$

and

$$\hat{\sigma}_g^2(n_1, n_2, k) = \frac{1}{L} \sum_{(i_1, i_2; \ell) \in \mathcal{S}_{n_1, n_2, k}} [g(i_1, i_2; \ell) - \hat{\mu}_g(n_1, n_2, k)]^2 \quad (14.34)$$

where L is the number of pixels in $\mathcal{S}_{n_1, n_2, k}$. Then

$$\hat{\sigma}_s^2(n_1, n_2, k) = \max[0, \hat{\sigma}_g^2(n_1, n_2, k) - \sigma_v^2(n_1, n_2, k)] \quad (14.35)$$

in order to avoid the possibility of a negative variance estimate.

Substituting these estimates into the filter expression (14.32), we have

$$\hat{s}(n_1,n_2,k) = \frac{\hat{\sigma}_s^2(n_1,n_2,k)}{\hat{\sigma}_s^2(n_1,n_2,k)+\hat{\sigma}_v^2(n_1,n_2,k)}\, g(n_1,n_2,k) + \frac{\hat{\sigma}_v^2(n_1,n_2,k)}{\hat{\sigma}_s^2(n_1,n_2,k)+\hat{\sigma}_v^2(n_1,n_2,k)}\, \hat{\mu}_g(n_1,n_2,k) \quad (14.36)$$

The adaptive nature of the filter can be observed from (14.36). When the spatio-temporal signal variance is much smaller than the noise variance, $\hat{\sigma}_s^2(n_1,n_2,k) \approx 0$, that is, the support $\mathcal{S}_{n_1,n_2,k}$ is uniform, the estimate approaches the spatio-temporal mean, $\hat{\mu}_g = \hat{\mu}_s$. In the extreme, when the spatio-temporal signal variance is much larger than the noise variance, $\hat{\sigma}_s^2(n_1,n_2,k) \gg \hat{\sigma}_v^2(n_1,n_2,k)$, due to poor motion estimation or the presence of sharp spatial edges in $\mathcal{S}_{n_1,n_2,k}$, the estimate approaches the noisy image value to avoid blurring.

A drawback of the adaptive LMMSE filter is that it turns the filtering down even if there is a single outlier pixel in the filter support, thus often leaving noisy spots in the filtered image. An alternative implementation, called "the switched LMMSE filter (LMMSE-SW)," may maintain the advantages of both the spatio-temporal and the temporal LMMSE filtering by switching between a selected set of temporal and spatio-temporal supports at each pixel, depending on which support is the most uniform [Ozk 93]. If the variance $\hat{\sigma}_g^2(n_1,n_2,k)$ computed over $\mathcal{S}_{n_1,n_2,k}$ is less than the noise variance, then the filtering is performed using this support; otherwise, the largest support over which $\hat{\sigma}_g^2(n_1,n_2,k)$ is less than the noise variance is selected. In the next section, we introduce an adaptive weighted averaging (AWA) filter which employs an implicit mechanism for selecting the most uniform subset within $\mathcal{S}_{n_1,n_2,k}$ for filtering.

14.3.2 Adaptive Weighted Averaging Filter

The adaptive weighted averaging (AWA) filter computes a weighted average of the image values within the spatio-temporal support along the motion trajectory. The weights are determined by optimizing a criterion functional, and they vary with the accuracy of motion estimation as well as the spatial uniformity of the region around the motion trajectory. In the case of sufficiently accurate motion estimation across the entire trajectory and spatial uniformity, image values within the spatio-temporal filter support attain equal weights, and the AWA filter performs direct spatio-temporal averaging. When the value of a certain pixel within the spatio-temporal support deviates from the value of the pixel to be filtered by more than a threshold, its weight decreases, shifting the emphasis to the remaining image values within the support that better matches the pixel of interest. The AWA filter is therefore particularly well suited for efficient filtering of sequences containing segments with varying scene contents due, for example, to rapid zooming and changes in the view of the camera.

The AWA filter can be defined by

$$\hat{s}(n_1, n_2, k) = \sum_{(i_1,i_2;\ell)\in\mathcal{S}_{n_1,n_2,k}} w(i_1, i_2; \ell)\; g(i_1, i_2; \ell) \tag{14.37}$$

where

$$w(i_1, i_2; \ell) \doteq \frac{K(n_1, n_2, k)}{1 + a\ \max\{\epsilon^2\ ,\ [g(n_1, n_2, k) - g(i_1, i_2; \ell)]^2\}} \tag{14.38}$$

are the weights within the support $\mathcal{S}_{n_1,n_2,k}$ and $K(n_1, n_2, k)$ is a normalization constant, given by

$$K(n_1, n_2, k) = \Big(\sum_{(i_1,i_2;\ell)\in\mathcal{S}_{n_1,n_2,k}} \frac{1}{1 + a\ \max\{\epsilon^2\ ,\ [g(n_1, n_2, k) - g(i_1, i_2; \ell)]^2\}} \Big)^{-1} \tag{14.39}$$

The quantities a ($a > 0$) and ϵ are the parameters of the filter. These parameters are determined according to the following principles:

1) When the differences in the intensities of pixels within the spatio-temporal support are merely due to noise, it is desirable that the weighted averaging reduces to direct averaging. This can be achieved by appropriately selecting the parameter ϵ^2. Note that, if the square of the differences are less than ϵ^2, then all the weights attain the same value $K/(1 + a\epsilon^2) = 1/L$, and $\hat{s}(n_1, n_2, k)$ reduces to direct averaging. We set the value of ϵ^2 equal to two times the value of the noise variance, i.e., the expected value of the square of the difference between two image values that differ due to the presence of noise only.

2) If the square of the difference between the values $g(n_1, n_2, k)$ and $g(i_1, i_2; \ell)$ for a particular $(i_1, i_2, \ell) \in \mathcal{S}_{n_1,n_2,k}$ is larger than ϵ^2, then the contribution of $g(i_1, i_2; \ell)$ is weighted down by $w(i_1, i_2; \ell) < w(n_1, n_2, k) = K/(1 + a\epsilon^2)$. The parameter a is a "penalty" parameter that determines the sensitivity of the weight to the squared difference $[g(n_1, n_2, k) - g(i_1, i_2; \ell)]^2$. The "penalty" parameter a is usually set equal to unity.

The effect of the penalty parameter a on the performance of the AWA filter can be best visualized considering a special case where one of the frames within the filter support is substantially different from the rest. In the extreme, when $a = 0$, all weights are equal. That is, there is no penalty for a mismatch, and the AWA filter performs direct averaging. However, for a large, the weights for the $2M$ "matching" frames are equal, whereas the weight of the nonmatching frame approaches zero. Generally speaking, the AWA filter takes the form of a "limited-amplitude averager," where those pixels whose intensities differ from that of the center pixel by not more than $\pm\epsilon$ are averaged. A similar algorithm is K nearest neighbor averaging [Dav 78], where the average of K pixels within a certain window whose values are closest to the value of the pixel of interest are computed.

14.4 Examples

We demonstrate the performance of the motion-compensated LMMSE and AWA filters with temporal and spatio-temporal supports using five frames of an image sequence scanned from motion picture film at a resolution of 1216 pixel $\times$ 832 lines. The central three frames of the sequence, degraded by actual film-grain noise, are shown in Figure 14.4.

The motion trajectories have been estimated as shown in Figure 14.3, by a hierarchical block-matching algorithm with five levels using log-D search, the MAD criterion, and 1/4 pixel accuracy. We have estimated the variance of the noise process over a uniform region in the frame to be filtered (selected interactively). In the LMMSE filter the noise variance appears directly in the filter expression, whereas in the AWA filter it is used in defining the filter parameter ϵ^2, which is typically set equal to twice the estimated noise variance. Figure 14.5 (b-f) show the results of applying the temporal LMMSE, spatio-temporal LMMSE, switched LMMSE, temporal AWA, and spatio-temporal AWA filters, respectively, on the noisy face section depicted in Figure 14.5 (a). Figure 14.6 shows the same set of results on the hand section. Inspection of the results suggests that the spatio-temporal filters provide better noise reduction than the respective temporal filters, at the expense of introducing some blurring. The switched LMMSE filter strikes the best balance between noise reduction and retaining image sharpness. The visual difference between the results of the LMMSE and AWA filters was insignificant for this particular sequence, because there were no sudden scene changes involved. It has been shown that the AWA filter outperforms the LMMSE filter, especially in cases of low input SNR and abruptly varying scene content [Ozk 93].

14.5 Exercises

1. In the Wiener filter (14.10), how would you estimate the power spectrum of the original image $P_{ss}(f_1, f_2)$ and the noise $P_{vv}(f_1, f_2)$?

2. Let $\mathbf{s}$ denote the lexicographic ordering of pixel intensities in an image. Show that convolution by an FIR filter kernel $h(n_1, n_2)$ can be expressed as $\mathbf{y} = \mathbf{H}\mathbf{s}$, where $\mathbf{H}$ is block-Toeplitz and block-circulant for linear and circular convolution, respectively. Write the elements of $\mathbf{H}$ in terms of $h(n_1, n_2)$.

3. Show that a block-circulant matrix can be diagonalized by a 2-D DFT.

4. Show a complete derivation of (14.23).

5. How do you compare motion-adaptive filtering and adaptive motion-compensated filtering in the presence of motion estimation errors?

6. How would you compare the motion-compensated adaptive LMMSE and AWA filters in the presence of a sudden scene change?

Figure 14.4: Three frames of a sequence digitized from motion picture film.

Figure 14.5: Comparison of the filters at the central frame (the face section): a) the noisy original (top-left), b) temporal LMMSE filtering (top-right), c) spatio-temporal LMMSE filtering (center-left), d) switched LMMSE filtering, (center-right) e) temporal AWA filtering (bottom-left), and f) spatio-temporal AWA filtering (bottom-right). (Courtesy Mehmet Ozkan)

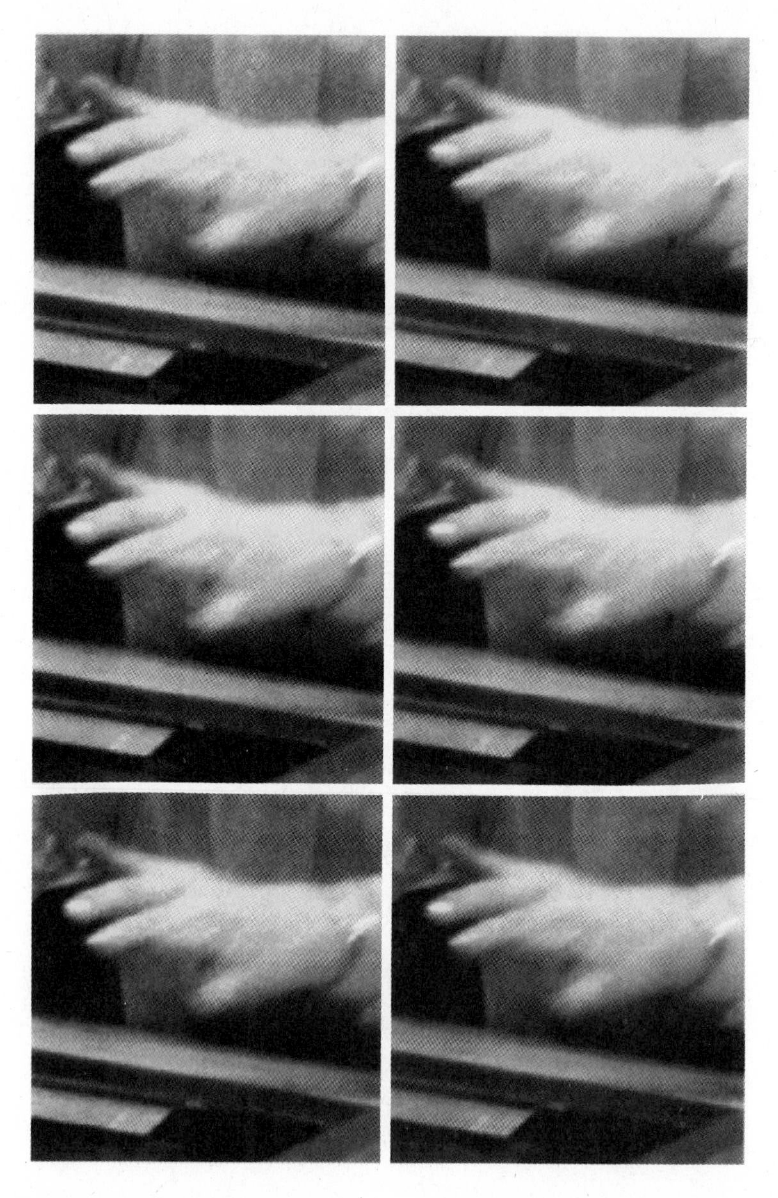

Figure 14.6: Comparison of the filters at the central frame (the hand section): a) the noisy original (top-left), b) temporal LMMSE filtering (top-right), c) spatio-temporal LMMSE filtering (center-left), d) switched LMMSE filtering, (center-right) e) temporal AWA filtering (bottom-left), and f) spatio-temporal AWA filtering (bottom-right). (Courtesy Mehmet Ozkan)

Bibliography

[Alp 90] B. Alp, P. Haavisto, T. Jarske, K. Oistamo, and Y. Neuvo, "Median based algorithms for image sequence processing," *Proc. SPIE Visual Comm. and Image Proc. '90*, Lausanne, Switzerland, pp. 122–134, Oct. 1990.

[Arc 86] G. R. Arce, N. C. Gallagher, and T. A. Nodes, "Median filters: Theory for one or two dimensional filters," in *Advances in Computer Vision and Image Processing*, T. S. Huang, ed., JAI Press, 1986.

[Arc 91] G. R. Arce, "Multistage order statistic filters for image sequence processing," *IEEE Trans. Signal Proc.*, vol. 39, pp. 1146–1163, 1991.

[Ata 80] E. Ataman, V. K. Aatre, and K. M. Wong, "A fast method for real-time median filtering," *IEEE Trans. Acoust. Speech and Sign. Proc.*, vol. 28, pp. 415–421, Aug. 1980.

[Boy 92] J. Boyce, "Noise reduction of image sequences using adaptive motion compensated frame averaging," *Proc. IEEE Int. Conf. Acoust. Speech and Sign. Proc.*, San Francisco, CA, pp. III-461–464, Mar. 1992.

[Cha 85] P. Chan and J. S. Lim, "One-dimensional processing for adaptive image restoration," *IEEE Trans. Acoust. Speech and Sign. Proc.*, vol. 33, pp. 117–126, Feb. 1985.

[Dav 75] L. S. Davis, "A survey of edge-detection techniques," *Comp. Graph. and Image Proc.*, vol. 4, pp. 248–270, 1975.

[Dav 78] L. S. Davis and A. Rosenfeld, "Noise cleaning by iterated local averaging," *IEEE Trans. Syst. Man and Cybern.*, vol. 8, pp. 705–710, 1978.

[Den 80] T. J. Dennis, "Non-linear temporal filter for television picture noise reduction," *IEE Proc.*, vol. 127G, pp. 52–56, 1980.

[Doy 86] T. Doyle and P. Frencken, "Median filtering of television images," *Proc. IEEE Int. Conf. on Consumer Elec.*, pp. 186–187, 1986.

[Dub 92] E. Dubois, "Motion-compensated filtering of time-varying images," *Multidim. Syst. Sign. Proc.*, vol. 3, pp. 211–239, 1992.

[Dub 84] E. Dubois and S. Sabri, "Noise reduction in image sequences using motion-compensated temporal filtering," *IEEE Trans. Comm.*, vol. COM-32, pp. 826–831, July 1984.

[Gon 92] R. C. Gonzalez and R. E. Woods, *Digital Image Processing*, Addison-Wesley, 1992.

[Hua 66] T. S. Huang, "Some notes on film-grain noise," in *Restoration of Atmospherically Degraded Images*, NSF Summer Study Rep., pp. 105–109, Woods Hole, MA, 1966.

[Hua 81] T. S. Huang and Y. P. Hsu, "Image sequence enhancement," in *Image Sequence Analysis*, T. S. Huang, ed., Berlin: Springer Verlag, 1981.

[Kat 91] A. K. Katsaggelos, R. P. Kleihorst, S. N. Efstratiadis, and R. L. Lagendijk, "Adaptive image sequence noise filtering methods," *Proc. SPIE Conf. Visual Comm. and Image Proc.*, Nov. 1991.

[Kua 85] D. T. Kuan, A. A. Sawchuk, T. C. Strand, and P. Chavel, "Adaptive noise smoothing filter for images with signal-dependent noise", *IEEE Trans. Patt. Anal. Mach. Intel.*, vol. PAMI-7, pp. 165–177, Mar. 1985.

[Lee 80] J. S. Lee, "Digital image enhancement and noise filtering by use of local statistics," *IEEE Trans. Patt. Anal. Mach. Intel.*, vol. PAMI-2, pp. 165–168, Mar. 1980.

[Mar 85] D. Martinez and J. S. Lim, "Implicit motion compensated noise reduction of motion video scenes," *Proc. IEEE Int. Conf. Acoust. Speech, Signal Proc.*, pp. 375–378, Tampa, FL, 1985.

[McM 79] R. H. McMann, S. Kreinik, J. K. Moore, A. Kaiser and J. Rossi, "A digital noise reducer for encoded NTSC signals," *SMPTE Jour.*, vol. 87, pp. 129–133, Mar. 1979.

[Ozk 93] M. K. Ozkan, M. I. Sezan, and A. M. Tekalp, "Adaptive motion-compensated filtering of noisy image sequences," *IEEE Trans. Circuits and Syst. for Video Tech.*, vol. 3, pp. 277–290, Aug. 1993.

[Rei 91] T. A. Reinen, "Noise reduction in heart movies by motion compensated filtering," *Proc. SPIE. Vis. Comm. and Image Proc. '91*, Boston, MA, pp. 755–763, Nov. 1991.

[Sam 85] R. Samy, "An adaptive image sequence filtering scheme based on motion detection," *SPIE*, vol. 596, pp. 135–144, 1985.

[Sez 91] M. I. Sezan, M. K. Ozkan, and S. V. Fogel, "Temporally adaptive filtering of noisy image sequences using a robust motion estimation algorithm," *Proc. IEEE Int. Conf. Acoust. Speech, Signal Processing*, pp. 2429–2432, Toronto, Canada, 1991.

[Uns 90] M. Unser and M. Eden, "Weighted averaging of a set of noisy images for maximum signal-to-noise ratio," *IEEE Trans. on Acoust., Speech and Sign. Proc.*, vol. ASSP-38, pp. 890–895, May 1990.

Chapter 15

RESTORATION

Restoration refers to spatial deblurring of video degraded by optical smearing, including linear motion or out-of-focus blurs. Optical blurring arises when a single object point spreads over several image pixels, which may be due to relative motion between the object and the camera, or out-of-focus imaging. The extent of the spatial spread is determined by the point spread function (PSF) of the imaging system. The chapter starts with modeling of spatially invariant and spatially varying blur formation in Section 15.1. Temporal blurring is not discussed because it is often negligible. In the discrete spatio-temporal domain, the restoration problem can be formulated as solving a set of simultaneous linear equations, which simplifies to a deconvolution problem in the case of spatially invariant blurs. We present some selected intraframe methods for the restoration of spatially invariant and spatially varying blurred images in Sections 15.2 and 15.3, respectively. Multiframe methods for image restoration are discussed in Section 15.4, where we make use of the temporal correlation and frame-to-frame motion information to better regularize the restoration process and/or identify the spatial extent of the blur PSF.

15.1 Modeling

Images suffer from optical blurring as well as noise. Spatial blurring due to motion and out-of-focus may be encountered in images acquired by fixed surveillance cameras, or cameras mounted on moving vehicles such as aircraft and robot-arms. Motion blurs are also common in still images of scenes with high activity, such as in freeze-frames of sports video. Temporal blurring in video is often negligible, and will not be considered here. In the following, we discuss modeling of shift-invariant and shift-varying spatial blurring, respectively. A vector-matrix model for the multiframe formulation, where each frame possibly suffers from a different spatial blur, is also presented.

15.1.1 Shift-Invariant Spatial Blurring

A spatially invariant blur, at a particular frame k, can be modeled by means of a 2-D convolution with the spatial PSF $h_k(n_1, n_2)$ given by

$$g_k(n_1, n_2) = h_k(n_1, n_2) ** s_k(n_1, n_2) + v_k(n_1, n_2) \quad k = 1, \ldots L \tag{15.1}$$

where $g_k(n_1, n_2)$, $s_k(n_1, n_2)$ and $v_k(n_1, n_2)$ denote the kth frame of the degraded video, ideal video, and noise, respectively. The frame index k appears as a subscript in (15.1), where $g_k(n_1, n_2) \doteq g(n_1, n_2, k)$, and so on, to emphasize the spatial nature of the degradation. Observe that Equation 15.1 represents spatio-temporally shift-invariant blurring if the 2-D PSF remains the same in every frame, that is, $h_k(n_1, n_2) = h(n_1, n_2)$.

The model (15.1) can be expressed in vector-matrix form as

$$\mathbf{g}_k = \mathbf{D}_k \mathbf{s}_k + \mathbf{v}_k$$

where $\mathbf{g}_k$, $\mathbf{s}_k$, and $\mathbf{v}_k$ denote the observed frame k, the original frame k, and the noise in frame k which are lexicographically ordered into $N^2 \times 1$ vectors (for an image of size $N \times N$), respectively, and the matrix $\mathbf{D}_k$ characterizes the PSF of the blur in frame k in operator form [Gon 92]. Note that for a spatially shift-invariant blur, $\mathbf{D}_k$ is block-Toeplitz.

Suppose we have L frames of video, each blurred by possibly a different spatially invariant PSF, $h_k(n_1, n_2)$, $k = 1, \ldots, L$. The vector-matrix model can be extended to multiframe modeling over L frames as

$$\mathbf{g} = \mathcal{D}\mathbf{s} + \mathbf{v} \tag{15.2}$$

where

$$\mathbf{g} \doteq \begin{bmatrix} \mathbf{g}_1 \\ \vdots \\ \mathbf{g}_L \end{bmatrix}, \quad \mathbf{s} \doteq \begin{bmatrix} \mathbf{s}_1 \\ \vdots \\ \mathbf{s}_L \end{bmatrix}, \quad \mathbf{v} \doteq \begin{bmatrix} \mathbf{v}_1 \\ \vdots \\ \mathbf{v}_L \end{bmatrix}$$

denote $N^2 L \times 1$ vectors representing the observed, ideal, and noise frames, respectively, stacked as multiframe vectors, and

$$\mathcal{D} \doteq \begin{bmatrix} \mathbf{D}_1 & \cdots & 0 \\ \vdots & \ddots & \vdots \\ 0 & \cdots & \mathbf{D}_L \end{bmatrix}$$

is an $N^2 L \times N^2 L$ matrix representing the multiframe blur operator. Observe that the multiframe blur matrix $\mathcal{D}$ is block diagonal, indicating no temporal blurring.

Based on the models presented in this section, we can pose the following video restoration problems:

(i) Intraframe restoration: Given the model (15.1), find an estimate $\hat{s}_k(n_1, n_2)$ of each frame, processing each frame independently.
(ii) Multiframe restoration: Given the model (15.2), find an estimate $\hat{s}(n_1, n_2, k)$ of the ideal video, processing all frames simultaneously.

Although we have only spatial blurring in both problems, the multiframe formulation allows us to utilize the temporal correlation among the frames of the original video. Temporal correlation between the ideal frames can be modeled either statistically by means of auto and cross power spectra, or deterministically by a motion model. Multiframe techniques based on the model (15.2) are discussed in Section 15.4.

15.1.2 Shift-Varying Spatial Blurring

Image formation in the presence of a linear shift-varying (LSV) blur can be modeled as the output of a 2-D LSV system, denoted by $\mathcal{L}_k$, where k is the frame index. We define the space-varying PSF of this imaging system as

$$h_k(n_1, n_2; i_1, i_2) = \mathcal{L}_k\{\delta(n_1 - i_1, n_2 - i_2)\} \tag{15.3}$$

which represents the image of a point source located at (i_1, i_2), and may have infinite extent. It is well-known that a discrete image can be represented as a weighted and shifted sum of point sources as

$$s_k(n_1, n_2) = \sum_{(i_1, i_2) \in \mathcal{S}} s_k(i_1, i_2)\delta(n_1 - i_1, n_2 - i_2) \tag{15.4}$$

where $s_k(i_1, i_2)$ denotes the samples of the kth frame, and $\mathcal{S}$ denotes the image support. Then, applying the system operator $\mathcal{L}_k$ to both sides of (15.4), using (15.3), and including the observation noise, $v_k(n_1, n_2)$, the blurred frame can be expressed as

$$g_k(n_1, n_2) = \sum_{(i_1, i_2) \in \mathcal{S}} s_k(i_1, i_2)h_k(n_1, n_2; i_1, i_2) + v_k(n_1, n_2) \tag{15.5}$$

From (15.5), it is clear that for an image of size $N_1 \times N_2$, assuming that the PSF is known, we can obtain $N_1 \times N_2$ observation equations for a total of $N_1 \times N_2$ unknowns, $s_k(i_1, i_2)$, where the equations are coupled with each other. Hence, the restoration problem can be formulated as a problem of solving a set of $N_1 \times N_2$ simultaneous equations. For the case of LSI blurs, the system matrix is block-Toeplitz; thus, the $N_1 \times N_2$ equations can be decoupled, under the block-circulant approximation, by applying the 2-D discrete Fourier transform. In the LSV case, however, the system matrix is not block-Toeplitz, and fast methods in the transform domain are not applicable. We will describe a nonlinear algorithm, based on convex projections, to solve this system of equations in Section 15.3.

15.2 Intraframe Shift-Invariant Restoration

Image restoration may be defined as the process of undoing image blurring (smearing) based on a mathematical model of its formation. Because the model (15.1) is generally not invertible, restoration algorithms employ regularization techniques by using *a priori* information about the solution (the ideal image). Many methods exist for image restoration depending on the amount and nature of the *a priori* information used about the ideal image. The most commonly used techniques are pseudo inverse filtering, Wiener and constrained least squares filtering, the MAP method, the constrained iterative method, and the POCS method. In this section, we briefly discuss pseudo inverse filtering and constrained least squares filtering. Wiener filtering is introduced as a special case of the constrained least squares filtering. The interested reader is referred to [And 77] for the fundamentals and to [Tek 94, Sez 90] for more recent overviews of image restoration methods.

15.2.1 Pseudo Inverse Filtering

If we neglect the presence of noise in the degradation model (15.1); that is, if the vector $\mathbf{g}_k$ lies in the column space of the matrix $\mathbf{D}_k$, and the matrix $\mathbf{D}_k$ is invertible, an exact solution can be found by direct inversion

$$\hat{\mathbf{s}}_k = \mathbf{D}_k^{-1}\mathbf{g}_k \tag{15.6}$$

where $\hat{\mathbf{s}}_k$ denotes an estimate of the ideal frame k. The operation (15.6) is known as inverse filtering [And 77]. However, because inverse filtering using (15.6) requires a huge matrix inversion, it is usually implemented in the frequency domain. Taking the 2-D (spatial) Fourier transforms of both sides of (15.1), we have

$$G_k(f_1, f_2) = H_k(f_1, f_2)S_k(f_1, f_2) + V_k(f_1, f_2) \tag{15.7}$$

Then, inverse filtering can be expressed in the frequency domain as

$$\hat{S}_k(f_1, f_2) = \frac{G_k(f_1, f_2)}{H_k(f_1, f_2)} \tag{15.8}$$

The implementation of (15.8) in the computer requires sampling of the frequency variables (f_1, f_2), which corresponds to computing the discrete Fourier transform (DFT) of the respective signals.

The relationship between the expressions (15.6) and (15.8) can be shown by diagonalizing the matrix operator $\mathbf{D}_k$. Because the $N^2 \times N^2$ matrix $\mathbf{D}_k$ is block-Toeplitz for spatially shift-invariant blurs, it can be diagonalized, under the block-circulant approximation [Gon 87], as

$$\mathbf{D}_k = \mathbf{W}\mathbf{\Lambda}_k\mathbf{W}^{-1} \tag{15.9}$$

where $\mathbf{W}$ is the $N^2 \times N^2$ 2-D DFT matrix which contains N^2 partitions of size $N \times N$. The imth partition of $\mathbf{W}$ is in the form $\mathbf{W}_{im} = \exp\left\{-j\frac{2\pi}{N}im\right\}\exp\left\{-j\frac{2\pi}{N}ln\right\}$,

$l, n = 0, \ldots, N-1$, and $\mathbf{\Lambda}_k$ is a diagonal matrix of the eigenvalues of $\mathbf{D}_k$ which are given by the 2-D DFT of $h_k(n_1, n_2)$ [Gon 87]. Then, premultiplying both sides of (15.6) by $\mathbf{W}^{-1}$, and inserting the term $\mathbf{W}\mathbf{W}^{-1}$ between $\mathbf{D}_k^{-1}$ and $\mathbf{g}_k$, we have

$$\begin{aligned} \mathbf{W}^{-1}\hat{\mathbf{s}}_k &= \mathbf{W}^{-1}\mathbf{D}_k^{-1}\mathbf{W}\mathbf{W}^{-1}\mathbf{g}_k, \quad \text{or} \\ \hat{S}_k(u_1, u_2) &= \frac{G_k(u_1, u_2)}{H_k(u_1, u_2)}, \quad \text{for all } (u_1, u_2) \end{aligned} \tag{15.10}$$

where multiplication by $\mathbf{W}^{-1}$ computes the 2-D DFT, and u_1 and u_2 denote the discretized spatial frequency variables; hence, the equivalence of (15.8) and (15.10).

However, inverse filtering has some drawbacks. First, the matrix $\mathbf{D}_k$ may be singular, which means at least one of its eigenvalues $H(u_1, u_2)$ would be zero, resulting in a division by zero in the 2-D DFT implementation (15.8). Even if the matrix $\mathbf{D}_k$ is nonsingular, i.e., $H(u_1, u_2) \neq 0$, for all (u_1, u_2), the vector $\mathbf{g}_k$ almost never lies in the column space of the matrix $\mathbf{D}_k$ due to presence of observation noise $\mathbf{v}_k$; thus, an exact solution (15.6) does not exist. As a result, one must resort to a least squares (LS) solution that minimizes the norm of the residual $\mathbf{g}_k - \mathbf{D}_k\mathbf{s}_k$. The LS solution, given by

$$\hat{\mathbf{s}}_k = (\mathbf{D}_k^T\mathbf{D}_k)^{-1}\mathbf{D}_k^T\mathbf{g}_k \tag{15.11}$$

exists whenever the columns of the matrix $\mathbf{D}_k$ are linearly independent. If this condition is not satisfied, we can define the minimum norm LS solution by means of the pseudo inverse $\mathbf{D}_k^{\#}$ (see, for example [Rus 87]). In the frequency domain, the corresponding solution is referred to as the pseudo-inverse filter, where division by zero is defined as zero.

Deconvolution by pseudo inversion is often ill-posed owing to the presence of observation noise [Tek 90]. This follows because the pseudo inverse of the blur transfer function usually has a very large magnitude at high frequencies and near those frequencies where the blur transfer function has zeros. This results in excessive amplification of the sensor noise at those frequencies. Regularized deconvolution techniques using information about the image generation process attempt to roll off the transfer function of the pseudo inverse filter at these frequencies in an attempt to limit the noise amplification. However, inevitably the regularized filter deviates from the exact inverse at these frequencies, which leads to other artifacts, known as regularization artifacts. Regularization of the inversion can be achieved either via deterministic constrained optimization techniques or by statistical estimation techniques. We discuss the constrained least squares and Wiener filtering in the following.

15.2.2 Constrained Least Squares and Wiener Filtering

The constrained least squares (CLS) filter is one of the most commonly used regularized deconvolution methods for image restoration. It is formulated as minimizing a functional of the image subject to a constraint imposed by the observation model, that is,

$$\begin{array}{ll} \min_{\mathbf{s}_k} & \mathbf{L}\mathbf{s}_k \\ \text{subject to} & ||g_k(n_1,n_2) - h_k(n_1,n_2) ** s_k(n_1,n_2)||^2 = \sigma_v^2 \end{array} \tag{15.12}$$

where $\mathbf{L}$ is called a regularization operator (matrix), and σ_v^2 denotes the spatial variance of the noise. The operator $\mathbf{L}$ is usually chosen to have high-pass characteristics so that the CLS method finds the smoothest image (that is, the image with the smallest high frequency content) which statistically complies with the observation model (15.1).

The solution to the above problem, known as the CLS filter, can be expressed as [Gon 87]

$$\hat{\mathbf{s}}_k = (\mathbf{D}_k^T \mathbf{D}_k + \mathbf{L}^T \mathbf{L})^{-1} \mathbf{D}_k^T \mathbf{g}_k \tag{15.13}$$

Again, in order to avoid a huge matrix inversion, we premultiply (15.13) by the 2-D DFT operator $\mathbf{W}^{-1}$ to arrive at the frequency domain filter expression

$$\hat{S}_k(u_1,u_2) = \frac{H_k^*(u_1,u_2)G_k(u_1,u_2)}{|H_k(u_1,u_2)|^2 + \alpha|L(u_1,u_2)|^2} \tag{15.14}$$

where $L(u_1,u_2)$ denotes the eigenvalues of the regularization operator $\mathbf{L}$. In practice, the operator $\mathbf{L}$ is defined in terms of a 2-D shift-invariant generating kernel (impulse response), and $L(u_1,u_2)$ is given by the 2-D DFT of this kernel.

Two common choices for the regularization operator are the identity operator and the Laplacian operator. In the former case, we have $\mathbf{L} = \mathbf{I}$, where $\mathbf{I}$ is the identity matrix. The 2-D generating kernel in this case is the Dirac delta function $\delta(x_1,x_2)$, and $L(u_1,u_2) = 1$, which is the 2-D DFT of the discrete representation given by the Kronecker delta function $\delta(n_1,n_2)$. In the latter case, the generating kernel is the Laplacian operator defined by

$$l(x_1,x_2) = \nabla \cdot \nabla(\cdot) = \frac{\partial^2(\cdot)}{\partial x_1^2} + \frac{\partial^2(\cdot)}{\partial x_2^2} \tag{15.15}$$

where ∇ is the gradient operator, and $\cdot$ denotes the vector dot product. Several discrete approximations exist for the Laplacian kernel. Two discrete Laplacian kernels $l(n_1,n_2)$ are shown in Figure 15.1. Then $L(u_1,u_2)$ is the 2-D DFT of a particular discrete approximation $l(n_1,n_2)$. The case of the Laplacian exemplifies frequency-dependent regularization, where the frequency response of the restoration filter rolls off at the high frequencies.

The reader can readily verify that the inverse filter is a special case of the CLS filter with $|L(u_1,u_2)| = 0$. Furthermore, the Wiener deconvolution filter, which can be derived by following the steps shown in Chapter 14 based on the observation model (15.1), is another special case with $\mathbf{L} = \mathbf{R}_s^{-1}\mathbf{R}_v$ or $|L(u_1,u_2)|^2 = P_{vv}(u_1,u_2)/P_{ss}(u_1,u_2)$, where $\mathbf{R}_s$ and $\mathbf{R}_v$ denote the covariance

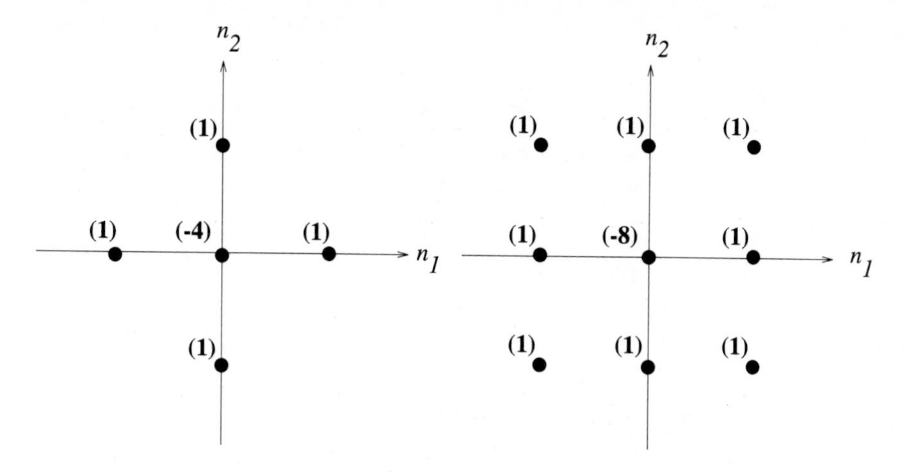

Figure 15.1: Discrete approximations for the Laplacian operator.

matrices of the ideal frame and the noise, and $P_{ss}(u_1, u_2)$ and $P_{vv}(u_1, u_2)$ denote their power spectra, respectively. Recall that the Wiener filter gives the linear minimum mean square error (LMMSE) estimate of the ideal frame given the observation model. However, it requires *a priori* information about the image and noise in the form of their power spectra.

15.3 Intraframe Shift-Varying Restoration

In comparison with the amount of work on linear space-invariant (LSI) image restoration, the literature reports only a few methods for the restoration of images degraded by linear shift-varying (LSV) blurs [Sez 90]. Nevertheless, in real-world applications the degradations are often space-varying; that is, the degrading system PSF changes with spatial location over the image. Examples of LSV blurs include the motion blur when the relative motion is not parallel to the imaging plane or contains acceleration, and the out-of-focus blur when the scene has depth. Space-varying restoration methods include coordinate transformations [Saw 74], sectional processing [Tru 78, Tru 92], and iterative methods [Sch 81]. The coordinate transformation approach is applicable only to a special class of LSV degradations that can be transformed into an LSI degradation. In sectional methods the image is sectioned into rectangular regions, where each section is restored using a space-invariant method, such as the maximum *a posteriori* filter [Tru 78] or the modified Landweber iterative filter [Tru 92].

Here, we present a relatively novel approach using the method of projections onto convex sets (POCS) [Ozk 94]. This POCS formulation, which is robust for any blur type and size, is different from those developed for the space-invariant image restoration problem [Tru 84, Sez 91]. In [Tru 84] Trussell and Civanlar use the variance of the residual constraint for restoration of space-invariant images. Unfortunately, this constraint cannot easily be extended for space-variant restoration since it involves inversion of huge matrices that would not be Toeplitz for space-varying blurs. In [Sez 91], Sezan and Trussell develop a general framework of prototype-image-based constraints. Although this framework is rather general, the specific constraints proposed in [Sez 91] are designed for the space-invariant blur case.

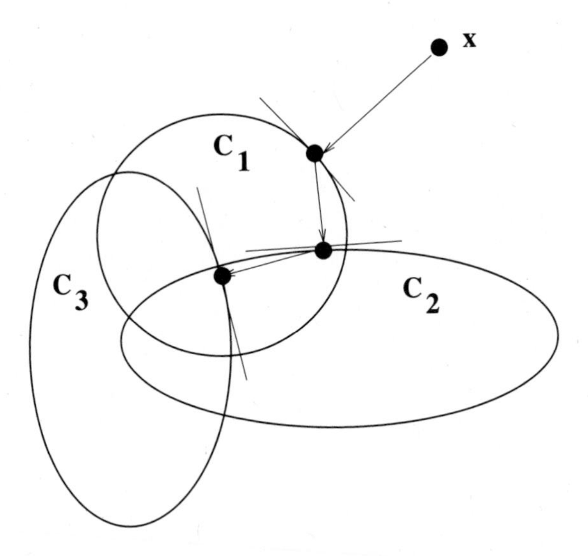

Figure 15.2: The method of projection onto convex sets.

15.3.1 Overview of the POCS Method

In the POCS method, the unknown signal, $\mathbf{s}$, is assumed to be an element of an appropriate Hilbert space $\mathcal{H}$. Each *a priori* information or constraint restricts the solution to a closed convex set in $\mathcal{H}$. Thus, for m pieces of information, there are m corresponding closed convex sets $C_i \in \mathcal{H}$, $i = 1, 2, ..., m$ and $\mathbf{s} \in C_0 \doteq \bigcap_{i=1}^m C_i$ provided that the intersection C_0 is nonempty. Given the sets C_i and their respective projection operators $\mathbf{P}_i$, the sequence generated by

$$\mathbf{s}_{k+1} = \mathbf{P}_m \mathbf{P}_{m-1} \ldots \mathbf{P}_1 \mathbf{s}_k; \quad k = 0, 1, \ldots \tag{15.16}$$

or, more generally, by

$$\mathbf{s}_{k+1} = \mathbf{T}_m \mathbf{T}_{m-1} \ldots \mathbf{T}_1 \mathbf{s}_k; \quad k = 0, 1, \ldots \tag{15.17}$$

where $\mathbf{T}_i \doteq (1-\lambda_i)\mathbf{I} + \lambda_i \mathbf{P}_i$, $0 < \lambda_i < 2$, is the relaxed projection operator, converges weakly to a feasible solution in the intersection C_0 of the constraint sets. The convex sets and the successive projections are illustrated in Figure 15.2.

Indeed, any solution in the intersection set is consistent with the *a priori* constraints and therefore is a feasible solution. Note that $\mathbf{T}_i$'s reduce to $\mathbf{P}_i$'s for unity relaxation parameters, i.e., $\lambda_i = 1$. The initialization $\mathbf{s}_0$ can be arbitrarily chosen from $\mathcal{H}$. It should be emphasized that the POCS algorithm is in general nonlinear, because the projection operations are in general nonlinear. For more detailed discussions of the fundamental concepts of the POCS, the reader is referred to [Sta 87].

15.3.2 Restoration Using POCS

The POCS method can readily be applied to the space-varying restoration problem using a number of space-domain constraints that are defined on the basis of the observed image, and *a priori* information about the space-varying degradation process, the noise statistics, and the ideal image itself.

Assuming that the space-varying blur PSF and the statistics of the noise process are known, we define the following closed, convex constraints (one for each observed blurred image pixel):

$$C_{n_1,n_2,k} = \{y_k(i_1,i_2) : \; |r^{(\mathbf{y}_k)}(n_1,n_2)| \leq \delta_0\}, \quad 0 \leq n_1 \leq N_1 - 1, \quad 0 \leq n_2 \leq N_2 - 1$$

where

$$r^{(\mathbf{y}_k)}(n_1,n_2) \doteq g_k(n_1,n_2) - \sum_{i_1=0}^{N_1-1} \sum_{i_2=0}^{N_2-1} y_k(i_1,i_2) h_k(n_1,n_2;i_1,i_2) \qquad (15.18)$$

is the residual associated with $\mathbf{y}_k$, which denotes an arbitrary member of the set. The quantity δ_0 is an *a priori* bound reflecting the statistical confidence with which the actual image is a member of the set $C_{n_1,n_2,k}$. Since $r^{(\mathbf{s}_k)}(n_1,n_2) = v_k(n_1,n_2)$, where $\mathbf{s}_k$ denotes the ideal frame, the statistics of $r^{(\mathbf{s}_k)}(n_1,n_2)$ is identical to that of $v_k(n_1,n_2)$. Hence, the bound δ_0 is determined from the statistics of the noise process so that the actual frame (i.e., the ideal solution) is a member of the set within a certain statistical confidence. For example, if the noise has Gaussian distribution with the standard deviation σ_v, δ_0 is set equal to $c\sigma_v$, where $c \geq 0$ is determined by an appropriate statistical confidence bound (e.g., $c = 3$ for 99% confidence). The bounded residual constraint enforces the estimate to be consistent with the observed image. In other words, at each iteration, the estimate is constrained such that at every pixel (n_1,n_2), the absolute value of the residual between the observed image value $g_k(n_1,n_2)$ and the pixel value obtained at (n_1,n_2) by simulating the imaging process using that estimate is required to be less than a predetermined bound.

From [Tru 84], the projection $z_k(i_1,i_2) \doteq \mathbf{P}_{n_1,n_2,k}[x_k(i_1,i_2)]$ of an arbitrary

$x_k(i_1, i_2)$ onto $C_{n_1,n_2,k}$ can be defined as:

$$\mathbf{P}_{n_1,n_2,k}[x_k(i_1,i_2)] = \begin{cases} x_k(i_1,i_2) + \frac{r^{(\mathbf{x}_k)}(n_1,n_2)-\delta_0}{\sum_{o_1}\sum_{o_2} h_k^2(n_1,n_2;o_1,o_2)} h_k(n_1,n_2;i_1,i_2) & \text{if } r^{(\mathbf{x}_k)}(n_1,n_2) > \delta_0 \\ x_k(i_1,i_2) & \text{if } -\delta_0 \le r^{(\mathbf{x}_k)}(n_1,n_2) \le \delta_0 \\ x_k(i_1,i_2) + \frac{r^{(\mathbf{x}_k)}(n_1,n_2)+\delta_0}{\sum_{o_1}\sum_{o_2} h_k^2(n_1,n_2;o_1,o_2)} h_k(n_1,n_2;i_1,i_2) & \text{if } r^{(\mathbf{x}_k)}(n_1,n_2) < -\delta_0 \end{cases} \tag{15.19}$$

Additional constraints, such as bounded energy, amplitude, and limited support, can be utilized to improve the results. For example, the amplitude constraint can be defined as

$$C_A = \{y_k(i_1,i_2) : \alpha \le y_k(i_1,i_2) \le \beta \text{ for } 0 \le i_1 \le N_1-1,\ 0 \le i_2 \le N_2-1\} \tag{15.20}$$

with amplitude bounds of $\alpha = 0$ and $\beta = 255$. The projection $\mathbf{P}_A$ onto the amplitude constraint C_A is defined as

$$\mathbf{P}_A[x_k(i_1,i_2)] = \begin{cases} 0 & \text{if } x_k(i_1,i_2) < 0 \\ x_k(i_1,i_2) & \text{if } 0 \le x_k(i_1,i_2) \le 255 \\ 255 & \text{if } x_k(i_1,i_2) > 255 \end{cases} \tag{15.21}$$

Given the above projections, an estimate $\hat{s}_k(i_1, i_2)$ of the ideal image $s_k(i_1, i_2)$ is obtained iteratively as

$$\hat{s}_k^{(j+1)}(i_1,i_2) = \mathbf{T}_A\ \mathbf{T}_{N_1-1,N_2-1,k}\ \mathbf{T}_{N_1-2,N_2-1,k} \cdots \mathbf{T}_{1,0,k}\ \mathbf{T}_{0,0,k}\ [\hat{s}_k^{(j)}(i_1,i_2)] \tag{15.22}$$

where $j = 0, 1, \ldots,\quad 0 \le i_1 \le N_1 - 1,\quad 0 \le i_2 \le N_2 - 1$, and $\mathbf{T}$ denotes the generalized projection operator defined in (15.17), and the observed blurred image $g_k(i_1, i_2)$ can be used as the initial estimate $\hat{s}_k^{(0)}(i_1, i_2)$. Note that in every iteration cycle j, each of the projection operators $\mathbf{T}_{n_1,n_2,k}$ is used once, but the order in which they are implemented is arbitrary.

15.4 Multiframe Restoration

The sequential nature of images in a video source can be used to both estimate the PSF of motion blurs and improve the quality of the restored images, making use of frame-to-frame image correlations. For example, in the case of linear motion blur, the extent of the spatial spread of a point source at a given pixel during the aperture time can be computed from an estimate of the frame-to-frame motion vector at that pixel, provided that the shutter speed of the camera is known [Tru 92]. In the remainder of this section, we address multiframe Wiener deconvolution to exploit the temporal correlation between the frames in the case of the spatially shift-invariant but temporally shift-varying blur model (15.2). We consider the extension of the spatially shift-varying restoration to a multiframe formulation in Chapter 17.

Applying the CLS filtering expression (15.13) to the observation equation (15.2) with $\mathbf{L} = \mathcal{R}_s^{-1}\mathcal{R}_v$, we obtain the Wiener estimate $\hat{\mathbf{s}}$ of the L frames $\mathbf{s}$ as

$$\hat{\mathbf{s}} = (\mathcal{D}^T\mathcal{D} + \mathcal{R}_s^{-1}\mathcal{R}_v)^{-1}\mathcal{D}^T\mathbf{g} \tag{15.23}$$

where

$$\hat{\mathbf{s}} \doteq \begin{bmatrix} \hat{\mathbf{s}}_1 \\ \vdots \\ \hat{\mathbf{s}}_L \end{bmatrix}, \quad \mathcal{R}_s \doteq \begin{bmatrix} \mathbf{R}_{s;11} & \cdots & \mathbf{R}_{s;1L} \\ \vdots & \ddots & \vdots \\ \mathbf{R}_{s;L1} & \cdots & \mathbf{R}_{s;LL} \end{bmatrix}, \quad \text{and } \mathcal{R}_v \doteq \begin{bmatrix} \mathbf{R}_{v;11} & \cdots & \mathbf{R}_{v;1L} \\ \vdots & \ddots & \vdots \\ \mathbf{R}_{v;L1} & \cdots & \mathbf{R}_{v;LL} \end{bmatrix}$$

in which $\mathbf{R}_{s;ij} \doteq \mathcal{E}\{\mathbf{s}_i\mathbf{s}_j^T\}$ and $\mathbf{R}_{v;ij} \doteq \mathcal{E}\{\mathbf{v}_i\mathbf{v}_j^T\}$, $i,j = 1,2,\ldots,L$. Note that if $\mathbf{R}_{s;ij} = \mathbf{0}$ for $i \neq j$, $i,j = 1,2,\ldots,L$, then the multiframe estimate becomes equivalent to stacking the L single-frame estimates obtained individually.

Again, direct computation of $\hat{\mathbf{s}}$ requires the inversion of the $N^2L \times N^2L$ matrix in (15.23). Because the blur PSF is not necessarily the same in each frame, and the image correlations are generally not shift-invariant in the temporal direction, the matrices $\mathcal{D}$, $\mathcal{R}_s$, and $\mathcal{R}_v$ are not block-Toeplitz; thus, a 3-D DFT would not diagonalize them. However, each $\mathbf{D}_k$ is block-Toeplitz. Furthermore, assuming each image and noise frame is wide-sense stationary in the 2-D plane, $\mathbf{R}_{s;ij}$ and $\mathbf{R}_{v;ij}$ are also block-Toeplitz. Approximating the block-Toeplitz submatrices $\mathbf{D}_i$, $\mathbf{R}_{s;ij}$, and $\mathbf{R}_{v;ij}$ by block-circulant submatrices, each submatrix can be diagonalized by separate 2-D DFT operations in an attempt to simplify matrix calculations.

To this effect, we define the $N^2L \times N^2L$ transformation matrix

$$\mathcal{W}^{-1} \doteq \begin{bmatrix} \mathbf{W}^{-1} & \cdots & 0 \\ \vdots & \ddots & \vdots \\ 0 & \cdots & \mathbf{W}^{-1} \end{bmatrix}$$

where the $N^2 \times N^2$ matrix $\mathbf{W}^{-1}$ is the 2-D DFT operator defined previously. Note that the matrix $\mathcal{W}^{-1}$, when applied to $\mathbf{s}$, stacks the vectors formed by 2-D DFTs of the individual frames, but it is not the 3-D DFT operator.

We premultiply both sides of (15.23) with $\mathcal{W}^{-1}$ to obtain

$$\begin{aligned} \mathcal{W}^{-1}\hat{\mathbf{s}} &= [\mathcal{W}^{-1}\mathcal{D}\mathcal{W}\mathcal{W}^{-1}\mathcal{D}^T\mathcal{W} + (\mathcal{W}\mathcal{R}_s\mathcal{W}^{-1})^{-1}\mathcal{W}^{-1}\mathcal{R}_v)\mathcal{W}]^{-1}\mathcal{W}^{-1}\mathcal{D}^T\mathcal{W}\mathcal{W}^{-1}\mathbf{g} \\ \hat{\mathcal{S}} &= \mathcal{Q}^{-1}\mathcal{H}^*\mathcal{G} \end{aligned} \tag{15.24}$$

where $\mathcal{G} = \mathcal{W}^{-1}\mathbf{g}$ and $\mathcal{P}_s = \mathcal{W}^{-1}\mathcal{R}_s\mathcal{W}$, $\mathcal{H}^* = \mathcal{W}^{-1}\mathcal{D}^T\mathcal{W}$ and $\mathcal{Q} = \mathcal{H}\mathcal{H}^* + \mathcal{P}_s$ are block matrices whose blocks are diagonal. The structure of these block matrices is depicted in Figure 15.3.

In the following, we propose two approaches for the computation of $\mathcal{Q}^{-1}$: a general algorithm called the cross-correlated multiframe (CCMF) Wiener filter, which requires the power and cross-power spectra of the frames; and a specific algorithm called the motion-compensated multiframe filter (MCMF), which applies to the special case of global, constant-velocity interframe motion when a closed-form solution that does not involve any matrix inversion becomes possible.

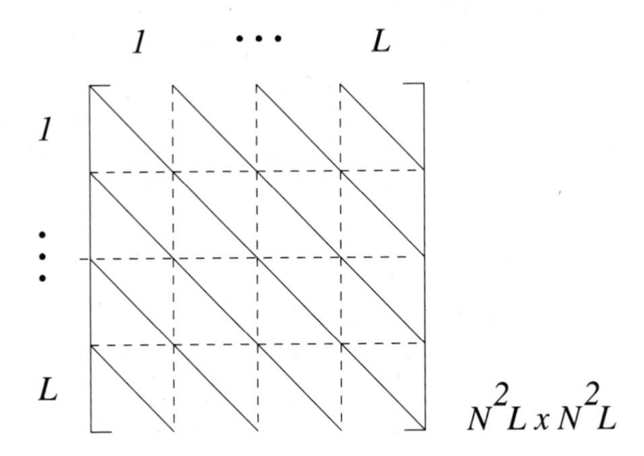

Figure 15.3: The structure of the matrices.

15.4.1 Cross-Correlated Multiframe Filter

It was shown in [Ozk 92] that $\mathcal{Q}^{-1}$ is also a block matrix with diagonal blocks, and the elements of $\mathcal{Q}^{-1}$ can be computed by inverting N^2 matrices each $L \times L$. The details of this derivation will not be given here; instead, in the following we will demonstrate the computation of $\mathcal{Q}^{-1}$ by means of an example. The inversion of $L \times L$ matrices can be performed in parallel. Furthermore, if L is sufficiently small, the $L \times L$ matrices can be inverted using an analytic inversion formula.

Example: The case of $N^2 = 4$ and $L = 2$.

Let

$$\mathcal{Q} = \left[\begin{array}{cccc|cccc} a_1 & 0 & 0 & 0 & b_1 & 0 & 0 & 0 \\ 0 & a_2 & 0 & 0 & 0 & b_2 & 0 & 0 \\ 0 & 0 & a_3 & 0 & 0 & 0 & b_3 & 0 \\ 0 & 0 & 0 & a_4 & 0 & 0 & 0 & b_4 \\ - & - & - & - & - & - & - & - \\ c_1 & 0 & 0 & 0 & d_1 & 0 & 0 & 0 \\ 0 & c_2 & 0 & 0 & 0 & d_2 & 0 & 0 \\ 0 & 0 & c_3 & 0 & 0 & 0 & d_3 & 0 \\ 0 & 0 & 0 & c_4 & 0 & 0 & 0 & d_4 \end{array} \right] \tag{15.25}$$

Then according to Lemma 1 in [Ozk 92],

$$\mathcal{Q}^{-1} = \left[\begin{array}{cccc|cccc} \alpha_1 & 0 & 0 & 0 & \beta_1 & 0 & 0 & 0 \\ 0 & \alpha_2 & 0 & 0 & 0 & \beta_2 & 0 & 0 \\ 0 & 0 & \alpha_3 & 0 & 0 & 0 & \beta_3 & 0 \\ 0 & 0 & 0 & \alpha_4 & 0 & 0 & 0 & \beta_4 \\ \hline \gamma_1 & 0 & 0 & 0 & \delta_1 & 0 & 0 & 0 \\ 0 & \gamma_2 & 0 & 0 & 0 & \delta_2 & 0 & 0 \\ 0 & 0 & \gamma_3 & 0 & 0 & 0 & \delta_3 & 0 \\ 0 & 0 & 0 & \gamma_4 & 0 & 0 & 0 & \delta_4 \end{array}\right] \quad (15.26)$$

where the elements α_i, β_i, γ_i and δ_i are computed by a series of four 2×2 matrix inversions, given by

$$\begin{bmatrix} a_i & b_i \\ c_i & d_i \end{bmatrix}^{-1} = \begin{bmatrix} \alpha_i & \beta_i \\ \gamma_i & \delta_i \end{bmatrix} \qquad i = 1, 2, 3, 4$$

The computation of the multiframe Wiener estimate requires the knowledge of the covariance matrices $\mathcal{R}_s$ and $\mathcal{R}_v$. We assume that the noise is spatio-temporally white; thus, the matrix $\mathcal{R}_v$ is diagonal with all diagonal entries equal to σ_v^2, although the formulation allows for any covariance matrix for the noise. The estimation of the multiframe ideal video covariance matrix $\mathcal{R}_s$ can be performed by either the periodogram method or 3-D AR modeling [Ozk 92]. Once $\mathcal{Q}^{-1}$ is determined using the scheme discussed, the Wiener estimate can be easily computed from (15.24).

15.4.2 Motion-Compensated Multiframe Filter

Assuming that there is global (circular) motion, the auto power spectra of all the frames are the same and the cross spectra are related by a phase factor related to the motion information. Given the motion vectors (one for each frame) and the auto power spectrum of the reference frame, the matrix $\mathcal{Q}^{-1}$ (hence, the Wiener estimate) can be computed analytically using the Sherman-Morrison formula without an explicit matrix inversion. The interested reader is referred to [Ozk 92] for further details.

15.5 Examples

In this section, we provide two examples of spatially invariant blurring, an out-of-focus blur and a linear motion blur, and an example of spatially varying out-of-focus blurring due to the depth-of-field problem. All of these examples demonstrate the restoration of real camera blurs, not simulated blurs.

Figure 15.4 (a) shows an out-of-focus text image that is 500 pixels $\times$ 480 lines. The out-of-focus blur is modeled by a uniform PSF with a circular support. We have identified the diameter of the circular PSF as 15 pixels in this case using a maximum likelihood blur identification algorithm [Pav 92]. The restored image obtained by Wiener filtering is shown in Figure 15.4 (b). The power spectrum of the sharp image has been derived from an AR model of a typical text image. The restored image is clearly legible, while the blurred one was not.

A 452 pixels $\times$ 200 lines segment of the picture of a moving train, which suffers from linear motion blur, is shown in Figure 15.5 (a). It is known that the train translates parallel to the image plane, so that the resulting blur PSF is space-invariant. Linear motion blur is modeled by a uniform rectangular (boxcar) PSF. Maximum likelihood estimate of the extent of this rectangular PSF has been found to be 32 pixels. Figure 15.5 (b) depicts the restored image obtained by Wiener filtering, where the power spectrum of the original image is approximated by that of the blurred image computed using the periodogram method. The result shows significant resolution improvement.

Finally, Figure 15.6 (a) shows a space-varying blurred image due to the depth-of-field problem, where the flowers, the film boxes, and the background are all at different depths. Because the camera was focused on the background, the flowers and the film boxes suffer from varying amounts of blur. In this case we have manually segmented the image into three regions, and the blur on the flowers and film boxes have been separately identified as 12 and 25 pixels, respectively. The image is then restored using the POCS approach. The restored image, shown in Figure 15.6 (b), looks much sharper.

15.6 Exercises

1. Suppose we have a 7×1 (pixels) uniform motion blur; that is, $h_k(n_1, n_2)$ is a 1-D boxcar function that is 7 pixels long. How many elements of $H_k(u_1, u_2)$ in (15.10) are exactly zero, assuming a 256×256 DFT? Repeat for an 8×1 (pixels) uniform motion blur. Show, in general, that exact zeros will be encountered when the DFT size is an integer multiple of the blur size.

2. Show that the CLS filter (15.14) becomes a Wiener filter when $\mathbf{L} = \mathbf{R}_s^{-1}\mathbf{R}_v$.

3. While the term $\alpha|L(u_1, u_2)|^2$ in (15.14) avoids excessive noise amplification, it generates a different kind of artifacts, known as "regularization artifacts" due to the deviation of the regularized filter from the exact inverse. Can you provide a quantitative analysis of the tradeoff between noise amplification and regularization artifacts? (Hint: see [Tek 90].)

4. Iterative signal and image restoration based on variations of the Landweber iteration has been heavily examined [Kat 89]. Discuss the relationship between the Landweber iterations and the POCS method. (Hint: see [Tru 85].)

5. Discuss the frequency response of the MCMF filter (Section 15.4.2) as it relates to the theory of motion-compensated filtering presented in Chapter 13.

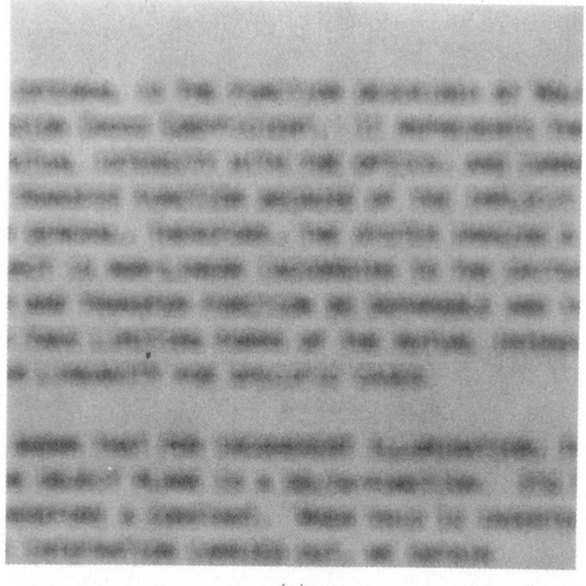

(a)

INTEGRAL IS THE FUNCTION DESCRIBED BY WOL
BION CROSS COEFFICIENT. IT REPRESENTS TH
UTUAL INTENSITY WITH THE OPTICS, AND CANN
TRANSFER FUNCTION BECAUSE OF THE IMPLICIT
GENERAL, THEREFORE, THE SYSTEM IMAGING A
ECT IS NON-LINEAR (ACCORDING TO THE CRITE
AND TRANSFER FUNCTION BE SEPARABLE AND I
TAKE LIMITING FORMS OF THE MUTUAL INTENS
M LINEARITY FOR SPECIFIC CASES.

SHOWN THAT FOR INCOHERENT ILLUMINATION, T
E OBJECT PLANE IS A DELTA-FUNCTION. ITS
EREFORE A CONSTANT. WHEN THIS IS INSERTE
INTEGRATION CARRIED OUT, WE OBTAIN

(b)

Figure 15.4: a) The out-of-focus text image; b) the text image restored by Wiener filtering. (Courtesy Gordana Pavlovic)

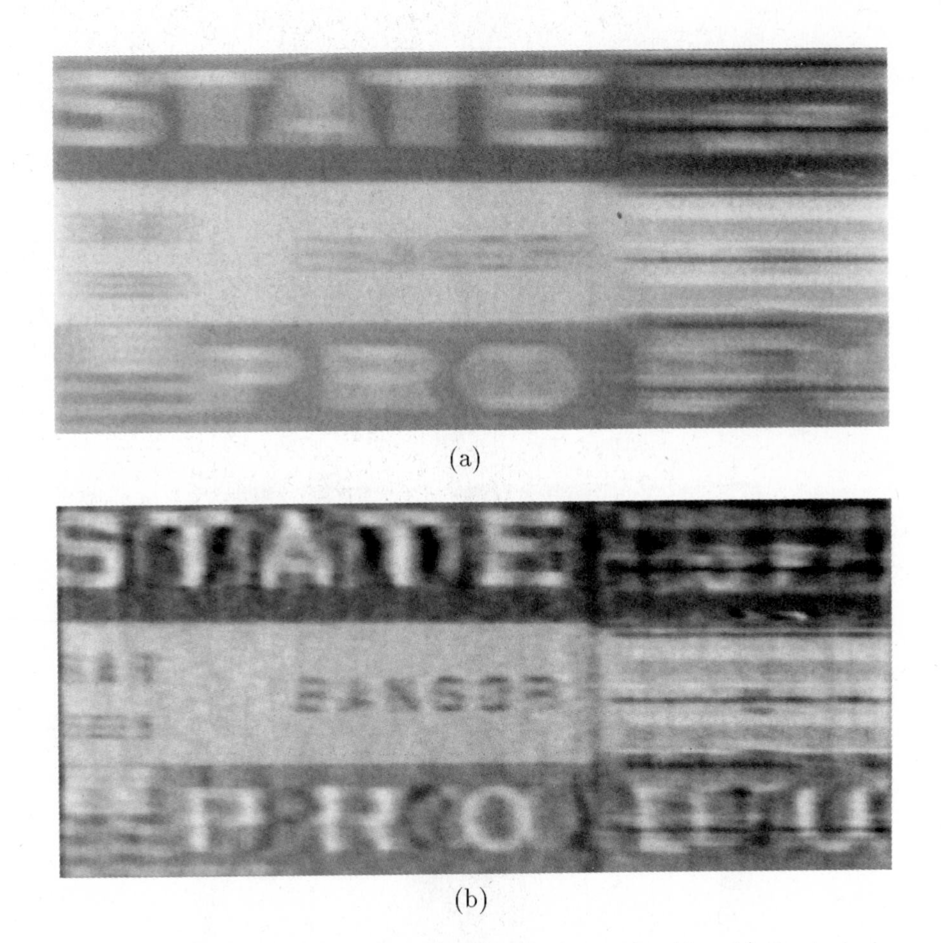

Figure 15.5: a) The train image degraded by linear motion blur; b) the train image restored by Wiener filtering. (Courtesy Gordana Pavlovic)

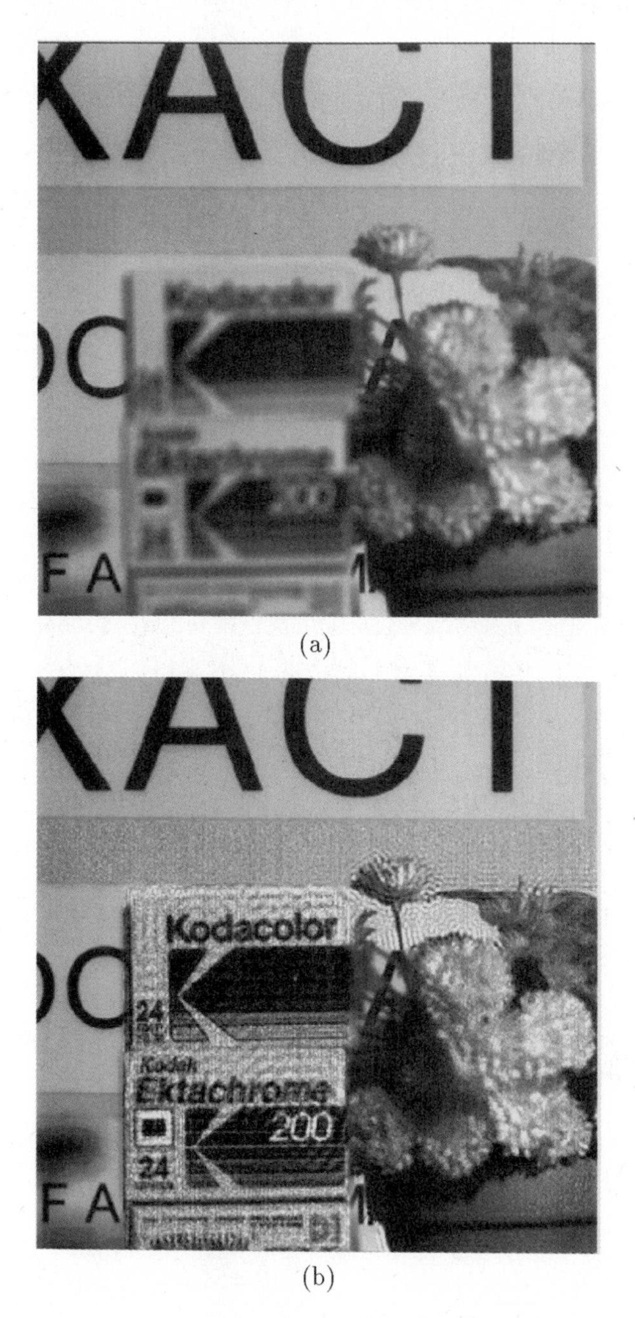

Figure 15.6: a) A space-varying blur due to the depth-of-field problem; b) the restored image using the POCS method. (Courtesy Mehmet Ozkan)

Bibliography

[And 77] H. C. Andrews and B. R. Hunt, *Digital Image Restoration*, Englewood Cliffs, NJ: Prentice-Hall, 1977.

[Gon 87] R. C. Gonzalez and P. Wintz, *Digital Image Processing*, Reading, MA: Addison-Wesley, 1987.

[Kat 89] A. K. Katsaggelos, "Iterative image restoration algorithms," *Optical Engineering*, vol. 28, pp. 735–748, 1989.

[Ozk 92] M. K. Ozkan, A. T. Erdem, M. I. Sezan, and A. M. Tekalp, "Efficient multiframe Wiener restoration of blurred and noisy image sequences," *IEEE Trans. Image Proc.*, vol. 1, pp. 453–476, Oct. 1992.

[Ozk 94] M. K. Ozkan, A. M. Tekalp, and M. I. Sezan, "POCS-based restoration of space-varying blurred images," *IEEE Trans. Image Proc.*, vol. 3, pp. 450–454, July 1994.

[Pav 92] G. Pavlovic and A. M. Tekalp, "Maximum likelihood parametric blur identification based on a continuous spatial domain model," *IEEE Trans. Image Proc.*, vol. 1, pp. 496–504, Oct. 1992.

[Rus 87] C. K. Rushforth, "Signal restoration, functional analysis, and fredholm integral equations of the first kind," in *Image Recovery: Theory and Application*, H. Stark, ed., Academic Press, 1987.

[Saw 74] A. Sawchuk, "Space-variant image restoration by coordinate transformations," *J. Opt. Soc. Am.*, vol. 60, pp. 138–144, Feb. 1974.

[Sch 81] R. W. Schafer, R. M. Mersereau, and M. A. Richards, "Constrained iterative restoration algorithms," *Proc. IEEE*, vol. 69, pp. 314–332, Apr. 1981.

[Sez 90] M. I. Sezan and A. M. Tekalp, "Survey of recent developments in digital image restoration," *Optical Engineering*, vol. 29, pp. 393–404, 1990.

[Sez 91] M. I. Sezan and H. J. Trussell, "Prototype image constraints for set-theoretic image restoration," *IEEE Trans. Signal Proc.*, vol. 39, pp. 2275–2285, Oct. 1991.

[Sta 87] H. Stark, ed., *Image Recovery: Theory and Application*, Academic Press, 1987.

[Tek 90] A. M. Tekalp and M. I. Sezan, "Quantitative analysis of artifacts in linear space-invariant image restoration," *Multidim. Syst. and Sign. Proc.*, vol. 1, no. 2, pp. 143–177, 1990.

[Tek 94] A. M. Tekalp and H. Kaufman, "Estimation techniques in image restoration," in *Digital Image Processing: Techniques and Applications*, Control and Dynamic Systems, vol. 67, Academic Press, 1994.

[The 89] C. W. Therrien and H. T. El-Shaer, "Multichannel 2-D AR spectrum estimation," *IEEE Trans. Acoust., Speech, and Signal Proc.*, vol. 37, pp. 1798–1800, 1989.

[Tru 78] H. J. Trussell and B. R. Hunt, "Image restoration of space-variant blurs by sectional methods," *IEEE Trans. Acoust., Speech and Signal Proc.*, vol. 26, pp. 608–609, 1978.

[Tru 79] H. J. Trussell and B. R. Hunt, "Improved methods of maximum a posteriori restoration," *IEEE Trans. Comput.*, vol. 27, no 1, pp. 57–62, 1979.

[Tru 84] H. J. Trussell and M. Civanlar, "Feasible solutions in signal restoration," *IEEE Trans. Acoust. Speech Sign. Proc.*, vol. 32, pp. 201–212, 1984.

[Tru 85] H. J. Trussell and M. Civanlar, "The Landweber iteration and projection onto convex sets," *IEEE Trans. Acoust. Speech Sign. Proc.*, vol. 33, pp. 1632–1634, Dec. 1985.

[Tru 92] H. J. Trussell and S. Fogel, "Identification and restoration of spatially variant motion blurs in sequential images," *IEEE Trans. Image Proc.*, vol. 1, pp. 123–126, Jan. 1992.

Chapter 16

STANDARDS CONVERSION

Various digital video systems, ranging from all-digital high-definition TV to videophone, have different spatio-temporal resolution requirements, leading to the emergence of different format standards to store, transmit, and display digital video signals. The task of converting digital video from one standard to another is referred to as standards conversion, which includes frame/field rate down-/up-conversion and interlacing/de-interlacing. Frame and field rate up-conversion increase the temporal sampling rate in progressive and interlaced video, respectively. De-interlacing refers to conversion from interlaced to progressive video. These problems are illustrated in Figure 16.1. Note that in field rate doubling two fields, one even and one odd, need to be inserted between existing fields.

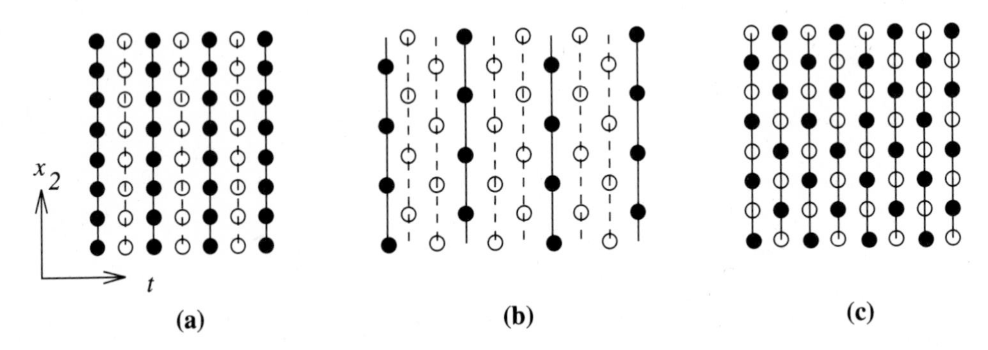

Figure 16.1: Different up-conversion problems: a) Frame rate up-conversion; b) field rate up-conversion; c) de-interlacing.

Standards conversion enables the exchange of information among various digital video systems employing different standards to ensure their interoperability. Some examples of commonly used video standards employed in such applications as motion pictures, workstations, and NTSC TV are shown in Figure 16.2. For example, video recorded by motion picture equipment can be used as high-quality source material for HDTV after proper frame rate conversion, and spatio-temporal down-conversion helps data compression. In addition to facilitating exchange of information, frame rate up-conversion generally yields higher visual quality through better motion rendition and less flicker, and de-interlacing provides improved spatial resolution in the vertical direction.

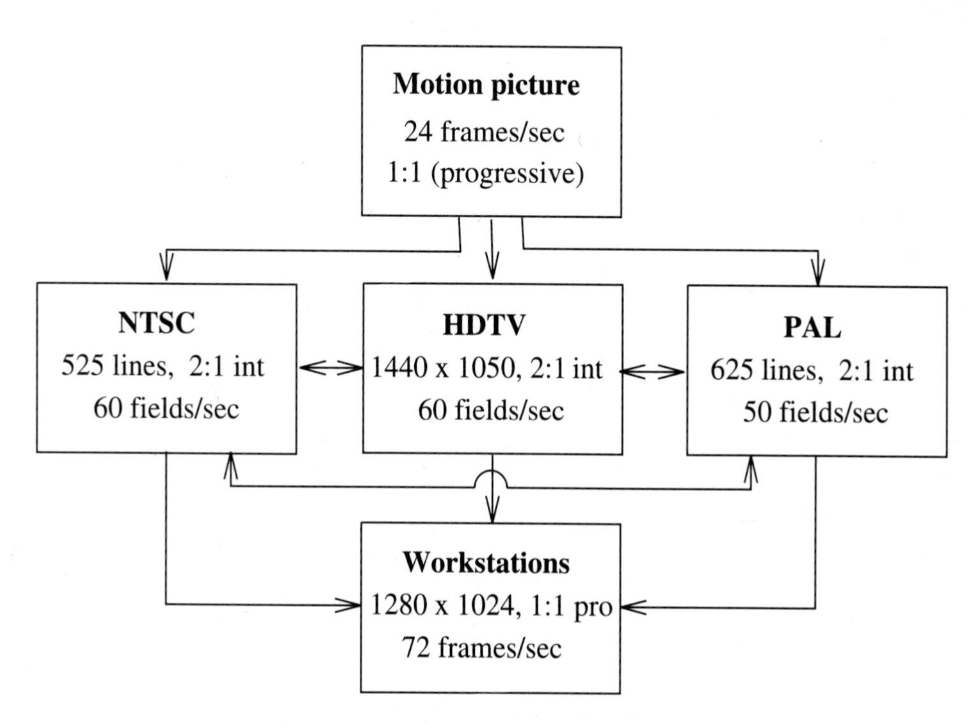

Figure 16.2: Typical standards conversion applications.

All standards conversion problems essentially deal with sampling structure conversion, which was introduced in Chapter 4. In Chapter 4, the sampling structure conversion problem has been discussed for arbitrary spatio-temporal signals within the framework of linear filtering. It is the goal of this chapter to extend this framework by incorporating motion models, following the analysis in Chapter 13, which characterize temporal variations of the video signal. The down-conversion problem is studied in Section 16.1, where we discuss methods with or without using an anti-alias filter. A variety of up-conversion methods exist in the literature, which can be organized in a number of different ways. One way would be to organize them as frame/field rate up-conversion and de-interlacing algorithms. However, because the main distinction between these two problems is the structure of the input and output lattices, and the underlying principles are the same, here we choose to classify them as intraframe filtering, motion-adaptive filtering, and motion-compensated filtering approaches similar to Chapter 14. Intraframe and motion-adaptive algorithms, discussed in Section 16.2, are practical but suboptimal algorithms that do not require explicit estimation of the motion trajectories. On the other hand, motion-compensated algorithms, presented in Section 16.3, are based on modeling and estimation of the motion trajectories. They yield the best results provided that the underlying motion models and the estimated motion trajectories are sufficiently accurate.

16.1 Down-Conversion

Spatio-temporal down-conversion is commonly used in video image communications over very-low-bitrate channels and in real-time transfer of uncompressed video data to and from storage devices such as hard disks and CD-ROMs. Frame/field rate down-conversion and interlacing are special cases of spatio-temporal down-conversion. For example, in videophone over existing copper networks at about 10 kbps, we can afford to transmit at most 5-10 frames/sec even with the use of state-of-the-art lossy compression schemes. Spatial down-conversion, another special case, refers to subsampling on a 2-D lattice to reduce the spatial resolution of individual frames or fields.

Recall from Chapter 4 that the down-conversion process refers to anti-alias filtering (optional) followed by downsampling (subsampling). First, let's review our model for downsampling. Suppose we downsample from an M-D input lattice Λ_1 with the sampling matrix $\mathbf{V}_1$ to an M-D output lattice Λ_2 with the sampling matrix $\mathbf{V}_2$. We define an $M \times M$ subsampling matrix

$$\mathbf{S} = \mathbf{V}_1^{-1}\mathbf{V}_2 \tag{16.1}$$

such that $\mathbf{V}_2 = \mathbf{V}_1\mathbf{S}$. Since $\mathbf{V}_1$ is always invertible, a subsampling matrix can always be defined. Then the sites of the output lattice Λ_2 can be expressed as

$$\mathbf{y} = \mathbf{V}_1\mathbf{S}\mathbf{n}, \quad \mathbf{n} \in \mathbf{Z}^M \tag{16.2}$$

For example, spatio-temporal subsampling according to the matrix

$$\mathbf{S} = \begin{bmatrix} 2 & 0 & 1 \\ 0 & 1 & 0 \\ 0 & 0 & 1 \end{bmatrix}$$

corresponds to discarding even and odd columns in alternating frames, respectively. The relative merits of down-conversion with or without anti-alias filtering are discussed in the following.

16.1.1 Down-Conversion with Anti-Alias Filtering

Should the Fourier spectrum of the video occupy the entire unit cell of the reciprocal lattice $\mathbf{V}_1^*$, an anti-alias filter is needed before downsampling in order to avoid aliasing errors. The anti-alias filter attenuates all frequencies outside the unit cell of the reciprocal lattice $\mathbf{V}_2^* = (\mathbf{V}_1\mathbf{S})^*$ so that the spectral replications centered about the sites of the reciprocal lattice $\mathbf{V}_2^*$ do not overlap with each other. However, because of this anti-alias filtering, the high-frequency information in the original signal is permanently lost. That is, if the video signal needs to be up-converted at the receiver for display purposes, the original spectral content of the video cannot be recovered even by using ideal motion-compensated reconstruction filters. This is illustrated by an example in the following (see Figure 16.4). Clearly, in this case successive frames contain no new information, and a simple low-pass filter is as good as a motion-compensated reconstruction filter.

16.1.2 Down-Conversion without Anti-Alias Filtering

Down-conversion without anti-alias filtering, which refers to simply discarding a subset of the input samples specified by a subsampling matrix $\mathbf{S}$, may preserve the frequency content of the original signal except in the case of critical velocities. Provided that we have global, constant-velocity motion, linear, shift-invariant motion-compensated filtering can be employed for subsequent up-conversion of the signal without any loss of resolution. This is clarified with the aid of the following example.

> *Example: Comparison of down-conversion with or without anti-aliasing*
>
> For the sake of simplicity, let's consider a single spatial coordinate and a temporal coordinate. Suppose that the output lattice is obtained by

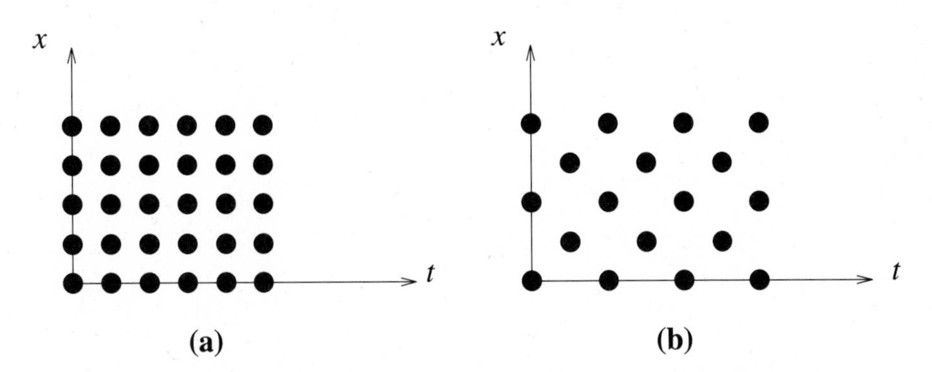

Figure 16.3: Two-dimensional a) input and b) output lattices.

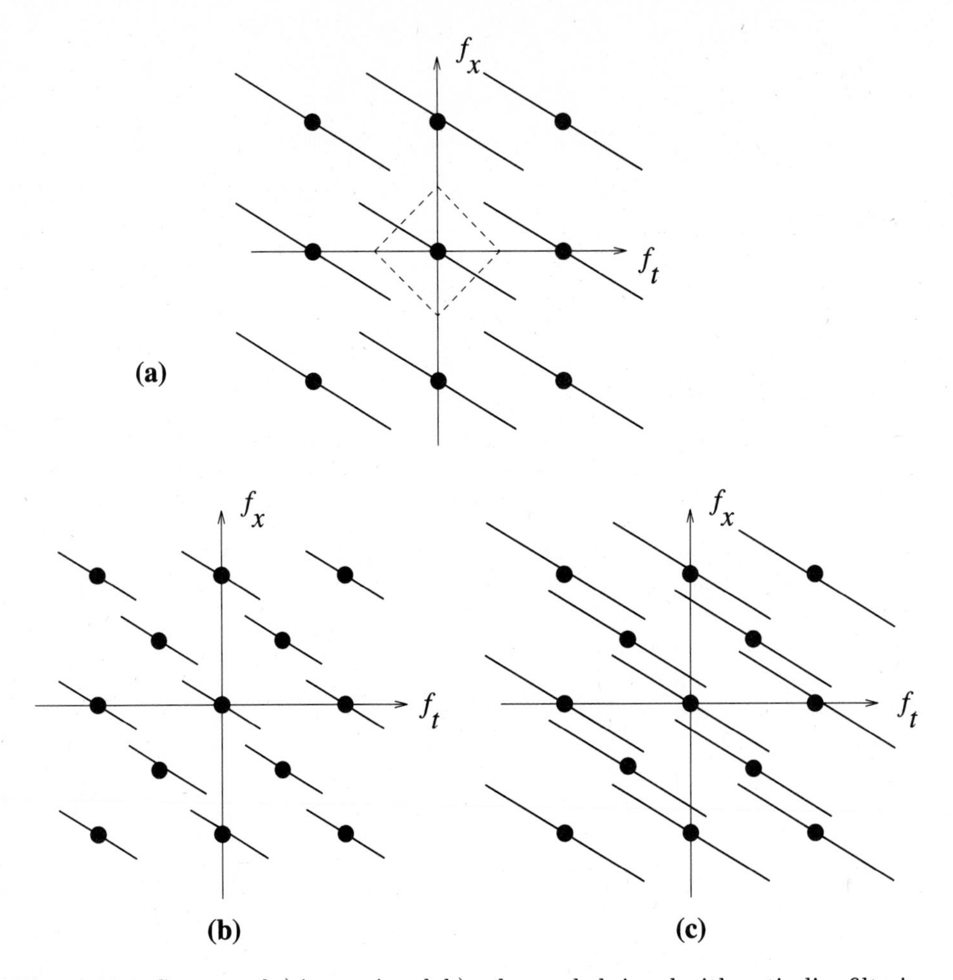

Figure 16.4: Spectra of a) input signal; b) subsampled signal with anti-alias filtering; c) subsampled signal without anti-alias filtering.

subsampling a progressive input lattice according to the subsampling matrix

$$\mathbf{S} = \begin{bmatrix} 2 & 1 \\ 0 & 1 \end{bmatrix}$$

which corresponds to discarding even and odd spatial samples in alternating time samples. The 2-D input and output lattices in this case are depicted in Figure 16.3.

The spectrum of the input video, assuming global, constant-velocity motion, is depicted in Figure 16.4 (a), where the solid dots indicate the sites of the reciprocal lattice $\mathbf{V}_1^*$. The dotted lines denote the support of an ideal anti-alias filter that would be used in the case of down-conversion with anti-alias filtering. The spectra of the down-converted signal with and without anti-alias filtering are shown in Figures 16.4 (b) and (c), respectively. The loss of high-frequency information when an anti-alias filter has been employed can be clearly observed. Further observe that the spectrum of the down-converted signal in Figure 16.4 (c) retains the high-frequency content of the original signal, provided that the subsampling structure does not lead to a critical velocity.

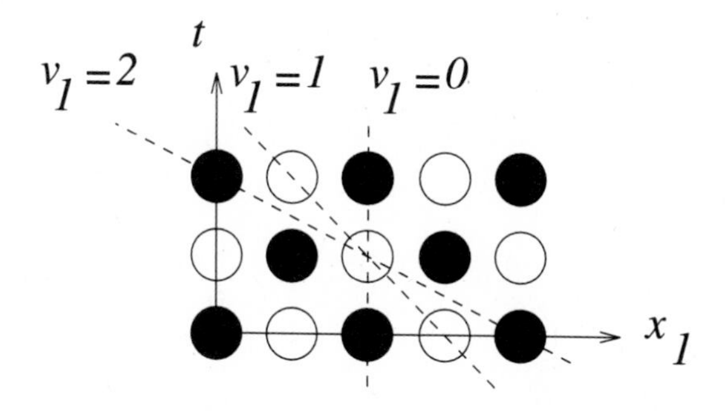

Figure 16.5: Critical velocities in the spatio-temporal domain.

It can easily be seen that for the subsampling matrix in this example, the velocities $v_c = 2i + 1,\ \ i \in Z$ constitute critical velocities in the absence of anti-alias filtering. For these velocities, the motion trajectory passes through either all existing or non-existing pixel sites at every frame, as shown in Figure 16.5. Note that with the use of proper anti-alias filtering, the replicas in the frequency domain can never overlap; hence, there are no critical velocities.

In conclusion, down-conversion to sub-Nyquist rates followed by up-conversion without loss of resolution is possible only if no anti-alias filtering has been used in the down-conversion process, and we do not have a critical velocity. Assuming global, constant-velocity motion, this process leads to motion-adaptive downsampling, where the subsampling matrix is selected to avoid the critical velocities given an estimate of the motion vector. Up-conversion methods are studied in the next three sections.

16.2 Practical Up-Conversion Methods

Standards conversion problems of common interest include frame/field rate up-conversion and de-interlacing. Examples of frame rate conversion include conversion from a digitized motion picture recorded with a temporal rate of 24 frames/sec to NTSC format, which requires 60 fields/sec. Conversion from NTSC to PAL format requires field (scan) rate conversion from 60 to 50 fields/s. Field rate doubling and de-interlacing are generally employed in improved-definition TV receivers to provide better visual quality.

Frame rate up-conversion refers to the case where the input and output sequences are both progressive. Suppose we represent a sequence of input frames as

$$\ldots, \mathbf{s}(t-1), \mathbf{s}(t), \mathbf{s}(t+1), \mathbf{s}(t+2), \ldots$$

Then the output sequence in the case of up-conversion by an integer factor L is given by

$$\ldots, \mathbf{s}(t), \underbrace{\mathbf{s}(t^*), \ldots, \mathbf{s}(t^*)}_{\text{L-1 times}}, \mathbf{s}(t+1), \ldots$$

where t^* denotes the time instances where new frames are inserted.

Field rate, or scan rate, up-conversion, where both the input and output sequences are interlaced, by an arbitrary factor is complicated by the fact that the ordering of even and odd fields must be preserved. If we let

$$\ldots, \mathbf{s}((t-1)_o), \mathbf{s}((t-1)_e), \mathbf{s}(t_o), \mathbf{s}(t_e), \mathbf{s}((t+1)_o), \mathbf{s}((t+1)_e), \ldots$$

denote a sequence of input fields, where $\mathbf{s}(t_o)$ and $\mathbf{s}(t_e)$ are the odd and even fields, respectively, then the output sequence

$$\ldots, \mathbf{s}(t_o), \mathbf{s}(t_e), \mathbf{s}(t_o^*), \mathbf{s}(t_e^*), \mathbf{s}((t+1)_o), \ldots$$

corresponds to the special case of doubling the scan rate. Note in this case that an even number of fields must be inserted between two existing fields to preserve the ordering of the fields. Conversion by an arbitrary factor can generally be modeled by de-interlacing each field, then frame rate conversion to the desired temporal rate, followed by interlacing.

De-interlacing refers to up-conversion from an interlaced to a progressive sampling lattice at the same temporal rate. It is often necessary to view interlaced video on progressively scanned display devices. In addition, proper de-interlacing reduces line flicker and improves the vertical resolution of the displayed images.

Both frame/field rate up-conversion and de-interlacing are based on the same principles of sampling structure conversion. Indeed, the only difference between frame/field rate up-conversion and de-interlacing is the structure of the input and output lattices. Standards up-conversion algorithms can be classified, in the order of increasing complexity, as intrafield, motion-adaptive, and motion-compensated filtering techniques. Although the motion-compensated approach is optimal, provided that a motion model is given and the trajectory can be accurately estimated, it is highly sensitive to errors. This fact, along with the cost of hardware implementation, have led to the development of suboptimal but more robust algorithms which do not require explicit estimation of the motion information. We first discuss some of these algorithms in the following.

16.2.1 Intraframe Filtering

Intraframe filtering refers to algorithms that need no more than a single framestore. Frame/field repetition algorithms for frame/field rate conversion and spatial interpolation methods for de-interlacing are examples of intraframe filtering.

Frame/Field Rate Conversion

The most obvious and simple way to increase the frame or the field rate is to repeat the existing information. Common examples of frame/field repetition are the "3 to 2 pull-down" method used for conversion of a motion picture source to NTSC video, and doubling of the field rate, which has been adopted by several TV manufacturers in Europe for commercial 100-Hz receivers.

In the "3 to 2 pull-down" method, each odd frame of the digitized motion picture is repeated three times and each even frame is repeated twice, yielding a 60 Hz field rate. That is, the output is a sequence of fields given by

$$\ldots, \mathbf{s}(t-1), \mathbf{s}((t-1)^*), \mathbf{s}((t-1)^*), \mathbf{s}(t), \mathbf{s}(t^*), \mathbf{s}(t+1), \mathbf{s}((t+1)^*), \mathbf{s}((t+1)^*), \mathbf{s}(t+2), \mathbf{s}((t+2)^*), \ldots$$

In order to characterize the 3:2 pull-down method in the temporal frequency domain, we describe it through the steps illustrated in Figure 16.6. They are: (i) insert four zero frames in between every existing frame, (ii) apply a zero-order hold filter whose impulse response is depicted, and (iii) downsample the resulting sequence of frames by two. The 3:2 pull-down method introduces temporal aliasing because the zero-order hold filter is a poor low-pass filter. The jerky motion rendition due to this aliasing is hardly visible at the spatio-temporal resolution provided by current NTSC receivers. However, with bigger displays and high-resolution video formats, there is need for more sophisticated frame/field rate conversion algorithms.

For scan rate doubling, there exist more than one way to repeat the fields. We provide two algorithms in the following.

Algorithm 1: In the first approach, the existing odd frame is repeated to form the next even frame, and the existing even frame is repeated for the next odd frame. This can be expressed as

$$\begin{aligned} \mathbf{s}(t_e^*) &= \mathbf{s}(t_o) \\ \mathbf{s}(t_o^*) &= \mathbf{s}(t_e) \end{aligned} \quad (16.3)$$

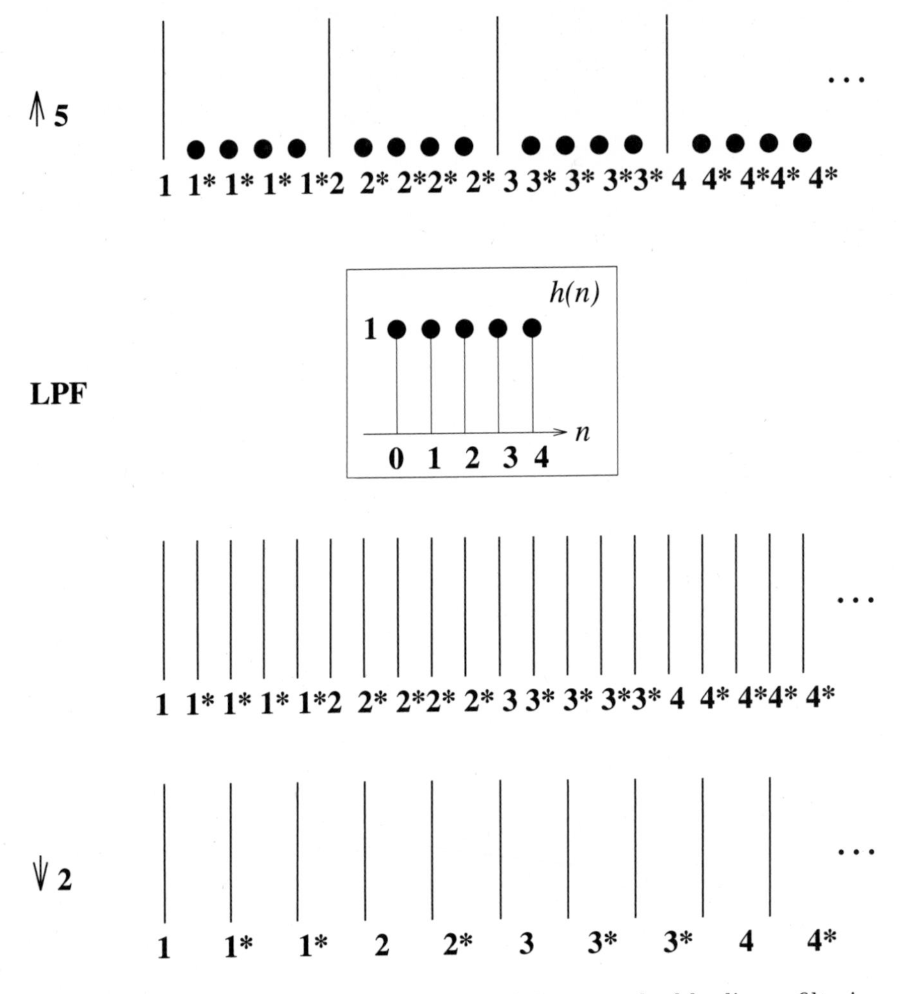

Figure 16.6: Interpretation of the 3 to 2 pull-down method by linear filtering.

Algorithm-1 has reasonably good performance with moving scenes, but poor results in stationary regions is inevitable.

Algorithm 2: The second approach repeats an even field for the next even field, and an odd field for the next odd field, which can be expressed as

$$\begin{aligned} \mathbf{s}(t_e^*) &= \mathbf{s}(t_e) \\ \mathbf{s}(t_o^*) &= \mathbf{s}(t_o) \end{aligned} \tag{16.4}$$

This strategy is optimal for stationary scenes but fails for the moving parts. Obviously, neither algorithm alone is sufficient for both stationary and moving regions

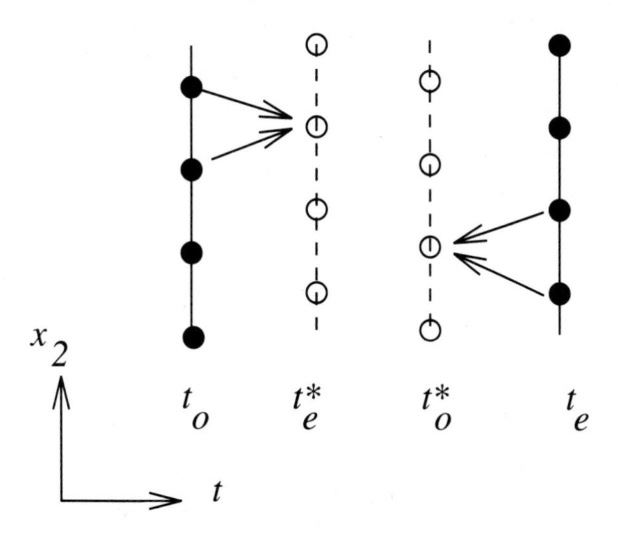

Figure 16.7: Two-point line averaging.

of the scene, which motivates the need for motion-adaptive or motion-compensated filtering schemes.

An interesting application which requires scan rate conversion by a rational factor is conversion from NTSC to PAL and vice versa. The NTSC standard employs 60 fields/sec and 262.5 lines/field, whereas the PAL standard uses 50 fields/sec and 312.5 lines/field. The conversion from NTSC to PAL may be achieved by dropping a complete frame (an even and an odd field) every five frames, and spatially interpolating for the missing lines. On the other hand, PAL to NTSC conversion requires dropping some extra lines per frame, and replicating a complete frame every five frames.

Alternatively, one can use a line averaging filter, defined by

$$\begin{aligned} s(x_1, x_2, t_e^*) &= \frac{1}{2}\left[s(x_1, x_2 - 1, t_o) + s(x_1, x_2 + 1, t_o)\right] \\ s(x_1, x_2, t_o^*) &= \frac{1}{2}\left[s(x_1, x_2 - 1, t_e) + s(x_1, x_2 + 1, t_e)\right] \end{aligned} \tag{16.5}$$

for field rate up-conversion. The solid pixels in Figure 16.7 denote the input pixels, and the open circles pointed by the arrows denote the output pixels. The line-averaging filter provides reasonable performance in the moving regions. However, in stationary image parts it introduces blurring because it is purely a spatial filter. Notice that while the repetition algorithms yield jagged edges, averaging algorithms provide blurred edges. A compromise between the performances of these two classes of algorithms may be reached by using interframe filtering.

De-interlacing

Intraframe de-interlacing techniques can employ line doubling (line repetition), line averaging, or more sophisticated nonlinear spatial interpolation methods within a single frame. Let $s(x_1, x_2, t_i)$, $i = e, o$, denote the even or the odd field, respectively, at time t, such that $s(x_1, x_2, t_e)$ is zero for odd values of x_2, and $s(x_1, x_2, t_o)$ is zero for even values of x_2. The projection of two frames of an interlaced video on the (x_2, t) coordinates is shown in Figure 16.8, where each circle denotes the cross-section of a complete line of video. The shaded circles denote lines that are available, and the open circles show the lines to be interpolated.

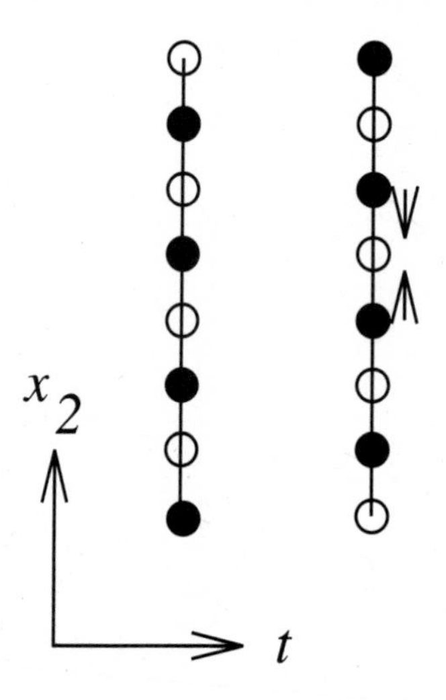

Figure 16.8: Intrafield filtering.

Assuming that the index of the first line of each field is zero, the line repetition algorithm can be described by

$$s(x_1, x_2, t_e) = s(x_1, x_2 - 1, t_e) \quad \text{for } x_2 \text{ odd} \tag{16.6}$$

and

$$s(x_1, x_2, t_o) = s(x_1, x_2 + 1, t_o) \quad \text{for } x_2 \text{ even} \tag{16.7}$$

The line-averaging algorithm is given by

$$s(x_1, x_2, t_i) = \frac{1}{2}[s(x_1, x_2 - 1, t_i) + s(x_1, x_2 + 1, t_i)] \quad \text{for } i = e, o \tag{16.8}$$

Similar to the case of frame/field rate up-conversion, the line repetition algorithm results in jagged edges, while the line-averaging algorithm causes undesired blurring of the edges.

Edge-adaptive spatial interpolation methods to overcome these problems were proposed by Isnardi [Isn 86], Martinez [Mar 86], and Lim [Lim 90]. In the edge-adaptive approach, each line of video in a given frame t_0 is modeled as a horizontally displaced version of the previous line in the same frame given by

$$s(x_1 - d/2, x_2 - 1, t_0) = s(x_1 + d/2, x_2 + 1, t_0) \tag{16.9}$$

where d denotes the horizontal displacement between the two consecutive even or odd lines. This model suggests a 1-D motion compensation problem where x_2 takes the place of the time variable. The displacement d can be estimated on a pixel-by-pixel basis either by using symmetric line segment matching about (x_1, x_2) in order to minimize the summed absolute difference (SAD) given by [Isn 86]

$$SAD(d) = \sum_{j=-1}^{1} |s(x_1 - d/2 + j, x_2 - 1, t_0) - s(x_1 + d/2 + j, x_2 + 1, t_0)| \tag{16.10}$$

or through the relation [Mar 86, Lim 90]

$$d/2 \frac{\partial s(x_1, x_2, t_0)}{\partial x_1} + \frac{\partial s(x_1, x_2, t_0)}{\partial x_2} = 0 \tag{16.11}$$

which is the equivalent of the optical flow equation in this case. Then an edge-adaptive contour interpolation filter can be defined as

$$s(x_1, x_2, t_i) = \frac{1}{2}[s(x_1 - d/2, x_2 - 1, t_i) + s(x_1 + d/2, x_2 + 1, t_i)], \quad i = e, o \tag{16.12}$$

This filter seeks those two pixels in the two neighboring lines that most likely belong to the same image structure, i.e., on the same side of the edge, and averages them. The fact that it is capable of preserving a 45 degree edge, unlike the linear averaging filter, is demonstrated in Figure 16.9. The crucial step here is the accurate estimation of the local displacement values.

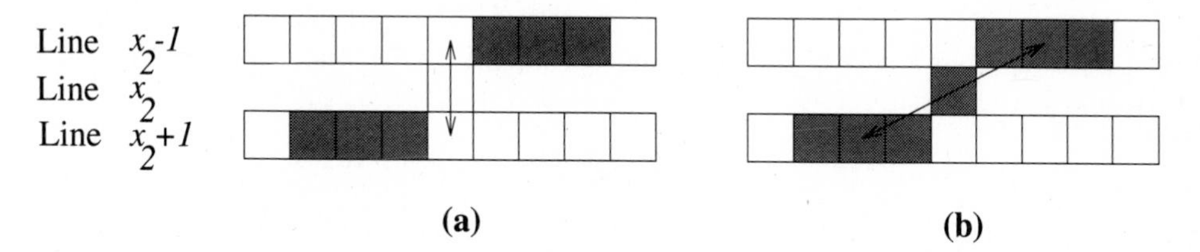

Figure 16.9: Demonstration of a) linear vs. b) edge-adaptive interpolation.

Intraframe filtering methods lead to simple hardware realizations. However, they are not well suited to de-interlacing in stationary regions, where spatial averaging usually causes blurring of image details; hence, the need for interframe filtering.

16.2.2 Motion-Adaptive Filtering

Motion-adaptive filtering refers to employing different filtering strategies in the presence and absence of motion without motion estimation. For example, in de-interlacing, in the absence of motion, the best strategy is to merge even and odd fields which simply doubles the vertical resolution. However, in the presence of motion this technique would suffer from motion artifacts. Motion-adaptive algorithms may employ explicit or implicit motion adaptation schemes. Explicit schemes make use of a motion-detection function and apply different filters depending on the value of this function. In implicitly adaptive filtering, such as median filtering, the motion adaptivity is inherent in the structure of the algorithm. We elaborate on some motion adaptive frame/field rate up-conversion and de-interlacing algorithms in the following.

Frame/Field Rate Conversion

A rudimentary linear shift-invariant interframe filtering strategy for frame rate up-conversion would be frame averaging, where a missing frame is replaced by the average of its two neighboring frames. Frame averaging improves the SNR in the stationary regions, and hence it is superior to simple frame repetition. However, it introduces ghost artifacts in the moving parts. Similarly, a naive linear shift-invariant filtering scheme for field rate up-conversion would be a three-point averaging filter, whose support is depicted in Figure 16.10. One way to avoid ghosting is to employ a shift-varying linear filter where the filter impulse response is determined locally based on a motion detection function. For example, we perform averaging only in the stationary image regions, and replicate pixels or compute a weighted average in the moving regions. The boundaries of the moving regions can be estimated by using a motion detection function, which may simply be the frame difference as in change detection. Because no optimal strategy exists to determine the filter weights in terms of the motion detection function in the moving areas, several researchers have suggested the use of spatio-temporal median filtering.

Median filtering is known to be edge-preserving in still-frame image processing. Considering the effect of motion as a temporal edge, one would expect spatio-temporal median filtering to provide motion adaptivity. In scan rate up-conversion we employ a three-point spatio-temporal median filter that can be described by

$$
\begin{aligned}
s(x_1, x_2, t_e^*) &= \text{Med}\,\{s(x_1, x_2 - 1, t_o), s(x_1, x_2 + 1, t_o), s(x_1, x_2, t_e)\} \\
s(x_1, x_2, t_o^*) &= \text{Med}\,\{s(x_1, x_2 - 1, t_e), s(x_1, x_2 + 1, t_e), s(x_1, x_2, t_o)\} \qquad (16.13)
\end{aligned}
$$

where "Med" denotes the median operation. The three pixels within the support of the filter are shown in Figure 16.10 for cases of even and odd field estimation.

The median filter is motion-adaptive such that in moving regions the filter output generally comes from one of the two pixels that is 1/2 pixel above or below the pixel to be estimated, and in the stationary regions it comes from the pixel that is in the same vertical position. Thus, it yields a reasonable performance in both the stationary and moving regions. The three-point median filter has already been used in prototype improved-definition TV receivers for field rate doubling. Several modifications including a combination of averaging and median filters have also been proposed for performance improvements [Haa 92].

In median filtering, each sample in the filter support is given an equal emphasis. Alternatively, one can employ weighted median filtering, given by

$$\begin{aligned} s(x_1, x_2, t_e^*) &= \text{Med}\,\{w_1 \diamond s(x_1, x_2 - 1, t_o), w_2 \diamond s(x_1, x_2 + 1, t_o), \\ &\quad w_3 \diamond s(x_1, x_2, t_e)\} \\ s(x_1, x_2, t_o^*) &= \text{Med}\,\{w_1 \diamond s(x_1, x_2 - 1, t_e), w_2 \diamond s(x_1, x_2 + 1, t_e), \\ &\quad w_3 \diamond s(x_1, x_2, t_o)\} \end{aligned} \tag{16.14}$$

where w_i denotes the weight of the ith sample, and $\diamond$ is the replication operator. That is, each sample i is replicated w_i times to affect the output of the median operation. To obtain the best possible results, the weights should be chosen as a function of a motion-detection signal. An example of a motion-detection signal for scan rate up-conversion and weight combinations as a function of this motion detection signal has been provided by Haavisto and Neuvo [Haa 92].

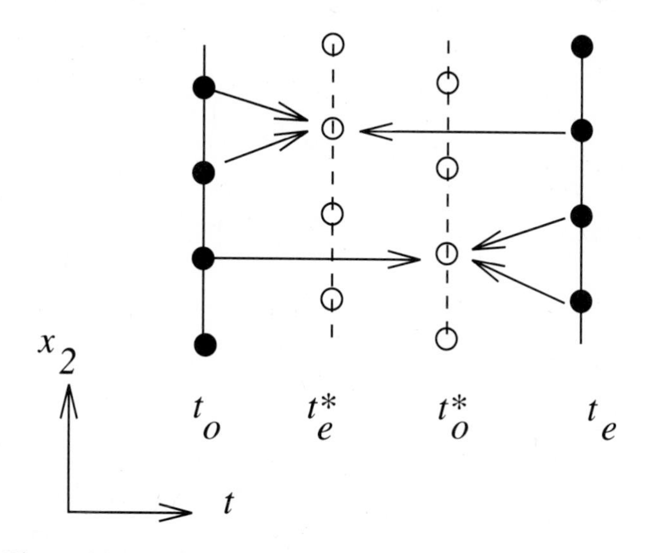

Figure 16.10: Three-point filtering for field rate doubling.

De-interlacing

The simplest interframe de-interlacing algorithm can be obtained by merging the even and odd fields, that is, copying all samples from the previous field, as shown by

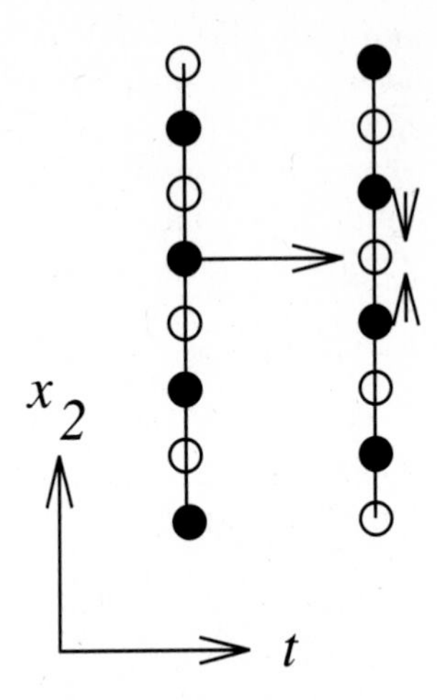

Figure 16.11: Two-field filtering for de-interlacing.

the horizontal arrow in Figure 16.11. The resulting frames are known as composite frames. Composite frames provide perfect resolution in stationary image regions, but they suffer from serious motion artifacts.

In order to obtain an acceptable performance in both the moving and stationary regions, we may consider motion-adaptive interframe filtering, which switches between merging and intraframe interpolation [Isn 86] or linearly blends them [Sch 87], based on a motion detection function. Two examples of a three-point motion-adaptive filter, whose support is depicted in Figure 16.11, are the three-point weighted averaging filter,

$$\begin{aligned} s(x_1, x_2, t_i) &= \alpha\, s(x_1 - d/2, x_2 - 1, t_i) + (1 - \alpha)\, s(x_1 + d/2, x_2 + 1, t_i) \\ &\quad + \beta\, s(x_1, x_2, t_i - 1) \end{aligned} \tag{16.15}$$

and the three-point median filter [Cap 90, Haa 92]

$$\begin{aligned} s(x_1, x_2, t_i) &= \text{Med}\,\{\, s(x_1 - d/2, x_2 - 1, t_i), \\ &\quad s(x_1 + d/2, x_2 + 1, t_i),\ s(x_1, x_2, t_i - 1) \,\} \end{aligned} \tag{16.16}$$

where α and β are determined based on the value of a motion detection function, and the parameter d, computed from Equation (16.10) or (16.11), allows for edge-adaptive intraframe interpolation.

Median filtering is attractive because of its computational simplicity and its edge-preserving property. In the case of the weighted averaging filter, we can employ a three- or four-field motion detection function. The three-field motion detection function is obtained by thresholding the difference between two fields of the same polarity (even-even or odd-odd), whereas the four-field motion detection function takes the logical OR of the thresholded differences of the respective even-even and odd-odd fields. The coefficients α and β may be given by

$$\alpha = 0.5,\ \beta = 0 \quad \text{if motion is detected, and}$$
$$\alpha = 0,\ \beta = 1 \quad \text{if no motion is detected.}$$

Motion adaptive methods provide satisfactory results provided that the scene does not contain fast-moving objects. In the presence of fast-moving objects some sort of motion-compensated filtering, as described next, is needed for the best results.

16.3 Motion-Compensated Up-Conversion

Motion-compensated filtering is the optimal method for standards up-conversion, provided that the motion trajectories can be accurately estimated. This follows from the derivation of the ideal motion-compensated reconstruction filter in Chapter 13. In the case of global, constant-velocity motion, the motion-compensated filter is linear and shift-invariant. There are, in general, two approaches to describe the theory of up-conversion: the digital approach, where we zero-fill the input signal and then apply a digital interpolation filter to the "filled-in" signal, and the analog approach, where we first conceptually reconstruct the analog signal and then re-sample it on the desired output lattice. In this section, we adopt the latter approach, since zero filling requires quantization of the motion estimates, limiting the accuracy of the motion trajectory.

16.3.1 Basic Principles

The basic concept of motion-compensated filtering is the same for frame/field rate up-conversion and de-interlacing, which is to perform filtering along the motion trajectories passing through the missing pixels [Reu 89]. The two problems differ only in the spatio-temporal locations of the missing samples. The procedure consists of three steps: i) motion estimation, ii) post processing of the motion vectors, and iii) filter design, which are described in the following.

Motion Estimation

Forward or backward block matching, illustrated in Figure 16.12 (a) and (b), respectively, are commonly used to estimate motion vectors between two frames. In motion-compensated up-conversion, we need to estimate motion trajectories that pass through missing pixel locations. Several motion estimators have been proposed for use in standards conversion [Bie 86, Tub 93]. Among these, a simple approach is *symmetric block matching*, which is illustrated in Figure 16.12 (c). In this scheme, blocks in the two existing neighboring frames/fields k and $k-1$ are moved symmetrically so that the line connecting the centers of these two blocks always passes through the missing pixel of interest (x_1, x_2) [Tho 89].

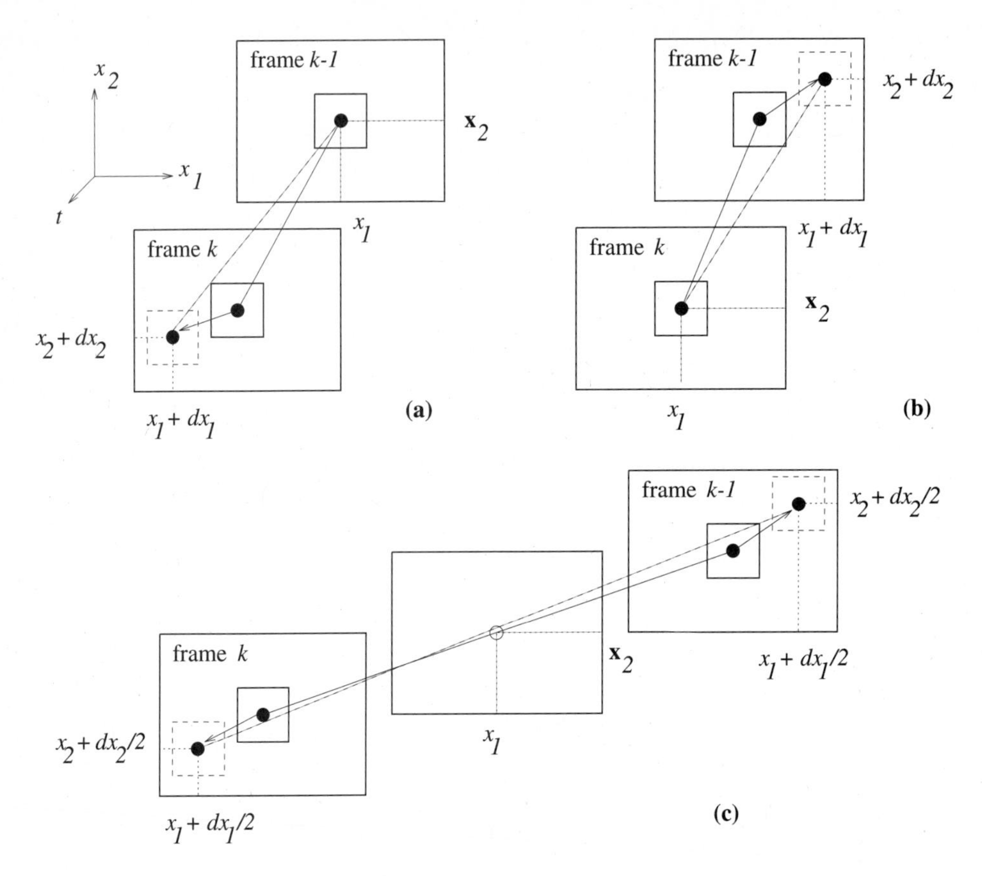

Figure 16.12: a) Forward, b) backward, and c) symmetric block matching.

The accuracy of the motion estimates is probably the most important factor in the effectiveness of motion-compensated interpolation. Thus, some kind of postprocessing is usually applied to the estimated motion vectors to improve their accuracy.

Postprocessing of Motion Estimates

The accuracy of the motion estimates can be improved by postprocessing techniques. Here we present two approaches to test the accuracy of the motion estimates: occlusion detection method, and displaced frame difference (DFD) method. Note that these methods are applied to the two nonzero frames used in symmetrical block matching.

The occlusion detection method is based on the assumption that a corona of motion vectors is usually located around moving objects. This assumption, coupled with the observation that the correct motion vectors from frame k to frame $k+1$

always map into the "changed region" $CD(k, k+1)$, leads to the following occlusion detection procedure, which was also used in Chapter 11. If an estimated motion vector from frame k to frame $k+1$ maps to outside of the changed region, it indicates a pixel in frame k that will be covered in frame $k+1$. Likewise, if a motion vector from frame $k+1$ to frame k maps to outside of the changed region, it indicates a pixel in frame $k+1$ that is uncovered. Of course, such motion vectors are not reliable and have to be discarded. An alternative method to detect unreliable motion vectors is to test the DFD between the frames k and $k+1$. In this scheme, all vectors yielding a DFD that is above a prespecified threshold are discarded.

The unreliable motion vectors can be replaced by a set of candidate motion vectors if any of the candidate vectors yields a DFD that is less than the threshold. The candidate vectors can be determined based on the analysis of the histogram of the reliable motion vectors, where the dominant peaks of the histogram indicate background motion or the motion of large objects in the scene [Lag 92]. If no reliable replacement motion vector can be found at a pixel, it is marked as a motion estimation failure, where a motion-adaptive or an intraframe filter is employed.

Filter Design

Given the velocity estimates, the motion-compensated low-pass filter is defined by

$$H(f_1, f_t) = \begin{cases} 1 & \text{for } \; -B_1 < f_1 < B_1 \\ & \text{and } \; -v_1 f_1 - B_t < f_t < -v_1 f_1 + B_t \\ 0 & \text{otherwise} \end{cases}$$

whose support is depicted in Figure 16.13. Because this filter is unrealizable, it needs to be approximated, usually by an FIR filter, which poses a trade-off between filter length and passband width. In practice, there is a limitation to the number of frame-stores that can be used in hardware implementations, which necessitates the use of short temporal filters. This causes a wider than desired transition band in the temporal frequency direction. Thus, the replications due to sub-Nyquist sampling should not come closer than the transition band of the filter frequency response.

For example, in the interest of real-time implementation with a small number of frame-stores, a simple zero-order hold filter may be employed along the motion trajectory for interpolation filtering. Thus, we attempt to find the missing pixel values from the past fields using the information provided by the motion vectors. In an attempt to find a match with one of the existing sample values, Woods and Han [Woo 91] carry this search over two previous fields, as shown in Figure 16.14. For those pixels whose motion vectors do not point to an existing pixel location in either of the past two fields, the values are taken to be the average of the two nearest existing pixels in the previous two fields. A number of other filter design issues for motion-compensated sampling structure conversion are addressed in [Dub 92, Bel 93, Gir 93]. Application of motion-compensated filtering to standards conversion from NTSC to PAL or vice versa has been discussed in [Yam 94].

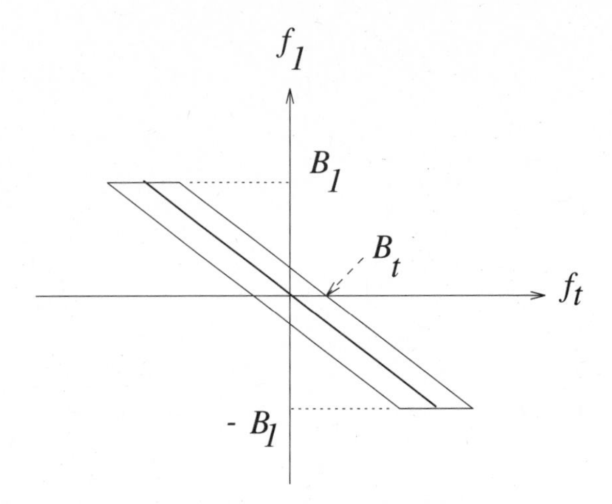

Figure 16.13: Ideal motion-compensated low-pass filtering.

Clearly, motion-compensated up-conversion is capable of producing video free from aliasing even if the original source is spatio-temporally aliased, such as shown in Figure 16.4, provided that i) the motion model is accurate, ii) the motion estimates are accurate, and iii) we do not have a critical velocity. Furthermore, neglecting the sensor PSF and any blurring due to relative motion between the object and the camera, motion-compensated up-conversion provides higher resolution than that of the input video. Assuming our model is at least locally accurate, the performance

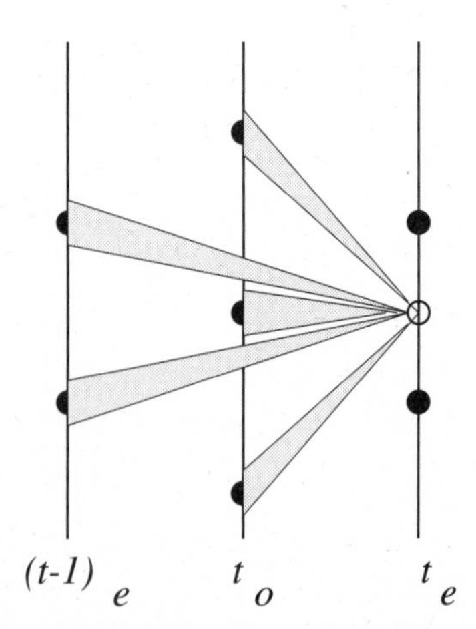

Figure 16.14: Zero-order hold filtering.

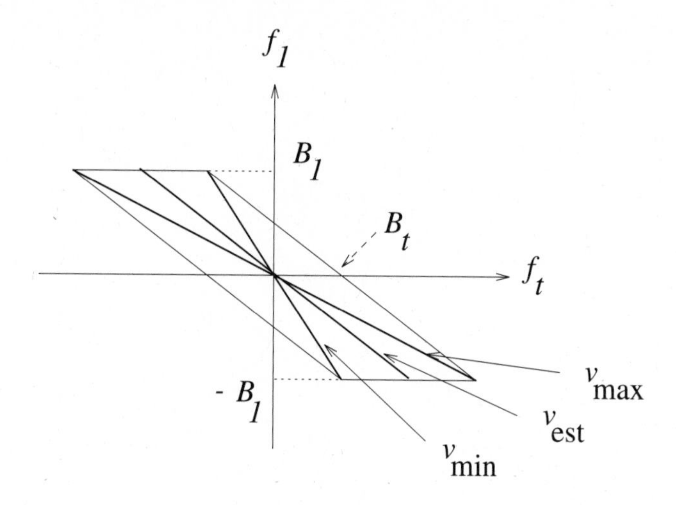

Figure 16.15: Sensitivity to motion estimation errors.

of a motion-compensated filter is closely related to the accuracy of the motion estimates and how well we deal with the critical velocities, which are discussed in the following.

Sensitivity to Errors in Motion Estimation

Because the motion-compensated ideal low-pass filter has a finite passband width B_t, as shown in Figure 16.15, an ideal filter based on an estimated velocity v_{est} is capable of eliminating the replicas successfully provided that the actual velocity is within the range $[v_{min}, v_{max}]$ [Gir 85].

A filter with a wider passband yields a larger tolerance to motion estimation errors, but may not suppress the spectral replications sufficiently. A smaller passband means better suppression of the spectral replications, but smaller error tolerance, particularly at higher frequencies.

Critical Velocities

Given the subsampling matrix and the bandwidth of the video signal, we can easily determine the critical velocities. Then if the estimated velocity vector at a particular frame or a block is close to this critical velocity, we need to do one of the following before subsampling [Bel 93]:

Anti-alias filtering: We apply anti-alias filtering with the proper passband. That is, the spatial resolution is sacrified at a particular frame or block of pixels when estimated motion is close to the critical velocity.

Adaptive subsampling: We change the subsampling lattice at a particular frame or block if a critical velocity is detected.

16.3.2 Global-Motion-Compensated De-interlacing

Motion-compensated filtering, in general, requires a different motion trajectory at each pixel. However, in practice, reliable estimation of these vectors at each pixel poses serious problems. To this effect, we propose a hybrid de-interlacing method, where we compensate for a single global motion, which is due to camera pan or shake, and then employ a motion-adaptive filter on the globally compensated image to account for any residual motion. The block diagram of a three-field hybrid de-interlacing filter is depicted in Figure 16.16, where the three consecutive fields are assumed to be an even field E_1, an odd field O_1, and an even field E_2.

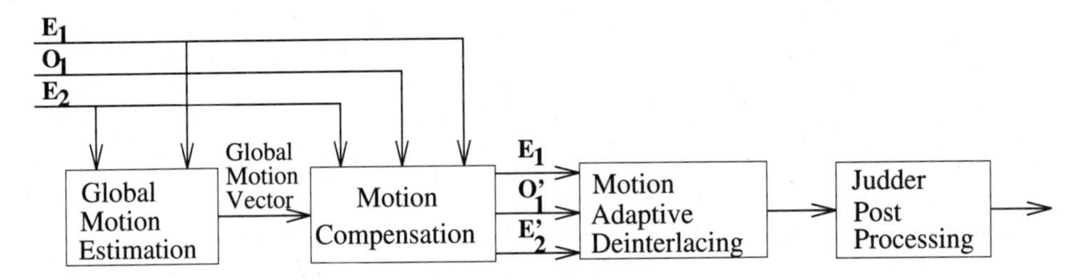

Figure 16.16: Motion-compensated/adaptive de-interlacing.

In the first stage, a global-motion vector between the fields E_1 and E_2 is estimated using the phase correlation method over four rectangular windows that are located near the borders of the fields, so that they are most likely affected by global motion only. Next, the fields O_1 and E_2 are motion-compensated with respect to E_1 to generate O_1' and E_2', respectively. The motion compensation step aims to create three consecutive fields, E_1, O_1', and E_2', that represent the interlaced video if no global motion were present. Subsequently, the three-field motion-adaptive weighted averaging filter, described in Section 16.2.2, is applied to the field sequence E_1, O_1' and E_2'. Finally, a judder post-processing step, proposed by Zaccarin and Liu [Zac 93], has been included. Judder, illustrated in Figure 16.17, refers to edge misalignment artifacts caused by incorrect motion vectors. In the postprocessing stage, the motion vectors at the pixels where judder is detected are deemed unreliable, and the corresponding pixels are replaced by spatially interpolated values.

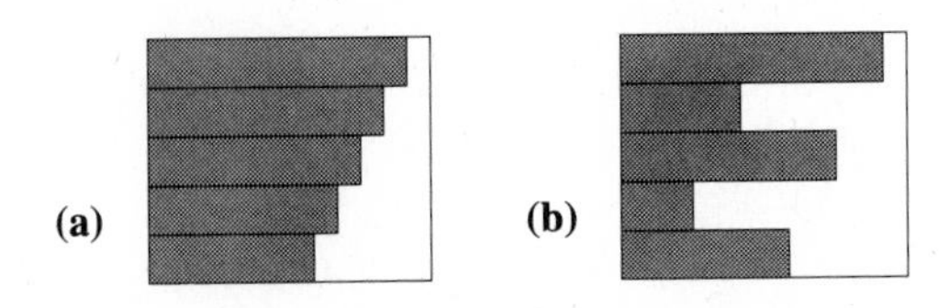

Figure 16.17: Illustration of judder: a) no judder; b) judder present.

16.4 Examples

In this section, we illustrate the performances of several different de-interlacing filters using four fields of the NTSC Train sequence, where each field is 720 pixels × 244 lines. Two different sequences have been captured, one with the camera mounted on a tripod and one with a hand-held camera. The former sequence contains a train and a small board that is attached to the train in motion against a still background, whereas the latter sequence exhibits an additional global motion due to camera shake.

Figure 16.18 (a) shows a composite frame obtained by merging an even and an odd field, when the camera was mounted on the tripod. Notice the motion artifacts in the form of jagged edges about the moving train. The simplest way to overcome the motion artifacts is to discard one of the fields and fill in the missing lines by 1-D linear interpolation in the vertical direction. Although the resulting image, shown in Figure 16.18 (b), has lower vertical resolution in the stationary areas, the disturbing motion artifacts disappear. In order to obtain higher vertical resolution by means of intraframe filtering, edge-adaptive techniques such as contour-oriented interpolation can be used. The result of edge-adaptive de-interlacing, where the edge orientation is determined by using the SAD criterion (16.10), is depicted in Figure 16.19 (a). Inspection of the results indicate that the edge-adaptive method strikes a good balance between retaining sharp edges and reducing motion artifacts.

Next, we illustrate the performances of motion-adaptive and global-motion-compensated de-interlacing filters. The result of two-field de-interlacing using three-point median filtering with the support depicted in Figure 16.11 is shown in Figure 16.19 (b). It can be easily seen that three-point median filtering did not perform well on the letters on the moving board. A more sophisticated approach is to perform adaptive filtering controlled by a motion-detection function. Figs. 16.20 and 16.21 depict the results of motion-adaptive filtering where the motion detection function has been estimated from three and four consecutive fields, respectively, as explained in Section 16.2.2. In both figures, the picture on the top shows the motion map, where white pixels denote the presence of motion. It can be seen that the four-field approach yields a wider motion map, and hence the four-field motion-adaptive algorithm performs slightly better in the vicinity of small text.

The composite frame in the case of the hand-held camera is depicted in Figure 16.22 (a). Notice the motion artifacts in the background as well as the moving parts in this case. The de-interlaced image using the globally motion-compensated and locally motion-adaptive algorithm, described in Section 16.3.2, is shown in Figure 16.22 (b). Observe that the image is almost free of motion artifacts without a noticable loss in vertical resolution.

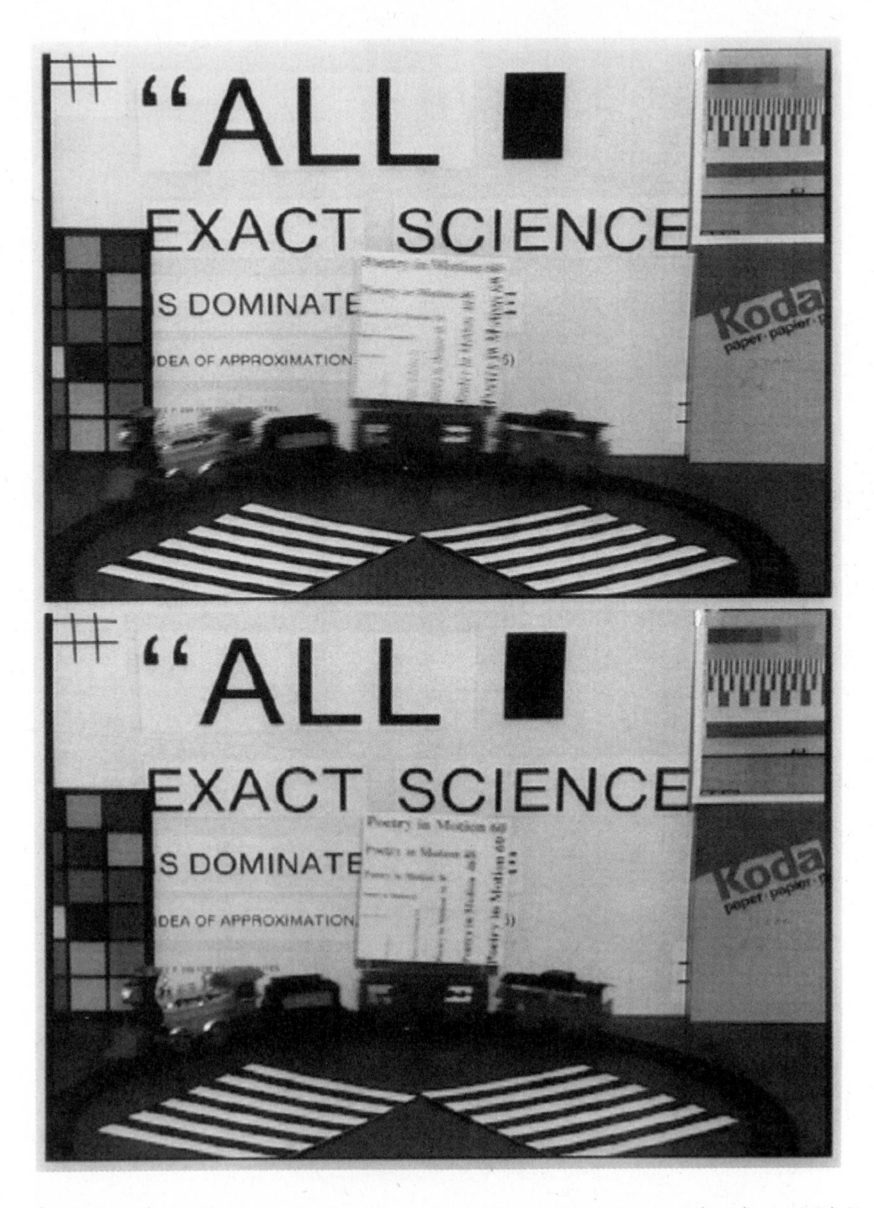

Figure 16.18: a) Composite frame when recording with a tripod (top), and b) linear interpolation of a single field image (bottom). (Courtesy Andrew Patti)

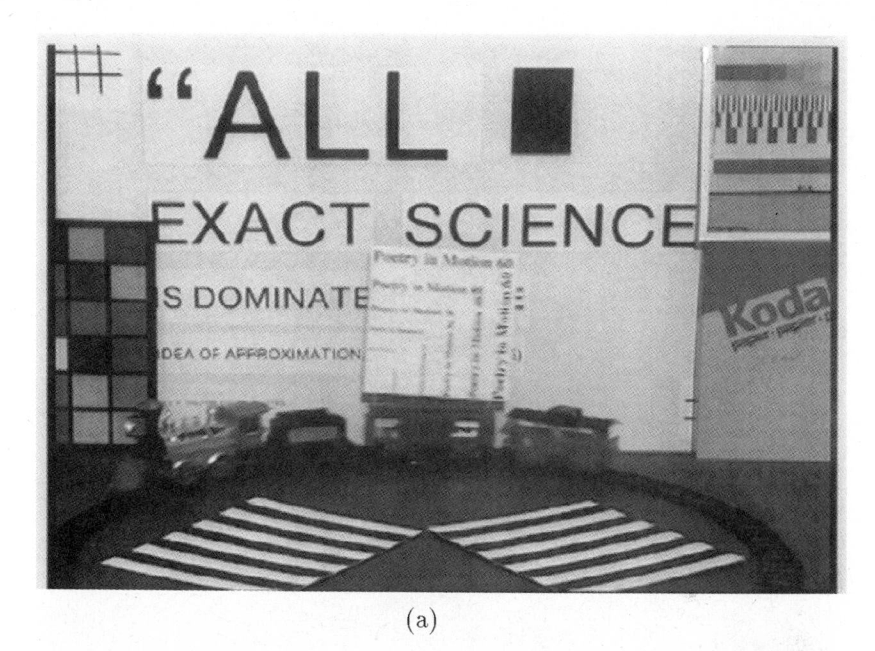

(a)

(b)

Figure 16.19: a) Edge-oriented interpolation of a single-field, and b) two-field median filtering. (Courtesy Andrew Patti)

Figure 16.20: Motion-adaptive de-interlacing with three fields: a) the motion detection map (top), and b) the de-interlaced picture (bottom). (Courtesy Andrew Patti)

Figure 16.21: Motion-adaptive de-interlacing with four fields: a) the motion detection map (top), and b) the de-interlaced picture (bottom). (Courtesy Andrew Patti)

(a)

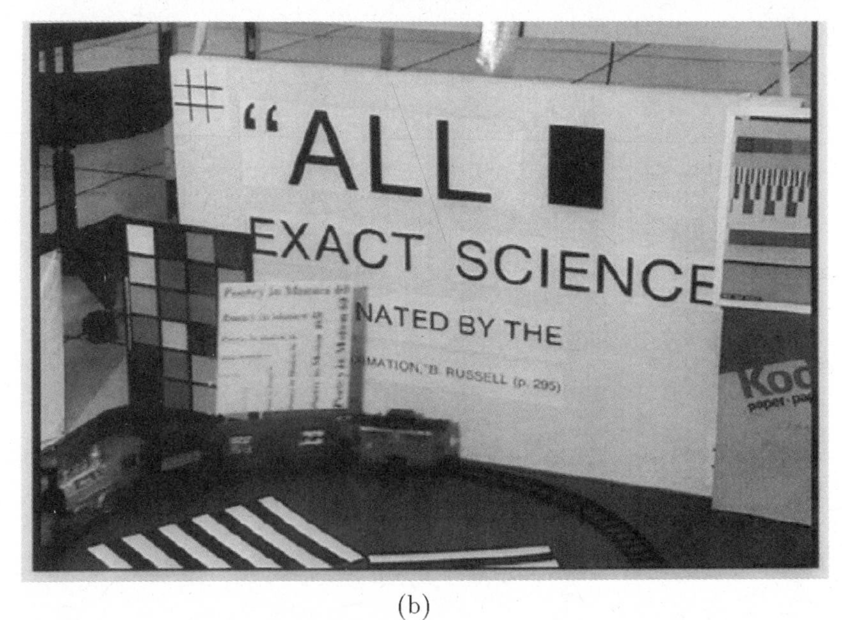

(b)

Figure 16.22: a) Composite frame when recording with a hand-held camcorder. b) Image obtained by the global motion-compensated, three-field de-interlacing algorithm. (Courtesy Andrew Patti)

16.5 Exercises

1. Show that the spatio-temporal impulse response of the filter that performs "merging" of even and odd fields to form a composite frame is given by $h(x_2, t) = \delta(x_2)\delta(t) + \delta(x_2)\delta(t+T)$ where T is the field interval. Find the frequency response of this filter. Discuss the frequency domain interpretation of merging for stationary and moving image regions.

2. Let the horizontal and vertical bandwidths of a video signal with an unknown global constant velocity be 10^6 cyc/mm. Suppose it is sampled on a vertically aligned 2:1 interlaced lattice with the parameters $\Delta x_1 = 100$ microns, $\Delta x_2 = 100$ microns, and $\Delta t = 1/60$sec. Find all critical velocities, if any.

3. Compare the relative advantages and disadvantages of two-, three-, and four-frame motion detection algorithms from interlaced video.

4. How would you deal with motion estimation errors in motion-compensated up-conversion? Can you modify the adaptive motion-compensated noise filtering structures presented in Chapter 14 for motion-compensated up-conversion?

Bibliography

[Bel 93] R. A. F. Belfor, R. L. Lagendijk, and J. Biemond, "Subsampling of digital image sequences using motion information," in *Motion Analysis and Image Sequence Proc.*, M. I. Sezan and R. L. Lagendijk, eds., Kluwer, 1993.

[Bie 86] M. Bierling and R. Thoma, "Motion compensating field interpolation using a hierarchically structured displacement estimator," *Signal Proc.*, vol. 11, pp. 387–404, 1986.

[Caf 90] C. Cafforio, F. Rocca, and S. Tubaro, "Motion compensated image interpolation," *IEEE Trans. Comm.*, vol. COM-38, pp. 215–222, Feb. 1990.

[Cap 90] L. Capodiferro, "Interlaced to progressive conversion by median filtering," *Signal Proc. of HDTV II*, L. Chiariglione, ed., Elsevier, 1990.

[Dub 92] E. Dubois, "Motion-compensated filtering of time-varying images," *Multidim. Syst. Sign. Proc.*, vol. 3, pp. 211–239, 1992.

[Gir 85] B. Girod and R. Thoma, "Motion-compensated field interpolation from interlaced and non-interlaced grids," *Proc. SPIE Conf. Image Coding*, pp. 186–193, 1985.

[Gir 93] B. Girod, "Motion compensation: Visual aspects, accuracy, and fundamental limits," in *Motion Analysis and Image Sequence Processing*, M. I. Sezan and R. L. Lagendijk, eds., Norwell, MA: Kluwer, 1993.

[Haa 92] P. Haavisto and Y. Neuvo, "Motion adaptive scan rate up-conversion," *Multidim. Syst. Sign. Proc.*, vol. 3, pp. 113–130, 1992.

[Isn 86] M. A. Isnardi, *Modeling the Television Process*, Ph.D. Thesis, Mass. Inst. of Tech., 1986.

[Lag 92] R. L. Lagendijk and M. I. Sezan, "Motion compensated frame rate conversion of motion pictures," *IEEE Int. Conf. Acoust. Speech, Sig. Proc.*, San Francisco, CA, 1992.

[Lim 90] J. S. Lim, *Two-Dimensional Signal and Image Processing*, Englewood Cliffs, NJ: Prentice Hall, 1990.

[Mar 86] D. M. Martinez, *Model-Based Motion Interpolation and its Application to Restoration and Interpolation of Motion Pictures*, Ph.D. Thesis, Mass. Inst. of Tech., 1986.

[Reu 89] T. Reuter, "Standards conversion using motion compensation," *Signal Proc.*, vol. 16, pp. 73–82, 1989.

[Sch 87] G. Schamel, "Pre- and post-filtering of HDTV signals for sampling rate reduction and display up-conversion," *IEEE Trans. Circuits and Syst.*, vol. 34, pp. 1432–1439, Nov. 1987.

[Tho 89] R. Thoma and M. Bierling, "Motion compensating interpolation considering covered and uncovered background," *Signal Proc.: Image Comm.*, vol. 1, pp. 191–212, 1989.

[Tub 93] S. Tubaro and F. Rocca, "Motion field estimators and their application to image interpolation," in *Motion Analysis and Image Sequence Processing*, M. I. Sezan and R. L. Lagendijk, eds., Norwell, MA: Kluwer, 1993.

[Wan 90] F.-M. Wang, D. Anastassiou, and A. N. Netravali, "Time-recursive deinterlacing for IDTV and pyramid coding," *Signal Proc.: Image Comm.*, vol. 2, pp. 365–374, 1990.

[Woo 91] J. W. Woods and S.-C. Han, "Hierarchical motion-compensated deinterlacing," *SPIE Vis. Comm. Image Proc.*, vol. 1605, pp. 805–810, 1991.

[Yam 94] T. Yamauchi, N. Ouchi, and H. Shimano, "Motion-compensated TV standards converter using motion vectors computed by an iterative gradient method," *Signal Proc.: Image Comm.*, vol. 6, pp. 267–274, 1994.

[Zac 93] A. Zaccarin and B. Liu, "Block motion compensated coding of interlaced sequences using adaptively de-interlaced fields," *Signal Proc.: Image Comm.*, vol. 5, pp. 473–485, 1993.

Chapter 17

SUPERRESOLUTION

Superresolution refers to obtaining video at a resolution higher than that of the camera (sensor) used in recording the image. Because most images contain sharp edges, they are not strictly bandlimited. As a result, digital images usually suffer from aliasing due to undersampling, loss of high-frequency detail due to low-resolution sensor point spread function (PSF), and possible optical blurring due to relative motion or out-of-focus. Superresolution involves up-conversion of the input sampling lattice as well as reducing or eliminating aliasing and blurring. For example, in order to obtain a high-quality print from an interlaced video of a football game, the video needs to be de-interlaced and deblurred (to remove motion blur). This would yield a sharper print that is free from motion artifacts. Other applications include detection of small targets in military or civilian surveillance imaging, and detection of small tumors in cine medical imaging.

Superresolution from a single low-resolution, and possibly blurred, image is known to be highly ill-posed. However, when a sequence of low-resolution frames is available, such as those obtained by a video camera, the problem becomes more manageable. It is evident that the 3-D spatio-temporal sampling grid contains more information than any 2-D still-frame sampling grid. Interframe superresolution methods exploit this additional information, contained in multiple frames, to reconstruct a high-resolution still image or a sequence of high-resolution images. As we proceed, it will become clear that both image restoration (Chapter 15) and standards conversion (Chapter 16) constitute special cases of the superresolution problem. Motion-compensated standards conversion is a special case where the optical blur and sensor PSF are taken as delta functions, and thus we are concerned with eliminating aliasing only; while in the restoration problem, the input and output lattices are identical, and hence we need only eliminate optical blurring.

This chapter starts with modeling and formulation of the superresolution problem in Section 17.1. Section 17.2 is devoted to two-step interpolation-restoration methods, whereas a frequency-domain method is presented in Section 17.3. Finally, a robust spatio-temporal domain algorithm is described in Section 17.4.

17.1 Modeling

An important limitation of electronic imaging today is that most available still-frame or video cameras can only record images at a resolution lower than desirable. This is related to certain physical limitations of the image sensors, such as finite cell area and finite aperture time. Although higher-resolution imaging sensors are being advanced, these may be too expensive and/or unsuitable for mobile imaging applications. In the following, we first present a model that relates the low-resolution sampled video to the underlying continuous video. Next, the low-resolution sampled video is expressed in terms of a higher-definition sampled video. The superresolution problem is then posed as the reconstruction of this higher-definition video from a number of observed low-resolution frames.

17.1.1 Continuous-Discrete Model

We present a comprehensive model of video acquisition, which includes the effects of finite sensor cell area and finite aperture time. Low-resolution images suffer from a combination of blurring due to spatial integration at the sensor surface (due to finite cell area), modeled by a shift-invariant spatial PSF $h_a(x_1, x_2)$, and aliasing due to sub-Nyquist sampling. In addition, relative motion between the objects in a scene and the camera during the aperture time gives rise (due to temporal integration) to shift-varying spatial blurring. The relative motion may be due to camera motion, as in camera pan or a camera mounted on a moving vehicle, or the motion of objects in a scene, which is especially significant in high-action movies and sports videos. Recall that it is often this motion which gives rise to temporal variation of the spatial intensity distribution. Images may also suffer from out-of-focus blur, which will be modeled by a shift-invariant spatial PSF $h_o(x_1, x_2)$ (ignoring the depth-of-field problem), and additive noise.

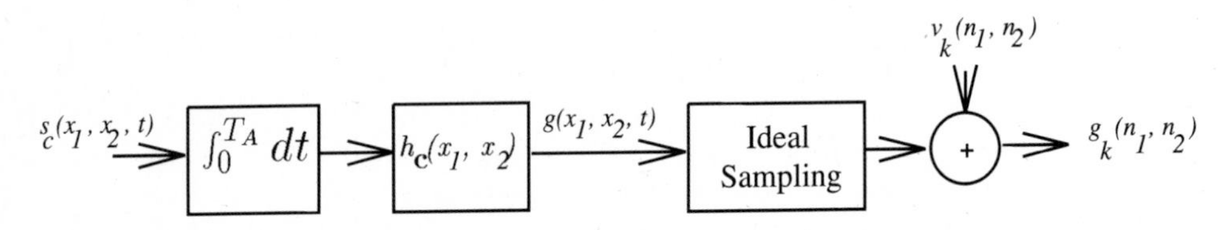

Figure 17.1: Block diagram of video acquisition.

A block diagram of the continuous-input, discrete-output model of a low-resolution video acquisition process is depicted in Figure 17.1. Here, the first subsystem represents the temporal integration during the aperture time T_A, resulting in a blurred video (due to relative motion) given by

$$g_m(x_1, x_2, t) = \frac{1}{T_A} \int_0^{T_A} s_c(x_1, x_2, t - \eta) d\eta \tag{17.1}$$

where $s_c(x_1, x_2, t)$ denotes the ideal continuous video that would be formed by an instantaneous aperture. The model (17.1) can be best interpreted by considering a stationary camera and mapping the effect of any camera motion to a relative scene motion. The next subsystem models the combined effects of integration at the sensor surface and any shift-invariant out-of-focus blur, resulting in a further blurred video given by

$$g(x_1, x_2, t) = h_c(x_1, x_2) * * g_m(x_1, x_2, t) \tag{17.2}$$

where $h_c(x_1, x_2) = h_a(x_1, x_2) * * h_o(x_1, x_2)$ is the combined shift-invariant spatial PSF, and ** denotes 2-D convolution.

The integration over time in the model (17.1) can equivalently be written as a spatial integration at a single time instant τ considering our ideal video source model

$$s_c(x_1, x_2, t) = s_c(c_1(\tau; x_1, x_2, t), c_2(\tau; x_1, x_2, t), \tau) \tag{17.3}$$

where $\mathbf{c}(\tau; x_1, x_2, t) = (c_1(\tau; x_1, x_2, t), c_2(\tau; x_1, x_2, t))$ is the motion trajectory function that is defined in Chapter 13. Then, under the assumption that the reference time τ is within the temporal span of all motion trajectories passing through $(x_1, x_2, t - \eta)$, $0 < \eta < T_A$, all (x_1, x_2) within the time interval $(t - T_A, t)$ can be traced onto the plane at time τ using the source model (17.3) to perform a spatial integration over these points. The corresponding spatial PSF is given by

$$h_m(u_1, u_2, \tau; x_1, x_2, t) = \begin{cases} \frac{1}{T_A} J(u_1, u_2, t) \delta(u_2 - c_2(\tau, x_1, x_2, c_1^{-1}(\tau, x_1, x_2, u_1))) \\ \quad \text{if } c_1(\tau; x_1, x_2, t - T_A) < u_1 < c_1(\tau; x_1, x_2, t) \\ 0 \quad \text{otherwise} \end{cases} \tag{17.4}$$

where $J(u_1, u_2, t)$ denotes the Jacobian of the change of variables, and $c_1^{-1}(\tau, x_1, x_2, u_1)$ returns the time $t - T_A < t_0 < t$ when $u_1 = c_1(\tau; x_1, x_2, t_0)$. (See [Pat 94] for a complete derivation of this PSF.) The spatial support of $h_m(u_1, u_2, \tau; x_1, x_2, t)$, which is a shift-variant PSF, is a path at the reference time τ, that is depicted in Figure 17.2. The path, mathematically expressed by the 1-D delta function within the interval $c_1(\tau; x_1, x_2, t - T_A) < u_1 < c_1(\tau; x_1, x_2, t)$ in (17.4), is obtained by mapping all points $(x_1, x_2, t - \eta)$, $0 < \eta < T_A$ (shown by the solid vertical line) onto time τ.

Therefore, combining the video acquisition model in Figure 17.1 with the source model (17.3), we can establish the relationship

$$g(x_1, x_2, t) = \int \int s_c(u_1, u_2, \tau) h(u_1, u_2, \tau; x_1, x_2, t) du_1 du_2 \tag{17.5}$$

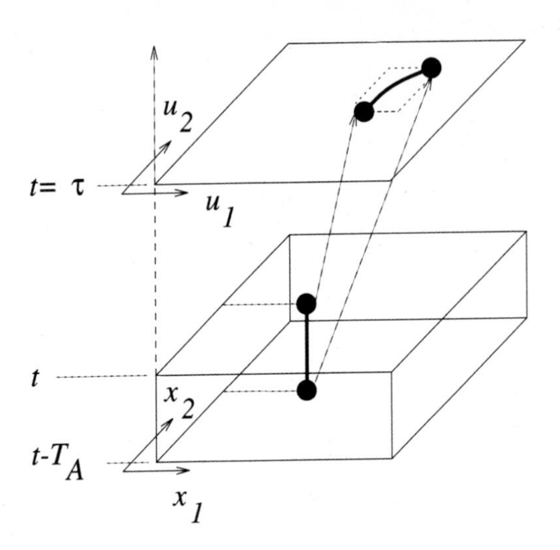

Figure 17.2: Representation of the effect of motion during the aperture time (temporal integration) by a spatial PSF (spatial integration).

between the observed intensity $g(x_1, x_2, t)$ and the ideal intensities $s_c(u_1, u_2, \tau)$ along the path $(u_1, u_2) = c(\tau; x_1, x_2, t)$ at time τ, where

$$h(u_1, u_2, \tau; x_1, x_2, t) = h_c(x_1, x_2) * *h_m(u_1, u_2, \tau; x_1, x_2, t) \tag{17.6}$$

denotes the effective shift-variant spatial PSF of the composite model. A block diagram of this composite model is shown in Figure 17.3.

The degraded video is then ideally sampled on a 3-D low-resolution lattice Λ_l, and noise is added to form the observed discrete video

$$g_k(n_1, n_2) = g(x_1, x_2, t)|_{[x_1\ x_2\ t]^T = \mathbf{V}_l[n_1\ n_2\ k]^T} + v_k(n_1, n_2) \tag{17.7}$$

where $\mathbf{V}_l$ is the sampling matrix of the low-resolution lattice Λ_l. If we assume that $g(x_1, x_2, t)$ is bandlimited, but the sampling intervals in the x_1, x_2, and t

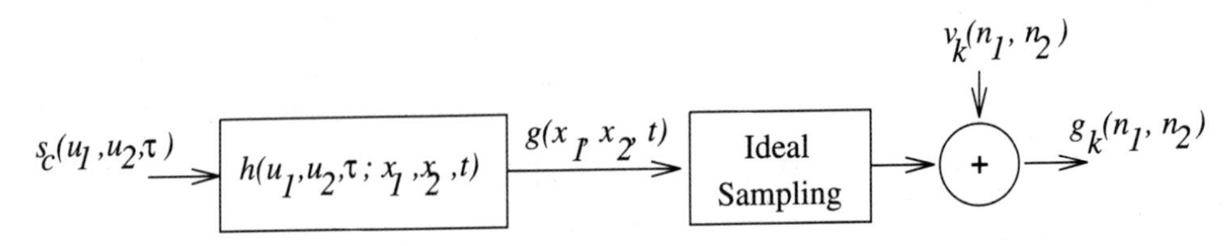

Figure 17.3: Block diagram of system model.

directions are larger than the corresponding Nyquist intervals, aliasing may result. It is essential that no anti-alias filtering be used prior to sampling in order to achieve superresolution, which follows from the discussion in Section 16.1.

17.1.2 Discrete-Discrete Model

Next, we relate a set of low-resolution observations, $g_k(n_1, n_2)$, to the desired high-resolution frame(s) to be reconstructed, which are defined as

$$s_i(m_1, m_2) = s_c(x_1, x_2, t)|_{[x_1 \ x_2 \ t]^T = \mathbf{V}_h [m_1 \ m_2 \ i]^T} \tag{17.8}$$

where $\mathbf{V}_h$ is the sampling matrix of the high-resolution sampling grid.

Let's assume that the high-resolution video $s_i(m_1, m_2)$ is sampled above the Nyquist rate, so that the continuous intensity pattern is more or less constant within each high-resolution pixel (cells depicted in Figure 17.4). Then, for any frame i within the temporal span of the motion trajectory passing through (n_1, n_2, k), we have, from (17.7) and (17.5),

$$g_k(n_1, n_2) \approx s_i(m_1, m_2) \int\int h(u_1, u_2, \tau; x_1, x_2, t) du_1 du_2 \tag{17.9}$$

where $[x_1 \ x_2 \ t]^T = \mathbf{V}_l [n_1 \ n_2 \ k]^T$, $[u_1 \ u_2 \ \tau]^T = \mathbf{V}_h [m_1 \ m_2 \ i]^T$, and $(u_1, u_2) = c(\tau; x_1, x_2, t)$. Next, we define

$$h_{ik}(m_1, m_2; n_1, n_2) = \int\int h(u_1, u_2, \tau; x_1, x_2, t) du_1 du_2 \tag{17.10}$$

to arrive at our discrete-input (high-resolution video), discrete-output (observed low-resolution video) model, given by

$$g_k(n_1, n_2) = \sum_{m_1} \sum_{m_2} s_i(m_1, m_2) h_{ik}(m_1, m_2; n_1, n_2) + v_k(n_1, n_2) \tag{17.11}$$

where the support of the summation over the high-resolution grid (m_1, m_2) at a particular low-resolution sample (n_1, n_2, k) is depicted in Figure 17.4. The size of the support in Figure 17.4 depends on the relative velocity of the scene with respect to the camera, the size of the support of the low-resolution sensor PSF $h_a(x_1, x_2)$ (depicted by the solid line, assuming no out-of-focus blur) with respect to the high resolution grid, and whether there is any out-of-focus blur. Because the relative positions of low- and high-resolution pixels in general vary from pixel to pixel, the discrete sensor PSF is space-varying.

The model (17.11) establishes a relationship between any high-resolution frame i and observed low-resolution pixels from all frames k which can be connected to the frame i by means of a motion trajectory. That is, each low-resolution observed pixel (n_1, n_2, k) can be expressed as a linear combination of several high-resolution pixels from the frame i, provided that (n_1, n_2, k) is connected to frame i by a motion trajectory. Note that both the continuous-discrete and discrete-discrete models are invalid in case of occlusion. We assume that occlusion regions can be detected *a priori* using a proper motion estimation/segmentation algorithm.

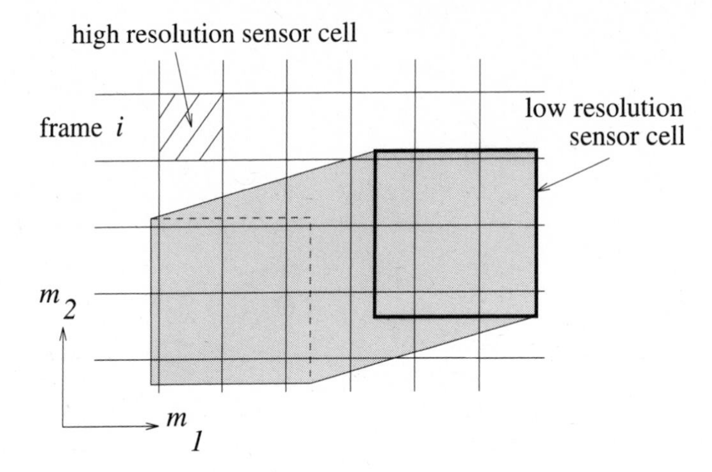

Figure 17.4: Illustration of the discrete system PSF.

17.1.3 Problem Interrelations

The superresolution problem stated by (17.5) and (17.11) is a superset of the filtering problems that were discussed in the previous chapters. This can be seen as follows:

i) Noise filtering: The input and output lattices are identical ($\Lambda_l = \Lambda_h$, $h_a(x_1, x_2) = \delta(x_1, x_2)$), the camera aperture time is negligible ($T_A = 0$), and there is no out-of-focus blur ($h_o(x_1, x_2) = \delta(x_1, x_2)$).

ii) Multiframe/intraframe image restoration: The input and output lattices are identical ($\Lambda_l = \Lambda_h$, $h_a(x_1, x_2) = \delta(x_1, x_2)$).

iii) Standards conversion: The input and output lattices are different; however, sensor PSF is not taken into account ($\Lambda_l \neq \Lambda_h$, $h_a(x_1, x_2) = \delta(x_1, x_2)$), the camera aperture time is negligible ($T_A = 0$), there is no out-of-focus blur ($h_o(x_1, x_2) = \delta(x_1, x_2)$), and there is no noise.

In Section 17.4, we present a single algorithm based on the theory of convex projections (POCS) that is capable of solving the general superresolution problem as well as each of these special cases.

17.2 Interpolation-Restoration Methods

Accurate image expansion ("digital zooming") is desirable in many applications, including surveillance imaging to track small targets and medical imaging to identify anomalies. In this section, we discuss intraframe interpolation and multiframe interpolation/restoration methods, respectively.

17.2.1 Intraframe Methods

It is well-known that no new high-frequency information can be generated by linear, shift-invariant interpolation techniques, including ideal band-limited interpolation (see Chapter 4) and its approximations, such as bilinear or cubic-spline interpolation. To this effect, several nonlinear and/or shift-varying interpolation methods have been proposed for improved definition image interpolation or superresolution [Wan 91, Aya 92, Sch 94].

Superresolution from a single observed image is known to be a highly ill-posed problem. Clearly, in the absence of any modeling assumptions, there exist infinitely many expanded images which are consistent with the original data based on the low-resolution image recording model (17.11) with $i = k$. The proposed intraframe improved-definition zooming techniques have been based on structural or statistical image and/or pattern models. Wang and Mitra [Wan 91] proposed directional filters to preserve oriented image features based on their representation by oriented polynomials. Ayazifar [Aya 92] proposes an edge-preserving interpolation technique that interpolates not only along linear edges but also curved contours. Schultz and Stevenson [Sch 94] modeled the high-resolution image by a discontinuity-preserving Huber-Markov random field model and computed its maximum *a posteriori* (MAP) estimate. Although these methods yield images that are sharper than can be obtained by linear shift-invariant interpolation filters, they do not attempt to remove either the aliasing or the blurring present in the observed data due to low-resolution sampling.

17.2.2 Multiframe Methods

Superresolution from multiple frames with subpixel motion is a better posed problem, since each low-resolution observation from neighboring frames potentially contains novel information about the desired high-resolution image, as suggested by the model (17.11). Several multiframe methods have been proposed which are in the form of two-stage interpolation-restoration algorithms. They are based on the premise that all pixels from available frames can be mapped back onto the reference frame, based on the motion trajectory information, to obtain an upsampled frame. However, unless we assume global, constant-velocity motion, the upsampled reference frame contains nonuniformly spaced samples. In order to obtain a uniformly spaced upsampled image, an interpolation onto a uniform sampling grid needs to be performed. Some methods include a postprocessing step, where image restoration is applied to the upsampled image to remove the effect of the sensor PSF blur. (See Chapter 15, [And 77], or [Gon 87] for a discussion of image restoration.)

Sauer and Allebach [Sau 87] propose an iterative method to reconstruct band-limited images from nonuniformly spaced samples. Their method estimates image intensities on a uniformly spaced grid using a projections-based method. In [Kom 93], Aizawa *et al.* use a nonuniform sampling theorem proposed by Clark *et al.* [Cla 85] to transform nonuniformly spaced samples (acquired by a stereo camera) onto a single uniform sampling grid. Recently, Ur and Gross [Ur 92] have proposed a nonuniform interpolation scheme based on the generalized sampling theorem of Papoulis to obtain an improved-resolution blurred image. They acknowledge the need for the restoration step after nonuniform interpolation. An important drawback of these algorithms is that they do not address removal of aliasing artifacts. Furthermore, the restoration stage neglects the errors in the interpolation stage.

17.3 A Frequency Domain Method

The frequency domain method, first proposed by Tsai and Huang [Tsa 84], exploits the relationship between the continuous and discrete Fourier transforms of the undersampled frames in the special case of global motion. If we let $s_0(x_1, x_2) \doteq s_c(x_1, x_2, 0)$ denote the reference frame, and assume zero aperture time ($T_A = 0$) and rectangular sampling with the sampling interval Δ in both directions, then the continuous-input, discrete-output model (17.5) simplifies as

$$\begin{aligned} g_k(n_1, n_2) &= \sum_{n_1=0}^{N-1} \sum_{n_2=0}^{N-1} [s_0(x_1 - \alpha_k, x_2 - \beta_k) * *h_c(x_1, x_2)]\ \delta(x_1 - n_1\Delta, x_2 - n_2\Delta) \\ &\quad + v_k(n_1, n_2) \qquad (17.12) \end{aligned}$$

where α_k and β_k denote the x_1 and x_2 components of the displacement of the kth frame with respect to the reference frame, $\delta(x_1, x_2)$ denotes the 2-D Dirac delta function, and the low-resolution frames are assumed to be $N \times N$. Because of the linearity of convolution, Equation (17.12) can be equivalently expressed as

$$\begin{aligned} g_k(n_1, n_2) &= \sum_{n_1=0}^{N-1} \sum_{n_2=0}^{N-1} [s_0(x_1, x_2) * *h_c(x_1 - \alpha_k, x_2 - \beta_k)]\ \delta(x_1 - n_1\Delta, x_2 - n_2\Delta) \\ &\quad + v_k(n_1, n_2) \qquad (17.13) \end{aligned}$$

Suppose we wish to reconstruct an $M \times M$ high-resolution sampled version of the reference frame $s_0(x_1, x_2)$, where M is an integer multiple of N, i.e., $R = M/N$ is an integer. Assuming that $s_0(x_1, x_2)$ is bandlimited, such that

$$| S_0(F_1, F_2) |= 0, \quad \text{for} \quad | F_1, F_2 |> R\frac{1}{2\Delta} \qquad (17.14)$$

the model (17.13) undersamples $s_0(x_1, x_2)$ by a factor R in both the x_1 and x_2 directions. Taking the Fourier transform of both sides of the observation model (17.13), we obtain

$$\begin{aligned} G_k(f_1, f_2) = \sum_{i_1=0}^{R-1} \sum_{i_2=0}^{R-1} \Bigg[& \frac{1}{\Delta^2} S_0(\frac{f_1 - i_1}{\Delta}, \frac{f_2 - i_2}{\Delta}) \\ & H_c(\frac{f_1 - i_1}{\Delta}, \frac{f_2 - i_2}{\Delta})\ e^{-j\frac{2\pi}{\Delta}\{(f_1 - i_1)\alpha_k + (f_2 - i_2)\beta_k\}} \Bigg] + V_k(f_1, f_2) \qquad (17.15) \end{aligned}$$

where $S_0(f_1/\Delta, f_2/\Delta)$ and $H_c(f_1/\Delta, f_2/\Delta)$ denote the 2-D continuous Fourier transform of the reference frame $s_0(x_1, x_2)$ and the sensor PSF $h_c(x_1, x_2)$, respectively, and $G_k(f_1, f_2)$ and $V_k(f_1, f_2)$ denote the 2-D discrete Fourier transform of $g_k(n_1, n_2)$ and $v_k(n_1, n_2)$, respectively. We note that the motion blur can be incorporated into the simplified model (17.15) (that is, the assumption $T_A = 0$ may be relaxed) only if we have global, constant-velocity motion, so that the frequency response of the motion blur remains the same from frame to frame.

In order to recover the unaliased spectrum $S_0(f_1/\Delta, f_2/\Delta)$ at a given frequency pair (f_1, f_2), we need at least R^2 equations (17.15) at that frequency pair, which could be obtained from $L > R^2$ low-resolution frames. The set of equations (17.15), $k = 1, \ldots, L$, at any frequency pair (f_1, f_2) are decoupled from the equations that are formed at any other frequency pair. The formulation by Tsai and Huang [Tsa 84] ignores the sensor blur and noise, hence proposes to set up $L = R^2$ equations in as many unknowns at each frequency pair to recover the alias-free spectrum $S_0(f_1/\Delta, f_2/\Delta)$. This procedure is illustrated by means of an example in the following. Taking the sensor blur into account, we first recover the product $S_0(f_1/\Delta, f_2/\Delta)\ H_c(f_1/\Delta, f_2/\Delta)$ similarly. Subsequently, $S_0(f_1/\Delta, f_2/\Delta)$ can be estimated by inverse filtering or any other regularized deconvolution technique.

Example: 1-D case

Consider two low-resolution observations of a 1-D signal ($L = 2$), which are shifted with respect to each other by a subsample amount α, such that

$$\begin{aligned} g_1(n) &= \sum_{n=0}^{N-1} s_0(x)\delta(x - n\Delta) + v_1(n) \\ g_2(n) &= \sum_{n=0}^{N-1} s_0(x-\alpha)\delta(x - n\Delta) + v_2(n) \end{aligned}$$

where N is the number of low-resolution samples, and the sensor PSF is assumed to be a delta function.

Assuming that $g_k(n)$, $k = 1, 2$, are sampled at one half of the Nyquist rate ($R = 2$), and taking the Fourier transform of both sides, we have

$$\begin{aligned} G_1(f) &= \frac{1}{\Delta} S_0(\frac{f}{\Delta}) + \frac{1}{\Delta} S_0(\frac{f-1}{\Delta}) \\ G_2(f) &= \frac{1}{\Delta} S_0(\frac{f}{\Delta}) e^{-j\frac{2\pi}{\Delta} f\alpha} + \frac{1}{\Delta} S_0(\frac{f-1}{\Delta}) e^{-j\frac{2\pi}{\Delta}(f-1)\alpha} \end{aligned}$$

There is only one aliasing term in each expression, because we assumed sampling at half the Nyquist rate.

The aliasing in the spectrum of the low-resolution sampled signal $G_1(f)$ is illustrated in Figure 17.5. We can solve for the two unknowns

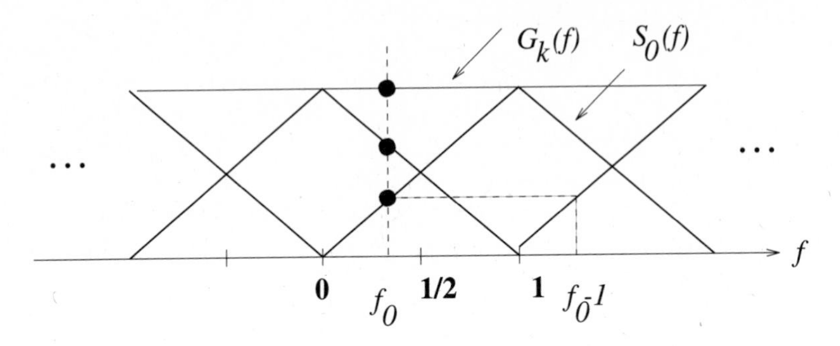

Figure 17.5: Demonstration of the frequency domain method.

$S_0(f_0/\Delta)$ and $S_0((f_0 - 1)/\Delta)$, which are marked on the figure, from these two equations given $G_1(f)$ and $G_2(f)$. Repeating this procedure at every frequency sample f, the spectrum of the alias-free continuous signal can be recovered.

The following remarks about the frequency domain method are in order: (i) In practice, the frequencies in the range $0 \leq f_1, f_2 \leq 1$ are discretized by K samples along each direction, where $K \geq M$, and M is the number of high-resolution signal samples. Then the samples of the high-resolution reference frame can be computed by a $K \times K$ inverse 2-D DFT. (ii) It is easy to see that if the image $s_c(x_1, x_2, 0)$ were passed through a perfect anti-alias filter before the low-resolution sampling, there would be only one term in the double summation in (17.15), and recovery of a high-resolution image would not be possible no matter how many low-resolution frames were available. Thus, it is the aliasing terms that make the recovery of a high-resolution image possible.

Note that the frequency domain method has some drawbacks: (i) The set of equations (17.15) at each frequency pair (f_1, f_2) can best be solved in the least squares sense due to the presence of observation noise. Thus, in practice, more equations, and hence more frames, than L^2 are needed to solve for the L^2 unknowns at each frequency pair (f_1, f_2). (ii) The set of equations (17.15) may be singular depending on the relative positions of the subpixel displacements, α_k and β_k, or due to zero-crossings in the Fourier transform $H_c(f_1, f_2)$ of the sensor PSF. In particular, if there are more than L frames with shifts (α_k, β_k) on a line parallel to either the x_1 or x_2 axis, or if there are more than $L(L^2 - L - 1)/2$ pairs of frames with shifts (α_k, β_k) which are symmetric with respect to the line $x_1 = x_2$, the system of equations becomes singular. This fact is stated by a theorem in [Kim 90]. Furthermore, it is clear from (17.15) that any zeros in $H_c(f_1, f_2)$ result in a column of zeros in the coefficient matrix of the system of equations. In this case, regularization techniques need to be employed which limit the resolution improvement.

This approach has been extended by Kim *et. al.* [Kim 90] to take into account noise and blur in the low-resolution images, by posing the problem in the least squares sense. In their approach, the blur and noise characteristics had to be the same for all frames of the low-resolution data, and impulse sampling was assumed for the low-resolution images (i.e., the low-resolution sensor had no physical size). This method is further refined by Kim and Su [Kim 93] to take into account blurs that are different for each frame of low-resolution data, by using a Tikhonov regularization for the previously reported least squares approach. The resulting algorithm does not treat the formation of blur due to motion or sensor size, and suffers from convergence problems. In the following, we discuss a more robust and general spatio-temporal domain method based on the theory of convex projections.

17.4 A Unifying POCS Method

In this section, we present a POCS-based algorithm to address the general superresolution problem, including the special cases of multiframe shift-varying restoration and standards conversion. Inspection of (17.11) suggests that the superresolution problem can be stated in the spatio-temporal domain as the solution of a set of simultaneous linear equations. Suppose that the desired high-resolution frames are $M \times M$, and we have L low-resolution observations, each $N \times N$. Then, from (17.11), we can set up at most $L \times N \times N$ equations in M^2 unknowns to reconstruct a particular high-resolution frame i. These equations are linearly independent provided that all displacements between the successive frames are at subpixel amounts. (Clearly, the number of equations will be reduced by the number of occlusion labels encountered along the respective motion trajectories.) In general, it is desirable to set up an overdetermined system of equations, i.e., $L > R^2 = M^2/N^2$, to obtain a more robust solution in the presence of observation noise. Because the impulse response coefficients $h_{ik}(n_1, n_2; m_1, m_2)$ are spatially varying, and hence the system matrix is not block-Toeplitz, fast methods to solve them are not available. In the following, we present a POCS-based method to solve the set of simultaneous linear equations (17.11). A similar but more limited solution was also proposed by Stark and Oskoui [Sta 89].

The POCS formulation presented in the following is similar to the one proposed in Section 15.3 for intraframe restoration of shift-varying blurred images. Here, we define a different closed, convex set for each observed low-resolution pixel (n_1, n_2, k) (which can be connected to frame i by a motion trajectory $c(i; n_1, n_2, k)$) as follows

$$C_{n_1,n_2;i,k} = \{x_i(m_1, m_2) : |r_k^{(\mathbf{x}_i)}(n_1, n_2)| \leq \delta_0\},$$
$$0 \leq n_1, n_2 \leq N-1, \quad k = 1, \ldots, L \tag{17.16}$$

where

$$r_k^{(\mathbf{x}_i)}(n_1, n_2) \doteq g_k(n_1, n_2) - \sum_{m_1=0}^{M-1} \sum_{m_2=0}^{M-1} x_i(m_1, m_2) h_{ik}(m_1, m_2; n_1, n_2)$$

and δ_0 represents the confidence that we have in the observation and is set equal to $c\sigma_v$, where σ_v is the standard deviation of the noise and $c \geq 0$ is determined by an appropriate statistical confidence bound. These sets define high-resolution images which are consistent with the observed low-resolution frames within a confidence bound that is proportional to the variance of the observation noise.

The projection $y_i(m_1, m_2) \doteq P_{n_1,n_2;i,k}[x_i(m_1, m_2)]$ of an arbitrary $x_i(m_1, m_2)$ onto $C_{n_1,n_2;i,k}$ is defined as

$$P_{n_1,n_2;i,k}[x_i(m_1,m_2)] = \tag{17.17}$$
$$\begin{cases} x_i(m_1,m_2) + \frac{r_k^{(\mathbf{X}_i)}(n_1,n_2)-\delta_0}{\sum_o \sum_p h_{ik}^2(o,p;n_1,n_2)} h_{ik}(m_1,m_2;n_1,n_2) & \text{if } r_k^{(\mathbf{X}_i)}(n_1,n_2) > \delta_0 \\ x_i(m_1,m_2) & \text{if } -\delta_0 < r_k^{(\mathbf{X}_i)}(n_1,n_2) < \delta_0 \\ x_i(m_1,m_2) + \frac{r_k^{(\mathbf{X}_i)}(n_1,n_2)+\delta_0}{\sum_o \sum_p h_{ik}^2(o,p;n_1,n_2)} h_{ik}(m_1,m_2;n_1,n_2) & \text{if } r_k^{(\mathbf{X}_i)}(n_1,n_2) < -\delta_0 \end{cases}$$

where

$$r_k^{(\mathbf{X}_i)}(n_1,n_2) \doteq g_k(n_1,n_2) - \sum_{m_1=0}^{M-1} \sum_{m_2=0}^{M-1} x_i(m_1,m_2) h_{ik}(n_1,n_2;m_1,m_2)$$

Additional constraints, such as amplitude and/or finite support constraints, can be utilized to improve the results. The amplitude constraint is given by

$$C_A = \{x_i(m_1,m_2) : 0 \leq x_i(m_1,m_2) \leq 255 \;\text{ for }\; 0 \leq m_1, m_2 \leq M-1\} \tag{17.18}$$

The projection $\mathbf{P}_A$ onto the amplitude constraint C_A is defined as

$$\mathbf{P}_A[x_i(m_1,m_2)] = \begin{cases} 0 & \text{if } x_i(m_1,m_2) < 0 \\ x_i(m_1,m_2) & \text{if } 0 \leq x_i(m_1,m_2) \leq 255 \\ 255 & \text{if } x_i(m_1,m_2) > 255 \end{cases} \tag{17.19}$$

Given the above projections, an estimate $\hat{s}_i(m_1, m_2)$ of the high-resolution image $s_i(m_1, m_2)$ is obtained iteratively as

$$\hat{s}_i^{(j+1)}(m_1,m_2) = \mathbf{T}_A \prod_{\substack{(0,0) \leq (n_1,n_2) \leq (N-1,N-1) \\ k=1,\ldots,L^2}} \mathbf{T}_{n_1,n_2;k} \, [\hat{s}_i^{(j)}(m_1,m_2)]$$
$$j = 0,1,\ldots, \quad 0 \leq m_1 \leq M-1, \quad 0 \leq m_2 \leq M-1 \tag{17.20}$$

where $\mathbf{T}$ denotes the generalized projection operator. The low-resolution reference frame can be interpolated by a factor of M/N in both directions, using bilinear or bicubic interpolation, to use as an initial estimate $\hat{s}_i^{(0)}(m_1, m_2)$. Excellent reconstructions have been reported using this procedure [Ozk 92, Pat 94].

A few observations about the POCS method are in order:

i) While certain similarities exist between the POCS iterations and the Landweber-type iterations [Ira 91, Ira 93, Kom 93], the POCS method can adapt to the amount

of the observation noise, while the latter generally cannot.
ii) The proposed POCS method can also be applied to shift-varying multiframe restoration and standards conversion problems by specifying the input and output lattices and the shift-varying system PSF $h_{ik}(m_1, m_2; n_1, n_2)$ appropriately (as stated in Section 17.1.3).
iii) The POCS method finds a feasible solution, that is, a solution consistent with all available low-resolution observations. Clearly, the more the low-resolution observations (more frames with reliable motion estimation), the better the high-resolution reconstructed image $\hat{s}_i(m_1, m_2)$ will be. In general, it is desirable that $L > M^2/N^2$. Note however that, the POCS method generates a reconstructed image with any number L of available frames. The number L is just an indicator of how large the feasible set of solutions will be. Of course, the size of the feasible set can be further reduced by employing other closed, convex constraints in the form of statistical or structural image models.

17.5 Examples

The feasibility of the unifying POCS approach presented in this chapter has been demonstrated by means of a set of six low-resolution still images of a stationary target taken by an Apple Quick-Take 100 camera, each from a slightly different view. The green components of the low-resolution images, 130 pixels $\times$ 142 lines each, are shown in Figure 17.6 (a-f). The camera uses a color filter array (CFA) that samples the green component using a diamond lattice. We assume that the focal plane of the camera is completely covered by the CFA array, and all elements have equal size and uniform response. A conversion from the diamond lattice to progressive, and further upsampling of the progressive grid by a factor of two, is carried out simultaneously using the proposed algorithm. The focal blur was assumed to be Gaussian, with a variance of 1, and a 5×5 support (units are relative to the high-resolution spatial sampling period). Motion was computed between the low-resolution pictures using five-level hierarchical block matching with quarter-pixel accuracy (relative to a progressive low-resolution grid). The aperture time was assumed to be zero.

Figure 17.7 (a) shows the first frame enlarged by using bilinear interpolation. Clearly, no new frequency components can be generated as a result of bilinear interpolation; hence, there is no resolution improvement. Next, we use the POCS method, which is initialized by the result of the bilinear interpolation, to reconstruct the first frame using data from all six low-resolution frames. The result after 20 iterations, depicted in Figure 17.7 (b), demonstrates noticeable resolution improvement over that of the bilinear interpolation. We remark that the larger the number of available low-resolution frames, the smaller in size the feasible set of solutions will be. An analysis of the number of equations and the number of unknowns suggests that we need a minimum of 16 low-resolution frames in order to hope to have a unique solution in this case.

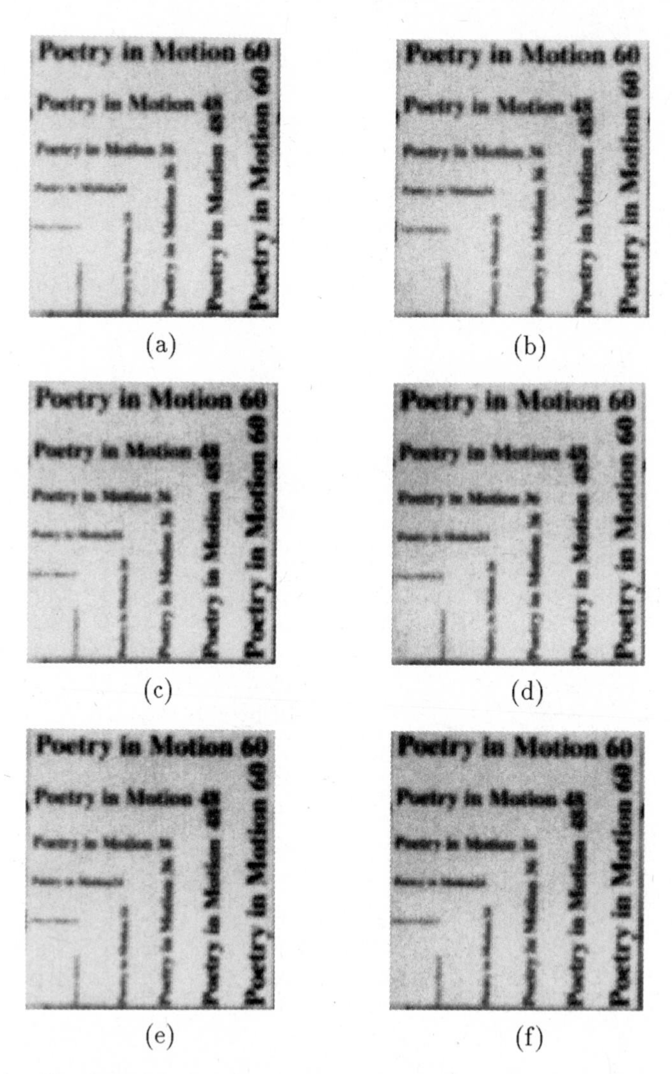

Figure 17.6: Six frames of a low-resolution image sequence taken by an Apple QuickTake 100 camera. (Courtesy Andrew Patti)

Figure 17.7: The first frame a) interpolated by 2 using bilinear interpolation, and b) reconstructed by the POCS algorithm. (Courtesy Andrew Patti)

17.6 Exercises

1. Consider global, constant-velocity motion, where the velocity is modeled as constant during each aperture time (piecewise constant-velocity model). Let its value during ith aperture time (acquisition of the ith frame) be given by $\mathbf{v}_i = [v_{1i}\ v_{2i}]^T$. Show that the Jacobian $J(u_1, u_2, t)$ in (17.4) is equal to $\frac{1}{|v_{1i}|}$.

2. Suppose we have four low-resolution images that are globally translating with respect to a reference frame with velocities $v_{1i} = v_{2i} = 8.25, 8.5, 9$, and 10 in units of pixels per high-resolution sampling grid for $i = 1, \ldots, 4$, respectively. Assume that each side of the low-resolution sensor cell is four times that of the high-resolution cell with rectangular sensor geometry shown in Figure 17.2. How would you calculate the discrete-discrete PSF $h_{ik}(m_1, m_2; n_1, n_2)$ in (17.11)?

3. Derive (17.15).

4. Derive (17.17).

5. Discuss the relationship between the POCS reconstruction method discussed in Section 17.4 and the backprojection iterations presented in [Ira 93].

Bibliography

[And 77] H. C. Andrews and B. R. Hunt, *Digital Image Restoration*, Englewood Cliffs, NJ: Prentice-Hall, 1977.

[Aya 92] B. Ayazifar, *Pel-Adaptive Model-Based Interpolation of Spatially Subsampled Images*, M.S. Thesis, Mass. Inst. of Tech., 1992.

[Cla 85] J. J. Clark, M. R. Palmer, and P. D. Lawrence, "A transformation method for the reconstruction of functions from nonuniformly spaced samples," *IEEE Trans. Acoust. Speech Sign. Proc.*, vol. 33, pp. 1151–1165, Oct. 1985.

[Gon 87] R. C. Gonzalez and P. Wintz, *Digital Image Processing*, Reading, MA: Addison-Wesley, 1987.

[Ira 91] M. Irani and S. Peleg, "Improving resolution by image registration," *CVGIP: Graphical Models and Image Proc.*, vol. 53, pp. 231–239, May 1991.

[Ira 93] M. Irani and S. Peleg, "Motion analysis for image enhancement: Resolution, occlusion and transparency," *J. Vis. Comm. and Image Rep.*, vol. 4, pp. 324–335, Dec. 1993.

[Kim 90] S. P. Kim, N. K. Bose, and H. M. Valenzuela, "Recursive reconstruction of high-resolution image from noisy undersampled frames," *IEEE Trans. Acoust., Speech and Sign. Proc.*, vol. 38, pp. 1013–1027, June 1990.

[Kim 93] S. P. Kim and W.-Y. Su, "Recursive high-resolution reconstruction of blurred multiframe images," *IEEE Trans. Image Proc.*, vol. 2, pp. 534–539, Oct. 1993.

[Kom 93] T. Komatsu, T. Igarashi, K. Aizawa, and T. Saito, "Very high-resolution imaging scheme with multiple different aperture cameras," *Signal Proc.: Image Comm.*, vol. 5, pp. 511–526, Dec. 1993.

[Ozk 94] M. K. Ozkan, A. M. Tekalp, and M. I. Sezan, "POCS-based restoration of space-varying blurred images," *IEEE Trans. Image Proc.*, vol. 3, pp. 450–454, July 1994.

[Pat 94] A. Patti, M. I. Sezan, and A. M. Tekalp, "High-resolution image reconstruction from a low-resolution image sequence in the presence of time-varying motion blur," *Proc. IEEE Int. Conf. Image Proc.*, Austin, TX, Nov. 1994.

[Sau 87] K. D. Sauer and J. P. Allebach, "Iterative reconstruction of bandlimited images from non-uniformly spaced samples," *IEEE Trans. Circ. and Syst.*, vol. 34, pp. 1497–1505, 1987.

[Sch 94] R. R. Schultz and R. L. Stevenson, "A Bayesian approach to image expansion for improved definition," *IEEE Trans. Image Proc.*, vol. 3, pp. 233–242, May 1994.

[Sta 89] H. Stark and P. Oskoui, "High-resolution image recovery from image plane arrays using convex projections," *J. Opt. Soc. Amer. A*, vol. 6, pp. 1715–1726, 1989.

[Tek 92] A. M. Tekalp, M. K. Ozkan, and M. I. Sezan, "High-resolution image reconstruction from lower-resolution image sequences and space-varying image restoration," *IEEE Int. Conf. Acoust. Speech and Sign. Proc.*, San Francisco, CA, vol. III, pp. 169–172, March 1992.

[Tsa 84] R. Y. Tsai and T. S. Huang, "Multiframe image restoration and registration," in *Advances in Computer Vision and Image Processing*, vol. 1, T. S. Huang, ed., pp. 317-339, Greenwich, CT: Jai Press, 1984.

[Ur 92] H. Ur and D. Gross, "Improved resolution from subpixel shifted pictures," *CVGIP: Graphical Models and Image Processing*, vol. 54, pp. 181–186, March 1992.

[Wan 91] Y. Wang and S. K. Mitra, "Motion/pattern adaptive interpolation of interlaced video sequences," *Proc. IEEE Int. Conf. Acoust. Speech Sign. Proc.*, pp. 2829–2832, Toronto, Canada, May 1991.

Chapter 18

LOSSLESS COMPRESSION

The need for effective data compression is evident in almost all applications where storage and transmission of digital images are involved. For example, an 8.5×11 in document scanned at 300 pixels/in with 1 bit/pixel generates 8.4 Mbits data, which without compression requires about 15 min transmission time over a 9600 baud line. A 35 mm film scanned at 12 micron resolution results in a digital image of size 3656 pixels $\times$ 2664 lines. With 8 bits/pixel per color and three color channels, the storage required per picture is approximately 233 Mbits. The storage capacity of a CD is about 5 Gbits, which without compression can hold approximately 600 pages of a document, or 21 color images scanned from 35 mm film. Several world standards for image compression, such as ITU (formerly CCITT) Group 3 and 4 codes, and ISO/IEC/CCITT JPEG, have recently been developed for efficient transmission and storage of binary, gray-scale, and color images.

Compression of image data without significant degradation of the visual quality is usually possible because images contain a high degree of i) spatial redundancy, due to correlation between neighboring pixels, ii) spectral redundancy, due to correlation among the color components, and iii) psychovisual redundancy, due to properties of the human visual system. The higher the redundancy, the higher the achievable compression. This chapter introduces the basics of image compression, and discusses some lossless compression methods. It is not intended as a formal review of the related concepts, but rather aims to provide the minimum information necessary to follow the popular still-image and video compression algorithms/standards, which will be discussed in the subsequent chapters. Elements of an image compression system as well as some information theoretic concepts are introduced in Section 18.1. Section 18.2 discusses symbol coding, and in particular entropy coding, which is an integral part of lossless compression methods. Finally, three commonly used lossless compression algorithms are presented in Section 18.3.

18.1 Basics of Image Compression

In this section, we first present the elements of a general image compression system, then summarize some results from the information theory which provide bounds on the achievable compression ratios and bitrates.

18.1.1 Elements of an Image Compression System

In information theory, the process of data compression by redundancy reduction is referred to as source encoding. Images contain two types of redundancy, statistical (spatial) and pyschovisual. Statistical redundancy is present because certain spatial patterns are more likely than others, whereas psychovisual redundancy originates from the fact that the human eye is insensitive to certain spatial frequencies. The block diagram of a source encoder is shown in Figure 18.1. It is composed of the following blocks:

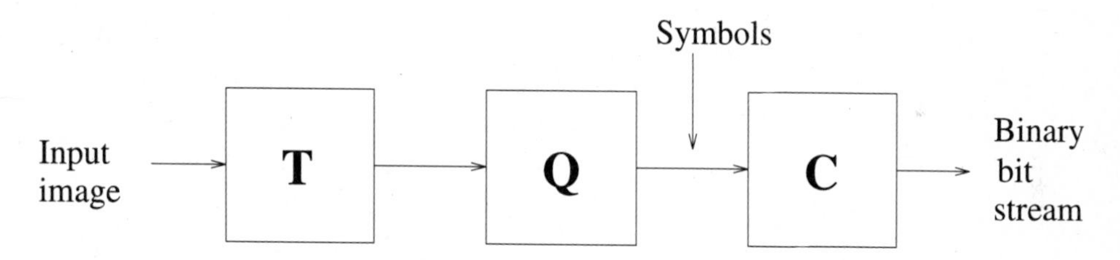

Figure 18.1: Block diagram of an image compression system.

i) Transformer (T) applies a one-to-one transformation to the input image data. The output of the transformer is an image representation which is more amenable to efficient compression than the raw image data. Typical transformations are linear predictive mapping, which maps the pixel intensities onto a prediction error signal by subtracting the predictible part of the pixel intensities; unitary mappings such as the discrete cosine transform, which pack the energy of the signal to a small number of coefficients; and multiresolution mappings, such as subband decompositions and the wavelet transform.
ii) Quantizer (Q) generates a limited number of symbols that can be used in the representation of the compressed image. Quantization is a many-to-one mapping which is irreversible. It can be performed by scalar or vector quantizers. Scalar quantization refers to element-by-element quantization of the data, whereas quantization of a block of data at once is known as vector quantization.
iii) Coder (C) assigns a codeword, a binary bitstream, to each symbol at the output of the quantizer. The coder may employ fixed-length or variable-length codes. Variable-length coding (VLC), also known as entropy coding, assigns codewords in such a way as to minimize the average length of the binary representation of the

symbols. This is achieved by assigning shorter codewords to more probable symbols, which is the fundamental principle of entropy coding.

Different image compression systems implement different combinations of these choices. Image compression methods can be broadly classified as:
i) Lossless (noiseless) compression methods, which aim to minimize the bitrate without any distortion in the image.
ii) Lossy compression methods, which aim to obtain the best possible fidelity for a given bitrate, or to minimize the bitrate to achieve a given fidelity measure.
The transformation and encoding blocks are lossless. However, quantization is lossy. Therefore, lossless methods, which only make use of the statistical redundancies, do not employ a quantizer. In most practical cases a small degradation in the image quality must be allowed to achieve the desired bitrate. Lossy compression methods make use of both the statistical and psychovisual redundancies.

In the following, we first briefly review some results from the information theory which gives bounds on the achievable bitrates in both lossless and lossy compression. In Section 18.2, we present techniques for symbol coding (the third box in Figure 18.1). The discussion of the second box, the quantizer, is deferred until the next chapter. Finally, some commonly used lossless image compression methods are presented in Section 18.3.

18.1.2 Information Theoretic Concepts

A source $\mathcal{X}$ with an alphabet $\mathcal{A}$ is defined as a discrete random process (a sequence of random variables X_i, $i = 1, \ldots$) in the form $\mathcal{X} = X_1 \, X_2 \, \ldots$, where each random variable X_i takes a value from the alphabet $\mathcal{A}$. In the following, we assume that the alphabet contains a finite number (M) of symbols, i.e., $\mathcal{A} = \{a_1, a_2, \ldots, a_M\}$. Here, we introduce two source models, a discrete memoryless source (DMS) and a Markov-K source.

A DMS is such that successive symbols are statistically independent. It is completely specified by the probabilities $p(a_i) = p_i$, $i = 1, \ldots, M$ such that $p_1 + \ldots + p_M = 1$. In VLC, the optimum length of the binary code for a symbol is equal to the information (in bits) that the symbol conveys. According to information theory, the information content of a symbol is related to the extent that the symbol is unpredictable or unexpected. If a symbol with low probability occurs, a larger amount of information is transferred than in the occurence of a more likely symbol. This quantitative concept of surprise is formally expressed by the relation

$$I(a_i) = \log_2 (1/p(a_i)), \quad \text{for } a_i \in \mathcal{A} \tag{18.1}$$

where $I(a_i)$ is the amount of information that the symbol a_i with probability $p(a_i)$ carries. The unit of information is bit when we use logarithm with base-2. Observe that if $p = 1$, then $I = 0$ as expected, and as $p \to 0$, $I \to \infty$. In practice, the probability of occurrence of each symbol is estimated from the histogram of a specific source, or a training set of sources.

The entropy $H(\mathcal{X})$ of a DMS $\mathcal{X}$ with an alphabet $\mathcal{A}$ is defined as the average information per symbol in the source, given by

$$\begin{aligned} H(\mathcal{X}) &= \sum_{a \in \mathcal{A}} p(a) \log_2 (1/p(a)) \\ &= -\sum_{a \in \mathcal{A}} p(a) \log_2 p(a) \end{aligned} \tag{18.2}$$

The more skewed the probability distribution of the symbols, the smaller the entropy of the source. The entropy is maximized for a flat distribution, that is, when all symbols are equally likely. It follows that a source where some symbols are more likely than others has a smaller entropy than a source where all symbols are equally likely.

> *Example:* Entropy of raw image data
>
> Suppose an 8-bit image is taken as a realization of a DMS $\mathcal{X}$. The symbols i are the gray levels of the pixels, and the alphabet $\mathcal{A}$ is the collection of all gray levels between 0 and 255. Then the entropy of the image is given by
>
> $$H(\mathcal{X}) = -\sum_{i=0}^{255} p(i) \log_2 p(i)$$
>
> where $p(i)$ denotes the relative frequency of occurrence of the gray level i in the image. Note that the entropy of an image consisting of a single gray level (constant image) is zero.

Most realistic sources can be better modeled by Markov-K random processes. That is, the probability of occurence of a symbol depends on the values of K preceding symbols. A Markov-K source can be specified by the conditional probabilities $p(X_j = a_i | X_{j-1}, \ldots, X_{j-K})$, for all j, $a_i \in \mathcal{A}$. The entropy of a Markov-K source is defined as

$$H(\mathcal{X}) = \sum_{S^K} p(X_{j-1}, \ldots, X_{j-K}) H(\mathcal{X} | X_{j-1}, \ldots, X_{j-K}) \tag{18.3}$$

where S^K denotes all possible realizations of $X_{j-1}, \ldots, X_{j-K}$, and

$$H(\mathcal{X} | X_{j-1}, \ldots, X_{j-K}) = \sum_{a \in \mathcal{A}} p(a_i | X_{j-1}, \ldots, X_{j-K}) \log p(a_i | X_{j-1}, \ldots, X_{j-K})$$

In the following, we present two fundamental theorems, the Lossless Coding Theorem and the Source Coding Theorem, which are used to measure the performance of lossless coding and lossy coding systems, respectively.

Lossless Coding Theorem: [Shannon, 1948] The minimum bitrate that can be achieved by lossless coding of a discrete memoryless source $\mathcal{X}$ is given by

$$\min\{R\} = H(\mathcal{X}) + \epsilon \text{ bits/symbol} \tag{18.4}$$

where R is the transmission rate, $H(\mathcal{X})$ is the entropy of the source, and ϵ is a positive quantity that can be made arbitrarily close to zero.

The Lossless Coding Theorem establishes the lower bound for the bitrate necessary to achieve zero coding-decoding error. In the case of a DMS, we can approach this bound by encoding each symbol independently. For sources with memory, we need to encode blocks of N source symbols at a time to come arbitrarily close to the bound. For example, a Markov-M source should be encoded M symbols at a time. In the next section, we introduce two coding techniques, Huffman coding and arithmetic coding, that approach the entropy bound.

In lossy coding schemes, the achievable minimum bitrate is a function of the distortion that is allowed. This relationship between the bitrate and distortion is given by the rate distortion function [Ber 71].

Source Coding Theorem: There exists a mapping from the source symbols to codewords such that for a given distortion D, $R(D)$ bits/symbol are sufficient to enable source reconstruction with an average distortion that is arbitrarily close to D. The actual rate R should obey

$$R \geq R(D) \tag{18.5}$$

for the fidelity level D. The function $R(D)$ is called the rate-distortion function. Note that $R(0) = H(\mathcal{X})$.

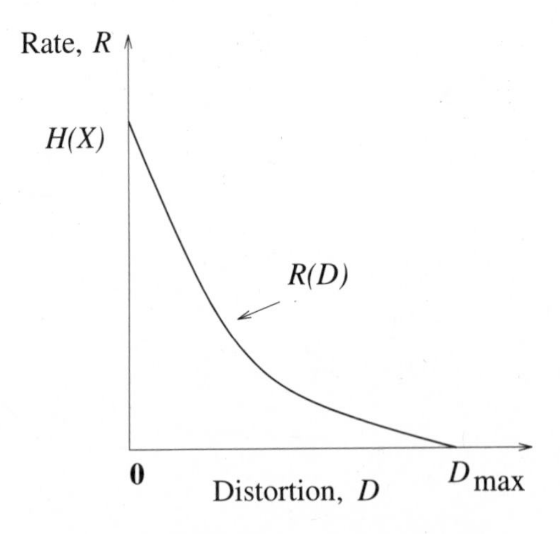

Figure 18.2: Rate distortion function.

A typical rate-distortion function is depicted in Figure 18.2. The rate distortion function can be computed analytically for simple source and distortion models. Computer algorithms exist to compute $R(D)$ when analytical methods fail or are unpractical [Ber 71]. In general, we are interested in designing a compression system to achieve either the lowest bitrate for a given distortion or the lowest distortion at a given bitrate. Note that the source coding theorem does not state how to design algorithms to achieve these desired limits. Some well-known lossy coding algorithms are discussed in Chapters 19, 20, and 21.

18.2 Symbol Coding

Symbol coding is the process of assigning a bit string to individual symbols or to a block of symbols comprising the source. The simplest scheme is to assign equal-length codewords to individual symbols or a fixed-length block of symbols, which is known as fixed-length coding. Because compression is generally achieved by assigning shorter-length codewords to more probable symbols, we next describe two variable-length coding, also known as entropy coding, schemes. The first, Huffman coding, assigns variable-length codes to a fixed-length block of symbols, where the block length can be one. In Huffman coding, the length of the codewords is proportional to the information (in bits) of the respective symbols or block of symbols. The latter, arithmetic coding, assigns variable-length codes to a variable-length block of symbols.

18.2.1 Fixed-Length Coding

In fixed-length coding, we assign equal-length code words to each symbol in the alphabet $\mathcal{A}$ regardless of their probabilities. If the alphabet has M different symbols (or blocks of symbols), then the length of the code words is the smallest integer greater than $\log_2 M$. Two commonly used fixed-length coding schemes are natural codes and Gray codes, which are shown in Table 18.1 for the case of a four-symbol source. Notice that in Gray coding, the consecutive codewords differ in only one bit position. This property of the Gray codes may provide an advantage in error detection. We will see in Section 18.3 that Gray codes are also better suited for run-length encoding of bit-planes.

Table 18.1: Fixed-length codes for a four-symbol alphabet.

Symbol	Natural code	Gray code
a_1	00	00
a_2	01	01
a_3	10	11
a_4	11	10

It can easily be shown that fixed-length coding is optimal only when:

1) the number of symbols is equal to a power of 2, and
2) all the symbols are equiprobable.

Only then would the entropy of the source be equal to the average length of the codewords, which is equal to the length of each codeword in the case of fixed-length coding. For the example shown in Table 18.1, both the entropy of the source and the average codeword length is 2, assuming all symbols are equally likely. Most often, some symbols are more probable than others, where it would be more advantageous to use entropy coding. Actually, the goal of the transformation box in Figure 18.1 is to obtain a set of symbols with a skew probability distribution, to minimize the entropy of the transformed source.

18.2.2 Huffman Coding

Huffman coding yields the optimal integer prefix codes given a source with a finite number of symbols and their probabilities. In prefix codes, no codeword is a prefix of another codeword. Such codes are uniquely decodable since a given binary string can only be interpreted in one way. Huffman codes are optimal in the sense that no other integer-length VLC can be found to yield a smaller average bitrate. In fact, the average length of Huffman codes per codeword achieves the lower bound, the entropy of the source, when the symbol probabilities are all powers of 2.

Huffman codes can be designed by following a very simple procedure. Let $\mathcal{X}$ denote a DMS with the alphabet $\mathcal{A}$ and the symbol probabilities $p(a_i)$, $a_i \in \mathcal{A}$, $i = 1, \ldots, M$. Obviously, if $M = 2$, we must have

$$c(a_1) = 0 \quad \text{and} \quad c(a_2) = 1 \tag{18.6}$$

where $c(a_i)$ denotes the codeword for the symbol a_i, $i = 1, 2$. If $\mathcal{A}$ has more than two symbols, the Huffman procedure requires a series of source reduction steps. In each step, we find and merge the two symbols with the smallest probabilities, which results in a new source with a reduced alphabet. The probability of the new symbol in the reduced alphabet is the sum of the probabilities of the two "merged" symbols from the previous alphabet. This procedure is continued until we reach a source with only two symbols, for which the codeword assignment is given by (18.6). Then we work backwards towards the original source, each time splitting the codeword of the "merged" symbol into two new codewords by appending it with a zero and one, respectively. The following examples demonstrate this procedure.

Example: Symbol probabilities are powers of 2

Let the alphabet $\mathcal{A}$ consist of four symbols, shown in Table 18.2. The probabilities and the information content of the symbols in the alphabet are also listed in the table. Note that all symbol probabilities are powers of 2, and consequently the symbols have integer information values.

Table 18.2: An alphabet where the symbol probabilities are powers of 2.

Symbol	Probability	Information
a_1	0.50	1 bit
a_2	0.25	2 bits
a_3	0.125	3 bits
a_4	0.125	3 bits

The Huffman coding procedure is demonstrated for this alphabet in Table 18.3. The reduced alphabet in Step 1 is obtained by merging the symbols a_3 and a_4 in the original alphabet which have the lowest two probabilities. Likewise, the reduced alphabet in Step 2 is obtained by merging the two symbols with the lowest probabilities after Step 1. Since the reduced alphabet in Step 2 has only two symbols, we assign the codes 0 and 1 to these symbols in arbitrary order. Next, we assign codes to the reduced alphabet in Step 1. We recall that the symbol 2 in Step 2 is obtained by merging the symbols 2 and 3 in Step 1. Thus, we assign codes to symbols 2 and 3 in Step 1 by appending the code for symbol 2 in Step 2 by a zero and one in arbitrary order. The appended zero and one are shown by bold fonts in Table 18.3. Finally, the codes for the original alphabet are obtained in a similar fashion.

Table 18.3: Illustration of alphabet reduction.

Original Alphabet		Reduced Alphabet Step 1		Reduced Alphabet Step 2	
p	c	p	c	p	c
0.50	0	0.50	0	0.50	**0**
0.25	10	0.25	1**0**	0.50	**1**
0.125	11**0**	0.25	1**1**		
0.125	11**1**				

This procedure can alternatively be described by the tree diagram shown in Figure 18.3.

Observe that in this case, the average codeword length is

$$\bar{R} = 0.5 \times 1 + 0.25 \times 2 + 0.125 \times 3 + 0.125 \times 3 = 1.75$$

and the entropy of the source is

$$H = -0.5 \ln 0.5 - 0.25 \ln 0.25 - 0.125 \ln 0.125 - 0.125 \ln 0.125 = 1.75$$

which is consistent with the result that Huffman coding achieves the entropy of the source when the symbol probabilities are powers of 2.

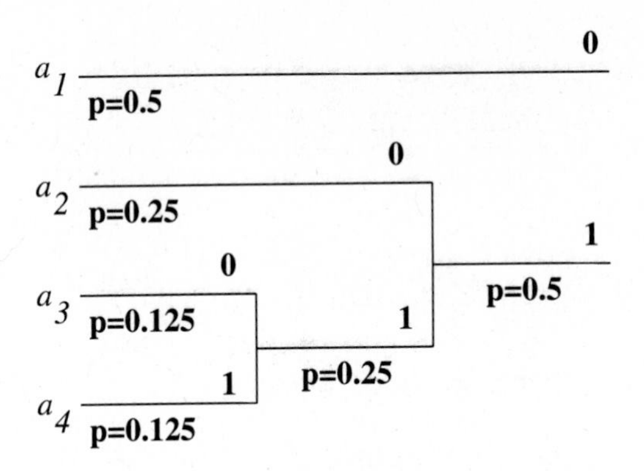

Figure 18.3: Tree-diagram for Huffman coding.

Next, we present an example in which the symbol probabilities are not powers of 2.

Example 2: Symbol probabilities are not powers of 2

The information content of each symbol is a real number, as shown in Table 18.4, when the probabilities of the symbols are not powers of 2.

Table 18.4: An alphabet with arbitrary symbol probabilities.

Symbol	Probability	Information
a_1	0.40	1.32 bits
a_2	0.25	2.00 bits
a_3	0.15	2.73 bits
a_4	0.15	2.73 bits
a_5	0.05	4.32 bits

Since the length of each codeword must be an integer, it is not possible to design codewords whose lengths are equal to the information of the respective symbols in this case. Huffman code design for the alphabet in Table 18.4 is shown in Table 18.5. It can be easily seen that for this example the average length of codewords is 2.15, and entropy of the source is 2.07.

Notice that Huffman codes are uniquely decodable, with proper synchronization, because no codeword is a prefix of another. For example, a received binary string

$$001101101110000\ldots$$

can be decoded uniquely as

$$a_3\ a_1\ a_2\ a_1\ a_2\ a_1\ a_1\ a_4\ \ldots$$

Table 18.5: Huffman coding when probabilities are not powers of 2.

Original Alphabet		Step 1		Step 2		Step 3	
p	c	p	c	p	c	p	c
0.40	1	0.40	1	0.40	1	0.60	**0**
0.25	01	0.25	01	0.35	0**0**	0.40	**1**
0.15	001	0.20	00**0**	0.25	0**1**		
0.15	000**0**	0.15	00**1**				
0.05	000**1**						

Huffman coding can also be used as a block coding scheme where we assign codewords to combinations of L symbols from the original alphabet at a time. Of course, this requires building a new block alphabet with all possible combinations of the L symbols from the original alphabet and computing their respective probabilities. Huffman codes for all possible combinations of the L symbols from the original alphabet can be formed using the above design procedure with the new block alphabet. Thus, Huffman coding is a block coding scheme, where we assign variable-length codes to fixed-length (L) blocks of symbols. The case $L = 1$ refers to assigning an individual codeword to each symbol of the original alphabet, as shown in the above two examples. It has been shown that for sources with memory, the coding efficiency improves as L gets larger, although the design of Huffman codes gets more complicated.

18.2.3 Arithmetic Coding

In arithmetic coding a one-to-one correspondence between the symbols of an alphabet $\mathcal{A}$ and the codewords does not exist. Instead, arithmetic coding assigns a single variable-length code to a source $\mathcal{X}$, composed of N symbols, where N is variable. The distinction between arithmetic coding and block Huffman coding is that in arithmetic coding the length of the input sequence, i.e., the block of symbols for which a single codeword is assigned, is variable. Thus, arithmetic coding assigns variable-length codewords to variable-length blocks of symbols. Because arithmetic coding does not require assignment of integer-length codes to fixed-length blocks of symbols, in theory it can achieve the lower bound established by the noiseless coding theorem.

Arithmetic coding associates a given realization of $\mathcal{X}$, $\mathbf{x} = \{x_1, \ldots, x_N\}$, with a subinterval of $[0, 1)$ whose length equals the probability of the sequence $p(\mathbf{x})$. The encoder processes the input stream of symbols one by one, starting with $N = 1$, where the length of the subinterval associated with the sequence gets smaller as N increases. Bits are sequentially sent to the channel starting from the most significant bit towards the least significant bit as they are determined according to a procedure, which is presented in an algorithmic form in the following. At the end of the transmission, the transmitted bitstream is a uniquely decodable codeword representing the source, which is a binary number pointing to the subinterval associated with this sequence.

The Procedure

Consider an alphabet $\mathcal{A}$ that has M symbols a_i, $i = 1, \ldots, M$, with the probabilities $p(a_i) = p_i$, such that $p_1 + \ldots + p_M = 1$. The procedure starts with assigning each individual symbol in the alphabet a subinterval, within 0 to 1, whose length is equal to its probability. It is assumed that this assignment is known to the decoder.

1. If the first input symbol $x_1 = a_i$, $i = 1, \ldots, M$, then define the initial subinterval as $I_1 = [l_1, r_1) = [p_{i-1}, p_{i-1} + p_i)$, where $p_0 = 0$. Set $n = 1$, $L = l_1$, $R = r_1$, and $d = r_1 - l_1$.

2. Obtain the binary expansions of L and R as

$$L = \sum_{k=1}^{\infty} u_k 2^{-k}, \qquad \text{and} \qquad R = \sum_{k=1}^{\infty} v_k 2^{-k}$$

where u_k and v_k are 0 or 1.

Compare u_1 and v_1. If they are not the same, send nothing to the channel at this time, and go to step 3.

If $u_1 = v_1$, then send the binary symbol u_1, and compare u_2 and v_2. If they are not the same, go to step 3.

If $u_2 = v_2$, also send the binary symbol u_2, and compare u_3 and v_3, and so on, until the next two corresponding binary symbols do not match, at which time go to step 3.

3. Increment n, and read the next symbol. If the nth input symbol $x_n = a_i$, then subdivide the interval from the previous step as

$$I_n = [l_n, r_n) = [l_{n-1} + p_{i-1}d, l_{n-1} + (p_{i-1} + p_i)d).$$

Set $L = l_n$, $R = r_n$, and $d = r_n - l_n$, and go to step 2.

Note that the decoder may decode one binary symbol into several source symbols, or it may require several binary symbols before it can decode one or more source symbols. The arithmetic coding procedure is illustrated by means of the following example.

Example

Suppose we wish to determine an arithmetic code to represent a sequence of symbols,

$$a_2 \; a_1 \; a_3 \; \ldots$$

from the source shown in Table 18.2. Because we have four symbols in the alphabet, the interval from 0 to 1 is initially subdivided into 4, where the lengths of the subintervals are equal to 0.5, 0.25, 0.125 and 0.125, respectively. This is depicted in Figure 18.4.

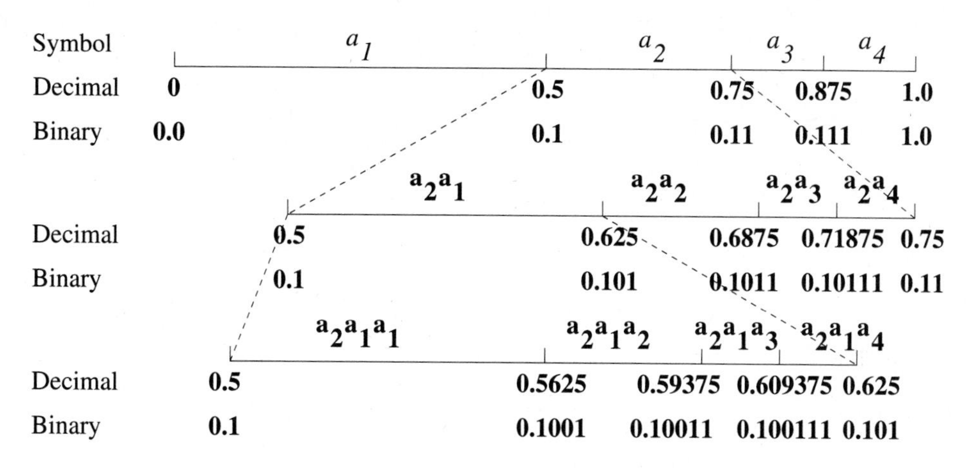

Figure 18.4: Illustration of the concept of arithmetic coding.

The first symbol defines the initial interval as $I_1 = [0.5, 0.75)$, where the binary representations of the left and right boundaries are $L = 2^{-1} = 0.1$ and $R = 2^{-1} + 2^{-2} = 0.11$, respectively. According to step 2, $u_1 = v_1 = 1$; thus, 1 is sent to the channel. Noting that $u_2 = 0$ and $v_2 = 1$, we read the second symbol, a_1. Step 3 indicates that $I_2 = [0.5, 0.625)$, with $L = 0.10$ and $R = 0.101$. Now that $u_2 = v_2 = 0$, we send 0 to the channel. However, $u_3 = 0$ and $v_3 = 1$, so we read the third symbol, a_3. It can be easily seen that $I_3 = [0.59375, 0.609375)$, with $L = 0.10011$ and $R = 0.100111$. Note that $u_3 = v_3 = 0$, $u_4 = v_4 = 1$, and $u_5 = v_5 = 1$, but $u_6 = 0$ and $v_6 = 1$. At this stage, we send 011 to the channel, and read the next symbol. A reserved symbol usually signals the end of a sequence.

Let's now briefly look at how the decoder operates, which is illustrated in Figure 18.5. The first bit restricts the interval to $[0.5, 1)$. However,

Received Bit	Interval	Symbol
1	$[0.5, 1)$	-
0	$[0.5, 0.75)$	a_2
0	$[0.5, 0.609375)$	a_1
1	$[0.5625, 0.609375)$	-
1	$[0.59375, 0.609375)$	a_3
⋮	⋮	

Figure 18.5: The operation of the decoder.

three symbols are within this range; thus, the first bit does not contain sufficient information. After receiving the second bit, we have 10 which points to the interval [0.5, 0.75). All possible combinations of two symbols pointing to this range start with a_2. Hence, we can now decode the first symbol as a_2. The information that becomes available after the receipt of each bit is summarized in Figure 18.5.

In practice, two factors cause the performance of the arithmetic encoder to fall short of the theoretical bound: the addition of an end-of-message indicator, and the use of finite precision arithmetic. Practical implementations of the arithmetic coder overcome the precision problem by a scaling and a rounding strategy.

18.3 Lossless Compression Methods

Error-free coding is the only acceptable means of compression in some applications for various reasons. For example, in the transmission or archival of medical images lossy compression is not allowed for legal reasons. Recalling the elements of a compression system, lossless coding schemes do not employ a quantizer. They consist of a transformation, which generates symbols whose probability distribution is highly peaked, followed by an entropy coder. The transformation aims to minimize the entropy of its output, so that significant compression becomes possible by variable-length coding of the generated symbols.

In this section, we present three popular methods for lossless compression: i) Lossless predictive coding, where an integer predictive mapping is employed, followed by entropy coding of the integer prediction errors. ii) Run-length coding of bit-planes, where the image is decomposed into individual bit-planes (binary images), and the run-lengths of zeros and ones in these planes are entropy coded. iii) Ziv-Lempel coding, which is a deterministic coding procedure, where the input bit string is parsed into blocks of variable length to form a dictionary of blocks (symbols), each of which is represented by a fixed-length codeword. The achievable compression ratio using lossless coding methods ranges between 2:1 to 5:1, depending on the characteristics of the input image.

18.3.1 Lossless Predictive Coding

The first step in lossless predictive coding is to form an integer-valued prediction of the next pixel intensity to be encoded based on a set of previously encoded neighboring pixels. Then the difference between the actual intensity of the pixel and its prediction is entropy coded. Assuming each pixel is integer-valued, then the prediction errors are also integer-valued, which facilitates lossless compression. The block diagram of a simple predictor is shown in Figure 18.6, where the prediction is taken as the intensity of the previous pixel encoded. Some other commonly used integer predictors are shown in the following example.

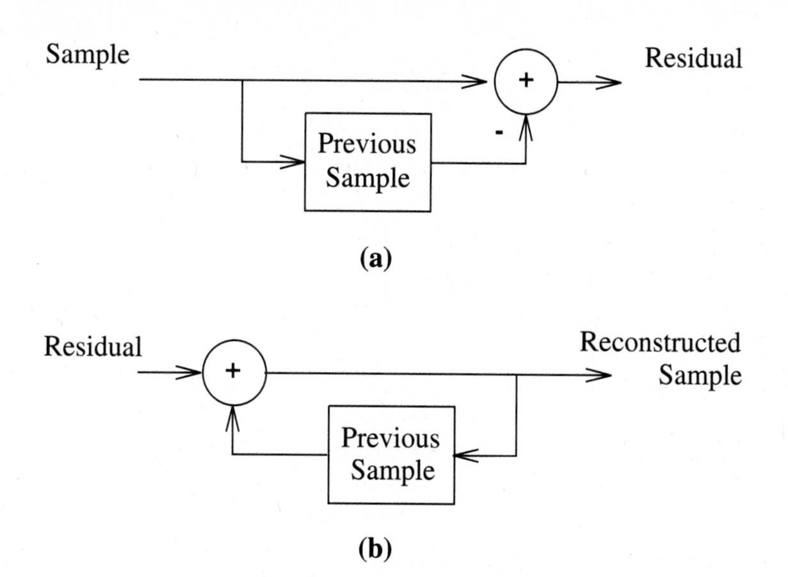

Figure 18.6: Block diagram of a) an encoder, and b) a decoder using a simple predictor.

Example: Integer prediction

In order for the decoder to be able to duplicate the prediction step, the prediction operation must be based on already-encoded pixels. In 2-D, such prediction models are called recursively computable. The support of a recursively computable predictor is shown in Figure 18.7, where the coefficients a, b, c, and d denote the intensities of the respective pixels.

Two simple predictors based on this model can be written as

$$\hat{x} = \text{int}\{(a+b)/2\} \tag{18.7}$$

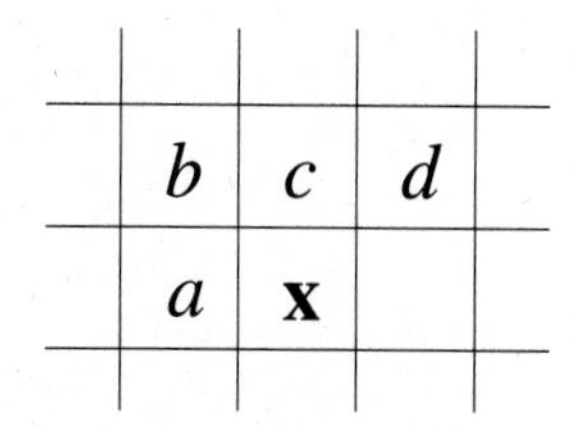

Figure 18.7: Prediction schemes.

and

$$\hat{x} = \text{int}\{(a + b + c + d)/4\} \tag{18.8}$$

where $\hat{x}$ denotes the predicted value for the pixel x. In both expressions, the prediction is rounded to the nearest integer to ensure an integer prediction error. We note that many other forms of prediction, such as edge-adaptive prediction, also exist.

If the input image intensity x has a dynamic range of (0,255), then the prediction error $x - \hat{x}$ has a theoretical dynamic range of (-255,255). The reader may have noticed that the predictive mapping in fact results in an expansion of the dynamic range of the signal to be encoded. However, inspection of the histogram (probability density) of the prediction error shows that it is highly peaked about 0, as compared to the histogram of the actual image intensities, as shown in Figure 18.8. Therefore, the prediction error always has much smaller entropy than the original intensity values, which implies that the prediction process removes a great deal of interpixel (statistical) redundancy.

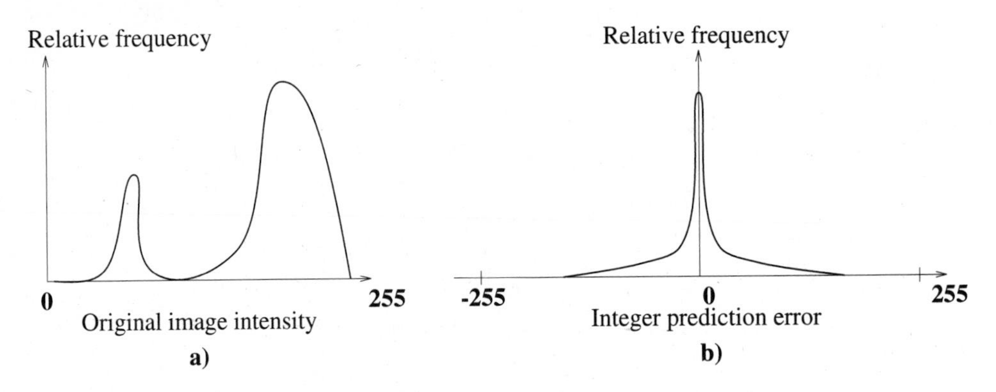

Figure 18.8: Histograms of a) the original image intensity and b) integer prediction error.

For lossless coding, every possible difference value in the range (-255,255) needs to be taken as a different symbol. Binary codewords for these symbols can be assigned by entropy coding, such as Huffman coding or arithmetic coding. However, because it is usually very costly to assign a different codeword to 513 different symbols, we usually assign a unique codeword to every difference value in the range (-15,16). In addition, codewords are assigned to a shift up (SU) symbol and a shift down (SD) symbol, which shifts a given difference value up or down by 32, respectively. Using these codewords, every possible difference value can be uniquely coded by using an appropriate number of SU or SD operators followed by a difference code in the range (-15,16). For example, a difference value of 100 can be represented by cascading the codes for the symbols SD, SD, SD, and 4. This scheme results in a slightly higher bitrate than does designing 513 different codes, but offers significant reduction in complexity. The probabilities of each of these symbols are estimated by analyzing a histogram of the prediction errors obtained from a training set of images.

18.3.2 Run-Length Coding of Bit-Planes

Bit-plane decomposition refers to expressing a multilevel (monochrome or color) image by a series of binary images, one for each bit used in the representation of the pixel intensities. Let the gray levels of an m-bit gray-scale image be represented as

$$a_{m-1}2^{m-1} + a_{m-2}2^{m-2} + \ldots + a_1 2^1 + a_0 2^0 \tag{18.9}$$

where a_i, $i = 0, \ldots, m-1$ are either 0 or 1. The zeroth-order bit-plane is generated by collecting the a_0 bits of each pixel, while the $(m-1)$st-order bit-plane contains the a_{m-1} bits. For example, for the case of an 8-bit image, a pixel in the most significant bit-plane is represented by a 1 if the corresponding pixel intensity is equal to or greater than 128. Observe that a binary image can be represented by a single bit-plane. Bit-plane decomposition is illustrated in Fig, 18.9.

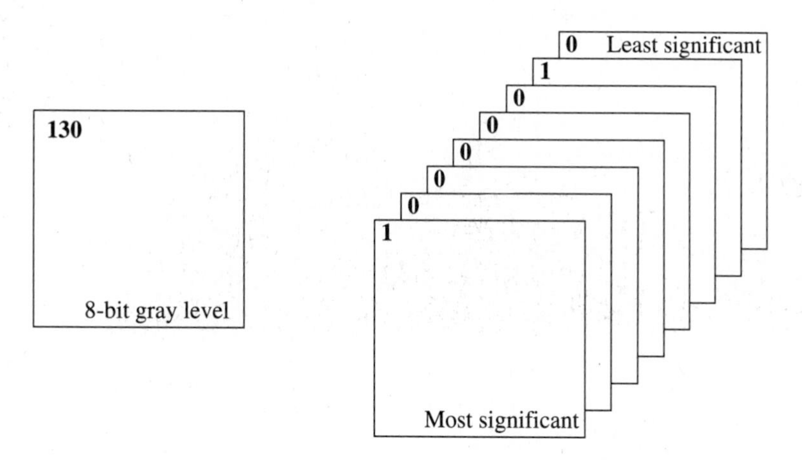

Figure 18.9: Bit-plane decomposition of an 8-bit image.

A disadvantage of the above bit-plane representation is that small changes in gray level, such as variations due to noise, may cause edges in all bit planes. For example, the binary representation for 127 is 01111111, and for 128, it is 10000000. To reduce the effect of such small gray-level variations in the bit-planes, we may choose to represent the pixel intensities by an m-bit Gray code where successive codewords differ only in one bit position. The m-bit Gray code $g_{m-1} \ldots g_2 g_1 g_0$ is given by

$$\begin{aligned} g_i &= a_i \oplus a_{i+1} \quad \text{for} \quad 0 \leq i \leq m-2 \\ g_{m-1} &= a_{m-1} \end{aligned} \tag{18.10}$$

where $\oplus$ denotes the exclusive OR operation.

An effective approach to encode bit-planes is to employ run-length coding (RLC), which is often used for binary image compression. RLC algorithms can be classified as 1-D RLC and 2-D RLC. In 1-D RLC, the length of each contiguous group of 0's or 1's encountered in a left-to-right scan of a row of bit-plane is entropy coded. In most implementations, the length of each run is limited by the number of pixels in a line. For unique decodability, we need to establish a convention to either specify the first run of each row or assume that each row begins with a white run (i.e., the first symbol in each row is 1), whose run length may be zero. A different variable-length code is designed for each possible value of runs. Because the statistics of 0-runs and 1-runs are usually different, we design different codes for the white-runs and black-runs. Once again, the statistics of these runs are estimated from a training set of images. RLC has been adopted in the international standards for fax transmission, such as ITU (formerly CCITT) Group 3 and Group 4 codes. A more detailed discussion of 1-D and 2-D RLC is provided in Chapter 21.

18.3.3 Ziv-Lempel Coding

Ziv-Lempel coding is a block coding method which assigns fixed-length codes to variable-size blocks of input symbols by means of a table lookup using a dictionary of variable-length blocks of symbols [Ziv 94]. The input sequence of symbols is parsed into nonoverlapping blocks of variable length, whose length depends on the size of the blocks in the present dictionary, while updating the dictionary of blocks of symbols according to the following algorithm.

The length of the next block to be parsed, L, is defined to be equal to that of the longest word that is already in the dictionary. The initial dictionary is set equal to the list of all symbols in the alphabet $A = \{a_1, a_2, \ldots, a_M\}$. Thus, initially $L = 1$. If the next parsed block, w, is already in the dictionary, the encoder sends to the channel a fixed-length code for the index of this block. Before continuing parsing, w is concatenated with the next input symbol and added to the dictionary. If w is not in the dictionary, then the encoder sends a fixed-length code for the first $L-1$ symbols of w (which must be in the dictionary) and adds w to the dictionary. This process is repeated until the entire input sequence is coded. The Ziv-Lempel procedure is demonstrated by an example in the following.

Example

We demonstrate the Ziv-Lempel coding procedure for the case of a binary alphabet, that is, an alphabet which contains only two symbols, 0 and 1. The initial dictionary in this case contains only two symbols, 0 and 1, and $L = 1$. The first parsed symbol in the sequence to be coded is 0, as shown in Table 18.6. Clearly, 0 is in the dictionary; therefore, its code, 0, is transmitted. Then we check the next symbol, which is a 1, and append the block 01 in the dictionary. Now that $L = 2$, we read the next block of two symbols in the input sequence, which is 11. The block 11 is not in the dictionary, so it is added to the dictionary, and the code for the first $L - 1$ symbols in this block, which is a 1, is transmitted. The operation of the algorithm is illustrated in Table 18.6, which shows a list of input symbols, the blocks that enter the dictionary, the index that is coded by fixed-length coding, and the output of the coder.

Observe from Table 18.6 that Ziv-Lempel codes have the so-called "last-first" property which states that the last symbol of the most recent word added to the table is the first symbol of the next parsed block.

Table 18.6: An example for Ziv-Lempel coding.

Input	Dictionary	Index	Output
	0 1	0 1	
0	01	2	0
1	11	3	1
1	10	4	1
0	00	5	0
0 1	011	6	2
1 0	100	7	4
0 1	010	8	2
0 1 1	0110	9	6
0 0 0	. . .	. . .	

Ziv-Lempel coding is noiseless, does not require probabilities of the source symbols to be known or estimated, and is optimum in the limit of unbounded dictionary size. In practice, one places a bound on the dictionary size. Once this size limit is reached, the encoder can no longer add codewords and must simply use the existing dictionary. However, methods exist to adapt the dictionary to varying input characteristics.

Ziv-Lempel coding has been succesfully used for compression of binary data files, with a compression ratio of approximately 2.5:1. Indeed, it forms the basis of the "compress" utility in UNIX and the "arc" program for the PC environment.

18.4 Exercises

1. Suppose we have a discrete memoryless source with the alphabet $\mathcal{A}$ and the symbol probabilities $p(a_i)$ for $a_i \in \mathcal{A}$ specified by the following table:

Symbol	a_0	a_1	a_2	a_3	a_4	a_5
Probability	0.3	0.2	0.2	0.1	0.1	0.1

   a) Find the entropy of this source.

   b) Design a Huffman code for this source.

   c) Find the average codeword length.

   d) How good is this code?

2. Let $\mathbf{X}$ be a binary, Markov-2 source with the alphabet $\mathcal{A} = \{\ 0,\ 1\ \}$. The source is modeled by the conditional probabilities

$$\begin{array}{ll} P(0|0,0) = 0.7 & P(1|0,0) = 0.3 \\ P(0|0,1) = 0.6 & P(1|0,1) = 0.4 \\ P(0|1,0) = 0.4 & P(1|1,0) = 0.6 \\ P(0|1,1) = 0.3 & P(1|1,1) = 0.7 \end{array}$$

   a) What is the entropy of this source?

   b) Design a block Huffman code for $N = 3$.

   c) What is the average codeword length? What would be the average codeword length if we design a scalar Huffman code?

3. Suppose we have an 8-bit gray-level image. How would you estimate the entropy of this image:

   a) Assuming that it is a discrete memoryless source?

   b) Assuming that it is a Markov-1 source?

   Which one do you guess will be smaller? Why?

4. Assume that the histogram of the differential image shown in Figure 18.8 can be modeled by a Laplacian distribution, given by

$$p(e) = \frac{1}{\sqrt{2}\sigma_e} \exp\left\{\frac{-\sqrt{2}|e|}{\sigma_e}\right\}$$

with $\sigma_e = 15$. Write an expression for the entropy of the differential image. Estimate its value. What is the expected compression ratio (using lossless differential encoding)?

5. How do you compare arithmetic coding with block Huffman coding?

6. What is the primary motivation of using the Gray codes in bit-plane encoding?

Bibliography

[Ber 71] T. Berger, *Rate Distortion Theory*, Englewood Cliffs, NJ: Prentice-Hall, 1971.

[Ger 92] A. Gersho and R. M. Gray, *Vector Quantization and Signal Compression*, Norwell, MA: Kluwer, 1992.

[Jai 81] A. K. Jain, "Image data compression: A review," *Proc. IEEE*, vol. 69, pp. 349–389, Mar. 1981.

[Jay 84] N. S. Jayant and P. Noll, *Digital Coding of Waveforms, Principles and Applications to Speech and Video*, Englewood Cliffs, NJ: Prentice Hall, 1984.

[Jay 92] N. S. Jayant, "Signal compression: Technology targets and research directions," *IEEE Trans. Spec. Areas Comm*, vol. 10, pp. 796–819, June 1992.

[Net88] A. N. Netravali and B. G. Haskell, *Digital Pictures: Representation and Compression*, New York, NY: Plenum Press, 1988.

[Rab 91] M. Rabbani and P. Jones, *Digital Image Compression Techniques*, Bellingham, WA: SPIE Press, 1991.

[Ziv 77] J. Ziv and A. Lempel, "A universal algorithm for sequential data compression," *IEEE Trans. Info. Theory*, vol. IT-23, pp. 337–343, 1977.

[Ziv 94] J. Ziv, "On universal data compression - An intuitive overview," *J. Vis. Comm. Image Rep.*, vol. 5, no. 4, pp. 317–321, Dec. 1994.

Chapter 19

DPCM AND TRANSFORM CODING

This chapter introduces two popular lossy compression schemes, differential pulse code modulation (DPCM) and discrete cosine transform (DCT) coding. Lossy compression methods compromise fidelity in exchange for increased compression. For example, DCT coding can easily provide a compression ratio of 25:1 without visually apparent degradations. The main difference between the lossy and lossless compression schemes is that the former employs a quantizer while the latter does not. To this effect, we start our discussion of lossy compression schemes with the process of quantization and quantizer design in Section 18.1. Next, DPCM is covered in Section 18.2. In DPCM, a predictive mapping transforms pixel intensities to prediction errors, which are then quantized and entropy coded. In transform coding, studied in Section 18.3, the image is first divided into blocks. Then the DCT coefficients for each block are zig-zag scanned, quantized, and run-length/entropy coded. Both DCT coding and DPCM play important roles in the international standards for still-frame and video compression. Thus, this chapter serves to provide the necessary background to understand the international compression standards.

19.1 Quantization

Quantization is the process of representing a set of continuous-valued samples with a finite number of states. If each sample is quantized independently, then the process is known as scalar quantization. A scalar quantizer $Q(\cdot)$ is a function that is defined in terms of a finite set of decision levels d_i and reconstruction levels r_i, given by

$$Q(s) = r_i, \quad \text{if } s \in (d_{i-1}, d_i], \quad i = 1, \ldots, L \tag{19.1}$$

where L is the number of output states. That is, the output of the quantizer is the reconstruction level r_i, if s, the value of the sample, is within the range $(d_{i-1}, d_i]$.

Vector quantization refers to representing a set of vectors, each formed by several continuous-valued samples, with a finite number of vector states. In image compression, scalar or vector quantization is usually applied to a transform-domain image representation. In the following, we discuss how to design a scalar quantizer, that is, how to determine d_i and r_i in (19.1). We will return to vector quantization in Chapter 21.

Let $\dot{s} = Q(s)$ denote the quantized variable. Then $e = s - \dot{s}$ is the quantization error. The performance of a quantizer may be quantified by a distortion measure D which is a function of e, and therefore d_i and r_i. If we treat s as a random variable, then the distortion measure may be selected as the mean square quantization error, that is, $D = E\{(s - \dot{s})^2\}$, where $E\{\cdot\}$ stands for the expectation operator. Given a distortion measure D and the input probability density function (pdf) $p(s)$, there are, in general, two different optimum scalar quantizer design criteria:

i) For a fixed number of levels L, find r_i, $i = 1, \ldots, L$, and d_i, $i = 0, \ldots, L$, in order to minimize D. Quantizers with a fixed number of levels which are optimal in the mean square error sense, $D = E\{(s - r_i)^2\}$, are known as Lloyd-Max quantizers [Llo 82, Max 60].

ii) For a fixed output entropy, $H(\dot{s}) = C$, where C is a constant, find r_i and d_i (L is unknown) in order to minimize D. Quantizers that minimize a distortion measure for a constant output entropy are known as entropy-constrained quantizers [Woo 69].

Recall from Chapter 18 that the rate R in the case of fixed-length coding is given by $R = \lfloor log_2 L \rfloor$, ($\lfloor \cdot \rfloor$ denotes the smallest integer greater than), while $R > H(\dot{s})$ in the case of variable-length encoding (VLC). Thus, Lloyd-Max quantization is more suitable for use with fixed-length coding, whereas entropy-constrained quantization is well-suited for VLC (Huffman or arithmetic coding). A detailed discussion of entropy-constrained quantization is beyond the scope of this book. We elaborate on the design of Lloyd-Max quantizers below.

19.1.1 Nonuniform Quantization

The determination of the Lloyd-Max quantizer requires minimization of

$$E\{(s - r_i)^2\} = \sum_{i=1}^{L} \int_{d_{i-1}}^{d_i} (s - r_i)^2 p(s) ds \tag{19.2}$$

with respect to r_i and d_i. It has been shown that the necessary conditions to minimize (19.2) are given by [Jai 90, Lim 90]

$$r_i = \frac{\int_{d_{i-1}}^{d_i} s p(s) ds}{\int_{d_{i-1}}^{d_i} p(s) ds}, \quad 1 \le i \le L \tag{19.3}$$

$$d_i = \frac{r_i + r_{i+1}}{2}, \quad 1 \le i \le L - 1 \tag{19.4}$$

$$d_0 = -\infty$$

$$d_L = \infty$$

The expression (19.3) states that the reconstruction level r_i is the centroid of $p(s)$ over the interval $d_{i-1} < s < d_i$, that is, the conditional expectation of s given the interval, while (19.4) states that the decision level d_i is the midpoint of two reconstruction levels except for d_0 and d_L.

The solution of the nonlinear equations (19.3) and (19.4) requires an iterative method, such as the Newton-Raphson algorithm. The resulting decision and reconstruction levels are not equally spaced, and hence we have a nonuniform quantizer, except when s has a uniform pdf over a finite interval. The optimal r_i and d_i, which depend on the pdf $p(s)$, are tabulated for some commonly encountered pdfs, such as the Gaussian and Laplacian, and for certain values of L in the literature [Jai 90, Lim 90] (see example below).

19.1.2 Uniform Quantization

A uniform quantizer is a quantizer where all reconstruction levels are equally spaced, that is,

$$r_{i+1} - r_i = \theta \quad 1 \leq i \leq L-1 \tag{19.5}$$

where θ is a constant called the stepsize.

It has been shown that the Lloyd-Max quantizer becomes a uniform quantizer when $p(s)$ is uniformly distributed over an interval $[A, B]$, given by

$$p(s) = \begin{cases} \frac{1}{B-A} & A \leq s \leq B \\ 0 & \text{otherwise.} \end{cases} \tag{19.6}$$

Designing a uniform quantizer in this case is a trivial task, where

$$\theta = \frac{B-A}{L} \tag{19.7}$$

$$d_i = A + i\theta, \quad 0 \leq i \leq L \tag{19.8}$$

and

$$r_i = d_{i-1} + \frac{\theta}{2}, \quad 1 \leq i \leq L \tag{19.9}$$

We can also design a uniform quantizer when $p(s)$ has an infinite tail, e.g., in the case of a Laplacian distribution. This can be achieved by incorporating (19.5) as an additional constraint in the formulation that was presented in the previous section. Then the nonlinear optimization can be performed with respect to a single variable θ.

If the input pdf $p(s)$ is even, then the decision and reconstruction levels have certain symmetry relations in both nonuniform and uniform quantization, depending on whether L is even or odd. We assume, without loss of generality, that the mean of s is zero; that is, $p(s)$ is even symmetric about zero. To simplify the notation for symmetric quantizers, we introduce a change of variables, for L even, as

$$\begin{aligned}\tilde{r}_i &= r_{i+L/2+1} \quad \text{for } i=-L/2,\ldots,1 \\ \tilde{r}_i &= r_{i+L/2} \quad \text{for } i=1,\ldots,L/2 \\ \tilde{d}_i &= d_{i+L/2} \quad \text{for } i=-L/2,\ldots,L/2\end{aligned}$$

and, for L odd, as

$$\begin{aligned}\tilde{r}_i &= r_{i+(L+1)/2} \quad \text{for } i=-(L-1)/2,\ldots,(L-1)/2 \\ \tilde{d}_i &= d_{i+(L+1)/2} \quad \text{for } i=-(L+1)/2,\ldots,1 \\ \tilde{d}_i &= d_{i+(L-1)/2} \quad \text{for } i=1,\ldots,(L+1)/2\end{aligned}$$

The relationship between the levels r_i and d_i in (19.3) and (19.4), and those of the symmetric quantizer, $\tilde{d}_i$ and $\tilde{r}_i$ are illustrated in Figure 19.1, for $L=4$.

The symmetry relations can then be easily expressed, for L even, as

$$\begin{aligned}\tilde{r}_i &= -\tilde{r}_{-i} \quad \text{for } i=1,\ldots,L/2 \\ \tilde{d}_i &= -\tilde{d}_{-i} \quad \text{for } i=1,\ldots,L/2 \\ \tilde{d}_0 &= 0 \quad \text{and} \quad \tilde{d}_{L/2}=\infty\end{aligned} \tag{19.10}$$

and, for L odd, as

$$\begin{aligned}\tilde{r}_i &= -\tilde{r}_{-i} \quad \text{for } i=1,\ldots,(L-1)/2 \\ \tilde{d}_i &= -\tilde{d}_{-i} \quad \text{for } i=1,\ldots,(L+1)/2 \\ \tilde{r}_0 &= 0 \quad \text{and} \quad \tilde{d}_{(L+1)/2}=\infty\end{aligned} \tag{19.11}$$

These relationships are generally incorporated into the quantizer design to reduce the number of unknowns. Symmetric nonuniform quantizers for L even and L odd are depicted in Figure 19.2 (a) and (b), respectively.

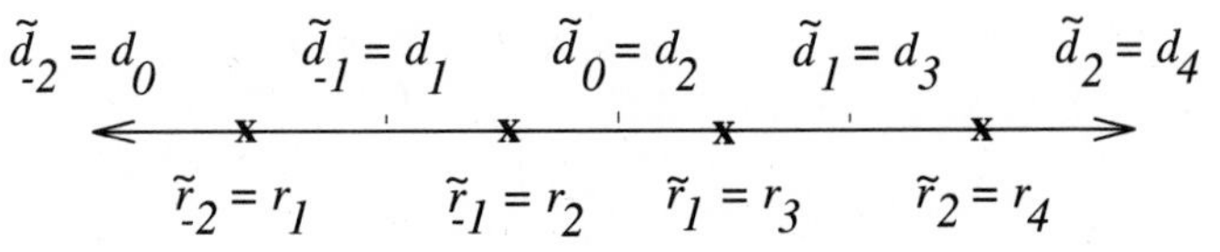

$\tilde{d}_{-2}=d_0 \quad \tilde{d}_{-1}=d_1 \quad \tilde{d}_0=d_2 \quad \tilde{d}_1=d_3 \quad \tilde{d}_2=d_4$

$\tilde{r}_{-2}=r_1 \quad \tilde{r}_{-1}=r_2 \quad \tilde{r}_1=r_3 \quad \tilde{r}_2=r_4$

Figure 19.1: The relationship between the levels for $L=4$.

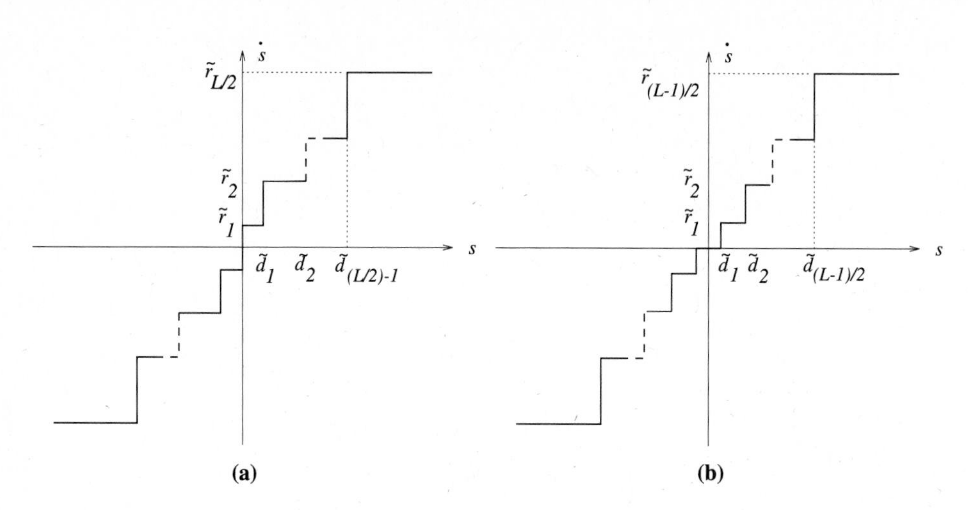

Figure 19.2: Typical nonuniform quantizers: a) L even; b) L odd.

Example: Lloyd-Max quantization of a Laplacian distributed signal.

The decision and reconstruction levels for uniform and nonuniform Lloyd-Max quantization of a Laplacian signal with unity variance are tabulated in Table 19.1. Because Laplacian is a symmetric pdf, only the positive levels are shown for the nonuniform quantizers. The uniform quantizer is defined by a single parameter θ in all cases. For example, in the case of nonuniform quantization with $L = 4$, the decision levels are $d_0 = \tilde{d}_{-2} = -\infty$, $d_1 = \tilde{d}_{-1} = -1.102$, $d_2 = \tilde{d}_0 = 0$, $d_3 = \tilde{d}_1 = 1.102$, and $d_4 = \tilde{d}_2 = \infty$, and the reconstruction levels are $r_1 = \tilde{r}_{-2} = -1.810$, $r_2 = \tilde{r}_{-1} = -0.395$, $r_3 = \tilde{r}_1 = 0.395$, and $r_4 = \tilde{r}_2 = 1.810$.

Table 19.1: The decision and reconstruction levels for Lloyd-Max quantizers.

Levels	2		4		8	
i	$\tilde{d}_i$	$\tilde{r}_i$	$\tilde{d}_i$	$\tilde{r}_i$	$\tilde{d}_i$	$\tilde{r}_i$
1	∞	0.707	1.102	0.395	0.504	0.222
2			∞	1.810	1.181	0.785
3					2.285	1.576
4					∞	2.994
θ	1.414		1.087		0.731	

The loss due to quantization, known as the quantization noise, is usually quantified by a signal-to-noise ratio (SNR). The next example demonstrates that a relationship can be established between the SNR and the number of quantization levels.

Example: Quantization noise

Suppose we have a memoryless Gaussian signal s with zero mean and variance σ^2. Let the distortion measure be the mean square error. What is the minimum number of levels, equivalently the rate R in bits/sample, to obtain 40 dB SNR, assuming uniform quantization?

We can express the mean square quantization noise as

$$D = \sum E\{(s - \dot{s})^2\} \tag{19.12}$$

Then the signal noise ratio in dB is given by

$$SNR = 10 \log_{10} \frac{\sigma^2}{D} \tag{19.13}$$

It can easily be seen that $SNR = 40$ dB implies $\sigma^2/D = 10,000$. Substituting this result into the rate-distortion function for a memoryless Gaussian source, given by

$$R(D) = \frac{1}{2} \log_2 \frac{\sigma^2}{D} \tag{19.14}$$

we have $R(D) \approx 7$ bits/sample. Likewise, we can show that quantization with 8 bits/sample yields approximately 48 dB SNR.

19.2 Differential Pulse Code Modulation

Pulse code modulation (PCM) is a commonly employed digital image representation method, which refers to quantizing the intensity of each pixel individually, neglecting the spatial correlation that exists between the intensities of neighboring pixels. One approach to exploit the inherent spatial redundancy (correlation) is to use differential pulse code modulation (DPCM), which first predicts the intensity of the next pixel based on its neighbors. The difference between the actual and predicted intensity is then quantized and coded. The principal difference between the lossless predictive coding and DPCM is that DPCM incorporates a quantizer which limits the allowed prediction error states to a small number. The reconstructed intensity values $\dot{s}(n_1, n_2)$ at the receiver (decoder) are then given by

$$\dot{s}(n_1, n_2) = \hat{s}(n_1, n_2) + \dot{e}(n_1, n_2) \tag{19.15}$$

where $\hat{s}(n_1, n_2)$ is the predicted intensity and $\dot{e}(n_1, n_2)$ denotes the quantized prediction error. The block diagrams of an encoder and decoder are shown in Figure 19.3.

Because of the quantization, the reconstructed intensities $\dot{s}(n_1, n_2)$ are not identical to the actual intensities $s(n_1, n_2)$. From the block diagram in Figure 19.3 (b), observe that the prediction at the decoder is based on reconstructed values of the received pixels. In order to prevent error build-up at the decoder, the prediction

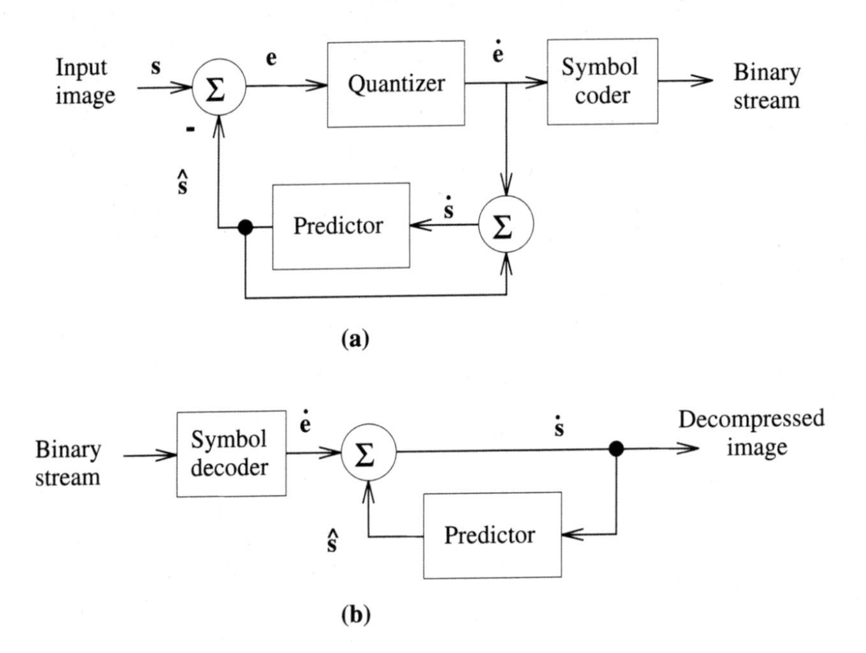

Figure 19.3: Block diagrams of a DPCM a) encoder and b) decoder.

at both the encoder and decoder must be identical. Therefore, the predictor at the encoder also uses the reconstructed values of the previously encoded samples in the prediction instead of their actual values. To this effect, the encoder simulates the decoder in the prediction loop, as shown in Figure 19.3 (a).

DPCM is relatively simple to implement; however, it provides only a modest compression. The prediction and quantization steps are discussed in detail in the following. Despite the apparent interaction between the prediction and quantization, e.g., the reconstructed values are used in prediction, the predictor is traditionally designed under the assumption of no quantization error, and the quantizer is designed to minimize its own error.

19.2.1 Optimal Prediction

Modeling the image by a stationary random field, a linear minimum mean square error (LMMSE) predictor, in the form

$$\begin{aligned}\hat{s}(n_1, n_2) &= c_1 s(n_1-1, n_2-1) + c_2 s(n_1-1, n_2) + c_3 s(n_1-1, n_2+1) \\ &\quad + c_4 s(n_1, n_2-1) \end{aligned} \tag{19.16}$$

can be designed to minimize the mean square prediction error

$$E\{[\mathbf{s} - \hat{\mathbf{s}}]^T [\mathbf{s} - \hat{\mathbf{s}}]\} \tag{19.17}$$

where $\mathbf{s}$ and $\hat{\mathbf{s}}$ denote lexicographic ordering of the pixels in the actual and predicted images, respectively.

The optimal coefficient vector $\mathbf{c} = [c_1\ c_2\ c_3\ c_4]^T$ in (19.16) can be found by solving the normal equation,

$$\phi = \mathbf{\Phi c} \tag{19.18}$$

where

$$\mathbf{\Phi} = \begin{bmatrix} R_s(0,0) & R_s(0,1) & R_s(0,2) & R_s(1,0) \\ R_s(0,-1) & R_s(0,0) & R_s(0,1) & R_s(1,-1) \\ R_s(0,-2) & R_s(0,-1) & R_s(0,0) & R_s(1,-2) \\ R_s(-1,0) & R_s(-1,1) & R_s(-1,2) & R_s(0,0) \end{bmatrix} \tag{19.19}$$

and

$$\phi = [R_s(1,1)\ R_s(1,0)\ R_s(1,-1)\ R_s(0,1)]^T \tag{19.20}$$

The function $R_s(i_1, i_2)$ stands for the autocorrelation function of the image, which is defined by an ensemble average

$$R_s(i_1, i_2) \quad = \quad E\{s(n_1, n_2)s(n_1 - i_1, n_2 - i_2)\} \tag{19.21}$$

where $E\{\cdot\}$ is the expectation operator. In practice, it is approximated by the sample average

$$R_s(i_1, i_2) \quad \approx \quad \frac{1}{N} \sum_{(n_1,n_2)} s(n_1, n_2)s(n_1 - i_1, n_2 - i_2) \tag{19.22}$$

where N is the number of pixels within the support of the summation. The matrix $\mathbf{\Phi}$ is almost always invertible, provided that the summation (19.22) is computed over a large enough support. Then

$$\mathbf{c} = \mathbf{\Phi}^{-1}\phi \tag{19.23}$$

Note that although these coefficients may be optimal in the LMMSE sense, they are not necessarily optimal in the sense of minimizing the entropy of the prediction error. Furthermore, images rarely obey the stationarity assumption. As a result, most DPCM schemes employ a fixed predictor.

19.2.2 Quantization of the Prediction Error

In standard DPCM, the difference signal is scalar quantized using K bits/sample, where K is an integer that is less than 8. Empirical studies show that the statistics of the prediction error can be well modeled by a Laplacian distribution with variance σ^2. Because the Laplacian distribution is heavily peaked about 0, nonuniform

Lloyd-Max quantization is employed in order to obtain the best image quality at a given bitrate.

The decision and reconstruction levels of a nonuniform Lloyd-Max quantizer for a unity variance Laplacian distribution have been tabulated [Jai 90]. The levels for the case $\sigma^2 \neq 1$ can easily be obtained by scaling the tabulated values by the standard deviation σ. Although reasonable-quality images can usually be obtained with $K = 2$ bits/pixel, i.e., using only 4 levels, better images can be obtained by using adaptive scalar quantization or vector quantization strategies.

19.2.3 Adaptive Quantization

Adaptive quantization refers to adjusting the decision and reconstruction levels according to the local statistics of the prediction error. Since the quantization levels are scaled by the standard deviation of the prediction error, many adaptive quantization schemes attempt to match the quantizer to the space-varying standard deviation (square root of the variance) of the prediction error. A reasonable strategy may be to estimate the variance of the prediction error on a block-by-block basis, hence, block-by-block adaptation of the quantizer. However, this would require encoding and transmission of the variance of each block as an overhead information.

An alternate strategy is the multiple model approach, where four scaled Lloyd-Max quantizers representing uniform regions, textured regions, moderate edges, and sharp edges, respectively, are predesigned. They are assumed to be known at both the encoder and decoder. We select one of these four quantizers at each block based on the variance of the block. In the multiple model scheme, the overhead information is thus reduced to 2 bits/block. The block diagram of the multiple model adaptive quantization scheme is shown in Figure 19.4 along with typical values of the variance of the prediction error assumed for designing each quantizer.

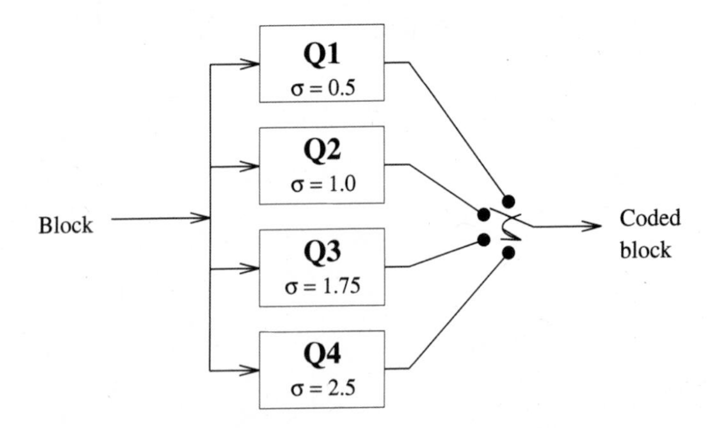

Figure 19.4: The block diagram of an adaptive quantizer.

19.2.4 Delta Modulation

Delta modulation (DM) refers to a simplified version of a DPCM system where the prediction is usually set equal to the reconstructed value of the most recently encoded sample, given by

$$\hat{s}_n = \alpha \dot{s}_{n-1} \tag{19.24}$$

The resulting prediction error is then quantized to just two states, given by

$$\dot{e}_n = \begin{cases} \zeta & \text{for } e_n > 0 \\ -\zeta & \text{otherwise} \end{cases} \tag{19.25}$$

As shown in Figure 19.5, the quantizer assigns a single level to all positive prediction errors and another to all negative prediction errors.

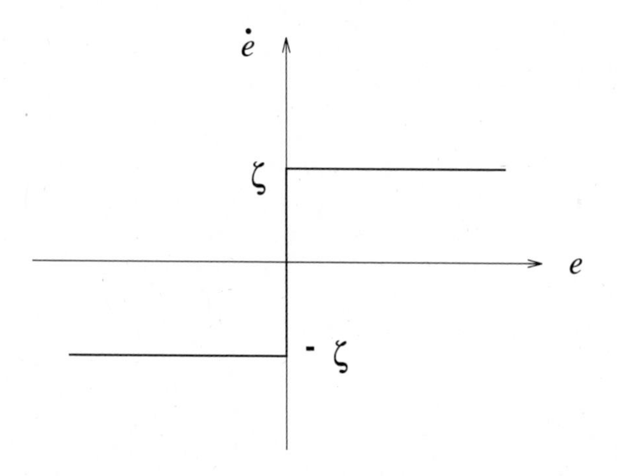

Figure 19.5: The quantizer for delta modulation.

Delta modulation corresponds to a staircase approximation of the source signal as depicted in Figure 19.6. This approximation may suffer from two drawbacks, slope overload and granularity. The reconstructed signal $\dot{s}$ can follow a slope of at most ζ/Δ, where Δ is the distance between the sampled pixels. If the pixel intensities vary faster than this rate, slope overload results. Slope overload manifests itself as blurring of sharp edges in images. One way to overcome slope overload is to increase ζ, which may, however, cause granularity in uniform-intensity regions. This suggests adaptive DM strategies where the quantizer step size ζ varies according to local image characteristics. If the local characteristics are computed based on a causal neighborhood, adaptivity can be achieved without transmitting overhead information.

Besides serving as a simple but crude source coding technique, DM also finds application as a simple analog-to-digital (A/D) conversion method, where a small ζ

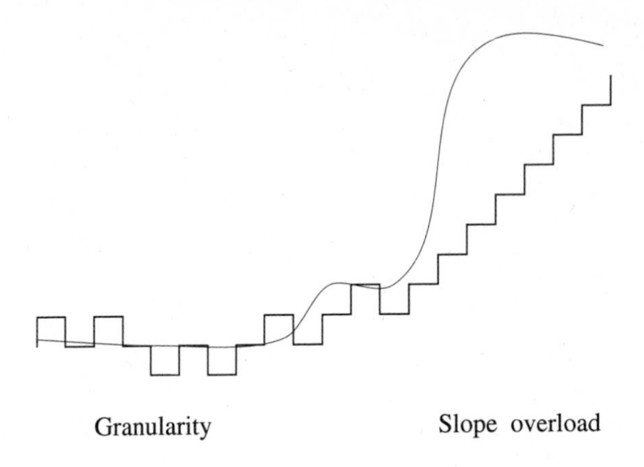

Figure 19.6: Illustration of granular noise and slope overload.

is usually preferred. In this case, the slope overload may be overcome by reducing Δ, which can be achieved by oversampling the signal approximately four to eight times. It is interesting to note that this scheme suggests a tradeoff between the number of quantization levels used and the sampling rate.

19.3 Transform Coding

Transform coding, developed more than two decades ago, has proven to be an effective image compression scheme, and is the basis of all world standards for lossy compression up to date. A basic transform coder segments the image into small square blocks. Each block undergoes a 2-D orthogonal transformation to produce an array of transform coefficients. Next, the transform coefficients are individually

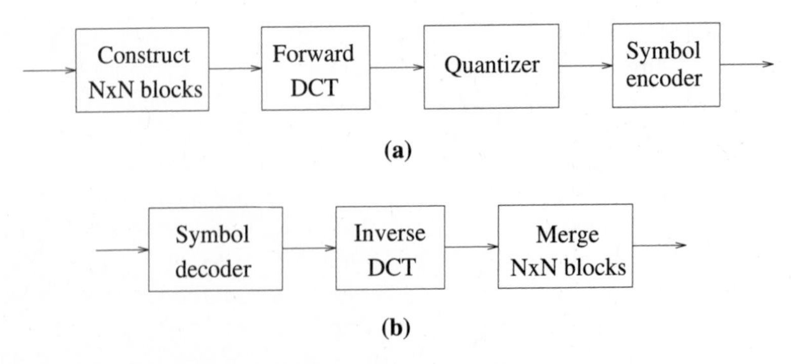

Figure 19.7: The block diagrams of a transform a) encoder and b) decoder.

quantized and coded. The coefficients having the highest energy over all blocks are most finely quantized, and those with the least energy are quantized coarsely or simply truncated. The encoder treats the quantized coefficients as symbols which are then entropy (VLC) coded. The decoder reconstructs the pixel intensities from the received bitstream following the inverse operations on a block-by-block basis. The block diagrams of a transform encoder and decoder are shown in Figure 19.7.

The best transformation for the purposes of effective quantization should produce uncorrelated coefficients, and pack the maximum amount of energy (variance) into the smallest number of coefficients. The first property justifies the use of scalar quantization. The second property is desirable because we would like to discard as many coefficients as possible without seriously affecting image quality. The transformation that satisfies both properties is the KLT. The derivation of the KLT assumes that each image block is a realization of a homogeneous random field, whose covariance matrix $\mathbf{R}_b = E\{\mathbf{b}\mathbf{b}^T\}$, where $\mathbf{b}$ is a vector obtained by lexicographic ordering of the pixels within the block, is known. It is well-known that any covariance matrix can be diagonalized as

$$\mathbf{R}_b = \mathbf{\Gamma}^T \mathbf{\Lambda} \mathbf{\Gamma} \tag{19.26}$$

where $\mathbf{\Gamma}$ is the matrix of the orthonormal eigenvectors of $\mathbf{R}_b$, and $\mathbf{\Lambda}$ is a diagonal matrix of the corresponding eigenvalues. It follows that $\mathbf{\Gamma}$ is a unitary matrix, such that $\mathbf{\Gamma}^T = \mathbf{\Gamma}^{-1}$. Then, the KLT of the block $\mathbf{b}$ is given by $\mathbf{y} = \mathbf{\Gamma}\mathbf{b}$. The properties of the KLT easily follows:

i) The coefficients of $\mathbf{y}$, the KLT, are uncorrelated since, $E\{\mathbf{y}\mathbf{y}^T\} = \mathbf{\Gamma}^T\mathbf{R}_b\mathbf{\Gamma} = \mathbf{\Lambda}$ is diagonal, by construction.

ii) Mean square quantization noise is preserved under the KLT, because

$$\begin{aligned} E\{||\mathbf{y} - \dot{\mathbf{y}}||^2\} &= E\{(\mathbf{b} - \dot{\mathbf{b}})^T \Gamma^T \Gamma (\mathbf{b} - \dot{\mathbf{b}})\} \\ &= E\{(\mathbf{b} - \dot{\mathbf{b}})^T (\mathbf{b} - \dot{\mathbf{b}})\} = E\{||\mathbf{b} - \dot{\mathbf{b}}||^2\} \end{aligned}$$

where $\dot{}$ denotes the quantized variable. It also follows that the mean square truncation error in the image domain is equal to the sum of the variances of the discarded coefficients [Win 72].

Despite its favorable theoretical properties, the KLT is not used in practice, because: i) its basis functions depend on the covariance matrix of the image, and hence they have to be recomputed and transmitted for every image, ii) perfect decorrelation is not possible, since images can rarely be modeled as realizations of homogeneous random fields, and iii) there are no fast computational algorithms for its implementation. To this effect, many other orthonormal transformations with data-independent basis functions, such as the discrete Fourier transform (DFT), discrete cosine transform (DCT), and Hadamard and Haar transforms have been used in image compression [Win 72]. Of these, DCT has been found to be the most effective with a performance close to that of the KLT.

19.3.1 Discrete Cosine Transform

The DCT is the most widely used transformation in transform coding. It is an orthogonal transform which has a fixed set of (image-independent) basis functions, an efficient algorithm for its computation, and good energy compaction and correlation reduction properties. Ahmed *et al.* [Ahm 74] first noticed that the KLT basis functions of a first-order Markov image closely resemble those of the DCT. They become identical as the correlation between adjacent pixel approaches one [Cla 85].

The DCT belongs to the family of discrete trigonometric transforms, which has 16 members [Mar 94]. The type-2 DCT of an $N \times N$ block is defined as

$$S(k_1, k_2) = \sqrt{\frac{4}{N^2}} C(k_1) C(k_2) \sum_{n_1=0}^{N-1} \sum_{n_2=0}^{N-1} s(n_1, n_2) \cos\left(\frac{\pi(2n_1+1)k_1}{2N}\right) \cos\left(\frac{\pi(2n_2+1)k_2}{2N}\right)$$

where $k_1, k_2, n_1, n_2 = 0, 1, \ldots N-1$, and

$$C(k) = \begin{cases} 1/\sqrt{2} & \text{for } k = 0, \\ 1 & \text{otherwise} \end{cases}$$

A significant factor in transform coding is the block size. The most popular sizes have been 8×8 and 16×16, both powers of 2 for computational reasons. It is no surprise that the DCT is closely related to the DFT. In particular, an $N \times N$ DCT of $s(n_1, n_2)$ can be expressed in terms of a $2N \times 2N$ DFT of its even-symmetric extension, which leads to a fast computational algorithm. Using the separability of the DFT and one of several fast Fourier transform (FFT) algorithms, it is possible to compute an $N \times N$ DCT by means of $\mathcal{O}(2N^2 \log_2 N)$ operations instead of $\mathcal{O}(N^4)$, where $\mathcal{O}(\cdot)$ stands for "order of." In addition, because of the even-symmetric extension process, no artificial discontinuities are introduced at the block boundaries, unlike the DFT, resulting in superior energy compaction. The fact that the computation of the DCT requires only real arithmetic facilitates its hardware implementation. As a result, DCT is widely available in special-purpose single-chip VLSI hardware, which makes it attractive for real-time use.

The energy compaction property of 8×8 DCT is illustrated in the following example.

Example: Energy Compaction Property of DCT

An 8×8 block from the seventh frame of the Mobile and Calendar sequence is shown below:

$$s(n_1, n_2) = \begin{bmatrix} 183 & 160 & 94 & 153 & 194 & 163 & 132 & 165 \\ 183 & 153 & 116 & 176 & 187 & 166 & 130 & 169 \\ 179 & 168 & 171 & 182 & 179 & 170 & 131 & 167 \\ 177 & 177 & 179 & 177 & 179 & 165 & 131 & 167 \\ 178 & 178 & 179 & 176 & 182 & 164 & 130 & 171 \\ 179 & 180 & 180 & 179 & 183 & 169 & 132 & 169 \\ 179 & 179 & 180 & 182 & 183 & 170 & 129 & 173 \\ 180 & 179 & 181 & 179 & 181 & 170 & 130 & 169 \end{bmatrix}$$

The DCT coefficients, after subtracting 128 from each element, are

$$\text{NINT}[S(k_1,k_2)] = \begin{bmatrix} 313 & 56 & -27 & 18 & 78 & -60 & 27 & -27 \\ -38 & -27 & 13 & 44 & 32 & -1 & -24 & -10 \\ -20 & -17 & 10 & 33 & 21 & -6 & -16 & -9 \\ -10 & -8 & 9 & 17 & 9 & -10 & -13 & 1 \\ -6 & 1 & 6 & 4 & -3 & -7 & -5 & 5 \\ 2 & 3 & 0 & -3 & -7 & -4 & 0 & 3 \\ 4 & 4 & -1 & -2 & -9 & 0 & 2 & 4 \\ 3 & 1 & 0 & -4 & -2 & -1 & 3 & 1 \end{bmatrix}$$

where NINT denotes nearest integer truncation. Observe that most of the high-frequency coefficients (around the lower right corner) are much smaller than those around the (0,0) frequency (at the upper left corner).

19.3.2 Quantization/Bit Allocation

Bit allocation refers to determination of which coefficients should be retained for coding, and how coarsely each retained coefficient should be quantized in order to minimize the overall mean square quantization noise. Bit allocation can be performed either globally, using zonal coding, or adaptively, using threshold coding.

Zonal Coding

Zonal coding is based on the premise that the transform coefficients of maximum variance carry the most picture information, in the information theoretic sense, and should be retained. The locations of the coefficients with the K largest variances are indicated by means of a zonal mask, which is the same for all blocks. A typical zonal mask, where the retained coefficients are denoted by a 1, is shown in Figure 19.8 (a). To design a zonal mask, variances of each coefficient can be calculated either over an ensemble of representative transformed subimage arrays (blocks) or based on a global image model, such as the Gauss-Markov model. Zonal masks can be customized for individual images, provided that they are stored with the encoded image data.

The Bit Allocation Problem: Because the dynamic range of each retained coefficient is usually different, the number of bits alloted to represent each coefficient should be proportional to the dynamic range of that coefficient. Various algorithms exist for bit allocation. A simple bit allocation strategy is to choose the number of bits proportional to the variance of each coefficient over the ensemble of blocks. Suppose a total of B bits are available to represent a block, and there are M retained coefficients with the variances σ_i^2, $i = 1, \ldots, M$. Then the number of bits allocated for each of these coefficients is given by

$$b_i = \frac{B}{M} + \frac{1}{2}\log_2 \sigma_i^2 - \frac{1}{2M}\sum_{i=1}^{M}\log_2 \sigma_i^2 \qquad (19.27)$$

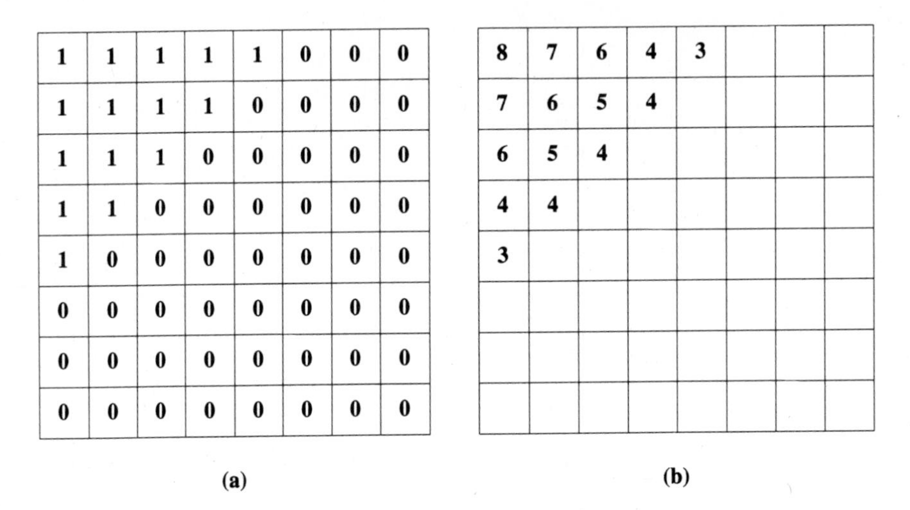

1	1	1	1	1	0	0	0
1	1	1	1	0	0	0	0
1	1	1	0	0	0	0	0
1	1	0	0	0	0	0	0
1	0	0	0	0	0	0	0
0	0	0	0	0	0	0	0
0	0	0	0	0	0	0	0
0	0	0	0	0	0	0	0

(a)

8	7	6	4	3			
7	6	5	4				
6	5	4					
4	4						
3							

(b)

Figure 19.8: Illustration of the a) zonal mask and b) zonal bit allocation.

Note that the results are rounded to the nearest integer such that $b_1 + \ldots + b_M = B$. The typical number of bits per each retained coefficient is shown in Figure 19.8 (b). Once the number of bits allocated for each coefficient is determined, a different Lloyd-Max quantizer can be designed for each coefficient.

Threshold Coding

In most images, different blocks have different spectral and statistical characteristics, necessitating the use of adaptive bit allocation methods. To obtain the best results, both the location of the retained coefficients and the average number of bits/pixel should vary from block to block. Threshold coding is an adaptive method, where only those coefficients whose magnitudes are above a threshold are retained within each block. In practice, the operations of thresholding and quantization can be combined by means of a quantization matrix as

$$\hat{S}(k_1, k_2) = NINT\left[\frac{S(k_1, k_2)}{T(k_1, k_2)}\right] \tag{19.28}$$

where $\hat{S}(k_1, k_2)$ is a thresholded and quantized approximation of $S(k_1, k_2)$, and $T(k_1, k_2)$ is the respective element of the quantization matrix (QM). A coefficient (k_1, k_2) is retained if $\hat{S}(k_1, k_2)$ is nonzero. The elements of the QM are 8-bit integers which determine the quantization step size for each coefficient location. The choice of the QM depends on the source noise level and viewing conditions. In general, coarser quantization, hence larger weights, is used for higher-frequency coefficients. A typical QM for luminance images is shown in the example below.

We can either use a single QM for all blocks, or scale the elements of the QM on a block-by-block basis as a function of the spatial activity within the block, which provides improved scene adaptivity. In the latter case, blocks containing more spatial activity are encoded using more bits than do low-energy blocks [Che 84].

Example: Threshold Coding of DCT Coefficients

Suppose the quantization matrix is given by

$$T(k_1, k_2) = \begin{bmatrix} 16 & 11 & 10 & 16 & 24 & 40 & 51 & 61 \\ 12 & 12 & 14 & 19 & 26 & 58 & 60 & 55 \\ 14 & 13 & 16 & 24 & 40 & 57 & 69 & 56 \\ 14 & 17 & 22 & 29 & 51 & 87 & 80 & 62 \\ 18 & 22 & 37 & 56 & 68 & 109 & 103 & 77 \\ 24 & 35 & 55 & 64 & 81 & 104 & 113 & 92 \\ 49 & 64 & 78 & 87 & 103 & 121 & 120 & 101 \\ 72 & 92 & 95 & 98 & 112 & 100 & 103 & 99 \end{bmatrix}$$

Applying the above QM to the DCT coefficients given in the previous example, we have

$$\hat{S}(k_1, k_2) = \begin{bmatrix} 20 & 5 & -3 & 1 & 3 & -2 & 1 & 0 \\ -3 & -2 & 1 & 2 & 1 & 0 & 0 & 0 \\ -1 & -1 & 1 & 1 & 1 & 0 & 0 & 0 \\ -1 & 0 & 0 & 1 & 0 & 0 & 0 & 0 \\ 0 & 0 & 0 & 0 & 0 & 0 & 0 & 0 \\ 0 & 0 & 0 & 0 & 0 & 0 & 0 & 0 \\ 0 & 0 & 0 & 0 & 0 & 0 & 0 & 0 \\ 0 & 0 & 0 & 0 & 0 & 0 & 0 & 0 \end{bmatrix}$$

In this example, 45 out of 64 DCT coefficients are truncated to zero. Coding of the remaining coefficients is discussed next.

19.3.3 Coding

In zonal coding, since the locations of the retained coefficients are assumed to be known at the receiver, they can be individually coded using entropy coding techniques, such as Huffman coding or arithmetic coding, and transmitted. However, in threshold coding, since the relative locations of the transmitted coefficients vary from block to block, more sophisticated coding strategies are required to encode the positions of these coefficients without an excessive amount of overhead information.

A common strategy is to zig-zag scan the transform coefficients, except the DC $(0,0)$ coefficient, in an attempt to order them in decreasing magnitude. Zig-zag scanning is illustrated in Figure 19.9. The coefficients along the zig-zag scan lines are mapped into symbols in the form [*run*, *level*], where *level* is the value of a nonzero

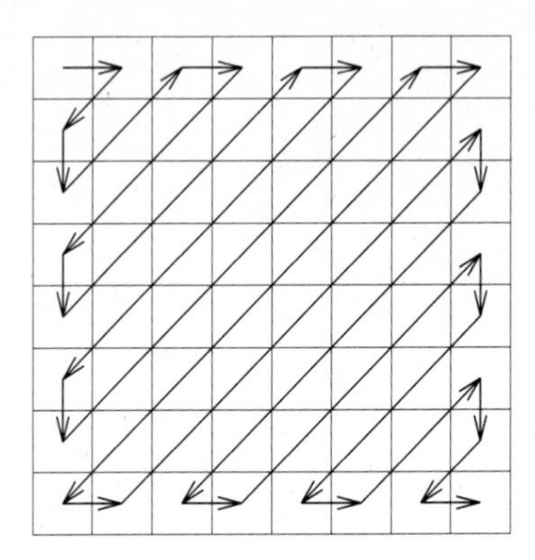

Figure 19.9: Illustration of the zig-zag scanning.

coefficient, and *run* is the number of zero coefficients preceding it. These symbols are then entropy coded for storage and/or transmission. (See Section 20.2.1 for an example.) It can easily be seen that zig-zag scanning increases the likelihood of grouping all nonzero coefficients together, and facilitates run-length coding of the zero coefficients.

It is crucial to transmit the DC coefficient of each block with full accuracy, because any mismatch of the mean level of the blocks generates severe blocking artifacts. To this effect, the DC coefficients of each block are DPCM encoded among

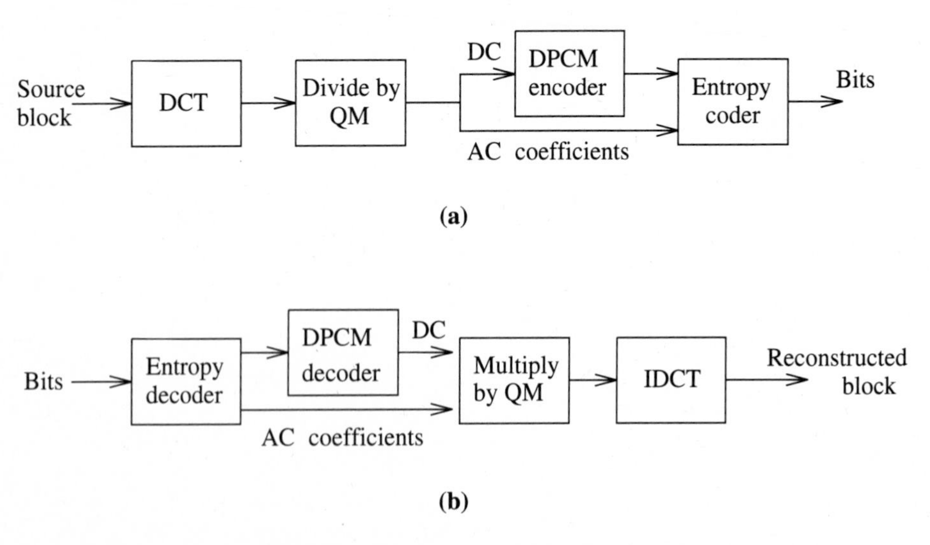

Figure 19.10: Threshold DCT a) encoder and b) decoder models.

themselves. Block diagrams of the complete encoder and decoder are depicted in Figure 19.10 (a) and (b), where QM and IDCT stand for the quantization matrix and inverse DCT, respectively.

19.3.4 Blocking Artifacts in Transform Coding

Compression artifacts in DCT coding can be classified as: i) blurring due to truncation of "high-frequency" coefficients, ii) graininess due to coarse quantization of some coefficients, and iii) blocking artifacts. Blocking artifacts, which refer to artificial intensity discontinuities at the borders of neighboring blocks, are the most disturbing. They appear because each block is processed independently, with different quantization strategies. The blocking artifacts become more noticeable at lower bitrates. See [Rab 91] for examples.

Several different methods were proposed for the reduction of blocking artifacts in DCT compression. In one of them, blocks are overlapped at the transmitter, which requires coding and transmission of the pixels in the overlapping regions more than once. These pixels are averaged at the receiver, providing a smoothing effect. Thus, reduction of blocking artifacts is achieved at the expense of increased bitrate. Using overlapping blocks at the encoder, without increasing the bitrate, has also been proposed by employing lapped orthogonal transforms (LOT) instead of the DCT. Because DCT is widely available in hardware, LOT is rarely used in practice. Finally, blocking artifacts may be reduced by postprocessing using image restoration techniques [Lim 90], including the method of projection onto convex sets (POCS) [Zak 92].

19.4 Exercises

1. Suppose a random variable X that is uniformly distributed between 0 and 10 is uniformly quantized with N levels.

   a) Find all decision and reconstruction levels.

   b) Calculate the mean square quantization error.

2. Let X_n, $n = \ldots, -1, 0, 1, \ldots$, denote a sequence of zero-mean scalar random variables with

$$E[X_n^2] = \sigma_X^2, \qquad \text{and} \qquad E[X_n X_{n-1}] = \alpha$$

   We define the prediction error sequence

$$\epsilon_n = X_n - \rho X_{n-1}$$

   where ρ is such that

$$E[(X_n - \rho X_{n-1})X_n] = 0 \qquad \text{for all } n.$$

a) Find the value of ρ.

b) Let D_1 and D_2 denote the distortion incurred by uniform quantization of X_n and ϵ_n, respectively, using r bits. Show that D_1 is always greater than or equal to D_2.

3. Suppose we have a differential image that is Laplacian distributed with the standard deviation 15. We wish to quantize it using 2 bits/sample.

   a) Find the decision and reconstruction levels of a uniform quantizer.

   b) Find the decision and reconstruction levels of the Lloyd-Max quantizer.

4. Explain why the transmitter in DPCM bases its prediction on previously reconstructed values rather than on the originals.

5. Compute the Karhunen-Loeve transform, DFT and DCT of the block shown below.

$$\begin{bmatrix} 16 & 11 & 10 & 16 & 24 & 40 & 51 & 61 \\ 12 & 12 & 14 & 19 & 26 & 58 & 60 & 55 \\ 14 & 13 & 16 & 24 & 40 & 57 & 69 & 56 \\ 14 & 17 & 22 & 29 & 51 & 87 & 80 & 62 \\ 18 & 22 & 37 & 56 & 68 & 109 & 103 & 77 \\ 24 & 35 & 55 & 64 & 81 & 104 & 113 & 92 \\ 49 & 64 & 78 & 87 & 103 & 121 & 120 & 101 \\ 72 & 92 & 95 & 98 & 112 & 100 & 103 & 99 \end{bmatrix}$$

   Compare these transforms on the basis of energy compaction (decorrelation), and similarity of the basis functions.

6. Suppose we have three scalar random variables, X_1, X_2, and X_3, with variances 2.5, 7, and 10. Propose a method to allocate a total of 20 bits to quantize each random variable independently to minimize the mean of the sum of the square quantization errors.

7. Explain the purpose of the zig-zag scanning in DCT compression. Do we need it with the zonal sampling strategy?

Bibliography

[Ahm 74] N. Ahmed, T, Natarajan, and K. R. Rao, "Discrete cosine transforms," *IEEE Trans. Comp.*, vol. 23, pp. 90–93, 1974.

[Che 84] W.-H. Chen and W. K. Pratt, "Scene adaptive coder," *IEEE Trans. Comm.*, vol. 32, pp. 225–232, Mar. 1984.

[Cla 85] R. J. Clark, *Transform Coding of Images*, Academic Press, 1985.

[Jai 81] A. K. Jain, "Image data compression: A review," *Proc. IEEE*, vol. 69, pp. 349–389, Mar. 1981.

[Jai 90] A. K. Jain, *Fundamentals of Digital Image Processing*, Englewood Cliffs, NJ: Prentice Hall, 1989.

[Jay 92] N. Jayant, "Signal compression: Technology targets and research directions," *IEEE J. Selected Areas in Comm.*, vol. 10, pp. 796–819, Jun. 1992.

[Lim 90] J. S. Lim, *Two-Dimensional Signal and Image Processing*, Englewood Cliffs, NJ: Prentice Hall, 1990.

[Llo 82] S. P. Lloyd, "Least squares quantization in PCM," *IEEE Trans. Info. The.*, vol. 28, pp. 129–137, Mar. 1982.

[Mar 94] S. Martucci, "Symmetric convolution and the discrete sine and cosine transforms," *IEEE Trans. Signal Proc.*, vol. 42, pp. 1038–1051, May 1994.

[Max 60] J. Max, "Quantizing for minimum distortion," *IRE Trans. Info. Theory*, vol. 6, pp. 7–12, Mar. 1960.

[Mus 85] H. G. Musmann, P. Pirsch, and H. J. Grallert, "Advances in picture coding," *Proc. IEEE*, vol. 73, pp. 523–548, Apr. 1985.

[Net 80] A. N. Netravali and J. O. Limb, "Picture coding: A review," *Proc. IEEE*, vol. 68, no. 3, pp. 366–406, Mar. 1980.

[Rab 91] M. Rabbani and P. Jones, *Digital Image Compression Techniques*, Bellingham, WA: SPIE Press, 1991.

[Rao 90] K. R. Rao and P. Yip, *Discrete Cosine Transform*, Academic Press, 1990.

[Win 72] P. A. Wintz, "Transform picture coding," *Proc. IEEE*, vol. 60, pp. 809–820, July 1972.

[Woo 69] R. C. Wood, "On optimum quantization," *IEEE Trans. Info. Theory*, vol. 15, no. 2, pp. 248–252, 1969.

[Zak 92] A. Zakhor, "Iterative procedures for reduction of blocking effects in transform image coding," *IEEE Trans. Circ. and Syst.: Video Tech.*, vol. 2, pp. 91–95, Mar. 1992.

Chapter 20

STILL IMAGE COMPRESSION STANDARDS

Recent developments in electronic imaging technology are creating both a capability and a need for digital storage and transmission of binary and continuous-tone color images and video. Image compression is an essential enabling technology behind these developments, considering the raw data rates associated with digital images. However, storage and transmission of digital images in compressed formats necessitates internationally accepted standards to ensure compatibility and interoperability of various products from different manufacturers.

Recognizing this need, several international organizations, including the International Standards Organization (ISO), the International Telecommunications Union (ITU) - formerly the International Consultative Committee for Telephone and Telegraph (CCITT), - and the International Electrotechnical Commission (IEC), have participated in programs to establish international standards for image compression in specific application areas. To this effect, expert groups, formed under the auspices of ISO, IEC, and ITU (CCITT), solicited proposals from companies, universities, and research laboratories. The best of those submitted have been selected on the basis of image quality, compression performance, and practical constraints. This chapter addresses the international standards on bilevel and continuous-tone color still image compression, which are summarized in Table 20.1. The bilevel image compression standards are discussed in Section 20.1, while the JPEG algorithm is presented in Section 20.2.

Table 20.1: International standards for still-image compression

Standard	Description
ITU (CCITT) Group 3 Group 4	Developed for bilevel image compression; primarily designed for fax transmission of documents; most widely used image compression standards.
JBIG	Joint (ITU-ISO/IEC) Bilevel Imaging Group; developed for improved bilevel image compression to handle progressive transmission and digital halftones.
JPEG	Joint (ITU-ISO/IEC) Photographic Expert Group; developed for color still-image compression.

20.1 Bilevel Image Compression Standards

Probably the best example of the relationship between standardization and market growth is the boom in the fax market that we experienced following the ITU (CCITT) standards for binary image compression. Today, undoubtedly the most widely used image compression standards are the ITU (CCITT) Group 3 (G3) and Group 4 (G4) codes for the compression of bilevel images.

ITU (CCITT) G3, also known as MR, modified READ (Relative Element Adress Designate), is based on nonadaptive, 1-D run-length coding in which the last $K-1$ lines of each group of K lines can be (optionally) 2-D run-length coded. Typical compression ratio has been found to be 15:1 on representative test documents. ITU (CCITT) G4 refers to a simplified version of G3 standard in which only 2-D run-length coding is allowed. Both standards use the same nonadaptive 2-D coding method and the same Huffman tables. G4 obtains 30% better compression by removing some error recovery from G3. ITU (CCITT) G3 and G4 standards have been vital in the world of fax machines and document storage applications.

It has been recognized that the ITU (CCITT) G3 and G4 codes cannot effectively deal with some images, especially digital halftones. JBIG is a new bilevel image compression standard developed to improve upon the performance of the G3 and G4 standards by using adaptive arithmetic coding, and adaptive templates for halftones. In the following, we discuss the 1-D and 2-D run-length coding methods which form the basis of ITU (CCITT) G3, G4, and JBIG. We conclude the section with a discussion of the new features in JBIG.

20.1.1 One-Dimensional RLC

In 1-D RLC, which is used in the ITU (CCITT) G3 algorithm, each line of a bilevel image is described as a series of alternating white and black runs using a left-to-right scan. The resulting run-lengths are Huffman coded by means of an appropriate combination of make-up codes and terminating codes. Run lengths

Table 20.2: ITU (CCITT) Terminating Codes [Gon 92].

Run Length	White Codeword	Black Codeword
0	00110101	0000110111
⋮	⋮	⋮
63	00110100	000001100111

Table 20.3: ITU (CCITT) Make-up Codes [Gon 92].

Run Length	White Codeword	Black Codeword
64	11011	0000001111
128	10010	000011001000
⋮	⋮	⋮
1728	010011011	0000001100101

Run Length	Codeword
1792	00000001000
1856	00000001100
⋮	⋮
2560	000000011111

Rafael C. Gonzalez and Richard E. Woods, DIGITAL IMAGE PROCESSING (pp. 391, 392, 395), ©1992 Addison Wesley Publishing Company, Inc. Reprinted by permission of the publisher.

between 0 and 63 are represented by a single codeword, called a terminating code. Run lengths that are greater than 63 are represented by a combination of a make-up code, which indicates the greatest multiple of 64 that is smaller than the actual run-length, followed by a terminating code indicating the difference. According to the ITU (CCITT) G3 standard, each line begins with a white run whose length may be zero, and ends with a unique end-of-line (EOL) codeword.

Since the statistics of white (1) and black (0) runs are generally different, we utilize different Huffman codes to represent the white and black run lengths. Some of the ITU (CCITT) terminating codes and make-up codes for white and black runs are listed in Tables 20.2 and 20.3, respectively. The reader is referred to [Gon 92] for a full listing of these codes. Observe that the make-up codes for run lengths greater than or equal to 1792 are the same for both white and black runs. The statistics needed to generate the listed Huffman codes were obtained from a number of training documents, including typed and handwritten documents in several languages.

20.1.2 Two-Dimensional RLC

Two-dimensional RLC is based on the principle of relative address coding, where the position of each black-to-white or white-to-black transition is coded with respect to the position of a *reference element* that is on the current coding line. The coding procedure has three modes: *pass mode*, *vertical mode*, and *horizontal mode*. The selection of the appropriate mode in coding a particular transition depends on the values of five parameters, the reference element a_0, and four *changing elements* a_1, a_2, b_1, and b_2.

The reference element is initially set equal to an imaginary white element to the left of the first pixel at the beginning of each new line. Subsequent values of the reference element are determined as indicated by the flowchart shown in Figure 20.2, where $||a_1, b_1||$ denotes the distance between a_1 and b_1. A changing element is defined as a pixel where a transition from black to white or vice versa occurs with respect to the previous element on the same line. In particular, a_1 is the location of the changing element to the right of a_0, and a_2 is the next to the right of a_1 on the coding line, b_1 is the location of the changing element, to the right of a_0 on the reference (previous) line, whose value is the opposite of a_0, and b_2 is the next changing element to the right of b_1 on the reference line. These parameters are illustrated by an example.

Example: Two-dimensional RLC

Suppose we have two lines of a binary image, shown in Figure 20.1. The initial positions of the parameters a_0, a_1, a_2, b_1, and b_2 are marked on the figure. Since, b_2 is to the right of a_1, and the distance between a_1 and b_1 is equal to 3, we are in the vertical mode, and the next value of $a_0 = a_1$ as marked.

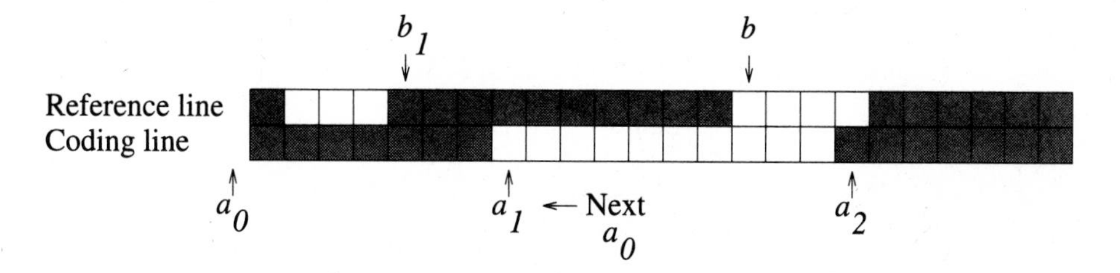

Figure 20.1: Parameters of the ITU (CCITT) 2-D run-length coding.

Once the compression mode is determined, the corresponding code word can be formed according to Table 20.4. The notation $M(a_i, a_j)$ indicates that the distance between a_i and a_j will be coded using the terminating codes and the make-up codes given in Tables 20.2 and 20.3, respectively. An extension code is often used to initiate an optional mode, such as transmission with no compression.

The algorithm can, then, be summarized as follows:

i) Initialize a_0.

ii) Determine the four changing elements, a_1, a_2, b_1, and b_2, according to their definitions given above.

iii) Determine the appropriate mode based on these values, using the flowchart in Figure 20.2, and form the codeword using Table 20.4.

iv) Establish the next *reference element* for the next coding cycle, according to the flowchart in Figure 20.2, and go to step ii. Go to step i if an end of line is encountered.

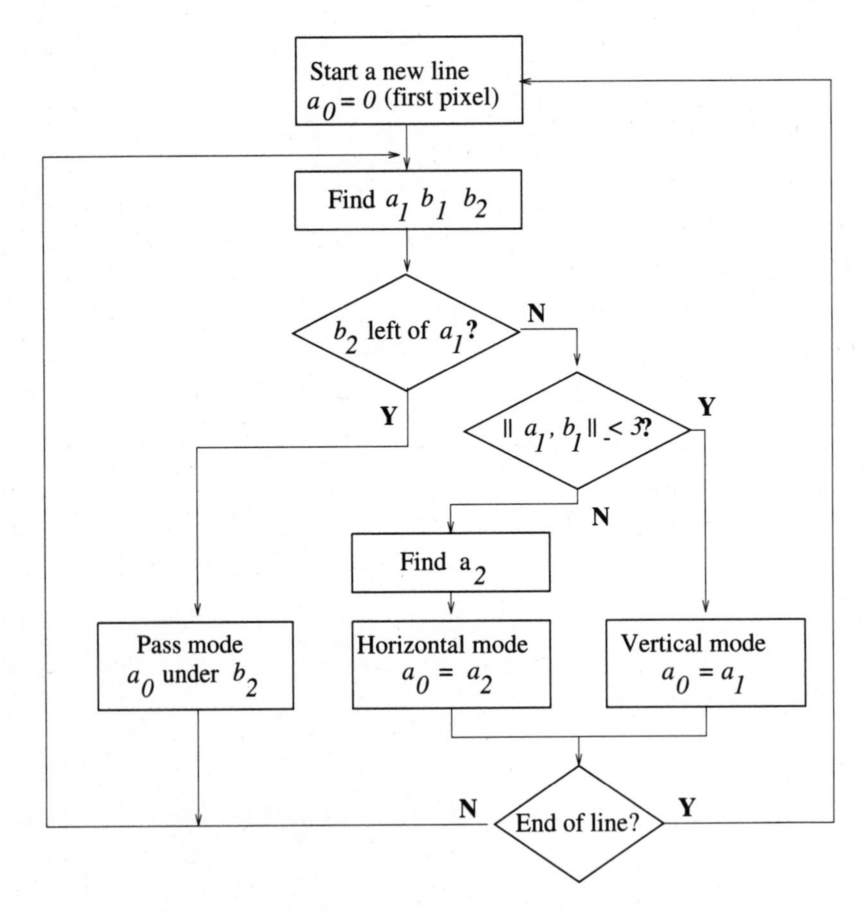

Figure 20.2: Flowchart of the ITU (CCITT) 2-D RLC algorithm.

Table 20.4: ITU (CCITT) two-dimensional run-length codes [Gon 92].

Mode	Codeword
Pass	0001
Horizontal	001+$M(a_0, a_1) + M(a_1, a_2)$
Vertical	
a_1 below b_1	1
a_1 one to the right of b_1	011
a_1 two to the right of b_1	000011
a_1 three to the right of b_1	0000011
a_1 one to the left of b_1	010
a_1 two to the left of b_1	000010
a_1 three to the left of b_1	0000010
Extension	0000001xxx

20.1.3 The JBIG Standard

JBIG is a new standard for lossless binary and low-precision gray-level (less than 6 bits/pixel) image compression proposed and developed by a joint committee between ISO/IEC and ITU (CCITT). The mission of the JBIG committee has been: (i) to develop adaptive lossless algorithms that will outperform the existing standards, and (ii) to accommodate progressive transmission.

In JBIG, binary pel values are directly fed into an arithmetic encoder, which utilizes a "sequential template" (a set of neighboring pixels that are already encoded) to create a "context" in the generation of each binary symbol. The context affects the decision probabilities used in the arithmetic encoder. The best performance can be obtained by using adaptive templates leading to adaptive arithmetic coding. Using adaptive arithmetic coding, JBIG improves upon the performance of the ITU (CCITT) G4 algorithm by another 30% for typical text images.

One of the primary reasons why JBIG has been formed is the performance improvement needed for binary halftones. Digital halftones have statistical properties that are quite different than those of the text documents. Because the Huffman tables used in ITU (CCITT) G3 and G4 have been based on a set of text documents, they are unsuitable for binary halftones, and have been found to cause data expansion in some cases. JBIG provides about 8:1 compression on typical digital halftones (using adaptive templates) which would have expanded with G3/G4.

In addition, JBIG defines a progressive mode in which a reduced-resolution starting layer image is followed by transmission of progressively higher-resolution layers. The templates used in the progressive mode may include pixels from previous layers as well as previously encoded pixels from the same layer. JBIG syntax also allows coding of images with more than one bit/pixel using bit-plane encoding.

20.2 The JPEG Standard

The JPEG standard describes a family of image compression techniques for continuous-tone (gray-scale or color) still images. Because of the amount of data involved and the psychovisual redundancy in the images, JPEG employs a lossy compression scheme based on transform coding. Activities leading to the JPEG standard got started in March 1986. In January 1988, JPEG reached a consensus that the adaptive DCT approach should be the basis for further refinement and enhancement. The JPEG Draft International Standard (DIS) document was submitted in October 1991. At the July 1992 meeting, the JPEG committee successfully responded to all of the comments, and editorial changes were incorporated into the DIS document in the process of converting it into the international standard document [Wal 91, Pen 93].

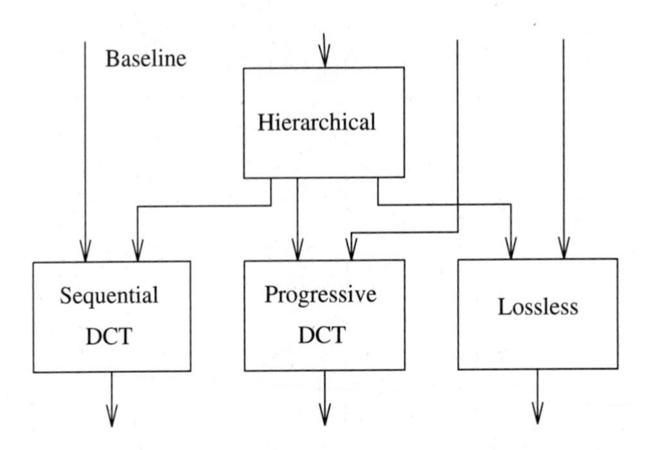

Figure 20.3: JPEG modes of operation.

JPEG provides four modes of operation, sequential (baseline), hierarchical, progressive, and lossless, as outlined in Figure 20.3. They are discussed below in detail. In general, the JPEG standard features:

i) Resolution independence: Arbitrary source resolutions can be handled. Images whose dimensions are not multiples of 8 are internally padded to multiples of 8 in DCT-based modes of operation.

ii) Precision: DCT modes of operation are restricted to 8 and 12 bits/sample precision. For lossless coding the precision can be from 2 to 16 bits/sample, although JBIG has been found to perform better below 4-5 bits/sample.

iii) No absolute bitrate targets: The bitrate/quality tradeoff is controlled primarily by the quantization matrix.

iv) Luminance-chrominance separability: The ability exists to recover the luminance-only image from luminance-chrominance encoded images without always having to decode the chrominance.

v) Extensible: There are no bounds on the number of progressive stages or, in the case of hierarchical progression, lower-resolution stages.

The reader should be aware that JPEG is not a complete architecture for image exchange. The JPEG data streams are defined in terms of what a JPEG decoder needs in order to decompress the data stream. No particular file format, spatial resolution, or color space model is specified as part of the standard. However, JPEG includes a minimal recommended file format, JPEG File Interchange Format (JFIF), which enables JPEG bit streams to be exchanged between a wide variety of platforms and applications. It uses a standard color space (CCIR 601-1). In addition, several commonly available image file formats, e.g., Tag Image File Format (TIFF), are JPEG-compatible. Note that in order to be JPEG-compatible, a product or system must include support for at least the baseline system.

20.2.1 Baseline Algorithm

The baseline algorithm, also called the sequential algorithm, is similar to the DCT compression scheme presented in Chapter 19. It can be summarized in three steps:

(1) DCT computation: The image is first subdivided into 8×8 blocks. Each pixel is level shifted by subtracting 2^{n-1}, where 2^n is the maximum number of gray levels. That is, for 8-bit images we subtract 128 from each pixel in an attempt to remove the DC level of each block. The 2-D DCT of each block is then computed. In the baseline system, the input and output data precision is limited to 8 bits, whereas the quantized DCT values are restricted to 11 bits.

(2) Quantization of the DCT coefficients: The DCT coefficients are threshold coded using a quantization matrix, and then reordered using zigzag scanning to form a 1-D sequence of quantized coefficients. The quantization matrix can be scaled to provide a variety of compression levels. The entries of the quantization matrix are usually determined according to psychovisual considerations, which are discussed below.

(3) Variable-length code (VLC) assignment: The nonzero AC coefficients are Huffman coded using a VLC code that defines the value of the coefficient and the number of preceding zeros. Standard VLC tables are specified. The DC coefficient of each block is DPCM coded relative to the DC coefficient of the previous block.

Color Images

Today most electronic images are recorded in color, in the R-G-B (red, green, blue) domain. JPEG transforms RGB images into a luminance-chrominance space, generally refered to as the Y-Cr-Cb domain, defined as

$$Y = 0.3R + 0.6G + 0.1B \tag{20.1}$$

$$Cr = \frac{B - Y}{2} + 0.5 \tag{20.2}$$

$$Cb = \frac{R - Y}{1.6} + 0.5 \tag{20.3}$$

Because, the human eye is relatively insensitive to the high-frequency content of the chrominance channels Cr and Cb (see Figure 20.5), they are subsampled by 2 in both directions. This is illustrated in Figure 20.4 where the chrominance channels contain half as many lines and pixels per line compared to the luminance channel.

Y1	**Y2**	**Y3**	**Y4**
Y5	**Y6**	**Y7**	**Y8**
Y9	**Y10**	**Y11**	**Y12**
Y13	**Y14**	**Y15**	**Y16**

Cr1	**Cr2**
Cr3	**Cr4**

Cb1	**Cb2**
Cb3	**Cb4**

Figure 20.4: Subsampling of the chrominance channels.

JPEG orders the pixels of a color image as either noninterleaved (three separate scans) or interleaved (a single scan). In reference to Figure 20.4, the noninterleaved ordering can be shown as

Scan 1: Y1, Y2, Y3, ..., Y16
Scan 2: Cr1, Cr2, Cr3, Cr4
Scan 3: Cb1, Cb2, Cb3, Cb4

whereas the interleaved ordering becomes

Y1, Y2, Y3, Y4, Cr1, Cb1, Y5, Y6, Y7, Y8, Cr2, Cb2, ...

Interleaving makes it possible to decompress the image, and convert from luminance-chrominance representation to RGB for display with a minimum of intermediate buffering. For interleaved data, the DCT blocks are ordered according to the parameters specified in the frame and scan headers.

Psychovisual Aspects

In order to reduce the psychovisual redundancy in images, JPEG incorporates the characteristics of the human visual system into the compression process through the specification of quantization matrices. It is well known that the frequency response of the human visual system drops off with increasing spatial frequency. Furthermore, this drop-off is faster in the two chrominance channels. The contrast sensitivity function depicted in Figure 20.5 demonstrates this effect. It implies that a small variation in intensity is more visible in slowly varying regions than in busier ones, and also more visible in luminance compared to a similar variation in chrominance.

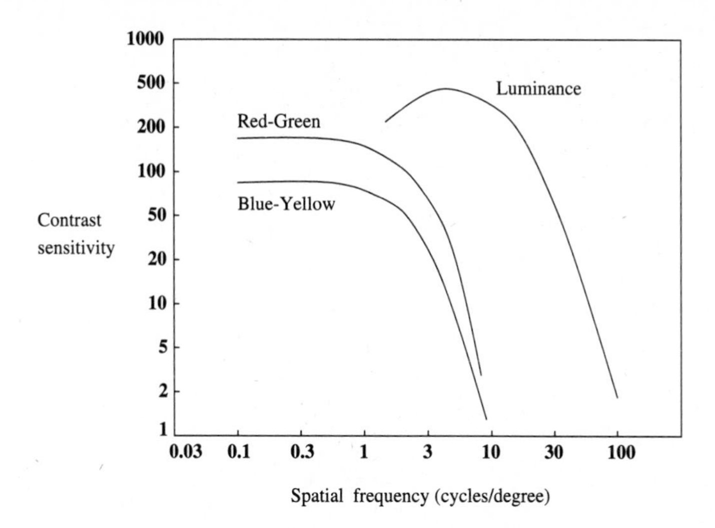

Figure 20.5: Visual response to luminance and chrominance variations [Mul 85].

As a result, JPEG allows specification of two quantization matrices, one for the luminance and another for the two chrominance channels to allocate more bits for the representation of coefficients which are visually more significant. Tables 20.5 and 20.6 show typical quantization matrices for the luminance and chrominance channels, respectively. The elements of these matrices are based on the visibility of individual 8×8 DCT basis functions with a viewing distance equal to six times the screen width. The basis functions were viewed with a luminance resolution of 720 pixels $\times$ 576 lines and a chrominance resolution of 360×576. The matrices suggest that those DCT coefficients corresponding to basis images with low visibility can be more coarsely quantized.

Table 20.5: Quantization Table for the Luminance Channel

16	11	10	16	24	40	51	61
12	12	14	19	26	58	60	55
14	13	16	24	40	57	69	56
14	17	22	29	51	87	80	62
18	22	37	56	68	109	103	77
24	35	55	64	81	104	113	92
49	64	78	87	103	121	120	101
72	92	95	98	112	100	103	99

Table 20.6: Quantization Table for the Chrominance Channel

17	18	24	47	99	99	99	99
18	21	26	66	99	99	99	99
24	26	56	99	99	99	99	99
47	66	99	99	99	99	99	99
99	99	99	99	99	99	99	99
99	99	99	99	99	99	99	99
99	99	99	99	99	99	99	99
99	99	99	99	99	99	99	99

Example: JPEG Baseline Algorithm

We demonstrate the application of the JPEG baseline algorithm on the 8×8 luminance block presented in Section 19.3.1. Following the same steps demonstrated on pages 380–383, the gray levels are first shifted by -128 (assuming that the original is an 8-bit image), and then a forward DCT is applied. The DCT coefficients divided by the quantization matrix, shown in Table 20.5, have been found as

$$\hat{S}(k_1, k_2) = \begin{bmatrix} 20 & 5 & -3 & 1 & 3 & -2 & 1 & 0 \\ -3 & -2 & 1 & 2 & 1 & 0 & 0 & 0 \\ -1 & -1 & 1 & 1 & 1 & 0 & 0 & 0 \\ -1 & 0 & 0 & 1 & 0 & 0 & 0 & 0 \\ 0 & 0 & 0 & 0 & 0 & 0 & 0 & 0 \\ 0 & 0 & 0 & 0 & 0 & 0 & 0 & 0 \\ 0 & 0 & 0 & 0 & 0 & 0 & 0 & 0 \\ 0 & 0 & 0 & 0 & 0 & 0 & 0 & 0 \end{bmatrix}.$$

Following a zigzag scanning of these coefficients, the 1-D coefficient sequence can be expressed as

$$[\,20,\ 5,\ -3,\ -1,\ -2,\ -3,\ 1,\ 1,\ -1,\ -1,\ 0,\ 0,\ 1,\ 2,\ 3,\ -2,\ 1,\ 1,\ 0,\ 0,\ 0,\ 0,\ 0,\ 0,\ 1,\ 1,\ 0,\ 1,\ EOB\,]$$

where EOB denotes the end of the block (i.e., all following coefficients are zero.)

The DC coefficient is coded using DPCM, as depicted in Figure 19.10. That is, the difference between the DC coefficient of the present block and that of the previously encoded block is coded. (Assume that the DC coefficient of the previous block was 29, so the difference is -9.) The AC coefficients are mapped into the following run-level pairs

(0, 5), (0, -3), (0, -1), (0, -2), (0, -3), (0, 1), (0, 1), (0, -1), (0, -1), (2, 1), (0, 2), (0, 3), (0, -2), (0, 1), (0, 1), (6, 1), (0, 1), (1, 1), EOB.

The codewords for the symbols formed from these coefficients can be found according to the Tables for JPEG Coefficient Coding Categories, JPEG Default DC Codes (Luminance/Chrominance), and JPEG Default AC Codes (Luminance/ Chrominance) [Gon 92, Pen 93]. For example, the DC difference -9 falls within the DC difference category 4. The proper base (default) code for category 4 is 011 (a 3-bit code), while the total length of a completely encoded category 4 coefficient is 7 bits. The remaining 4 bits will be the least significant bits (LSBs) of the difference value. Each default AC Huffman codeword depends on the number of zero-valued coefficients preceding the nonzero coefficient to be coded as well as the magnitude category of the coefficient. The remaining bits are found in the same way as in DC difference coding.

The decoder implements the inverse operations. That is, the received coefficients are first multiplied by the same quantization matrix to obtain

$$\tilde{S}(k_1,k_2) = \begin{bmatrix} 320 & 55 & -30 & 16 & 72 & -80 & 51 & 0 \\ -36 & -24 & 14 & 38 & 26 & 0 & 0 & 0 \\ -14 & -13 & 16 & 24 & 40 & 0 & 0 & 0 \\ -14 & 0 & 0 & 29 & 0 & 0 & 0 & 0 \\ 0 & 0 & 0 & 0 & 0 & 0 & 0 & 0 \\ 0 & 0 & 0 & 0 & 0 & 0 & 0 & 0 \\ 0 & 0 & 0 & 0 & 0 & 0 & 0 & 0 \\ 0 & 0 & 0 & 0 & 0 & 0 & 0 & 0 \end{bmatrix}$$

Performing an inverse DCT, and adding 128 to each element, we find the reconstructed block as

$$\tilde{s}(n_1,n_2) = \begin{bmatrix} 195 & 140 & 119 & 148 & 197 & 171 & 120 & 170 \\ 186 & 153 & 143 & 158 & 193 & 168 & 124 & 175 \\ 174 & 169 & 172 & 168 & 187 & 166 & 128 & 177 \\ 169 & 180 & 187 & 171 & 185 & 170 & 131 & 170 \\ 172 & 182 & 186 & 168 & 186 & 175 & 131 & 161 \\ 177 & 180 & 181 & 168 & 189 & 175 & 129 & 161 \\ 181 & 178 & 181 & 174 & 191 & 169 & 128 & 173 \\ 183 & 178 & 184 & 181 & 192 & 162 & 127 & 185 \end{bmatrix}$$

The reconstruction error varies ± 25 gray levels, as can be seen by comparing the result with the original block shown on page 380.

The compression ratio, and hence the quality of the reconstructed image, is controlled by scaling the quantization matrix as desired. Scaling the quantization matrix by a factor larger than 1 results in a coarser quantization, and a lower bitrate at the expense of higher reconstruction error.

20.2.2 JPEG Progressive

The progressive DCT mode refers to encoding of the DCT coefficients in multiple passes, where a portion of the quantized DCT coefficients is transmitted in each pass. Two complementary procedures, corresponding to different grouping of the DCT coefficients, have been defined for progressive transmission: the spectral selection and successive approximation methods. An organization of the DCT coefficients for progressive transmission is depicted in Figure 20.6 where MSB and LSB denote the most significant bit and the least significant bit, respectively.

In the spectral selection process, the DCT coefficients are ordered into spectral bands where the lower-frequency bands are encoded and sent first. For example, the DC coefficient of each block is sent in the initial transmission, yielding a rather blocky first image at the receiver. The image quality is usually acceptable after the first five AC coefficients are also transmitted. When all the DCT coefficients are eventually coded and transmitted, the image quality is the same as that of the sequential algorithm.

In the successive approximation method, the DCT coefficients are first sent with lower precision, and then refined in successive passes. The DC coefficient of each block is sent first with full precision to avoid mean level mismatch. The AC coefficients may be transmitted starting with the most significant bit (MSB) plane. Successive approximation usually gives better-quality images at lower bitrates. The two procedures may be intermixed by using spectral selection within successive approximations.

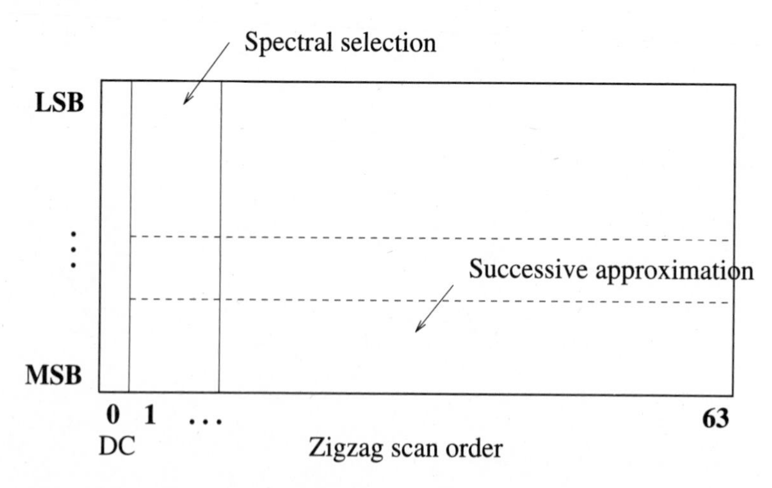

Figure 20.6: Arrangement of the DCT coefficients for progressive transmission.

20.2.3 JPEG Lossless

The lossless mode of operation uses a predictive coding technique which is similar to that described in Chapter 17. The neighboring samples denoted by a, b, and c, which are already encoded and transmitted, are used to predict the current sample x, shown in Figure 20.7. JPEG allows the use of eight different predictors, which are listed in Table 20.7. The selection value 1 is always used to predict the first line of an image. The symbols representing the integer prediction errors are VLC coded.

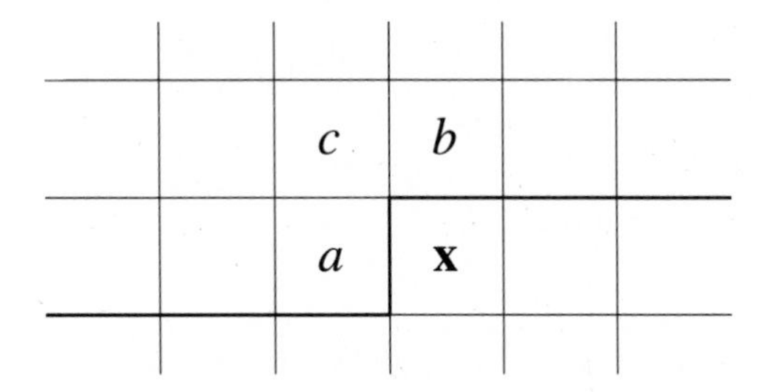

Figure 20.7: Support of the predictor.

Table 20.7: Possible predictors in JPEG lossless mode.

Selection	Predictor
0	Differential coding in hierarchical mode
1	a
2	b
3	c
4	a+b-c
5	a-(b-c)/2
6	b-(a-c)/2
7	(a+b)/2

20.2.4 JPEG Hierarchical

The hierarchical mode of operation may be considered as a special case of the progressive transmission, with increasing spatial resolution between the progressive stages. Hierarchical representation of an image in the form of a resolution pyramid is shown in Figure 20.8.

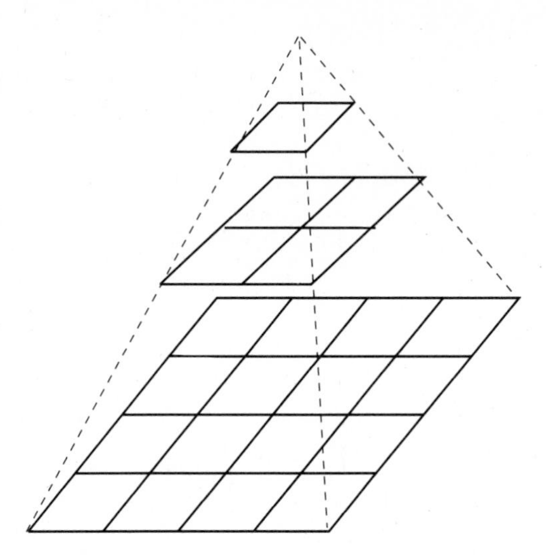

Figure 20.8: Hierarchical image representation.

In the first stage, the lowest-resolution image (top layer of the pyramid) is encoded using one of the sequential or progressive JPEG modes. The output of each hierarchical stage is then upsampled (interpolated) and is taken as the prediction for the next stage. The upsampling filters double the resolution horizontally, vertically, or in both dimensions. The bilinear interpolation filter is precisely specified in the DIS document. In the next stage, the difference between the actual second layer in the pyramid and the upsampled first-layer image is encoded and transmitted. The procedure continues until the residual of the highest-resolution image is encoded and transmitted.

In the hierarchical mode of operation, the image quality at extremely low bitrates surpasses any of the other JPEG modes, but this is achieved at the expense of a slightly higher bit rate at the completion of the progression.

20.2.5 Implementations of JPEG

JPEG is becoming increasingly popular for storage and transmission of full-color (24 bits) photographic images. At present, there are various software and hardware implementations of JPEG. For example, several video grab boards for popular workstation platforms host a JPEG chip for real-time (30 frames/s) image acquisition in JPEG-compressed form. In addition, software implementations exist for almost all workstations and PCs, including X-Windows, MS-DOS, Microsoft Windows, OS/2, and Macintosh, to view and exchange full-color images in JPEG-compressed form. However, because JPEG is a compression standard and does not specify a file format, JPEG images may come in many file formats, such as JFIF (JPEG File Exchange Format), TIFF/JPEG, and PICT/JPEG (for Macintosh). The reader is cautioned that exchanging JPEG-compressed images requires file format compatibility. In other words, the viewer or the application programs must recognize the specific file format in addition to being able to decode JPEG-compressed images. Most JPEG implementations give the user a choice to trade image size for image quality. This is achieved by selecting a value for the quality factor parameter, which adjusts the quantization levels. The range of values for the quality factor may vary from implementation to implementation. Up-to-date information about software implementations and other practical aspects can be found in the "JPEG-FAQ," which can be obtained over the Internet by ftp or e-mail. You can request the latest version by sending an e-mail to "mail-server@rtfm.mit.edu" containing the line "send usenet/news.answers/jpeg-faq." The "news.answers" archive is also accessible through the "www" and "gopher" services.

20.3 Exercises

1. Explain why the ITU (CCITT) Group 3 and Group 4 codes result in data expansion with halftone images. How does the JBIG standard address this problem?
2. How does the JPEG algorithm take advantage of spectral (color) redundancy?
3. How does the JPEG algorithm make use of perceptual (psychovisual) redundancy?

Bibliography

[Gon 92] R. C. Gonzalez and R. E. Woods, *Digital Image Processing*, Reading, MA: Addison-Wesley, 1992.

[Mul 85] K. T. Mullen, "The contrast sensitivity of human colour vision to red-green and blue-yellow chromatic gratings," *J. Physiol.*, 1985.

[Pen 93] W. B. Pennebaker and J. L. Mitchell, *JPEG: Still Image Data Compression Standard*, New York, NY: Van Nostrand Reinhold, 1993.

[Rab 91] M. Rabbani and P. W. Jones, *Digital Image Compression Techniques*, Bellingham, WA: SPIE Press, 1991.

[Wal 91] G. K. Wallace, "The JPEG still picture compression standard," *Commun. ACM*, vol. 34, pp. 30–44, 1991.

Chapter 21

VECTOR QUANTIZATION, SUBBAND CODING AND OTHER METHODS

This chapter is devoted to other promising image compression methods, including vector quantization, fractal compression, subband coding, wavelets, and the second-generation methods that have not been employed in the existing compression standards. Vector quantization, covered in Section 21.1, refers to quantization and coding of a block of pixels or transform coefficients at once, rather than element by element. Fractal compression, discussed in Section 21.2, can be considered as an extension of the vector quantization with a virtual codebook. In subband coding, which is presented in Section 21.3, the image is decomposed into several nonoverlapping frequency channels. The number of bits allocated for encoding each channel depends on the fidelity requirements of the particular application. The relationship between subband coding and wavelet transform coding is also discussed. Section 21.4 provides a brief introduction to the so-called second-generation methods, where the image is segmented into some primitives whose shape and texture are encoded. Our goal in this chapter is to introduce the essential concepts rather than implementational details.

21.1 Vector Quantization

Vector quantization (VQ), which refers to quantization of an array of samples into one of a finite number of array states, has been found to offer various advantages over scalar quantization of data for speech and image signals [Jay 84, Ger 92]. Because it can make effective use of the statistical dependency between the data samples to be quantized, VQ provides a lower distortion for a fixed number of reconstruction levels, or a lower number of reconstruction levels for a fixed distortion as compared to scalar quantization. Indeed, it can be shown that VQ approaches the rate-distortion bound as the dimension of the vectors (blocks of data) increases even if the data source is memoryless, i.e., consists of a sequence of independent random variables. However, this improved performance comes at the cost of increased computational complexity and storage requirements as will be seen below.

21.1.1 Structure of a Vector Quantizer

The first step in VQ is the decomposition of the image data to yield a set of vectors. In general, image data may refer to pixel intensities themselves or some appropriate transformation of them. Let $\mathbf{s} = [s_1, s_2, \ldots, s_N]^T$ denote an N-dimensional vector corresponding to a block of image data, which can be represented by a point in an N-dimensional space. In VQ, the N-dimensional space, consisting of 2^N possible blocks, is quantized into L regions $\mathcal{R}_i$, $i = 1, \ldots, L$, called Voronoi regions. The Voronoi regions are depicted in Figure 21.1 for the case $N = 2$. All vectors that fall into the region $\mathcal{R}_i$ are represented by a single vector, $\mathbf{r}_i = [r_{i1}, r_{i2}, \ldots, r_{iN}]^T$, called the codevector, which is analogous to a reconstruction level in scalar quantization. The set $\mathcal{C} = \{\mathbf{r}_1, \ldots, \mathbf{r}_L\}$ is called the *codebook*. The partitioning of the N-dimensional space into L Voronoi regions in such a way as to minimize the mean square quantization error is referred to as the codebook design problem.

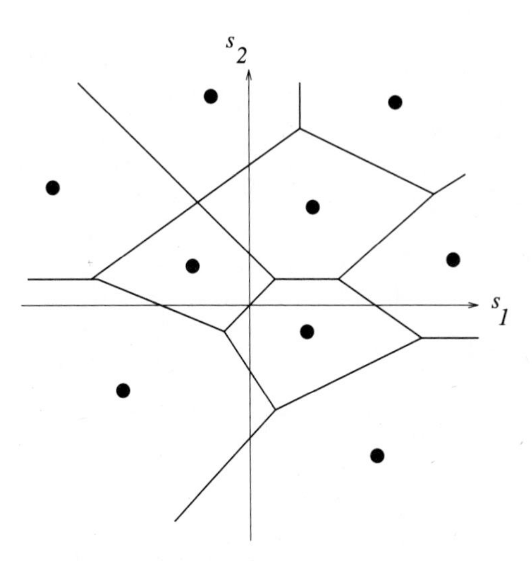

Figure 21.1: Partition of the 2-D space into Voronoi regions [Lim 90].

Given the codebook, a vector quantizer can be described by

$$\dot{\mathbf{s}} = VQ(\mathbf{s}) = \mathbf{r}_i, \quad \text{if } \mathbf{s} \in \mathcal{R}_i \tag{21.1}$$

That is, the encoder simply searches for the closest codevector from the codebook. The label of this reconstruction vector is then entropy coded for transmission or storage. The decoder then performs a table lookup using the label to find the respective reconstruction vector. The block diagrams of the encoder and decoder are shown in Figure 21.2. If we assume that the uncompressed image intensities are represented by 8 bits/pixel, and let L, the size of the codebook, be a power of 2, then the compression ratio (CR) can be defined as

$$\text{CR} = \frac{N \times 8}{\log_2 L} \tag{21.2}$$

Typical numbers are $N = 4 \times 4$ and $L = 1024$, so that $CR = 12.8$.

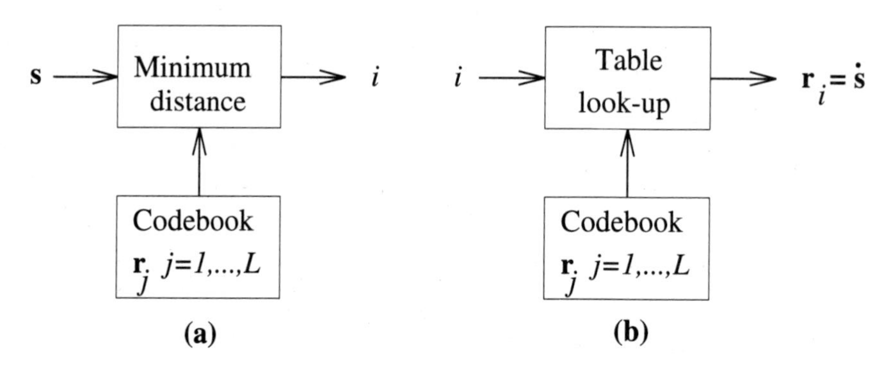

Figure 21.2: Block diagram of a VQ a) encoder and b) decoder.

The advantage of VQ over scalar quantization can best be appreciated by the following example.

Example: VQ vs. Scalar Quantization [Lim 90]

Assume that (s_1, s_2) are jointly uniformly distributed over the shaded region in Figure 21.3, so that the marginal pdfs of both s_1 and s_2 are uniform distributions between $(-a, a)$. As a result, if we quantize s_1 and s_2 individually and then plot the reconstruction levels for (s_1, s_2), we obtain four-levels, as shown in the bottom left. Observe that two of the four levels are wasted since scalar quantization of the components of the vector cannot make use of the apparent correlation between the components. In this case, VQ requires two reconstruction levels, shown in the bottom right, to provide the same distortion that scalar quantization yields with four levels.

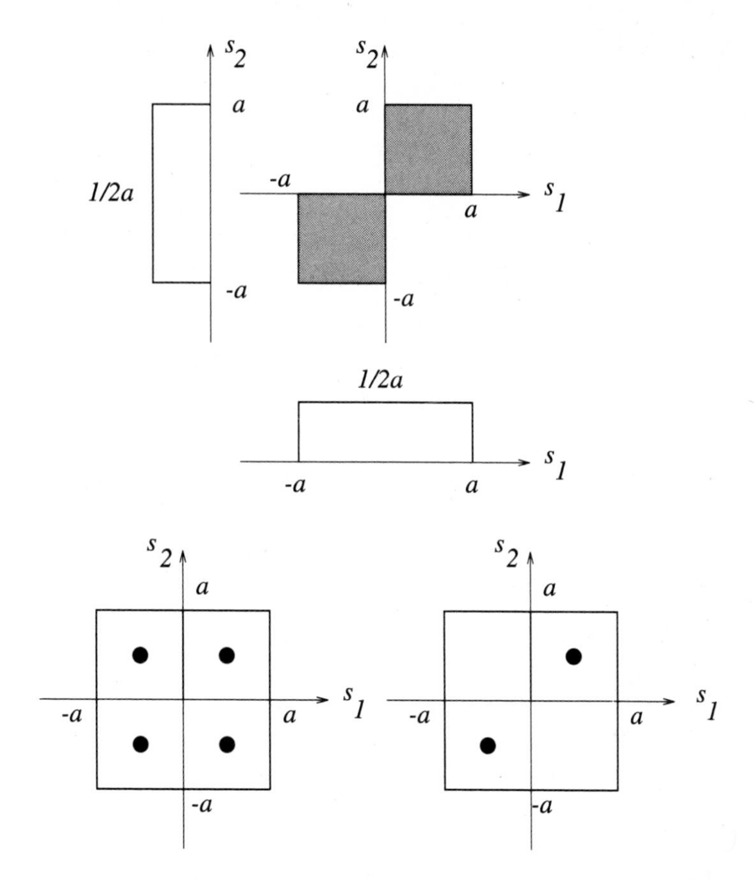

Figure 21.3: Comparison of vector vs. scalar quantization [Lim 90].

The above example clearly demonstrates that VQ is advantageous to scalar quantization of the components, when the components of the vector are correlated. It has also been shown that VQ carries a small advantage over scalar quantization of the components even when the components are statistically independent [Mak 85]. This advantage derives from the fact that scalar quantization of the components always provides rectangular-shaped decision regions in a hyperspace, whereas VQ can achieve almost spherical cells.

Another attractive feature of VQ is the simple structure of the decoder, which provides a distinct advantage in a broadcast environment where there is a single encoder but many decoders. However, the computational complexity of the encoder, which performs a search for the closest element in the codebook, generally prohibits its use in high-speed real-time applications. In the following, we first present an algorithm for the codebook design, and then discuss practical implementations of the VQ.

21.1.2 VQ Codebook Design

The better the codebook is suited for the types of images encoded, the better the performance of VQ. Because it is impractical to design a separate codebook for every image that is encoded, we usually design a single codebook for a class of images based on a training set of representative images.

The codebook design problem can then be stated as follows. Given a set of training data $\mathbf{s}$ and the size of the codebook L, choose $\mathcal{R}_i$ and $\mathbf{r}_i$, $i = 1, \ldots, L$ to minimize the overall distortion in the representation of the data. The distortion measure is typically taken as the mean square quantization error

$$d(\mathbf{s}, \mathbf{r}) = ||\mathbf{s} - \mathbf{r}||^2 \tag{21.3}$$

As a rule of thumb, for a good codebook, 100 training vectors are required per codevector, and typically $L < 5000$. A multidimensional extension of the K-means algorithm, commonly known as the LBG algorithm, was first proposed by Linde, Buzo, and Gray [Lin 80] for VQ codebook design.

The LBG algorithm is an extension of the scalar Lloyd-Max quantization (similar to the K-means algorithm presented in Appendix B) to the vector case, and consists of the following steps:

0) Initialization: Set $j = 0$. Choose an initial codebook $\mathcal{C}^{(j)} = \mathbf{r}_1, \ldots, \mathbf{r}_L$.

1) Determine the regions $\mathcal{R}_i$, $i = 1, \ldots, L$ using $\mathcal{C}^{(j)}$ and a minimum distance classifier.

2) If $\frac{D^{(j-1)} - D^{(j)}}{D^{(j)}} < \epsilon$, stop. Here $D^{(j)}$ denotes the overall distortion in data representation using $\mathcal{C}^{(j)}$.

3) Increment j. Update $\mathcal{C}^{(j)}$ by computing the centroids of each of $\mathcal{R}_i$, $i = 1, \ldots, L$.

We note that at each step of the algorithm the overall distortion decreases. However, the LBG algorithm, in general, converges to a local minimum of the distortion function. Therefore, the initial codebook is an important parameter of the algorithm. Several alternatives exist to initialize the codebook, including selecting the first L vectors formed from the training set, and splitting the vectors formed from the training set into L classes randomly and compute their centroids.

21.1.3 Practical VQ Implementations

The basic VQ structure and the LBG algorithm for codebook design discussed in the previous sections call for searching of the entire codebook to find the best matching vector for every input block, which is generally referred to as "the full-search VQ." The computational requirements of the full-search VQ algorithm can be reduced by employing suboptimal schemes, such as:

1) *Tree-Structured VQ*: Full search can be simplified to a tree search by imposing a tree structure on the codebook. In a binary tree, each node is split into two more nodes, whereas in a quad tree nodes are split to four from layer to layer. For

example, with a 10-layer binary tree we eventually have 1024 codevectors. The tree search reduces the computational complexity tremendously, since we need only 20 comparisons, as opposed to 1024 with a 10-layer binary tree. Mixed trees composed of more branches in the initial layers may avoid void nodes at the higher layers. The tree-structured codebook requires more storage to store all intermediate codevectors, which is well justified considering the reduction in computational complexity. We note that tree-structured VQ is suboptimal compared with full-search VQ, but in general yields acceptable performance.

2) *Product VQ*: Product VQ uses more than one codebook, which captures different aspects and properties of the vectors. The goal of a product VQ scheme is to reduce the variation between the input vectors by appropriate normalization or other preprocessing so that they be effectively represented by a smaller codebook. Most commonly used product codes are gain/shape VQ and mean-removed VQ. Shape is defined as the original input vector normalized by a gain factor, such as the energy of the block. In gain-shape VQ, first a unit energy "shape vector" is chosen from a shape codebook to best match a given block. Then a "gain constant" is selected from a scalar gain codebook. The optimal codevector is given by the product of the gain scalar and the shape vector. The indices of the shape and gain are stored or transmitted separately. In mean-removed VQ, instead of normalizing each block by a gain factor, the mean value of each block is subtracted and separately encoded using a scalar quantizer.

These VQ structures can be applied directly on the pixel intensities or to some other suitable representation of the image data, resulting in predictive-VQ, DCT-VQ, or subband-VQ (SB-VQ). In predictive VQ, the prediction error blocks rather than the actual pixel intensities are quantized similar to DPCM. Prediction error can be computed either via scalar prediction, as in DPCM, or via vector prediction, where the content of the present block is predicted as a linear combination of neighboring blocks. In DCT-VQ, the discrete cosine transform coefficients are vector quantized. The DCT coefficients having similar dynamic ranges can be arranged into smaller subvectors, with each subvector encoded using a separate codebook. In SB-VQ, the higher-frequency subbands are usually vector quantized, as will be discussed in the following.

21.2 Fractal Compression

An approach that is closely related to VQ is fractal coding, which exploits the self-similarity within images. Fractal objects are known to have high visual complexity but low information content. Fractal geometry became popular following the publication of the book *The Fractal Geometry of Nature* by Mandelbrot [Man 82], an IBM mathematician. It provided computer scientists with a powerful mathematical tool capable of generating realistic animations. Later, Barnsley published the book

Fractals Everywhere [Bar 88] in which he discussed the application of fractals to image compression using the theory of Iterated Function Systems (IFS). An IFS is a set of contractive transformations that map an image into smaller blocks. Often, affine transformations are used, which include translation, rotation, shear, and scaling. In addition, scaling and offset of gray level or color are also allowed. The representation of images by an IFS was formalized by the so-called Collage Theorem [Bar 88].

Because of practical difficulties with the initial approach, fractal-block coding (also known as a partitioned IFS), which utilizes self-similarity between pairs of blocks under a set of affine transformations, was later proposed [Jac 92, Lu 93]. Here, the transformations do not map from the whole image into the blocks, but from larger blocks to smaller blocks, to make use of local self-similarity. In Jacquin's notation [Jac 92], large blocks are called domain blocks and small blocks are called range blocks. It is necessary that every pixel of the original image belong to (at least) one range block. The pattern of range blocks is called the partitioning of the image. In fractal-block coding, an image is encoded in terms of a sequence of child-parent maps and a corresponding sequence of transformations chosen from a transformation codebook. In particular, the method consists of the following steps:

1. Partition the image into range blocks.
2. Form the set of domain blocks.
3. Choose a set of contractive transformations.
4. Select a distance metric between blocks.
5. Specify a search method for pairing range blocks to domain blocks.

The reconstruction process starts with a flat background. Observe that no domain block information is transmitted. The set of affine transformations is applied iteratively. The process converges, because all transformations are contractive, usually after a small number of iterations. Compression ratios in the range 25:1 to 50:1 are typical with fractal compression schemes. Note that because fractal images are scaleless, compression ratio must be calculated at the resolution of the input image.

The similarity of fractal compression to VQ can be seen as follows:

- In VQ, the codebook is computed and stored independent of the image being coded; in an IFS, strictly speaking, there is no codebook. However, the domain blocks can be viewed to define a "virtual codebook" which is adapted to each image.
- In VQ, domain blocks and range blocks are the same size, and the intensity within the domain blocks is copied directly; in an IFS, domain blocks are larger than range blocks, and each domain block undergoes a luminance scaling and offset in addition to an affine transformation.

21.3 Subband Coding

The basic idea of subband coding is to split the Fourier spectrum of an image into nonoverlapping bands, and then inverse transform each subband to obtain a set of bandpass images. Each bandpass image is then subsampled according to its frequency content, and encoded separately using a bitrate matched to the statistics and subjective visual perception requirements of the respective subband. The decoder reconstructs the original image by adding upsampled and appropriately filtered versions of the subimages. In the following, we first address the decomposition of the original image into subimages, and its reconstruction from the subimages, which is generally refered to as the analysis-synthesis problem. Methods for the encoding of the individual subimages are presented next.

21.3.1 Subband Decomposition

In order to understand the principles involved in the subband decomposition of images, we first consider the simple case of splitting a 1-D signal $s(n)$ into two equal-size subbands, called the lower and upper subbands. If we let the normalized sampling frequency set equal to $f = 1$, theoretically, we need an ideal low-pass filter with the passband $(0, 1/4)$ and an ideal high-pass filter with the passband $(1/4, 1/2)$ for the binary decomposition. The lower and upper subbands are inverse Fourier transformed, then subsampled by 2 to obtain the lower and upper subsignals. The block diagram of the analysis-synthesis filtering is shown in Figure 21.4.

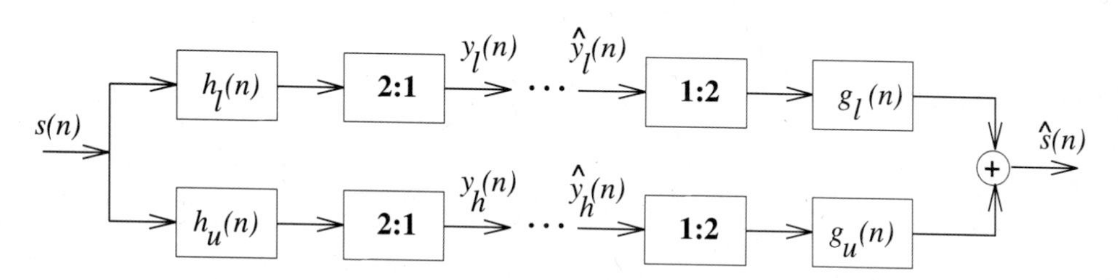

Figure 21.4: Block diagram of subband decomposition and reconstruction filtering.

Because ideal filters are unrealizable, we have to allow for an overlap between the passbands of the low-pass and high-pass filters in practice to avoid any frequency gaps. The frequency response of realizable binary subband filters are depicted in Figure 21.5. The fact that the fréquency response of the low-pass filter $H_l(f)$ extends into the band $(1/4, 1/2)$ and vice versa causes aliasing when each subband signal is decimated by 2.

Quadrature mirror filters (QMFs) which exhibit mirror symmetry about $f = 1/4$ have been introduced for alias-free analysis-synthesis filtering (assuming lossless coding). In QMF, the filters are designed in such a way that the aliasing introduced

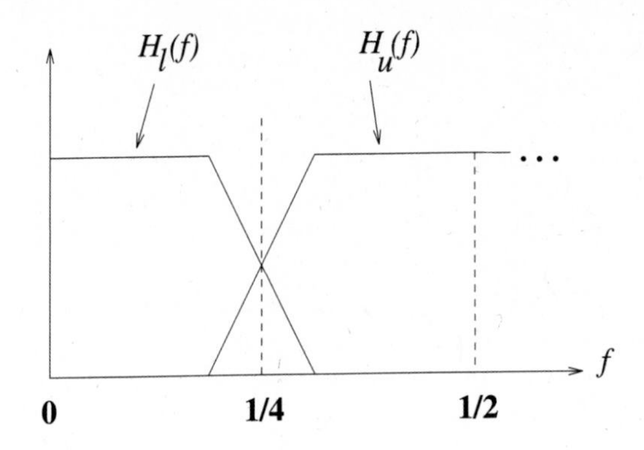

Figure 21.5: Frequency response of 1-D binary decomposition filters.

by the analysis part is exactly canceled by the synthesis part. In order to see how this is achieved, we express the Fourier transform of the subsignals $y_l(n)$ and $y_u(n)$, after subsampling by 2, as

$$\begin{aligned} Y_l(f) &= \frac{1}{2}[H_l(f/2)S(f/2) + H_l(-f/2)S(-f/2)] \\ Y_u(f) &= \frac{1}{2}[H_u(f/2)S(f/2) + H_u(-f/2)S(-f/2)] \end{aligned} \tag{21.4}$$

respectively. The reconstructed signal can be expressed as

$$\hat{S}(f) = G_l(f)\hat{Y}_l(2f) + G_u(f)\hat{Y}_u(2f) \tag{21.5}$$

Assuming lossless compression, $\hat{Y}_i(f) = Y_i(f)$, $i = l, u$, and substituting (21.4) into (21.5), we have

$$\begin{aligned} \hat{S}(f) &= \frac{1}{2}[H_l(f)G_l(f) + H_u(f)G_u(f)]S(f) + \\ & \frac{1}{2}[H_l(-f)G_l(f) + H_u(-f)G_u(f)]S(-f) \end{aligned} \tag{21.6}$$

The second term represents the aliased spectra and can be canceled by choosing

$$\begin{aligned} H_u(f) = H_l(-f) &= H_l(f + 1/2) \\ G_l(f) &= 2H_l(f) \\ G_u(f) &= -2H_u(f) \end{aligned} \tag{21.7}$$

which are the requirements for QMF. Note that the first condition is equivalent to

$$h_u(n) = (-1)^n h_l(n) \tag{21.8}$$

Substituting (21.7) into (21.5), the reconstructed signal becomes

$$\hat{S}(f) = [H_l^2(f) - H_u^2(f)]S(f) \tag{21.9}$$

It follows that the requirement for perfect reconstruction QMF also includes

$$H_l^2(f) - H_u^2(f) = 1, \quad \text{for all } f \tag{21.10}$$

It has been shown that the conditions (21.10) and (21.7) cannot be satisfied simultaneously by a linear phase FIR filter, except for a trivial two-tap filter. Thus, practical QMF allow for some amplitude distortion. The interested reader is referred to [Vai 87] and [Vet 89] for details of perfect reconstruction QMF design.

The 1-D decompositions can be extended to two dimensions by using separable filters, that is, splitting the image $s(n_1, n_2)$ first in the row and then in the column direction, or vice versa. Using a binary decomposition in each direction, we obtain four subbands called low (L) $y_L(n_1, n_2)$, horizontal (H) $y_H(n_1, n_2)$, vertical (V) $y_V(n_1, n_2)$, and diagonal (D) $y_D(n_1, n_2)$, corresponding to lower-lower, upper-lower, lower-upper, and upper-upper subbands, respectively. The decomposition can be continued by splitting all subbands or just the L-subband repetitively, as shown in Figure 21.6. Another alternative is a quad-tree-type decomposition where at each stage we split into four subbands followed by subsampling by four.

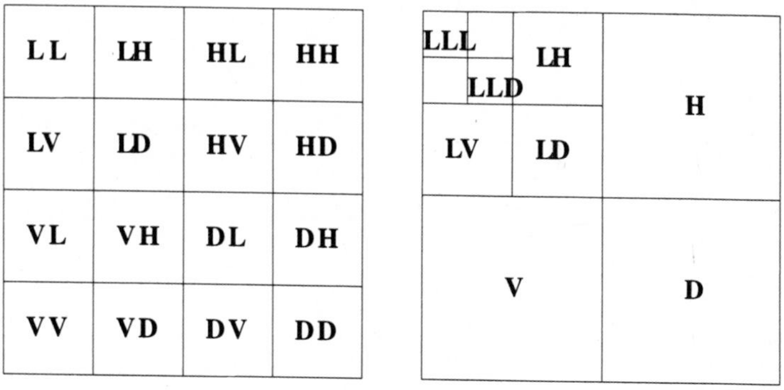

LL	LH	HL	HH
LV	LD	HV	HD
VL	VH	DL	DH
VV	VD	DV	DD

(a)

LLL
LH
LLD
H
LV
LD
V
D

(b)

Figure 21.6: Decomposition into frequency bands: a) Equal-size bands; b) binary-tree decomposition.

The total number of samples in all subimages $y_L(n_1, n_2)$, $y_V(n_1, n_2)$, $y_H(n_1, n_2)$, and $y_D(n_1, n_2)$ is the same as the number of samples in the input image $s(n_1, n_2)$ after subsampling of the subimages. Thus, the subband decomposition itself does not result in data compression or expansion. Observe that the image $y_L(n_1, n_2)$ corresponds to a low-resolution version of the image $s(n_1, n_2)$, while $y_D(n_1, n_2)$ contains the high-frequency detail information. Therefore, the subband decomposition is also known as a "multiscale" or "multiresolution" representation, and can be utilized in progressive transmission of images.

21.3.2 Coding of the Subbands

In subband coding, each subimage can be encoded using a method and bitrate most suitable to the statistics and the visual significance of that subimage. Experiments indicate that for typical images, more than 95% of the image energy resides in the (L) frequency band. The other bands contain high-frequency detail (residual) information which may be more coarsely represented. There are many approaches reported in the literature for the encoding of subimages. In most subband coding schemes, the (L) subimage is encoded by transform coding, DPCM or VQ. All other bands are either encoded by PCM or run-length coded after coarse thresholding. Some schemes apply VQ to high-frequency subimages using a separate codebook.

For quantization purposes, the statistics of the high-frequency subimages can be modeled by a generalized Gaussian distribution of the form

$$p(s) = Kexp\{-\eta(\alpha)|s|^{\alpha}\}, \quad \text{for} \quad \alpha > 0 \tag{21.11}$$

where

$$\eta(\alpha) = \left[\frac{\Gamma(3/\alpha)}{\Gamma(1/\alpha)}\right]^{1/2}$$

and for most images, $\alpha = 0.7$. The same pdf can also be used to model the statistics of the prediction error in DPCM or the DCT coefficients in the (L) subimage with

$$\begin{array}{ll} \alpha = 0.6 & \text{for the prediction residual, and} \\ \alpha = 0.6 - 1.0 & \text{for the DCT coefficients.} \end{array}$$

For heavy-tailed distributions, entropy coding results in significant reduction in bitrate over fixed-length coding.

21.3.3 Relationship to Transform Coding

One can easily establish a connection between subband coding and transform coding. In transform coding, recall that we compute the DCT of $N \times N$ image blocks. For example, if we order the DC coefficients of each block in an array, the resulting image corresponds to a realization of the lowest-resolution subband image with some choice of the filter parameters. Likewise, grouping intermediate frequency coefficients together would result in intermediate subband images (see Exercise 7). In the case of using a subband decomposition filter whose number of taps is larger than the decimation factor, the transform used in transform coding corresponds to the lapped orthogonal transform (LOT), where the transformation is applied to overlapping blocks of pixels. Since the LOT is primarily used to avoid blocking artifacts in transform coding, it should be of no surprise that blocking artifacts are not common in subband coding even at very low bitrates.

21.3.4 Relationship to Wavelet Transform Coding

Wavelet transform coding is closely related to subband coding in that it provides an alternative method for filter design. In subband coding, the decomposition of the image into subimages is usually accomplished by using perfect reconstruction QMF filters. The wavelet transform describes this multiresolution decomposition process in terms of expansion of an image onto a set of wavelet basis functions. The wavelet basis functions, unlike the basis function for DFT and DCT, are well localized in both space and frequency.

In 1-D wavelet analysis, there are two functions of interest: the mother wavelet ψ, and the scaling function ϕ. The dilated and translated versions of the scaling function, $\phi_{mn}(x) = 2^{-m/2}\phi(2^{-m}x - n)$, are associated with the design of the low-pass filter $H_l(f)$, and those of the mother wavelet are associated with the design of the high-pass filter $H_u(f)$. In 2-D wavelet analysis, we can define a separable scaling function,

$$\phi(x_1, x_2) = \phi(x_1)\phi(x_2) \tag{21.12}$$

and three directional wavelets,

$$\begin{aligned} \psi^H(x_1, x_2) &= \phi(x_1)\psi(x_2) \\ \psi^V(x_1, x_2) &= \psi(x_1)\phi(x_2) \\ \psi^D(x_1, x_2) &= \psi(x_1)\psi(x_2) \end{aligned} \tag{21.13}$$

where $\psi^H(x_1, x_2)$, $\psi^V(x_1, x_2)$, and $\psi^D(x_1, x_2)$ denote horizontal, vertical, and diagonal wavelets. These functions are then associated with a separable four-band decomposition of the image. The wavelet transform coefficients, then, correspond to pixels of the respective subimages. In most cases, the decomposition is carried out in multiple stages yielding a multiscale pyramidal decomposition. In this sense, the wavelet transform is also related to the well-known pyramid coding methods [Bur 83].

The wavelet-transform-based filters possess some regularity properties that not all QMF filters have, which give them improved coding performance over QMF filters with the same number of taps. For example, Antonini *et al.* [Ant 92] report that they can achieve a performance close to that Woods and O'Neil [Woo 86] by using only 7- and 9-tap filters, whereas the latter uses 32-tap Johnston filters. However, some more recent QMF designs may yield performance similar to that of the wavelet filters with the same number of taps [Ade 90].

21.4 Second-Generation Coding Methods

The compression methods described so far are generally known as waveform coding techniques because they operate on individual pixels or blocks of pixels based on some statistical image models. Efforts to utilize more complicated structural image models and/or models of the human visual system in an attempt to achieve higher compression ratios lead to a new class of coding methods which are referred to as "second generation coding techniques" [Kun 85]. Second-generation coding techniques can be grouped into two classes: local-operator-based techniques, and contour/texture-oriented techniques.

Local-operator-based techniques include pyramidal coding and anisotropic non-stationary predictive coding, which are discussed in [Kun 85]. Pyramid and subband coding have also been classified as "second-generation," because they are based on models that resemble the operation of the human visual system. For example, in subband coding, we use parallel bandpass filters, which can be chosen to resemble the direction-sensitive cells in the human visual system.

Contour/texture-oriented techniques attempt to describe an image in terms of structural primitives such as contours and textures. They segment the image into textured regions surrounded by contours such that the contours correspond, as much as possible, to those of the objects in the image. Two commonly employed methods are the region-growing method and the directional decomposition method. In these methods, the contour and texture information are encoded separately [Kun 85, Kun 87]. Contour/texture-based methods are capable of achieving very high compression ratios in comparison to classical waveform-based methods. Indeed, they form the basis of object-based very-low-bitrate video compression schemes, which are described in Chapter 24.

21.5 Exercises

1. Suppose we employ a mean-removed VQ, with 4×4 blocks, where we use 8 bits to encode the mean of each block, and the residual codebook size is $L = 512$. Calculate the compression ratio, assuming that we have encoded the indices k using fixed-length coding.

2. Show that $D^{(j-1)}$ is always greater than or equal to $D^{(j)}$ in the LBG algorithm.

3. Suppose that, in Figure 21.3, you are allowed to first transform the vectors by a linear transformation and then use scalar quantizers. (This is the principle of transform coding.)

   a) Describe a transformation that can match the performance of the VQ by using a single scalar quantizer. Explain clearly.

   b) Do you think that the difference in performance between scalar versus vector quantization would be more pronounced in waveform coding or transform coding? Explain why.

4. Elaborate on the similarities between VQ and fractal coding.

5. Verify that the requirements for perfect reconstruction are given by (21.7) and (21.10).

6. Show that a perfect reconstruction filter cannot satisfy all of the following requirements simultaneously:
   i) it is symmetric,
   ii) it is FIR, and
   iii) analysis and synthesis filters are equal-length.

7. Elaborate on the relationship between the subband coding and the DCT coding. In particular, consider a 64×64 image. Suppose we partition the image into 8×8 blocks, and compute their DCT.

   a) Show that we can construct 64 subband images, each of which is 8×8, by an ordering of the DCT coefficients.

   b) Discuss the relationship between DPCM encoding of the DC coefficients in JPEG, and DPCM encoding of the lowest subband in this case.

Bibliography

[Ade 90] E. H. Adelson and E. Simoncelli, "Non-separable extensions of quadrature mirror filters to multiple dimensions," *Proc. IEEE*, vol. 78, Apr. 1990.

[Ant 92] M. Antonini, M. Barlaud, P. Mathieu, and I. Daubechies, "Image coding using wavelet transform," *IEEE Trans. Image Proc.*, vol. 1, pp. 205–220, Apr. 1992.

[Bar 88] M. F. Barnsley, *Fractals Everywhere*, New York, NY: Academic Press, 1988.

[Bur 83] P. Burt and E. Adelson, "The Laplacian pyramid as a compact image code," *IEEE Trans. Comm.*, vol. 31, pp. 482–540, 1983.

[Ger 82] A. Gersho, "On the structure of vector quantizer," *IEEE Trans. Info. The.*, vol. IT-28, pp. 157-166, Mar. 1982.

[Ger 92] A. Gersho and R. Gray, *Vector Quantization and Signal Compression*, Norwell, MA: Kluwer, 1992.

[Gha 88] H. Gharavi and A. Tabatabai, "Subband coding of monochrome and color images," *IEEE Trans. Circ. and Syst.*, vol. 35, pp. 207–214, Feb. 1988.

[Gra 84] R. M. Gray, "Vector quantization," *IEEE ASSP Magazine*, vol. 1, pp. 4–29, April 1984.

[Jac 92] A. E. Jacquin, "Image coding based on a fractal theory of iterated contractive image transformations," *IEEE Trans. Image Proc.*, vol. 1, pp. 18–30, Jan. 1992.

[Jay 84] N. S. Jayant and P. Noll, *Digital Coding of Waveforms - Principles and Applications to Speech and Video*, Englewood Cliffs, NJ: Prentice Hall, 1984.

[Kun 85] M. Kunt, A. Ikonomopoulos and M. Kocher, "Second-generation image coding techniques," *Proc. IEEE*, vol. 73, 549–574, 1985.

[Kun 87] M. Kunt, M. Benard, R. Leonardi, "Recent results in high-compression image coding," *IEEE Trans. Circ. Systems*, Vol. 34, pp. 1306–1336, 1987.

[Lim 90] J. S. Lim, *Two-Dimensional Signal and Image Processing*, Englewood Cliffs, NJ: Prentice Hall, 1990.

[Lin 80] Y. Linde, A. Buzo, and R. M. Gray, "An algorithm for vector quantizer design," *IEEE Trans. Comm.*, vol. 28, pp. 84–89, Jan. 1980.

[Lu 93] G. Lu, "Fractal image compression," *Signal Processing: Image Comm.*, vol. 5, pp. 327–343, 1993.

[Mak 85] J. Makhoul, S. Roucos, and H. Gish, "Vector quantization in speech coding," *Proc. IEEE*, vol. 73, pp. 1551–1588, Nov. 1985.

[Man 82] B. Mandelbrot, *The Fractal Geometry of Nature*, Freeman, 1982.

[Nas 88] N. M. Nasrabadi and R. A. King, "Image coding using vector quantization: A review," *IEEE Trans. Comm*, vol. 36, pp. 957–971, Aug. 1988.

[Rab 91] M. Rabbani and P. W. Jones, *Digital Image Compression Techniques*, Bellingham, WA: SPIE Press, 1991.

[Vai 87] P. P. Vaidyanathan, "Quadrature mirror filter banks, M-band extensions, and perfect reconstruction technique," *IEEE Acoust. Speech Sign. Proc. Magazine*, vol. 4, pp. 4–20, 1987.

[Vet 84] M. Vetterli, "Multidimensional sub-band coding: Some theory and algorithms, *Signal Processing*, vol. 6, pp. 97–112, 1984.

[Vet 89] M. Vetterli and D. LeGall, "Perfect reconstruction FIR filter banks: Some properties and factorizations," *IEEE Trans. Acoust. Speech Sign. Proc,*, vol. 37, pp. 1057–1071, 1989.

[Wes 88] P. Westerink and J. W. Woods, "Subband coding of images using vector quantization," *IEEE Trans. Comm.*, vol. 36, pp. 713–719, June 1988.

[Woo 86] J. W. Woods and S. D. O'Neil, "Subband coding of images," *IEEE Trans. Acoust. Speech and Sign. Proc.*, vol. 34, pp. 1278–1288, Oct. 1986.

Chapter 22

INTERFRAME COMPRESSION METHODS

Video compression is a key enabling technology for desktop digital video. Without compression, digital transmission of an NTSC color video, with 720 pixels $\times$ 480 lines, 8 bits/pixel per color, and 30 frames/sec, requires a transmission capacity of 248 Mbps. Likewise, an HDTV color video, with 1920 pixels $\times$ 1080 lines, 8 bits/pixel per color, and 30 frames/sec, needs a channel capacity of 1.5 Gbps. A super-35 format motion picture is usually digitized with 4096 pixels $\times$ 3112 lines and 10 bits/pixel per color. At a rate of 24 frames/sec, one second of a color movie requires approximately 9 Gbits (1.15 Gbytes) storage space. These data rates suggest that a CD with a storage capacity of about 5 Gbits can hold, without compression, approximately 20 sec of NTSC video, 3 sec of HDTV video, and one-half second of a movie. A typical data transfer rate for a CD-ROM device is about 1.5 Mbps (although faster devices that can transfer up to 4 Mbps are appearing in the market). Then full-motion NTSC quality video can be played back from a CD (1.2 Mbps for video and 0.3 Mbps for stereo audio) with 200:1 compression of the video signal. At 200:1 compression, a single CD can hold 3400 sec or about one hour of video. The HDTV signal needs be broadcast over a 6 MHz channel which can support about 20 Mbps. As a result, a compression ratio of 75:1 is required for broadcasting HDTV signals. For transmission using fiber optic networks or satellite links, similar compression is needed to transmit multiple channels.

An elementary approach to video compression would be to employ any of the still-frame compression techniques discussed in Chapters 18-21 on a frame by frame basis. However, the compression that can be achieved by such an approach will be limited because each frame is treated as an independent image. Interframe compression methods exploit the temporal redundancies due to similarity between neighboring frames, in addition to the spatial, spectral, and pyschovisual redundancies to provide superior compression efficiency. Note, however, that some application specific requirements, such as random access capability at all frames, may dictate the use of intraframe compression rather than interframe methods in some cases. In general, interframe compression methods take advantage of temporal redundancies through i) 3-D waveform coding strategies, which are based on statistical signal models, ii) motion-compensated (MC) coding strategies, which use elementary motion models, or iii) object/knowledge based coding strategies, which utilize more sophisticated scene models. Three-dimensional waveform coding strategy is discussed in Section 22.1. Various motion-compensated compression schemes are presented in Section 22.2. Note that all international video compression standards, covered in Chapter 23, utilize the MC transform coding strategy. Section 22.3 provides an overview of model-based coding methods, which are studied in detail in Chapter 24.

22.1 Three-Dimensional Waveform Coding

The simplest way to extend still-frame image compression methods to interframe video compression is to consider 3-D waveform coding schemes, which include 3-D transform and 3-D subband coding. These methods exploit spatio-temporal redundancies in a video source through statistical signal models.

22.1.1 3-D Transform Coding

Three-dimensional DCT coding is a straightforward extension of the 2-D DCT coding method, where the video is divided into $M \times N \times J$ blocks (M, N, and J denote the horizontal, vertical, and temporal dimensions of the block, respectively). The transform coefficients are then quantized subject to zonal coding or threshold coding, and encoded similar to 2-D transform coding of still-frame images. Since the DCT coefficients are closely related to the frequency content of the blocks, for temporally stationary blocks the DCT coefficients in the temporal direction will be close to zero, and will be truncated in a threshold coding scheme. For most blocks, the DCT coefficients will be packed towards the low spatial and temporal frequency zone. The 3-D DCT coding scheme has the advantage that it does not require a separate motion estimation step. However, it requires J frame stores both in the transmitter and receiver. Therefore, J is typically chosen as 2 or 4, to allow for practical hardware implementations. Observe that random access to video is possible once for every J frames, as shown in Figure 22.1.

A related 3-D waveform coding approach is the hybrid transform/DPCM coding method, which has been proposed to overcome the multiple frame-store requirement [Roe 77, Nat 77]. This is an extension of the 2-D hybrid DCT/DPCM coding concept proposed by Habibi [Hab 74] to 3-D, where a 2-D orthogonal transform is performed on each spatial block within a given frame. A bank of parallel DPCM coders, each tuned to the statistics of a specific DCT coefficient, is then applied to

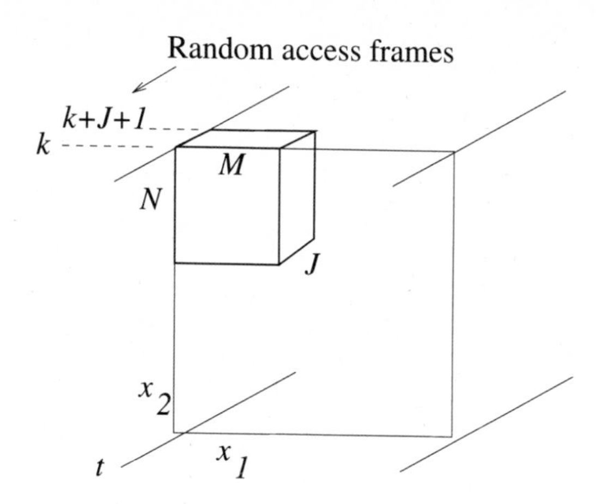

Figure 22.1: Three-dimensional transform coding.

the transform coefficients in the temporal direction. Thus, the differences in the respective DCT coefficients in the temporal direction are quantized and encoded, which eliminates the need for multiple frame-stores. This scheme generally requires adaptation of the DPCM quantizers to the temporal statistics of the 2-D DCT coefficients for results comparable to that of the 3-D DCT coding [Roe 77]. Note that neither 3-D DCT coding nor hybrid DCT/DPCM coding has been widely used in practical applications.

22.1.2 3-D Subband Coding

3-D subband coding, an extension of 2-D subband coding, has recently received increased attention [Vet 92, Bos 92, Luo 94] motivated by the following considerations: i) it is almost always free from blocking artifacts, which is a common problem with 3-D DCT and MC/DCT coding methods, especially at low bitrates, ii) unlike MC compression methods, it does not require a separate motion estimation stage, and iii) it is inherently scalable, both spatially and temporally. Scalability, which refers to availability of digital video at various spatial and temporal resolutions without having to decompress the entire bitstream, has become an important factor in recent years due to the growing need for storage and transmission of digital video that is comformable with various format standards. For example, standard TV and high-definition TV differ only in spatial resolution, whereas videophone systems offer lower spatial and temporal resolution.

In 3-D subband coding, the video is decomposed into various properly subsampled component video signals, ranging from a low spatial and temporal resolution component to various higher-frequency detail signal components. These various component video signals are encoded independently using algorithms adapted to the statistical and pyschovisual properties of the respective spatio-temporal frequency bands. Compression is achieved by appropriate quantization of the various components and entropy coding of the quantized values. Higher-resolution video, in both spatial and temporal coordinates, can be recovered by combining the decompressed low-resolution version with the decompressed detail components. Most 3-D subband decomposition schemes utilize 2- or 4-frame temporal blocks at a time due to practical implementation considerations.

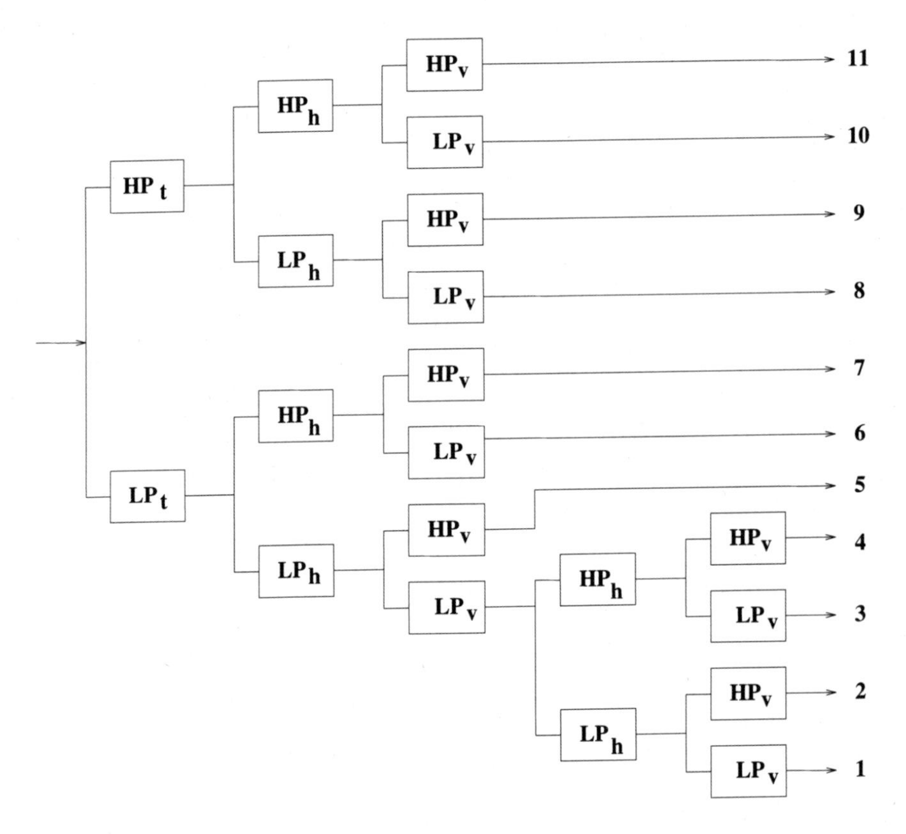

Figure 22.2: A typical 3-D subband decomposition.

An 11-band 3-D subband decomposition is illustrated in Figure 22.2. Typically, the temporal decomposition is based on a simple 2-tap Haar filterbank [Luo 94], which in the case of two frame blocks gives the average and the difference of the two frames for the low-pass (LP) and high-pass (HP) temporal components, respectively. This choice minimizes the number of frame-stores needed as well as the computational burden for the temporal decomposition. In the second stage, both the low and high temporal subbands are decomposed into low and high horizontal subbands, respectively. In the next stage, each of these bands are decomposed into low and high vertical subbands, as depicted in Figure 22.2. Subsequently, the low (temporal)-low (horizontal)-low (vertical) band is further decomposed into four spatial subbands to yield the 11-band decomposition. Note that longer filters can be applied for the spatial (horizontal and vertical) decompositions, since these filters can be operated in parallel and do not affect the frame-store requirements. To this effect, Luo *et al.* report using wavelet filterbanks for the spatial decompositions.

The resulting component video signals can be subsampled consistent with the spatio-temporal frequency characteristics of each band, which is illustrated in Figure 22.3. In this figure, the left- and right-hand side templates correspond to the low- and high-temporal-frequency components, respectively. For example, the component 1, which is subsampled by a factor of 2 in time and by a factor of 4 in each spatial direction, represents a spatio-temporally blurred version of a 2-frame block. As a result, a time-sequence composed of the component 1 for all 2 frame blocks constitutes a low-resolution (both spatially and temporally) version of the original video. Likewise, time sequences composed of the other components (e.g., 2-11) constitute auxilary video signals containing high-frequency detail information needed to reconstruct the original video from the low-resolution video. The reader is referred to [Vet 92, Bos 92, Luo 94] for various approaches offered to compress the individual component signals.

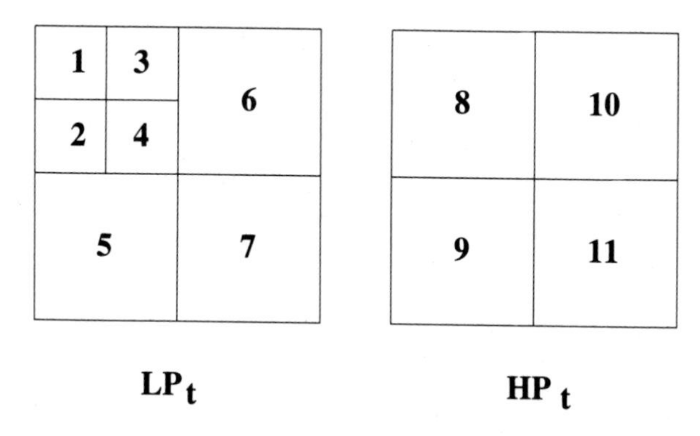

Figure 22.3: Representation of 3-D subband video.

The basic approach of 3-D DCT and subband coding is quite different from that of motion-compensated (MC) coding, which is presented next. MC techniques characterize the temporal correlation in a video by means of motion vectors rather than through the respective transform coefficients.

22.2 Motion-Compensated Waveform Coding

One of the earliest approaches in interframe image compression has been the so-called conditional replenishment technique, which is based on segmenting each frame into "changed" and "unchanged" regions with respect to the previous frame. Then information about the addresses and intensities of the pixels in the changed region would be transmitted using a bitrate that is matched to the channel rate. Intensities in the changed region are encoded by means of a DPCM method. Since the amount of changed information varies from frame to frame, the information to be transmitted needs to be buffered, and the quantization scheme is regulated according to the fullness of the buffer. A review of conditional replenishment algorithms can be found in [Has 72]. Note that conditional replenishment is a motion-detection based algorithm rather than a motion-compensated algorithm, since it does not require explicit estimation of the motion vectors.

The conditional replenishment was later extended to motion-compensated (MC) DPCM by encoding displaced frame difference values for those pixels in the changed area with respect to the previous frame [Has 78]. MC-DPCM yields more efficient compression provided that we can accurately estimate the displacement vectors. Most commonly used motion estimation methods in MC compression fall into pixel recursive algorithms or block-matching algorithms. Since these motion estimation methods were covered in detail in Chapters 5-8, here we assume that the motion vectors are known, and deal only with the encoding of the differential signal, also known as the temporal prediction error. We have already mentioned, in still-frame image compression, that transform coding and vector quantization both provide better compression efficiency compared with scalar DPCM. Thus, the next two sections are devoted to transform coding and vector quantization of the temporal prediction error.

22.2.1 MC Transform Coding

In MC transform coding, the temporal prediction error is 2-D transform coded by segmenting the displaced frame difference into blocks, and encoding the DCT coefficients of each block as in 2-D DCT coding. The temporal prediction aims at minimizing the temporal redundancy, while the DCT encoding makes use of the spatial redundancy in the prediction error. MC transform coding algorithms feature several modes to incorporate both progressive and interlaced inputs. These include intrafield, intraframe, and interfield and interframe prediction with or without motion compensation. In the two intra modes, the DCT blocks are formed by actual pixel intensities from a single field or from an entire frame. In the interfield and interframe modes the prediction is based on the previous field or frame, respectively. The MC transform coding is the basis of several world standards for video compression that are summarized in Table 22.1, with the possible exception of MPEG-4. The details of the MC transform coding will be covered in the next chapter where we discuss these world standards.

Table 22.1: World standards for video compression.

Standard	Description
H.261	ITU (CCITT) Expert Group on Visual Telephony; developed for ISDN applications at $p \times 64$ kbps ($p = 1, 2, \ldots, 30$); standardized in December 1990. *Application:* Videoconferencing and videophone using ISDN
MPEG	ISO Moving Picture Expert Group; PHASE 1: Storage and retrieval of digital video + audio at about 1.5 Mbps; draft finalized in June 1992. *Application:* Storage on CD-ROM and hard disk PHASE 2: Storage and retrieval of digital video + audio at about 10-20 Mbps; tech. spec. frozen in Mar. 1993. *Application:* Higher definition digital video including HDTV PHASE 4: Low-bitrate digital video + audio at below 64 kbps; embedded functionalities; just about to start. *Applications:* Videophone; database queries

The basic MC transform coding scheme employs block-based motion estimation and compensation. It has been argued that block motion models are not realistic for most image sequences, since moving objects hardly ever manifest themselves as rectangular blocks in the image plane. Recently, improved motion-compensation schemes, where more than a single motion vector per block are used without increasing the number of motion vectors to be transmitted, have been proposed [Orc 93] to circumvent this problem.

22.2.2 MC Vector Quantization

In MC-VQ the prediction error signal is encoded by vector quantization. The choice of using transform coding versus VQ for encoding the prediction error depends on several factors, including encoder and decoder complexity, cost of encoder and decoder, target bitrate, and real-time operation requirement. In MC transform coding, the complexity of the encoder and decoder is more or less symmetric, whereas in MC-VQ the encoder is significantly more complex than the decoder. Although the MC-VQ approach is capable of providing lower bitrates, a significant advantage the MC transform coding scheme enjoys is the availability of special-purpose hardware which enables encoding and decoding in real-time.

A typical MC-VQ scheme partitions pixels in the current frame into 4×4 or 8×8 blocks. First, a motion-detection test is applied to each block. The outcome of the motion-detection test determines one of three options: do nothing if no motion is detected, interframe VQ if motion is detected and the motion vector can be estimated with sufficient accuracy, or intraframe VQ if motion is detected but cannot be estimated within an acceptable accuracy limit. In the interframe VQ mode, the estimated motion vector, one for each block, and the displaced block difference are transmitted. The motion vectors for each block are generally DPCM encoded. The displaced block difference is VQ encoded using an interframe codebook. In the intraframe mode, the actual pixel intensities are VQ encoded using an intraframe codebook. The reader is referred to the literature for implementational details [Che 92, Mer 93]. It is also possible to replenish the intraframe and interframe codebooks at regular intervals to adapt the codebooks to the changing image and prediction error statistics [Gol 86].

22.2.3 MC Subband Coding

In MC subband coding the frame prediction error, also called the residual, is decomposed into 2-D subbands. The residual signal energy is typically unevenly distributed among the subbands which facilitates compression by simply truncating some of the subbands. Schemes have been proposed where the subbands are compressed using VQ [Mer 93]. Woods and Naveen [Woo 89] proposed integrating subband coding with hierarchical motion estimation. A motion-compensated subband decomposition technique which employs VQ has also been proposed [Nic 93].

22.3 Model-Based Coding

Three-dimensional and motion-compensated waveform coding provide satisfactory results with CIF images at bitrates over 1.5 Mbps. However, the quality of images that these techniques offer at very low bitrates, e.g. 10 kbps for videophone over existing telephone networks, is deemed unacceptable. In particular, decompressed images obtained by MC/DCT type methods generally suffer from blocking artifacts, which originate from the assumed translational block-motion model. To this effect, a variety of new motion-compensated coding schemes, generally known as model-based or analysis-synthesis coders, which are based on more realistic structural motion models, have recently been proposed for very-low-bitrate applications. An analysis-synthesis encoder can be characterized by the following steps:

- Image analysis: The frame to be encoded (present frame) is segmented into individually moving objects using the knowledge of the previously coded frame(s). Each object in the present frame is characterized by a set of shape (contour) and motion parameters.
- Image synthesis: The present frame is synthesized based on the estimated shape and motion parameters and the knowledge of the previously coded frame(s). The difference between the actual and the synthesized frame provides a measure of model compliance. Those regions where this difference is more than a threshold are labeled as "model failure" regions.
- Coding: The shape, motion and color (for the model failure regions) parameters are separately entropy encoded and transmitted.

Table 22.2: Overview of source models.

Method	Source Model	Encoded Information
MC/DCT	Translatory blocks	Motion vectors and color of blocks
Object-based	Moving unknown 2-D or 3-D objects	Shape, motion, and color of each moving object
Knowledge-based	Moving known objects	Shape, motion, and color of the known object
Semantic	Facial expressions	Action units

The image analysis and synthesis steps usually make extensive use of sophisticated computer vision and computer graphics tools, such as 3-D motion and structure estimation, contour modeling, and texture mapping [For 89]. Analysis-synthesis coding includes object-based, knowledge-based, and semantic coding approaches. Table 22.2 provides an overview of the source models used by various motion-compensated coding schemes, including the MC/DCT and several analysis-synthesis coding schemes. Note that MC/DCT coding, which forms the basis of several international video compression standards such as H.261, MPEG 1 and 2, is based on the overly simplistic source model of 2-D translatory blocks. Object-based, knowledge-based, and semantic coding methods are introduced in the following.

22.3.1 Object-Based Coding

Object-based coding (OBC) methods are based on structural image models derived from 3-D representation of a scene in terms of moving *unknown* (arbitrary) objects. The unknown objects can be treated as [Mus 89]: i) 2-D rigid or flexible objects with 2-D motion, ii) 2-D rigid objects with 3-D motion (affine or perspective mappings), or iii) 3-D rigid or flexible objects with 3-D motion.

Table 22.3: Expected bits per CIF frame (352 × 288) for different source models. r_s denotes bits/pixel for encoding the color information [Mus 93] (©1993 IEEE).

Source Model	Motion	Shape	Color
2-D rigid object 3-D motion	600	1300	15000 r_s
2-D flexible object 2-D motion	1100	900	4000 r_s
3-D rigid object 3-D motion	200	1640	4000 r_s

Efficiency of the source model can be measured by the data rate required for encoding the model parameters. The average bitrates for these object models for encoding a typical CIF format (386×288) frame are given in Table 22.3 [Mus 93]. It can be seen that the compression ratio increases as the complexity of the model increases.

22.3.2 Knowledge-Based and Semantic Coding

Knowledge-based coding deals with cases where we have some *a priori* information about the content of the video, since dealing with unknown arbitrary objects is, in general, quite difficult. For example, in videophone applications, head-and-shoulders-type images are common. Knowledge-based coding of facial image sequences using model-based techniques requires a common 3-D flexible wireframe model of the speaker's face to be present at both the receiver and transmitter sides. 3-D motion and structure estimation techniques are employed at the transmitter to track the global motion of the wireframe model and the changes in its structure from frame to frame. The estimated motion and structure (depth) parameters along with changing texture information are sent and used to synthesize the next frame in the receiver side. The *knowledge-based* approach can be summarized by the following source model and algorithm.

Source Model:

A moving known object which is characterized by
- A generic wireframe model to describe the shape of the object, e.g., the head and shoulders
- 3-D global motion parameters, e.g., to track the rotation and translation of the head from frame to frame
- 3-D local motion (deformation) parameters, e.g., to account for the motion of the eyes, lips, etc. due to facial expressions
- Color parameters to describe the model failure areas

Algorithm:

1. Detection of the boundaries of the object in the initial frame.
2. Adaptation of the generic wireframe model to the particular object, by proper scaling in the x_1, x_2, and x_3 directions.
3. Estimation of the 3-D global and local motion parameters, using 3-D motion and structure estimation methods.
4. Synthesis of the next frame.
5. Determination of model failure areas.
6. Coding of motion and color parameters.

A detailed presentation of knowledge-based coding schemes can be found in Chapter 24. While the knowledge-based scheme is generally successful in tracking the global motion of the head, the estimation of the local motion of the eyes, lips, and so on due to facial expressions is usually difficult. Errors in local motion estimation generally result in several model failure regions. *Semantic coding* is a subset of knowledge-based methods, which attempts to model the local motion in terms of a set of facial action units. After compensating for the global motion, the encoder estimates a combination of action units that best fits a given facial expression and encodes their indices along with the global motion and shape parameters. Semantic coding can then be summarized by the following source model and algorithm.

Source Model:

Facial expressions of a head, described by
- A wireframe model
- A limited set of action units

Coding Algorithm:

1. Detection of known object boundaries.
2. Adaptation of the wireframe model.
3. Estimation/compensation of the global motion.
4. Estimation of action units (AUs).
5. Synthesis of next frame.
6. Determination of model failure areas.
7. Coding of AUs, global motion, and color parameters.

In knowledge-based and semantic coding, we can effectively decrease the bitrate by conveying the information in a head-and-shoulders-type sequence in terms of a set of global motion and facial-action-unit parameters. Clearly, knowledge-based coding poses a trade-off between compression efficiency and the generality of the compression algorithm.

22.4 Exercises

1. Discuss the relative advantages and disadvantages of 3-D waveform coding versus motion-compensated coding methods.

2. It is well-known that the MC-DCT approach suffers from blocking artifacts, especially at low bitrates. This is mainly due to insufficiency of the assumed translational block motion model. There are, in general, two approaches to address this problem: i) improving the motion model by using spatial transformations (generalized block motion), and ii) 3-D subband coding. Evaluate the relative merits and demerits of both approaches, especially in terms of motion artifacts versus spatio-temporal blurring.

3. The generalized block motion model, coupled with a motion segmentation algorithm, results in a 2-D object-based image compression scheme. Compare 2-D versus 3-D object-based methods for image compression. What advantages do 3-D object-based methods offer over 2-D methods, if any?

Bibliography

[Bos 92] F. Bosveld, R. L. Lagendijk, and J. Biemond, "Compatible spatio-temporal subband encoding of HDTV," *Signal Proc.*, vol. 28, pp. 271–290, Sep. 1992.

[Che 92] H.-H. Chen, Y.-S. Chen, and W.-H. Hsu, "Low-rate sequence image coding via vector quantization," *Signal Proc.*, vol. 26, pp. 265–283, 1992.

[For 89] R. Forchheimer and T. Kronander, "Image coding-from waveforms to animation," *IEEE Trans. Acoust. Speech Sign. Proc.*, vol. 37, no. 12, pp. 2008–2023, Dec. 1989.

[Gha 91] H. Gharavi, "Subband coding of video signals," in *Subband Image Coding*, J. W. Woods, ed., Norwell, MA: Kluwer, 1991.

[Gol 86] M. Goldberg and H. Sun, "Image sequence coding using vector quantization," *IEEE Trans. Comm.*, vol. 34, pp. 703–710, July 1986.

[Hab 74] A. Habibi, "Hybrid coding of pictorial data," *IEEE Trans. Comm.*, vol. 22, no. 5, pp. 614–624, May 1974.

[Hab 81] A. Habibi, "An adaptive strategy for hybrid image coding," *IEEE Trans. Comm.*, vol. 29, no. 12, pp. 1736–1740, Dec. 1981.

[Has 72] B. G. Haskell, F. W. Mounts, and J. C. Candy, "Interframe coding of videotelephone pictures," *Proc. IEEE*, vol. 60, pp. 792–800, July 1972.

[Has 78] B. G. Haskell, "Frame replenishment coding of Television," in *Image Transmission Techniques,* W. K. Pratt, ed., Academic Press, 1978.

[Hot 90] M. Hotter, "Object oriented analysis-synthesis coding based on moving two-dimensional objects," *Signal Proc: Image Comm.*, vol. 2, no. 4, pp. 409–428, Dec 1990.

[Luo 94] J. Luo, C. W. Chen, K. J. Parker, and T. S. Huang, "Three dimensional subband video analysis and synthesis with adaptive clustering in high-frequency subbands," *Proc. IEEE Int. Conf. Image Processing*, Austin, TX, Nov. 1994.

[Mer 93] R. M. Mersereau, M. J. T. Smith, C. S. Kim, F. Kossentini, and K. K. Truong, "Vector quantization for video data compression," in *Motion Analysis and Image Sequence Processing*, M. I. Sezan and R. L. Lagendijk, eds., Norwell, MA: Kluwer, 1993.

[Mus 89] H. G. Musmann, M. Hotter, and J. Ostermann, "Object-oriented analysis synthesis coding of moving images," *Signal Proc: Image Comm.*, vol. 1, pp. 117–138, 1989.

[Mus 93] H. G. Musmann, "Object-oriented analysis-synthesis coding based on source models of moving 2D and 3D objects," *Proc. IEEE Int. Conf. Acoust. Speech and Sign. Proc.*, Minneapolis, MN, April 1993.

[Nat 77] T. R. Natarajan and N. Ahmed, "On interframe transform coding," *IEEE Trans. Comm.*, vol. 25, no. 11, pp. 1323–1329, Nov. 1977.

[Nic 93] A. Nicoulin, M. Mattavelli, W. Li, A. Basso, A. C. Popat, and M. Kunt, "Image sequence coding using motion compensated subband decomposition," in *Motion Analysis and Image Sequence Processing*, M. I. Sezan and R. L. Lagendijk, eds., Norwell, MA: Kluwer, 1993.

[Orc 93] M. Orchard, "Predictive motion-field segmentation for image sequence coding," *IEEE Trans. Circ. and Syst: Video Tech.*, vol. 3, pp. 54–70, Feb. 1993.

[Plo 82] A. Ploysongsang and K. R. Rao, "DCT/DPCM processing of NTSC composite video signal," *IEEE Trans. Comm.*, vol. 30, no. 3, pp. 541–549, Mar. 1982.

[Pur 91] A. Puri and R. Aravind, "Motion-compensated video coding with adaptive perceptual quantization," *IEEE Trans. Circ. Syst. Video Tech.*, vol. 1, pp. 351–361, 1991.

[Roe 77] J. A. Roese, W. K. Pratt, and G. S. Robinson, "Interframe cosine transform image coding," *IEEE Trans. Comm.*, vol. 25, no. 11, pp. 1329–1339, Nov. 1977.

[Sab 84] S. Sabri and B. Prasada, "Coding of broadcast TV signals for transmission over satellite channels," *IEEE Trans. Comm.*, vol. 32, no. 12, pp. 1323–1330, Dec. 1984.

[Saw 78] K. Sawada and H. Kotera, "32 Mbit/s transmission of NTSC color TV signals by composite DPCM coding," *IEEE Trans. Comm.*, vol. 26, no. 10, pp. 1432–1439, Oct. 1978.

[Vet 92] M. Vetterli and K. M. Uz, "Multiresolution coding techniques for digital television: A review," *Multidim. Syst. Signal Proc.*, vol. 3, pp. 161–187, 1992.

[Woo 89] J. W. Woods and T. Naveen, "Subband encoding of video sequences," *Proc. SPIE Visual Comm. Image Proc. IV*, vol. 1199, pp. 724–732, 1989.

Chapter 23

VIDEO COMPRESSION STANDARDS

Standardization of compressed digital video formats facilitates manipulation and storage of full-motion video as a form of computer data, and its transmission over existing and future computer networks, or over terrestrial broadcast channels. Envisioned areas of application for digital video compression standards include all-digital TV, videoconferencing, videophone, video mail, multimedia stations, digital movies, video games, other forms of entertainment, and education. In this chapter, we discuss the international video compression standards for videoconferencing (H.261), multimedia (MPEG-1), and all-digital TV (MPEG-2) applications.

23.1 The H.261 Standard

ITU (CCITT) Recommendation H.261 is a video compression standard developed to facilitate videoconferencing and videophone services over the integrated services digital network (ISDN) at $p \times 64$ kbps, $p = 1, \ldots, 30$. For example, 64 kbps ($p = 1$) may be appropriate for a low quality videophone service, where the video signal can be transmitted at a rate of 48 kbps, and the remaining 16 kbps is used for the audio signal. Videoconferencing services generally require higher image quality, which can be achieved with $p \geq 6$, i.e., at 384 kbps or higher. Note that the maximum available bitrate over an ISDN channel is 1.92 Mbps ($p = 30$), which is sufficeint to obtain VHS-quality (or better) images.

ITU (CCITT) Recommendation H.261 has emerged as a result of studies performed within the European Project COST (CoOperation in the field of Scientific and Technical research) 211bis during the period 1983-1990. In 1985, the COST 211bis videoconference hardware subgroup developed an initial codec operating at bit rates of $n \times 384$ kbps, $n = 1, \ldots, 5$, which was adopted in 1987 as ITU (CCITT) Recommendation H.120. Later, it became clear that a single standard can cover all ISDN rates, $p \times 64$ kbps, $p = 1, \ldots, 30$. The specifications of such a codec were completed in 1989, and the corresponding Recommendation H.261 was adopted by ITU (CCITT) in 1990.

In addition to forming a basis for the later video compression standards such as MPEG-1 and MPEG-2, the H.261 standard offers two important features: i) It specifies a maximum coding delay of 150 msec. because it is mainly intended for bidirectional video communication. It has been determined that delays exceeding 150 msec. do not give the viewer the impression of direct visual feedback. ii) It is amenable to low-cost VLSI implementation, which is rather important for widespread commercialization of videophone and teleconferencing equipment. The important aspects of the H.261 standard are summarized below. Further details can be found in [Lio 90, Lio 91, CCI 90].

23.1.1 Input Image Formats

To permit a single recommendation for use in and between regions using 625- and 525-line TV standards, the H.261 input picture format is specified as the so-called Common Intermediate Format (CIF). For lower-bitrate applications, a smaller format, QCIF, which is one-quarter of the CIF, has been adopted. The specifications of the CIF and QCIF formats are listed in Table 23.1, where the numbers in the parenthesis denote the modified specifications so that all four numbers are integer multiples of 8. Note that, at 30 frames/s, the raw data rate for the CIF is 37.3 Mbps, and for QCIF it is 9.35 Mbps. Even with QCIF images at 10 frames/s, 48:1 compression is required for videophone services over a 64 kbps channel. CIF images may be used when $p \geq 6$, that is, for videoconferencing applications. Methods for conversion to and from CIF/QCIF are not subject to recommendation.

Table 23.1: H.261 input image formats.

	CIF	QCIF
Number of active pels/line		
Lum (Y)	360 (352)	180 (176)
Chroma (U,V)	180 (176)	90 (88)
Number of active lines/pic		
Lum (Y)	288	144
Chroma (U,V)	144	72
Interlacing	1:1	1:1
Temporal rate	30, 15, 10, or 7.5	30, 15, 10 or 7.5
Aspect ratio	4:3	4:3

23.1.2 Video Multiplex

The video multiplex defines a data structure so that a decoder can interpret the received bit stream without any ambiguity. The video data is arranged in a hierarchical structure consisting of a picture layer, which is divided into several group-of-blocks (GOB) layers. Each GOB layer in turn consists of macroblocks (MB), which are made of blocks of pixels. Each layer has a header identifying a set of parameters used by the encoder in generating the bitstream that follows.

A macroblock is the smallest unit of data for selecting a compression mode. (The choices for the modes of compression are described below.) It consists of four 8 ×8 (i.e., 16 pixels by 16 lines) of Y (luminance) and the spatially corresponding 8 ×8 U and V (chrominance) blocks. Since the chrominance channels are subsampled in both directions, there is only one U and one V block for every four luminance blocks. The composition of a macroblock is shown in Figure 23.1.

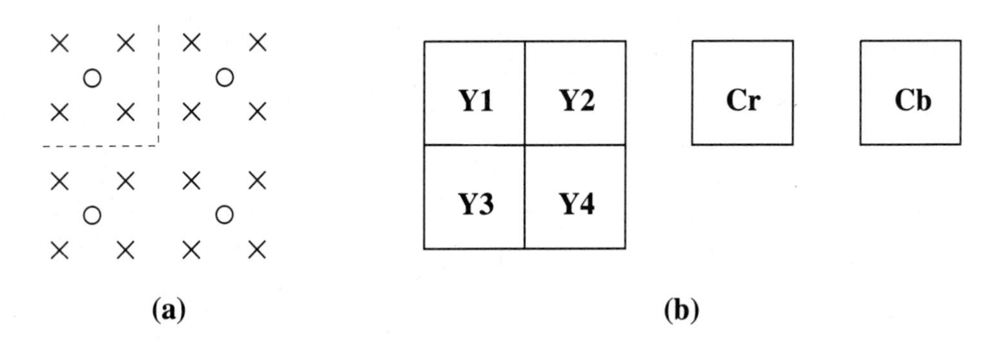

Figure 23.1: a) Positioning of luminance and chrominance pixels; b) the composition of a macroblock.

The GOB layer is always composed of 33 macroblocks, arranged as a 3 × 11 matrix, as depicted in Figure 23.2. Note that each MB has a header, which contains a MB address and the compression mode, followed by the data for the blocks.

Finally, the picture layer consists of a picture header followed by the data for GOBs. The picture header contains data such as the picture format (CIF or QCIF)

Table 23.2: The composition of the picture layer.

Format	Number of GOB in a frame	Number of MB in a GOB	Total number of MB in a frame
CIF	12	33	396
QCIF	3	33	99

1	2	3	4	5	6	7	8	9	10	11
12	13	14	15	16	17	18	19	20	21	22
23	24	25	26	27	28	29	30	31	32	33

(a)

MBA	MTYPE	MQUANT	MVD	CBP	Block data

(b)

Figure 23.2: a) Arrangement of macroblocks (MBs) in a GOB; b) structure of the MB layer [CCI 90].

and the frame number. Note that each CIF image has 12 GOBs, while the QCIF has only 3 GOBs, as shown in Table 23.2.

23.1.3 Video Compression Algorithm

The video compression scheme chosen for the H.261 standard has two main modes: the intra and inter modes. The *intra* mode is similar to JPEG still-image compression in that it is based on block-by-block DCT coding. In the *inter* mode, first a temporal prediction is employed with or without motion compensation (MC). Then the interframe prediction error is DCT encoded. Each mode offers several options, such as changing the quantization scale parameter, using a filter with MC, and so on. The algorithm can be summarized through the following steps:

1) Estimate a motion (displacement) vector for each MB. The standard does not specify a motion estimation method (in fact, MC is optional), but block matching based on 16 $\times$ 16 luminance blocks is generally used. The decoder accepts one integer-valued motion vector per MB whose components do not exceed $\pm$ 15.

2) Select a compression mode for each MB, based on a criterion involving the displaced block difference (dbd), which is defined by

$$dbd(\mathbf{x}, k) = b(\mathbf{x}, k) - b(\mathbf{x} - \mathbf{d}, k - 1) \tag{23.1}$$

where $b(.,.)$ denotes a block, $\mathbf{x}$ and k are pixel coordinates and time index, respectively, and $\mathbf{d}$ is the displacement vector defined for the kth frame relative to the $(k-1)$st frame. If $\mathbf{d}$ is set equal to zero, then dbd becomes a block difference (bd). Various compression modes are discussed in detail in the following.

3) Process each MB to generate a header followed by a data bitstream that is consistent with the compression mode chosen.

The motion estimation method, the criterion for the choice of a mode, and whether to transmit a block or not are not subject to recommendation. They are left as design parameters for a particular implementation. The choices that are presented below are those that are used in the Reference Model 8 (RM8), which is a particular implementation [RM8 89].

Compression Modes

Selecting a compression mode requires making several decisions, for each MB, including: i) should a motion vector be transmitted, ii) inter vs. intra compression, and iii) should the quantizer stepsize be changed? All possible compression modes are listed in Table 23.3, where "Intra," "Inter," "Inter+MC," and "Inter+MC+FIL" denote intraframe, interframe with zero motion vector, motion-compensated interframe, and motion-compensated interframe with loop-filtering, respectively. MQUANT stands for the quantizer step size, and an "x" in this column indicates that a new value for MQUANT will be transmitted. MVD stands for motion vector data, CBP for coded block pattern (a pattern number signifying those blocks in the MB for which at least one transform coefficient is transmitted), and TCOEFF for the transform coefficients that are encoded. Finally, VLC is the variable-length code that identifies the compression mode in the MB header.

In order to select the best compression mode, at each MB, the variance of the original macroblock, the macroblock difference (bd), and the displaced macroblock difference (dbd) with the best motion vector estimate are compared as follows:

Table 23.3: H.261 compression modes [CCI 90].

Prediction	MQUANT	MVD	CBP	TCOEFF	VLC
Intra				x	0001
Intra	x			x	0000 001
Inter			x	x	1
Inter	x		x	x	0000 1
Inter+MC		x			0000 0000 1
Inter+MC		x	x	x	0000 0001
Inter+MC	x	x	x	x	0000 0000 01
Inter+MC+FIL		x			001
Inter+MC+FIL		x	x	x	01
Inter+MC+FIL	x	x	x	x	0000 01

a) If the variance of *dbd* is smaller than *bd* as determined by a threshold, then the "Inter+MC" mode is selected, and the motion vector MVD needs to be transmitted as side information. The difference of the motion vector between the present and previous macroblocks is VLC coded for transmission. Observe from Table 23.3 that the transmission of the prediction error characterized by the DCT coefficients TCOEFF is optional.

b) Otherwise, a motion vector will not be transmitted, and a decision needs to be made between the "Inter" and "Intra" modes. If the original MB has a smaller variance, then the "Intra" mode is selected, where the DCT of each 8×8 block of the original picture elements are computed; otherwise, the "Inter" mode (with zero displacement vector) is selected. In both "Inter" and "Inter+MC" blocks the respective difference blocks (also called as the prediction error) are DCT encoded. The reader is referred to [CCI 90] for an exact specification of the various decision functions.

For MC blocks, the prediction error can be chosen to be modified by a 2-D spatial filter for each 8×8 block before the transformation by choosing the "Inter+MC+FIL" mode. The filter is separable, obtained by cascading two identical 1-D FIR filters. The coefficients of the 1-D filter are given by 1/4, 1/2, 1/4 except at the block boundaries should one of the taps fall outside the block, where they are modified as 0, 1, 0.

Thresholding

A variable thresholding is applied before quantization to increase the number of zero coefficients. The accuracy of the coefficients is 12 bits with dynamic range in [-2048,2047]. The flowchart of the variable thresholding algorithm is shown in Figure 23.3. In the flowchart, "g" refers to the quantizer step size, ξ is current value of the threshold, and "coef" is the value of the DCT coefficient. The variable thresholding scheme is demonstrated below by means of an example.

Quantization

The coefficient values after variable thresholding are quantized using a uniform quantizer. Within a macroblock the same quantizer is used for all coefficients except for the intra DC coefficient. The same quantizer is used for both luminance and chrominance coding. The intra DC coefficient is linearly quantized with a stepsize of 8 and no dead zone. Other coefficients are also linearly quantized, but with a central dead-zone about zero and with a stepsize MQUANT of an even value in the range 2 to 62 (31 stepsizes are allowed). This stepsize is controlled by the buffer state. To prevent overflow/underflow, a clipping of the image pixel values to within the range [0,255] is performed in both the encoder and decoder loops.

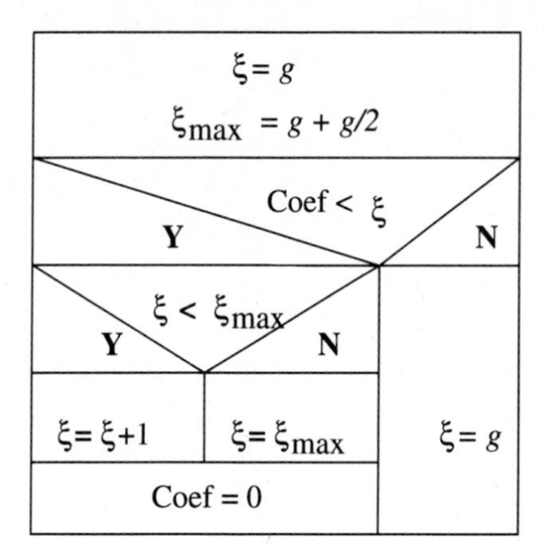

Figure 23.3: Flowchart of the thresholding algorithm [RM8 89].

Example: (with permission of PTT Research Labs., The Netherlands.)

An example of the variable thresholding and quantization is shown in Table 23.4 with $g = 32$. In this example, the threshold is incremented starting from 32 to 38, at which point the coefficient 40 is more than the value of the variable threshold. Thus, the threshold is reset to 32.

Table 23.4: An example to demonstrate variable thresholding [RM8 89].

Coefficients	50	0	0	0	33	34	0	40	33	34	10	32
Threshold ξ	32	32	33	34	35	36	37	38	32	32	32	33
New Coefficient	50	0	0	0	0	0	0	40	33	34	0	0
Quantized value	48	0	0	0	0	0	0	48	48	48	0	0

Coding

In order to increase the coding efficiency, the quantized coefficients are zigzag scanned, and events are defined which are then entropy coded. The events are defined as a combination of a run length of zero coefficients preceding a nonzero coefficient, and the level (value) of the nonzero coefficient, that is, $EVENT = (RUN, LEVEL)$. This is illustrated by an example [RM8 89].

Example: (with permission of PTT Research Labs., The Netherlands.)

For the block of transform coefficients shown in Figure 23.4, the events that represent this block, following a zigzag scan, are

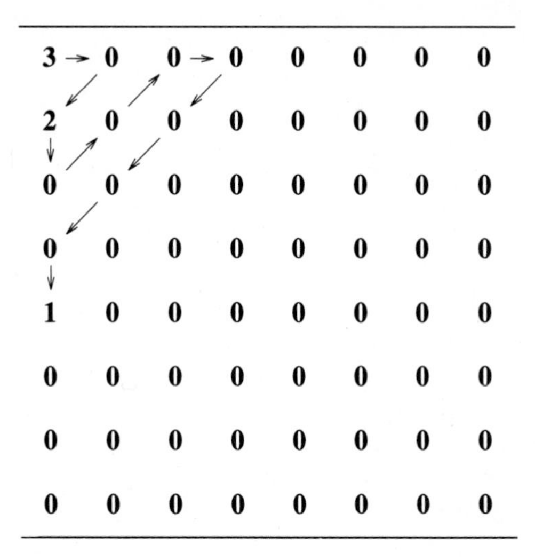

Figure 23.4: Illustration of the events [RM8 89].

(0,3) (1,2) (7,1) EOB

For 8 × 8 blocks, we have $0 \leq RUN < 64$, and the dynamic range of the nonzero coefficients, $LEVEL$, is [-128g, 127g], where g is the quantizer step size. These events are VLC coded. The VLC tables are specified in [CCI 90].

Rate/Buffer Control

Several parameters can be varied to control the rate of generation of coded video data. They are: i) processing prior to source coder, ii) the quantizer (step size), iii) block significance criterion, and iv) temporal subsampling (performed by discarding complete pictures). The proportions of such measures in the overall control strategy are not subject to recommendation.

In most implementations the quantizer step size is adjusted based on a measure of buffer fullness to obtain the desired bitrate. The buffer size is chosen not to exceed the maximum allowable coding delay (150 msec.) which also imposes a limit on the maximum bit count that can be generated by a single frame. Furthermore, the block significance criterion, which is employed to decide whether to transmit any data for a block, can be varied according to the desired bitrate.

Forced updating is used to control the accumulation of errors due to mismatch of the inverse DCT implementation at the encoder and decoder. The allowable bounds on the accuracy of the IDCT is specified in the standard. Forced updating refers to use of intra mode for a macroblock at least once every 132 times it is transmitted.

23.2 The MPEG-1 Standard

MPEG-1 is an ISO standard that has been developed for storage of CIF format video and its associated audio at about 1.5 Mbps on various digital storage media such as CD-ROM, DAT, Winchester disks, and optical drives, with the primary application perceived as interactive multimedia systems. The MPEG-1 algorithm is similar to that of H.261 with some additional features. The quality of MPEG-1 compressed/decompressed CIF video at about 1.2 Mbps (video rate) has been found to be similar (or superior) to that of VHS recorded analog video.

MPEG committee started its activities in 1988. Definition of the video algorithm (Simulation Model 1) was completed by September 1990. MPEG-1 was formally approved as an international standard by late 1992. Work is currently in progress to finalize the second phase algorithm, MPEG-2, for data rates up to 20 Mbps for high-definition video and associated audio. Efforts are also just underway for MPEG-4, which is concerned with very-low-bitrate compression (8-32 kbps) for videophone applications.

23.2.1 Features

MPEG-1 is a generic standard in that it standardizes a *syntax* for the representation of the encoded bitstream and a method of decoding. The syntax supports operations such as motion estimation, motion-compensated prediction, discrete cosine transformation (DCT), quantization, and variable-length coding. Unlike JPEG, MPEG-1 does not define specific algorithms needed to produce a valid data stream; instead, substantial flexibility is allowed in designing the encoder. Similar to H.261, MPEG-1 does not standardize a motion estimation algorithm or a criterion for selecting the compression mode. In addition, a number of parameters defining the coded bitstream and decoders are contained in the bitstream itself. This allows the algorithm to be used with pictures of a variety of sizes and aspect ratios and on channels or devices operating at a wide range of bitrates.

MPEG-1 also offers the following application-specific features: i) Random access is essential in any video storage application. It suggests that any frame should be decodable in a limited amount of time. In MPEG-1, this is achieved by allowing independent access points (I-frames) to the bitstream. ii) Fast forward/reverse search refers to scanning the compressed bit stream and to display only selected frames to obtain fast forward or reverse search. Reverse playback might also be necessary for some interactive applications. iii) Reasonable coding/decoding delay of about 1 sec to give the impression of interactivity in unidirectional video access. Recall that the coding delay in H.261 has been strictly limited to 150 msec to maintain bidirectional interactivity [Gal 92].

23.2.2 Input Video Format

MPEG-1 considers progressive (noninterlaced) video only. In order to reach the target bitrate of 1.5 Mbps, the input video is usually first converted into the MPEG standard input format (SIF). The (Y,Cr,Cb) color space has been adopted, as in CCIR Recommendation 601. In the MPEG-1 SIF, the luminance channel is 352 pixels × 240 lines and 30 frames/s. Luma and chroma components are represented by 8 bits/pixel, and the chroma components are subsampled by 2 in both the horizontal and vertical directions. The respective locations of the luma and chroma pixels are the same as in the H.261 standard.

While many video parameters, such as the picture size and temporal rate, can be specified in the syntax, and therefore are arbitrary, the following set of constrained parameters are specified to aid hardware implementations:

Maximum number of pixels/line: 720
Maximum number of lines/picture: 576
Maximum number of pictures/sec: 30
Maximum number of macroblocks/picture: 396
Maximum number of macroblocks/sec: 9900
Maximum bitrate: 1.86 Mbps
Maximum decoder buffer size: 376,832 bits.

Note, however, that the constrained parameter set does not suggest that a 720 pixels × 576 lines × 30 pictures/s video can be compressed artifact-free at 1.86 Mbps. For example, a CCIR 601 format video, 720 pixels × 488 lines × 30 pictures/sec, is usually downsampled to SIF before compression, which trades compression artifacts to spatial blurring in order to reach the target bitrate of 1.5 Mbps.

23.2.3 Data Structure and Compression Modes

Similar to H.261, the MPEG-1 bitstream also follows a hierarchical data structure, consisting of the following six layers, that enables the decoder to interpret the data unambiguously.

1) *Sequences* are formed by several group of pictures.

2) *Group of pictures* (GOP) are made up of pictures.

3) *Pictures* consist of slices. There are four picture types indicating the respective modes of compression: I-pictures, P-pictures, B-pictures, and D-pictures.

I-pictures are intra-frame DCT encoded using a JPEG-like algorithm. They serve as random access points to the sequence. There are two types of interframe encoded pictures, P- and B-pictures. In these pictures the motion-compensated prediction errors are DCT encoded. Only forward prediction is used in the P-pictures, which are always encoded relative to the preceding I- or P-pictures. The prediction of the B-pictures can be forward, backward, or bidirectional relative to other I- or P-pictures. D-pictures contain only the DC component of each block, and serve for browsing purposes at very low bitrates. The number of I, P, and B frames in a GOP are application-dependent, e.g., dependent on access time and bitrate requirements. The composition of a GOP is illustrated by an example.

Example

A GOP is shown in Figure 23.5 which is composed of nine pictures. Note that the first frame of each GOP is always an I-picture. In MPEG, the order in which the pictures are processed is not necessarily the same as their time sequential order. The pictures in Figure 23.5 can be encoded in one of the following orders:

0, 4, 1, 2, 3, 8, 5, 6, 7
or
0, 1, 4, 2, 3, 8, 5, 6, 7

since the prediction for P- and B-pictures should be based on pictures that are already transmitted.

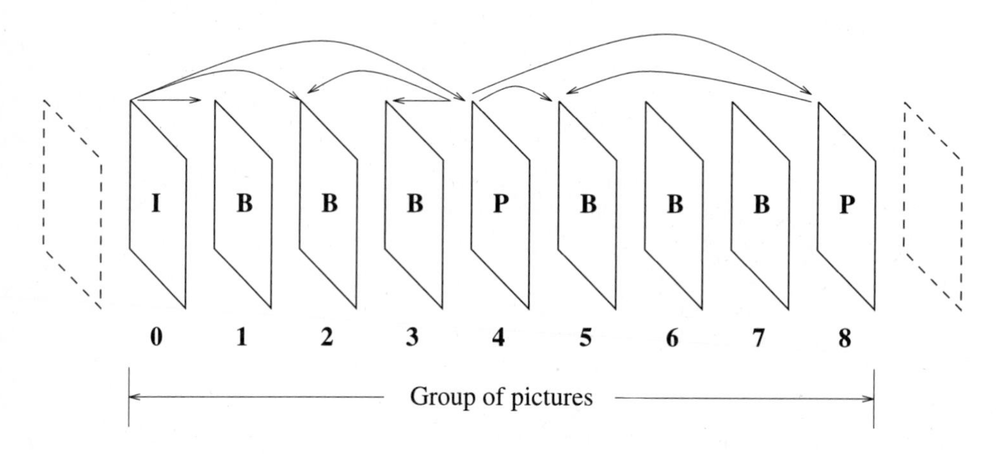

Figure 23.5: Group of pictures in MPEG-1.

4) *Slices* are made up of macroblocks. They are introduced mainly for error recovery.

5) The composition of *macroblocks* (MB) are the same as in the H.261 standard. Some compression parameters can be varied on a MB basis. The MB types depending on the choice of these parameters are listed in Table 23.5. We will take a closer look at each of these MB types in the following when we discuss the video compression algorithm.

6) *Blocks* are 8×8 pixel arrays. They are the smallest DCT unit.

Headers are defined for sequences, GOPs, pictures, slices, and MBs to uniquely specify the data that follows. For an extensive discussion of the MPEG-1 standard, the reader is referred to [ISO 91].

Table 23.5: Macroblock types in MPEG-1.

I-pictures	*P-pictures*	*B-pictures*
Intra	Intra	Intra
Intra-A	Intra-A	Intra-A
	Inter-D	Inter-F
	Inter-DA	Inter-FD
	Inter-F	Inter-FDA
	Inter-FD	Inter-B
	Inter-FDA	Inter-BD
	Skipped	Inter-BDA
		Inter-I
		Inter-ID
		Inter-IDA
		Skipped

23.2.4 Intraframe Compression Mode

The pixel intensity values are DCT encoded in a manner similar to JPEG and the intra mode of H.261. Compression is achieved by a combination of quantization and run-length coding of the zero coefficients.

Quantization

Assuming 8-bit input images, the DC coefficient can take values in the range [0,2040], and the AC coefficients are in the range [-1024,1023]. These coefficients are quantized with a uniform quantizer. The quantized coefficient is obtained by dividing the DCT coefficient value by the quantization step size and then rounding the result to the nearest integer. The quantizer step size varies by the frequency, ac-

Table 23.6: MPEG default intra quantization matrix.

8	16	19	22	26	27	29	34
16	16	22	24	27	29	34	37
19	22	26	27	29	34	34	38
22	22	26	27	29	34	37	40
22	26	27	29	32	35	40	48
26	27	29	32	35	40	48	58
26	27	29	34	38	46	56	69
27	29	35	38	46	56	69	83

cording to psycho-visual characteristics, and is specified by the *quantization matrix*. The MPEG default intra quantization matrix is shown in Table 23.6. According to this matrix, the DC coefficient is represented by 8 bits since its weight is 8. The AC coefficients can be represented with less than 8 bits using weights larger than 8.

MPEG allows for spatially-adaptive quantization by introducing a quantizer scale parameter $MQUANT$ in the syntax. As a result, there are two types of MBs in the I-pictures: **"Intra"** MBs are coded with the current quantization matrix. In **"Intra-A"** MBs, the quantization matrix is scaled by $MQUANT$, which is transmitted in the header. Note that $MQUANT$ can be varied on a MB basis to control the bitrate or for subjective quantization. Human visual system models suggest that MBs containing busy, textured areas can be quantized relatively coarsely. One of the primary differences between MPEG intra mode and JPEG is the provision of adaptive quantization in MPEG. It has been claimed that MPEG intra mode provides 30% better compression compared with JPEG due to adaptive quantization.

Coding

Redundancy among the quantized DC coefficients is reduced via DPCM. The resulting signal is VLC coded with 8 bits. The fixed DC Huffman table has a logarithmic amplitude category structure borrowed from JPEG. Quantized AC coefficients are zigzag scanned and converted into [run, level] pairs as in JPEG and H.261. A single Huffman-like code table is used for all blocks, independent of the color component to which they belong. There is no provision for downloading custom tables. Only those pairs which are highly probable are VLC coded. The rest of them are coded with an escape symbol followed by a fixed-length code to avoid extremely long codewords. The codebook is a superset of that of H.261, which is completely different from that of JPEG.

23.2.5 Interframe Compression Modes

In interframe compression modes, a temporal prediction is formed, and the resulting prediction error is DCT encoded. There are two types of temporal prediction modes allowed in MPEG-1: forward prediction (P-pictures) and bidirectional prediction (B-pictures).

P-Pictures

P-pictures allow motion-compensated forward predictive coding with reference to a previous I- or P-picture. The temporal prediction process is illustrated in Figure 23.6, where the prediction for a MB $\mathbf{b}$ is given by

$$\hat{\mathbf{b}} = \tilde{\mathbf{c}} \qquad (23.2)$$

and $\tilde{\mathbf{c}}$ denotes the MB corresponding to $\mathbf{b}$ in the "reconstructed" previous frame.

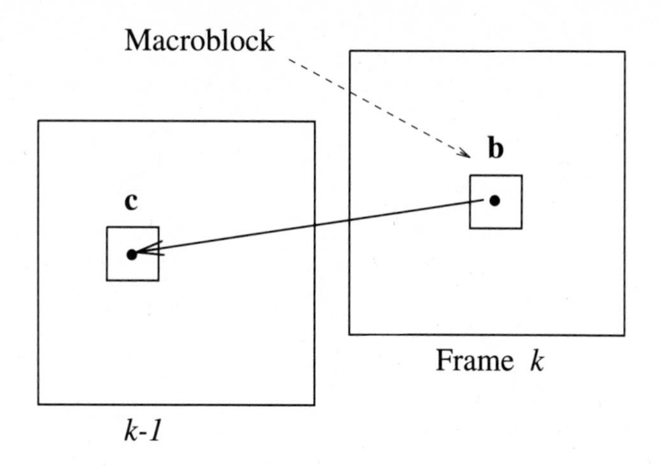

Figure 23.6: MPEG-1 forward prediction.

The mode of compression for each MB is selected by the encoder from the list of allowable modes for a P-picture shown in Table 23.5. **"Intra"** and **"Intra-A"** MBs in P-pictures are coded independently of any reference data just like MBs in the I-pictures. MBs classified as **"Inter"** are interframe coded, and the temporal prediction may use motion compensation (MC) and/or adaptive quantization. The subscript "D" indicates that the DCT of the prediction error will be coded, "F" indicates that forward MC is ON, and "A" indicates adaptive quantization (a new value of MQUANT is also transmitted). That is, if a MB is labeled **"Inter-F"** then the motion-compensated prediction $\hat{\mathbf{b}}$ is satisfactory, so we need to transmit just the motion vector $\mathbf{d}$ for that MB, **"Inter-FD"** denotes that we need to transmit a motion vector and the DCT coefficients of the prediction error, and **"Inter-FDA"** indicates that in addition to a motion vector and the DCT coefficients, a new value of MQUANT is also being transmitted for that MB. A macroblock may be **"Skipped"** if the block at the same position in the previous frame (without MC) is good enough, indicating a stationary area.

B-Pictures

B-pictures is a key feature of MPEG-1, which allows MC interpolative coding, also known as bidirectional prediction. The temporal prediction for the B-pictures is given by

$$\hat{\mathbf{b}} = \alpha_1 \tilde{\mathbf{c}}_1 + \alpha_2 \tilde{\mathbf{c}}_2 \qquad \alpha_1,\ \alpha_2 = 0,\ 0.5,\ 1 \qquad \alpha_1 + \alpha_2 = 1 \tag{23.3}$$

where $\tilde{}$ denotes "reconstructed" values. Then $\alpha_1 = 1$ and $\alpha_2 = 0$ yields forward prediction, $\alpha_1 = 0$ and $\alpha_2 = 0$ gives backward prediction, and $\alpha_1 = \alpha_2 = 0.5$

corresponds to bidirectional prediction. This is illustrated in Figure 23.7. Note that in the bidirectional prediction mode, two displacement vectors $\mathbf{d}_1$ and $\mathbf{d}_2$ and the corresponding prediction error $\mathbf{b} - \hat{\mathbf{b}}$ need to be encoded for each macroblock $\mathbf{b}$.

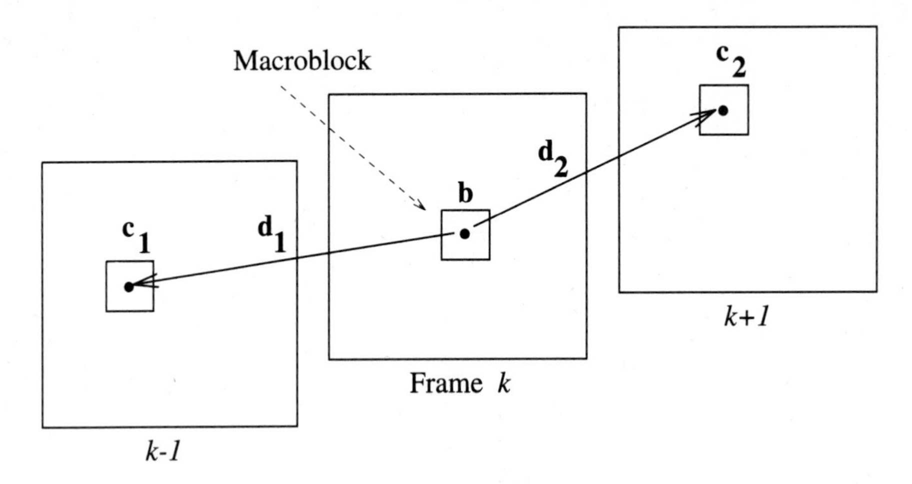

Figure 23.7: MPEG-1 bi-directional prediction.

The concept of bidirectional prediction or interpolative coding can be considered as a temporal multiresolution technique, where we first encode only the I- and P-pictures (typically 1/3 of all frames). Then the remaining frames can be interpolated from the reconstructed I and P frames, and the resulting interpolation error is DCT encoded. The use of B-pictures provides several advantages:

- They allow effective handling of problems associated with covered/uncovered background. If an object is going to be covered in the next frame, it can still be predicted from the previous frame or vice versa.
- MC averaging over two frames may provide better SNR compared to prediction from just one frame.
- Since B-pictures are not used in predicting any future pictures, they can be encoded with fewer bits without causing error propagation.

The trade-offs associated with using B-pictures are:

- Two frame-stores are needed at the encoder and decoder, since at least two reference (P and/or I) frames should be decoded first.
- If too many B-pictures are used, then i) the distance between the two reference frames increases, resulting in lesser temporal correlation between them, and hence more bits are required to encode the reference frames, and ii) we have longer coding delay.

The mode of compression for each MB in a B-picture is selected independently from the list of allowable modes shown in Table 23.5. Again, **"Intra"** and **"Intra-A"** MBs are coded independently of any reference frame. MBs classified as **"Inter"** have the following options: "D" indicates that the DCT of the prediction error will be coded, "F" indicates forward prediction with motion compensation, "B" indicates backward prediction with motion compensation, "I" indicates interpolated prediction with motion compensation, and "A" indicates adaptive quantization. A macroblock may be **"Skipped"** if the block from the previous frame is good enough as is; that is, no information needs to be sent.

Quantization and Coding

In the interframe mode, the inputs to the DCT are in the range [-255,255]; thus, all DCT coefficients have the dynamic range [-2048,2047]. The quantization matrix is such that the effective quantization is relatively coarser compared to those used for I-pictures. All quantized DCT coefficients, including the DC coefficient, are zigzag scanned to form $[run, level]$ pairs, which are then coded using VLC. Displacement vectors are DPCM encoded with respect to the motion vectors of the previous blocks. VLC tables are specified for the type of MB, the differential motion vector, and the MC prediction error. Different Huffman tables are defined for encoding the macroblock types for P- and B-pictures, whereas the tables for motion vectors and the DCT coefficients are the same for both picture types.

23.2.6 MPEG-1 Encoder and Decoder

An MPEG-1 encoder includes modules for motion estimation, selection of compression mode (MTYPE) per MB, setting the value of MQUANT, motion-compensated prediction, quantizer and dequantizer, DCT and IDCT, variable-length coding (VLC), a multiplexer, a buffer, and a buffer regulator. The dequantizer and the IDCT are needed in the encoder because the predictions are based on reconstructed data. The IDCT module at the encoder should match within a prespecified tolerance the IDCT module at the decoder to avoid propogation of errors in the prediction process. This tolerance is specified in IEEE Standard 1180-1990 for 64-bit floating-point IDCT implementations.

The relative number of I-, P- or B-pictures in a GOP is application-dependent. The standard specifies that one out of every 132 pictures must be an I-picture to avoid error propagation due to IDCT mismatch between the encoder and decoder. The use of B-pictures is optional. Neither the motion estimation algorithm nor the criterion to select MYTPE and MQUANT are part of the standard. In general, motion estimation is performed using the luminance data only. A single displacement vector is estimated for each MB. One-half (0.5) pixel accuracy is allowed for motion estimates. The maximum length of the vectors that may be represented can be changed on a picture-by-picture basis to allow maximum flexibility. Motion vectors that refer to pixels outside the picture are not allowed.

In summary, a typical MPEG encoder performs the following steps:

1. Decide on the labeling of I-, P- and B-pictures in a GOP.

2. Estimate a motion vector for each MB in the P- and B-pictures.

3. Determine the compression mode MTYPE for each MB from Table 23.5.

4. Set the quantization scale, MQUANT, if adaptive quantization is selected.

An MPEG-1 decoder reverses the operations of the encoder. The incoming bit stream (with a standard syntax) is demultiplexed into DCT coefficients and side information such as MTYPE, motion vectors, MQUANT, and so on. The decoder employs two frame-stores, since two reference frames are used to decode the B-pictures.

We conclude this section by summarizing the main differences between the H.261 and MPEG-1 standards in Table 23.7.

Table 23.7: Comparison of H.261 and MPEG-1 Standards

H.261	**MPEG-1**
Sequential access	Random access
One basic frame rate	Flexible frame rate
CIF and QCIF images only	Flexible image size
I and P frames only	I, P and B frames
MC over 1 frame	MC over 1 or more frames
1 pixel MV accuracy	1/2 pixel MV accuracy
121 filter in the loop	No filter
Variable threshold + uniform quantization	Quantization matrix
No GOF structure	GOF structure
GOB structure	Slice structure

23.3 The MPEG-2 Standard

The quality of MPEG-1 compressed video at 1.2 Mbps has been found unacceptable for most entertainment applications. Subjective tests indicate that CCIR 601 video can be compressed with excellent quality at 4-6 Mbps. MPEG-2 is intended as a compatible extension of MPEG-1 to serve a wide range of applications at various bitrates (2-20 Mbps) and resolutions. Main features of the MPEG-2 syntax are: i) it allows for interlaced inputs, higher-definition inputs, and alternative subsampling of the chroma channels, ii) it offers a scalable bitstream, and iii) it provides improved quantization and coding options.

Considering the practical difficulties with the implementation of the full syntax on a single chip, subsets of the full syntax have been specified under five "profiles," simple profile, main profile, SNR scalable profile, spatially scalable profile, and high profile. Furthermore, a number of "levels" have been introduced within these profiles to impose constraints on some of the video parameters [ISO 93]. It is important to note that the MPEG-2 standard has not yet been finalized. The syntax of the Main Profile was frozen in March 1993. However, work in other profiles is still ongoing. It is likely that the all-digital HDTV compression algorithm will conform with one of the profiles in MPEG-2.

In the following we discuss the MB structure in MPEG-2, how MPEG-2 handles interlaced video, the concepts related to scalability, and some extensions for encoding of higher-definition video along with a brief overview of the profiles and levels.

23.3.1 MPEG-2 Macroblocks

A macroblock (MB) refers to four 8 × 8 luminance blocks and the spatially associated chroma blocks. MPEG-2 allows for three chroma subsampling formats, 4:2:0 (same as MPEG-1), 4:2:2 (chroma subsampled in the horizontal direction only), and 4:4:4 (no chroma subsampling). The spatial locations of luma and chroma pixels for the 4:2:0 and 4:2:2 formats are depicted in Figure 23.8. Therefore, in MPEG-2 a MB may contain 6 (4 luma, 1 Cr, and 1 Cb), 8 (4 luma, 2 Cr, and 2 Cb), or 12 (4 luma, 4 Cr, and 4 Cb) 8 × 8 blocks.

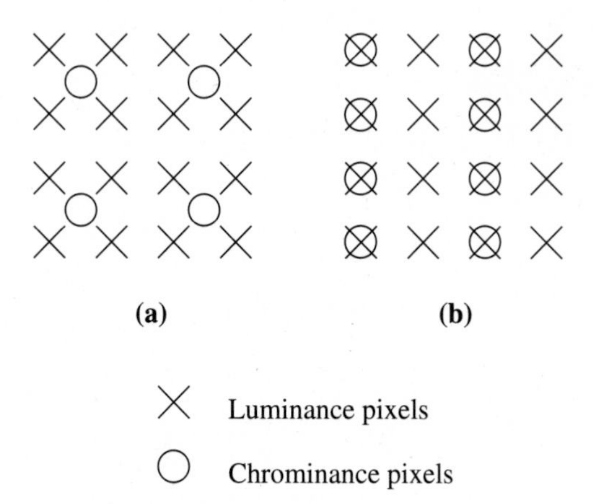

Figure 23.8: Chrominance subsampling options: a) 4:2:0 format; b) 4:2:2 format.

23.3.2 Coding Interlaced Video

MPEG-2 accepts both progressive and interlaced inputs. If the input is interlaced, the output of the encoder consists of a sequence of fields that are separated by the field period. There are two options in coding interlaced video: i) every field can be encoded independently (field pictures), or ii) two fields may be encoded together as a composite frame (frame pictures). It is possible to switch between frame pictures and field pictures on a frame-to-frame basis. Frame encoding is preferred for relatively still images; field encoding may give better results when there is significant motion. In order to deal with interlaced inputs effectively, MPEG-2 supports:

- two new picture formats: frame-picture and field-picture,
- field/frame DCT option per MB for frame pictures, and
- new MC prediction modes for interlaced video,

which are described in the following.

New Picture Types for Interlaced Video

Interlaced video is composed of a sequence of even and odd fields separated by a field period. MPEG-2 defines two new picture types for interlaced video. They are:

i) *Frame pictures*, which are obtained by interleaving lines of even and odd fields to form composite frames. Frame pictures can be I-, P-, or B-type. An MB of the luminance frame picture is depicted in Figure 23.9.

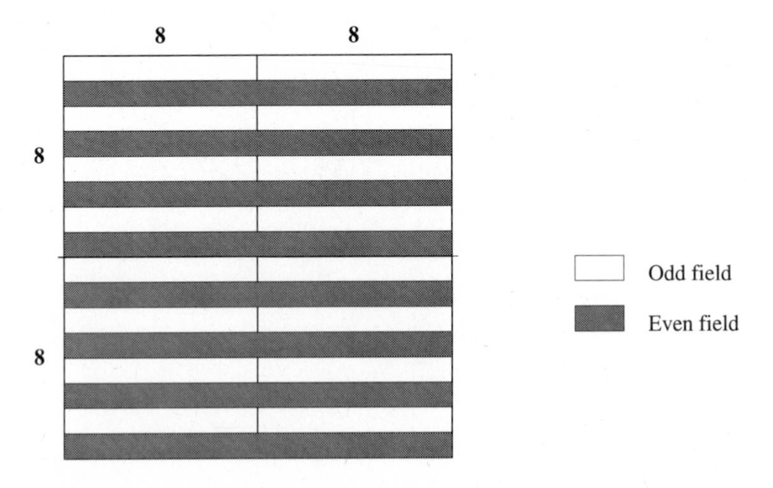

Figure 23.9: The luminance component of a MB of a frame picture.

ii) *Field pictures* are simply the even and odd fields treated as separate pictures. Each field picture can be I-, P- or B-type.

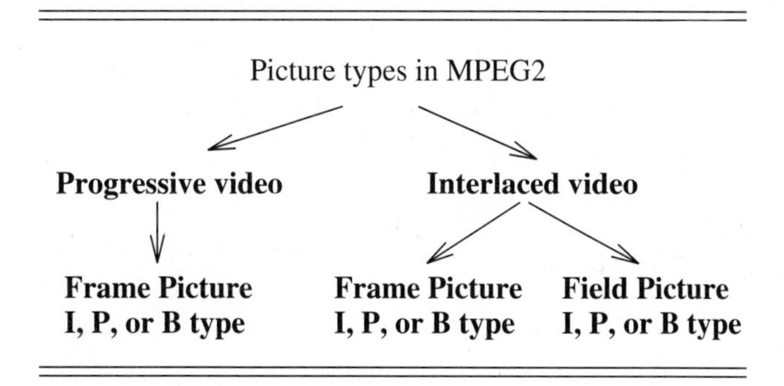

Figure 23.10: Summary of picture types in MPEG-2.

A summary of all picture types is shown in Figure 23.10. Clearly, in progressive video all pictures are frame pictures. A group of pictures can be composed of an arbitrary mixture of field and frame pictures. Field pictures always appear in pairs (called the top field and bottom field) which together constitute a frame. If the top field is a P- (B-) picture, then the bottom field must also be a P- (B-) picture. If the top field is an I-picture, then the bottom field can be an I- or a P-picture. A pair of field pictures are encoded in the order in which they should appear at the output. An example of a GOP for an interlaced video is shown in Figure 23.11.

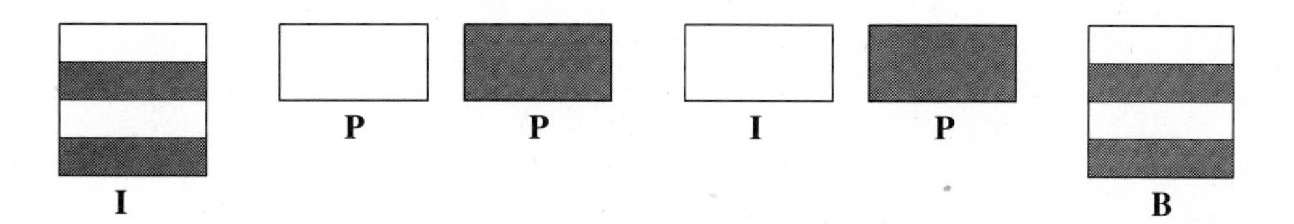

Figure 23.11: A GOP for an interlaced video.

Field/Frame DCT Option for Frame Pictures

MPEG-2 allows a field- or frame-DCT option for each MB in a frame picture. This allows computing DCT on a field-by-field basis for specific parts of a frame picture. For example, field-DCT may be chosen for macroblocks containing high motion, whereas frame-DCT may be appropriate for macroblocks with little or no motion but containing high spatial activity. The internal organization of a MB for frame (on the left) and field (on the right) DCT is shown in Figure 23.12. Note that in 4:2:0 sampling, only frame DCT can be used for the chroma blocks to avoid 8×4 IDCT.

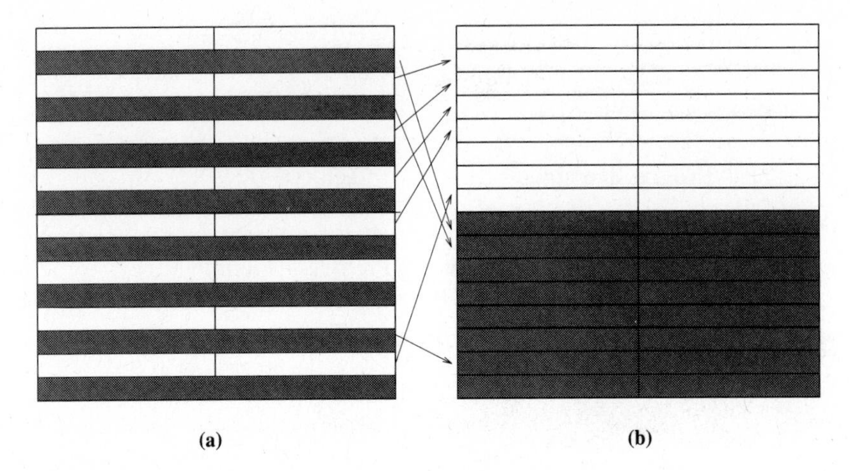

Figure 23.12: DCT options for interlaced frame pictures: a) Frame DCT and b) field DCT.

MC Prediction Modes for Interlaced Video

There are two main types of predictions: simple field and simple frame prediction. In simple field prediction, each field is predicted independently using data from one or more previously decoded fields. Simple frame prediction forms a prediction for an entire frame based on one or more previously decoded frames. Within a field picture only field predictions can be used. However, in a frame picture either field or frame prediction may be employed on an MB-by-MB basis. Selection of the best prediction mode depends on presence/absence of motion in an MB, since in the presence of motion, frame prediction suffers from strong motion artifacts, while in the absence of motion, field prediction does not utilize all available information.

There are also two other prediction modes: 16×8 MC mode and dual-prime mode. 16×8 MC mode is only used in field pictures, where two motion vectors are used per MB, one for the upper and the other for the lower 16×8 region, which belong to the top and bottom fields, respectively. In the case of bidirectional prediction, four motion vectors will be needed. In dual-prime mode, one motion vector and a small differential vector are encoded. In the case of field pictures, two motion vectors are derived from this information and used to form predictions from two reference fields, which are averaged to form the final prediction [ISO 93]. Dual-prime mode is used only for P-pictures.

23.3.3 Scalable Extensions

Scalability refers to ability to decode only a certain part of the bit-stream to obtain video at the desired resolution. It is assumed that decoders with different complexities can decode and display video at different spatio-temporal resolutions from the same bitstream. The minimum decodable subset of the bitstream is called the base layer. All other layers are enhancement layers, which improve the resolution of the base layer video. MPEG-2 syntax allows for two or three layers of video. There are different forms of scalability:

Spatial (pixel resolution) scalability provides the ability to decode video at different spatial resolutions without first decoding the entire frame and decimating it. The base layer is a low spatial resolution version of the video. Enhancement layers contain successively higher-frequency information. MPEG-2 employs a pyramidal coding approach. The base layer video is obtained by decimating the original input video. The enhancement layer is the difference of the actual input video and the interpolated version of the base-layer video.

SNR scalability offers decodability using different quantizer step sizes for the DCT coefficients. The base-layer video is obtained by using a coarse quantization of the DCT coefficients. It is at the same spatio-temporal resolution with the input video. The enhancement layer simply refers to the difference of the base layer and the original input video.

Temporal scalability refers to decodability at different frame rates without first decoding every single frame. *Hybrid scalability* refers to some combination of the above. An important advantage of scalability is that it provides higher resilience to transmission errors as the base-layer video is usually transmitted with better error correction capabilities.

23.3.4 Other Improvements

MPEG-2 also features some extensions in the quantization and coding steps for improved image quality in exchange to slightly higher bitrate. In particular, it allows for i) a new scanning scheme (alternate scan) in addition to the zigzag scanning of the DCT coefficients, ii) finer quantization of the DCT coefficients, iii) finer adjustment of the quantizer scale factor, and iv) a separate VLC table for the DCT coefficients for the intra macroblocks, which are explained in the following.

Alternate Scan

In addition to the zigzag scanning, MPEG-2 allows for an optional scanning pattern, called the "alternate scan." The alternate scan pattern, which is said to fit interlaced video better, is depicted in Figure 23.13.

Finer Quantization of the DCT Coefficients

In intra macroblocks, the quantization weight for the DC coefficient can be 8, 4, 2 or 1. That is, 11 bits (full) resolution is allowed for the DC coefficient. Recall that this weight is fixed to 8 in MPEG-1. AC coefficients are quantized in the range [-2048,2047], as opposed to [-256,255] in MPEG-1. In non-intra macroblocks, all coefficients are quantized into the range [-2048,2047]. This range was [-256,255] in MPEG-1.

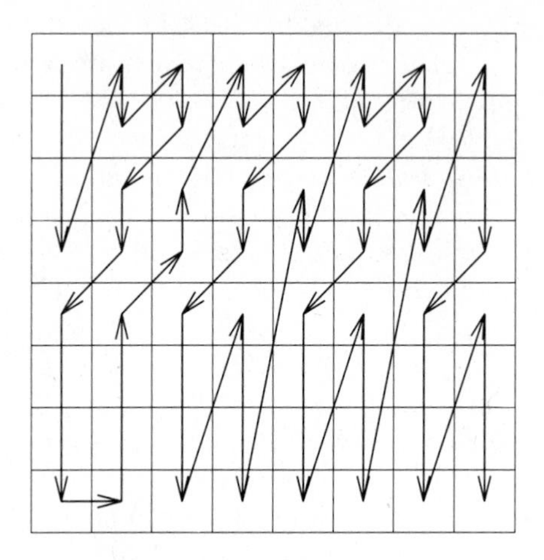

Figure 23.13: Alternate scan.

Finer Adjustment of MQUANT

In addition to a set of MQUANT values that are integers between 1 and 31, MPEG-2 allows for an optional set of 31 values that include real numbers ranging from 0.5 to 56. These values are listed in Table 23.8.

Table 23.8: Optional set of MQUANT values.

0.5	5.0	14.0	32.0
1.0	6.0	16.0	36.0
1.5	7.0	18.0	40.0
2.0	8.0	20.0	44.0
2.5	9.0	22.0	48.0
3.0	10.0	24.0	52.0
3.5	11.0	26.0	56.0
4.0	12.0	28.0	

23.3.5 Overview of Profiles and Levels

MPEG-2 full syntax covers a wide range of features and free parameters. The five MPEG-2 profiles define subsets of the syntax while the four levels impose constraints on the values of the free parameters for the purpose of practical hardware implementations. The parameter constraints imposed by the four levels are summarized in Table 23.9.

Table 23.9: Parameter constraints according to levels.

Level	Max. Pixels	Max. Lines	Max. Frames/s
Low	352	288	30
Main	720	576	30
High-1440	1440	1152	60
High	1920	1152	60

The Simple profile does not allow use of B-pictures and only support the Main level. The maximum bitrate for the Simple profile is 15 Mbps. The Main profile supports all four levels with upper bounds on the bitrates equal to 4, 15, 60, and 80 Mbps for the Low, Main, High-1440 and High levels, respectively. The Main profile does not include any scalability. The SNR Scalable profile supports Low and Main levels with maximum bitrates 4 (3) and 15 (10) Mbps, respectively. The numbers in parentheses indicate the maximum bitrate for the base layer. The Spatially Scalable profile supports only High-1440 level with a maximum bitrate of 60 (15) Mbps. The High profile includes Main, High-1440 and High levels with maximum bitrates of 20 (4), 80 (20), and 100 (25) Mbps, respectively.

23.4 Software and Hardware Implementations

There are several software and hardware implementations of the H.261 and MPEG algorithms. The Portable Video Research Group (PVRG) at Stanford University has public-domain source codes for both H.261 and MPEG standards, called PVRG-P64 and PVRG-MPEG, respectively. They can be obtained through anonymous ftp from "havefun.stanford.edu" - IP address [36.2.0.35] - (/pub/p64/P64v1.2.tar.Z) and (/pub/mpeg/MPEGv1.2.tar.Z). Other public-domain software includes the INRIA H.261 codec, which can be obtained from "avahi.inria.fr" (/pub/h261.tar.Z), and the Berkeley Plateau Research Group MPEG encoder, which can be obtained from "toe.cs.berkeley.edu" - IP address [128.32.149.117] - (/pub/multimedia/mpeg/mpeg-2.0.tar.Z).

Single- or multichip implementations of the video compression standards are available from many vendors, including

- C-Cube: CL-450, single-chip MPEG-1, SIF rates. CL-950, MPEG-2, CL-4000, single-chip, can code MPEG-1, H.261, and JPEG,
- SGS-Thomson: STi-3400, single-chip MPEG-1, SIF rates. STi-3500, the first MPEG-2 chip on the market,
- Motorola: MCD250, single-chip MPEG-1, SIF rates, and

- GEC Plassey: Chip sets for the H.261 algorithm.

For more complete listings, the reader is referred to MPEG-FAQ, available from "phade@cs.tu-berlin.de," and to Compression-FAQ (part 3), available by ftp from "rtfm.mit.edu" (/pub/usenet/news.answers/compression-faq/part[1-3].

Bibliography

[CCI 90] CCITT Recommendation H.261: "Video Codec for Audio Visual Services at $p \times 64$ kbits/s," COM XV-R 37-E, 1990.

[Che 93] C.-T. Chen, "Video compression: Standards and applications," *J. Visual Comm. Image Rep.*, vol. 4, no. 2, pp. 103–111, Jun. 1993.

[Chi 95] L. Chiariglione, "The development of an integrated audiovisual coding standard: MPEG," *Proc. IEEE*, vol. 83, no. 2, pp. 151–157, Feb. 1995.

[Gal 91] D. J. LeGall, "MPEG: A video compression standard for multimedia applications," *Commun. ACM*, vol. 34, pp. 46–58, 1991.

[Gal 92] D. J. LeGall, "The MPEG video compression algorithm," *Signal Proc.: Image Comm.*, vol. 4, pp. 129–140, 1992.

[ISO 91] ISO/IEC CD 11172: "Coding of Moving Pictures and Associated Audio - For Digital Storage Media at up to 1.5 Mbits/sec," Dec. 1991.

[ISO 93] ISO/IEC CD 13818-2: "Generic Coding of Moving Pictures and Associated Audio," Nov. 1993.

[Lio 91] M. L. Liou, "Overview of the $p \times 64$ kbits/sec video coding standard," *Commun. ACM*, vol. 34, pp. 59–63, 1991.

[Lio 90] M. L. Liou, "Visual telephony as an ISDN application," *IEEE Communications Magazine*, vol. 28, pp. 30–38, 1990.

[Oku 92] S. Okubo, "Requirements for high quality video coding standards," *Signal Proc.: Image Comm.*, vol. 4, pp. 141–151, 1992.

[Pen 93] W. B. Pennebaker and J. L. Mitchell, *JPEG: Still image data compression standard*, New York, NY: Van Nostrand Reinhold, 1993.

[RM8 89] COST211bis/SIM89/37: "Description of Reference Model 8 (RM8)," PTT Research Laboratories, The Netherlands, May 1989.

Chapter 24

MODEL-BASED CODING

Due to growing interest in very-low-bitrate digital video (about 10 kbps), significant research effort has recently been focused on new compression methods based on structural models, known as model-based analysis-synthesis coding. Scientists became interested in model-based coding because the quality of digital video provided by hybrid waveform encoders, such as the CCITT Rec. H.261 encoder, has been deemed unacceptable at these very low bitrates. The general principles of model-based coding, including general object-based, knowledge-based, and semantic coding, were presented in Chapter 22. These techniques employ structural models ranging from general purpose 2-D or 3-D object models [Mus 89, Die 93, Ost 93] to application-specific wireframe models [Aiz 89, Li 93, Boz 94]. In the following, we elaborate on general 2-D/3-D object-based coding in Section 24.1, and knowledge-based and semantic coding in Section 24.2.

24.1 General Object-Based Methods

A major deficiency of the MC/DCT compression is that it is based on a 2-D translatory block-motion model. This model is not adequate for a precise description of most motion fields, because it does not include rotation and zooming, and boundaries of moving objects hardly ever coincide with those of the rectangular blocks. Object-based methods aim to develop more realistic 2-D motion field models by considering affine, perspective, and bilinear spatial transformations and/or segmentation of the scene into individually moving objects. Typically, a frame is partitioned into an "unchanged region," an "uncovered background," and a number of moving objects which are either model-compliant (MC) or model-failure (MF) objects, as depicted in Figure 24.1. Segmentation is an integral part of these schemes, because different regions require different model parameters. We can classify object-based models as: i) piecewise planar or arbitrary 3-D surfaces with 3-D motion, ii) 2-D flexible models with 2-D translational motion, and iii) spatial transformations with triangular/rectangular patches, which are described in the following.

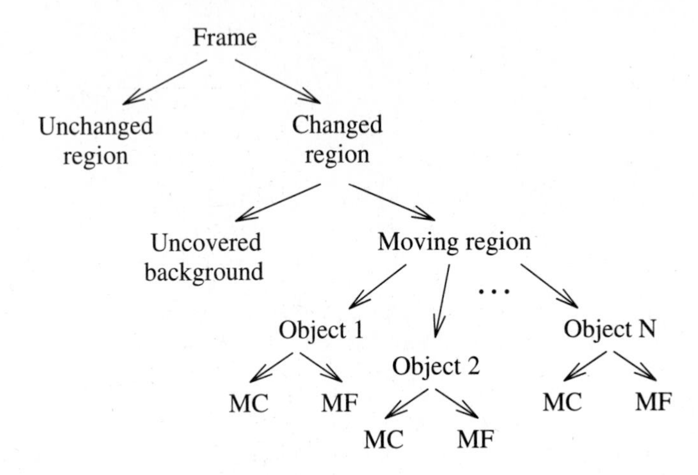

Figure 24.1: Segmentation of a frame into objects.

24.1.1 2-D/3-D Rigid Objects with 3-D Motion

In this approach, a 3-D model of a moving scene is estimated, which is consistent with a given sequence of frames. A realistic 2-D motion field model can then be obtained by projecting the resulting 3-D motion into the image plane. Two approaches are commonly used to represent 3-D surfaces: i) approximation by a piecewise planar model, also known as the case of 2-D rigid objects with 3-D motion [Hot 90, Die 93], and ii) estimation of a global surface model under a smoothness constraint, which leads to the so-called 3-D object with 3-D motion models [Mus 89, Ost 90, Ost 93, Mor 91, Koc 93]. Here, we provide a brief overview of the former approach.

We have seen in Chapter 9 that the orthographic and perspective projections of arbitrary 3-D motion of a rigid planar patch into the image plane yield the 6-parameter affine and the 8-parameter perspective models, respectively. It follows that, using the 2-D rigid object with 3-D motion approach, the 2-D motion field can be represented by a piecewise affine or a piecewise perspective field, where the boundaries of the patches and the respective model parameters can be estimated by a simultaneous motion estimation and segmentation algorithm (see Chapter 11). Then the parameter set for each independently moving object (motion parameters) together with the segmentation mask denoting the boundaries of each object in the image plane (shape parameters) constitute a complete description of the frame-to-frame pixel correspondences, which provide an MC-prediction of the next frame. Because of motion estimation/segmentation errors and problems related to uncovered background, the frame prediction error or the synthesis error (color parameters) also needs to be transmitted to improve upon image quality in the model failure regions.

An object-oriented analysis-synthesis encoder implements the following steps:

1. *Analysis:* Perform simultaneous motion parameter estimation and segmentation, between frame k and the reconstructed frame $k-1$, to find the best motion and shape parameters. Two specific algorithms that have been reported with successful results are those of Hotter *et al.* [Hot 90] and Diehl [Die 93]. They can be summarized using the flowchart shown in Figure 24.2, where the frame is initially segmented into a changed and an unchanged region. Each contiguous changed segment is assumed as an independent moving object, and a set of mapping parameters, such as a 6-parameter affine or an 8-parameter perspective mapping, is estimated for each moving object. The present frame (frame k) can then be synthesized from the previously reconstructed frame $k-1$ using these parameters. Those objects, where the synthesis error is

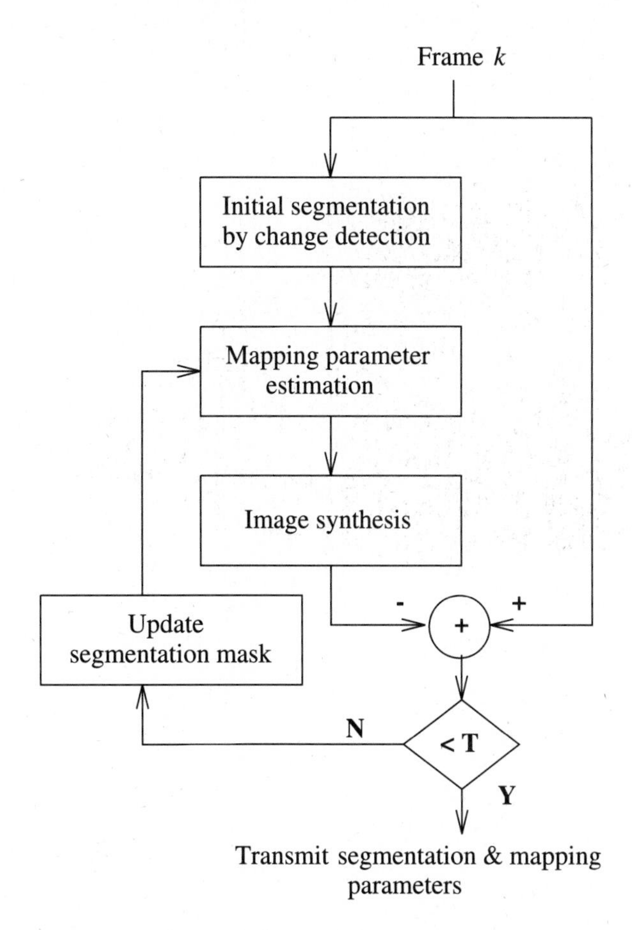

Figure 24.2: Flowchart of the mapping-parameter-based method.

above a threshold T, are classified as model-failure (MF) objects, which are then subdivided into more objects. The process is repeated until a predefined performance measure is satisfied.

2. *Synthesis:* Given the segmentation field and the motion parameters for each region, synthesize the present frame from the previous reconstructed frame. Compute the synthesis error, which specifies the color parameters.

3. *Coding:* In the case of the 8-parameter model, the parameters $(a_1, \ldots, a_8)$ are first normalized by a factor K, except for a_3 and a_6 which describe the horizontal and vertical translation of the object. The factor K is coded by 4 bits. The normalization aims at making the accuracy of the motion description independent of the spatial extent of the objects. The normalized parameters $(a_1, a_2, a_4, a_5, a_7, a_8)$ are coded with 6 bits each. The parameters a_3 and a_6 are quantized with quarter pixel accuracy, and coded with 7 bits each. Clearly, the amount of motion information to be transmitted depends on the number of independently moving objects. There exist several approaches for encoding the segmentation mask (shape) and the frame prediction error (color), which are described in [Mus 89, Die 93, Sch 93].

The difficulty with this approach is in the simultaneous estimation of the model parameters and scene segmentation, which is computationally demanding.

24.1.2 2-D Flexible Objects with 2-D Motion

Here, the source model is a flexible 2-D object with translational motion. This model, which forms the basis of the COST 211ter simulation model, has been proposed because it does not require complex motion estimation schemes, and hence leads to practical codec realizations. The main idea is to segment the current frame into three regions: i) unchanged or stationary areas, ii) moving areas, where each pixel corresponds to a pixel from the previous frame, and iii) uncovered background for which no corresponding pixel exists in the previous frame. Motion estimation is performed using a hierarchical block matching algorithm on a sparse grid of pixels within the moving region. A dense motion field is then computed via bilinear interpolation of the estimated motion vectors. Model-compliant (MC) moving objects are encoded by the respective 2-D motion vectors and their shape parameters. Shape and color information is encoded and transmitted for model-failure (MF) moving areas and the uncovered background.

The resulting algorithm, called the object-based analysis-synthesis coder (OBASC), whose flowchart is shown in Figure 24.3, can be summarized as follows:

1. Compute the Change Detection Mask (use 3×3 averaging, thresholding, 5×5 median filtering, and/or morphological operations to eliminate small changed/unchanged regions) The change detection mask is a binary mask that marks the changed and unchanged regions of the current picture with respect to the previous reconstructed picture.

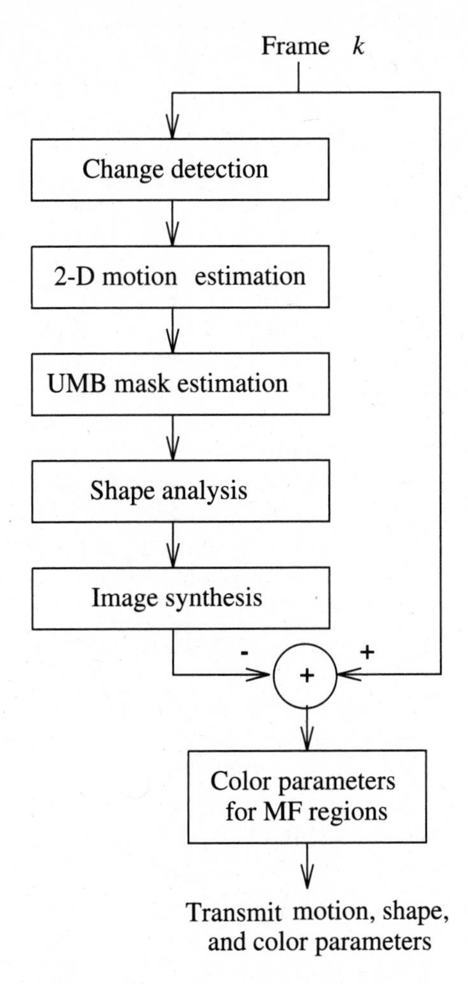

Figure 24.3: Flowchart of the 2-D translatory flexible model method.

2. Estimate motion vectors for those pixels within the changed region using three-level hierarchical block matching (HBM). The parameters used in HBM is tabulated in [Ger 94].

3. Compute the ternary UMB mask that marks unchanged areas, moving areas, and the uncovered background in the current picture. In order to distinguish moving pixels from the uncovered background, for each pel in the CHANGED region, we invert its motion vector. If the pel pointed to by the inverse of the motion vector is not in the CHANGED region, it is said to belong to the UNCOVERED background.

4. Approximate the shape of moving regions using the combined polygon/spline approximation method.

5. Synthesize moving regions in the present picture using motion and shape information. All analysis is performed on the luma component only, but synthesis must be performed for both luma and chroma components.

6. Determine Model Failure (MF) Regions.

7. Code pel data in MF and uncovered background regions. Several methods for coding of pel (color) data have been discussed in [Sch 93].

8. Code motion and shape parameters for Model Compliance (MC) regions. The sparse motion field is encoded using DPCM. The x_1 and x_2 components of the motion vectors are predicted separately using three-point spatial predictors whose coefficients are encoded using 6 bits each. The prediction error is usually run-length encoded [Hot 90]. The effectiveness of 2-D object-based coding methods depends strongly on how efficiently and accurately the shape (segmentation) information can be encoded. Several contour coding methods, including Fourier descriptors and polygon approximations, have been tested [Hot 90]. A combination of polygon approximation and spline-based representation of contours has been found to be most effective. In this approach the vertices of the polygon approximation are used to fit a spline to represent the segment boundary. The spline approximation is constrained to be within a prespecified distance from the actual boundary. The vertices of the polygon approximation are coded by MC-temporal DPCM. That is, first the vertices from the previous frame are translated by the estimated motion vectors. The MC-predicted vertices are tested for whether to accept or reject them. Additional vertices may need be inserted to account for the flexible nature of the object model. The positions of the newly inserted vertices are encoded by relative addressing.

9. Buffer regulation ensures smooth operation of the encoder at a fixed bitrate.

24.1.3 Affine Transformations with Triangular Meshes

Because of the computational complexity of simultaneous mapping parameter estimation and segmentation, and shape analysis algorithms, simpler motion-compensation schemes using spatial transformations based on a predetermined partition of the image plane into triangular or rectangular patches have recently been proposed [Nak 94]. These methods, although based on patches with a predetermined shape, provide results that are superior to those of block-based MC/DCT approach, because the spatial transformations accomodate rotation and scaling in addition to translation, and the implicit continuity of the motion field alleviates blocking artifacts even at very low bitrates.

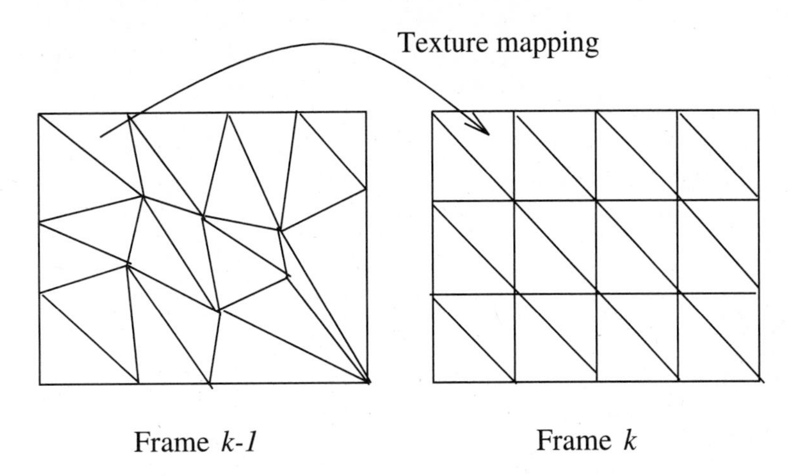

Figure 24.4: Motion compensation using triangular patches.

Nakaya and Harashima [Nak 94] propose using affine, bilinear, or perspective transformations, along with some new motion estimation algorithms, for improved motion compensation. In the case of the affine transformation, the present frame is segmented into triangular patches, because the affine transform has six free parameters which can uniquely be related to the (x_1, x_2) coordinates of the vertices of a triangular patch. Note that both the bilinear and perspective transforms have eight free parameters which can be related to the vertices of a quadrilateral patch. Assuming continuity of the motion field across the patches, it is sufficient to estimate the motion vectors at the vertices of the patches in frame k, called the grid points. Then texture within warped triangles in frame $k-1$, whose boundaries are determined by the estimated motion vectors, is mapped into the respective triangles in frame k, as depicted in Fig, 24.4.

An encoder using affine motion compensation (AFMC) implements the following steps:

1. Partition the present frame into triangular patches.
2. Estimate the motion vectors at the grid points in frame k. At the first stage, a rough estimate of the motion vectors can be obtained by using a standard block matching algorithm with rectangular blocks centered at the grid points. These estimates are then refined by the hexagonal search, which is a connectivity-preserving search procedure [Nak 94].
3. Determine the affine mapping parameters for each triangle given the displacement vectors at its vertices. Synthesize the present frame by mapping the color information from the previous reconstructed frame onto the corresponding patches in the present frame. Compute the synthesis error.
4. Encode both the motion vectors at the grid points and the synthesis error.

Note that no shape information needs to be transmitted in this approach. The number of grid points and the bit allocation for the transmission of synthesis error (color information) vary with the available bitrate. For example, at very low bitrates (under 10 kbps) no color information is usually transmitted. Segmentation of the current frame with adaptive patch boundaries that coincide with the boundaries of the moving objects (which would necessitate transmission of additional shape information) is left as a future research topic.

24.2 Knowledge-Based and Semantic Methods

Estimation of 3-D motion and structure of an unknown 3-D object from two frames (the case of 2-D/3-D rigid objects with 3-D motion) is a difficult problem, in that it requires high computational complexity, and the solution is very sensitive to noise. On the other hand, methods based on the simpler 2-D models, that is, the 2-D flexible model with translational motion and the affine transform using triangular patches, make certain assumptions that may not be satisfied. For example, in the former case motion may not be entirely translational, and in the latter, some triangles may cover two different objects. Knowledge-based 3-D modeling offers a compromise. It is applicable to cases where we have a priori information about the content of the scene in the form of a wireframe model, which is a mesh model composed of a set of triangular planar patches that are connected. However, the solution is considerably simpler than the case of unknown 3-D objects.

In knowledge-based coding, it is assumed that generic wireframe models have been designed off-line for certain objects of interest, which are available at both the transmitter (encoder) and the receiver (decoder). The encoder selects a suitable wireframe model for a particular scene, which is then scaled according to the size of the object in the reference frame. The motion of the object can then be described by the displacement of the vertices of the wireframe model. The objects can be modeled as moving 3-D rigid objects, where the global motion of all vertices on the same object can be characterized by a single set of six rigid motion parameters, or moving 3-D flexible objects (flexibly connected rigid components), where the wireframe model undergoes local motion deformations. Local motion deformations can be described by a set of motion parameters for each individual patch, where the vertices may move semi-independently under the geometrical constraints of a connected mesh model [Boz 94]. Alternatively, in the case of facial images, they can be described by semantic modeling techniques [Cho 94].

One of the main applications of model-based coding has been videophone, where scenes are generally restricted to head-and-shoulder types. In designing a wireframe model, the first step is to obtain the depth map of the speaker's head and shoulders, usually by scanning the speaker using collimated laser light. Once the depth map is obtained, the 3-D wireframe model is obtained through a triangularization procedure where small triangles are used in high-curvature areas and larger ones at low-curvature areas. The wireframe model is stored in the computer as a set of linked arrays. One set of arrays lists the X_1, X_2, X_3 coordinates of each vertex and another set gives the addresses of the vertices forming each triangle. There are several wireframe models used by different research groups. An extended version of the CANDIDE model [Ryd 87] is shown in Figure 24.5.

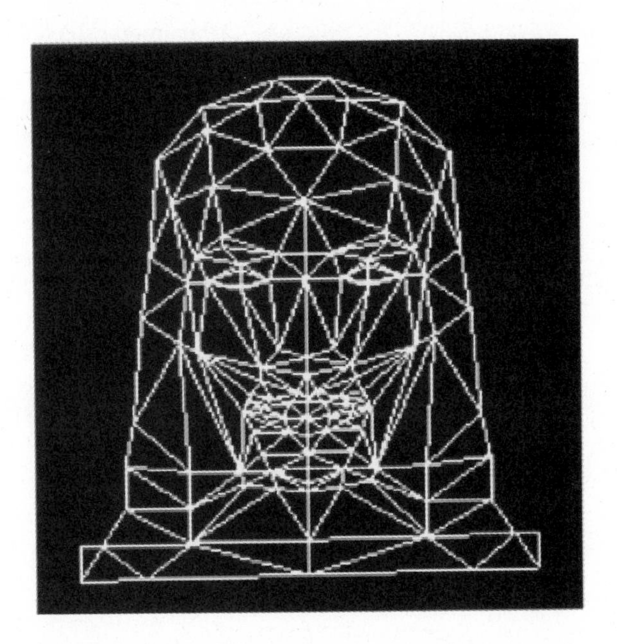

Figure 24.5: Wireframe model of a typical head-and-shoulder scene [Wel 91].

In the following, we first discuss the basic principles of the knowledge-based approach. Then we present two specific algorithms, the MBASIC algorithm [Aiz 93], and a more sophisticated adaptive scaling and tracking algorithm [Boz 94].

24.2.1 General Principles

A block diagram of the 3-D knowledge-based coding scheme is shown in Figure 24.6. The encoder is composed of four main components: i) image analysis module, which includes scaling of the 3-D wireframe model, global and local motion estimation, ii) image synthesis module, which includes texture mapping, iii) model update, which is updating of the coordinates of the wireframe model and the texture information, and iv) parameter coding. Some of these steps are described below.

1. *Wireframe model fitting*

   The accuracy of tracking the motion of the wireframe model from frame to frame strongly depends on how well the wireframe model matches the actual

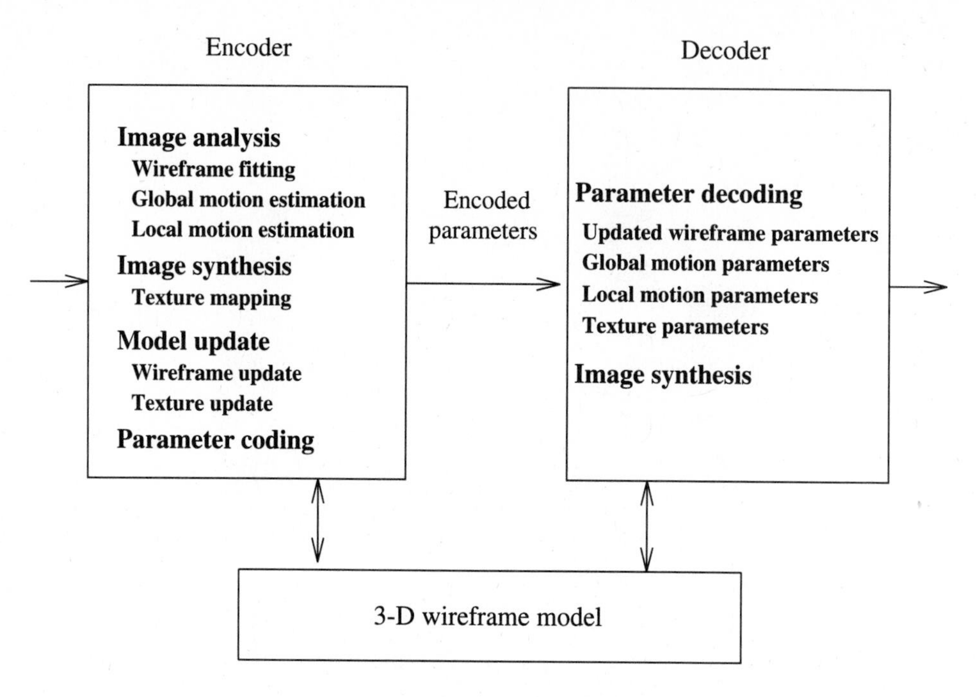

Figure 24.6: Block diagram of knowledge-based coding.

speaker in the scene. Since the size and shape of the head and the position of the eyes, mouth, and nose vary from person to person, it is necessary to modify the 3-D wireframe model to fit the actual speaker. Thus, the first step in 3-D knowledge-based coding of a facial image sequence is to adapt a generic wireframe model to the actual speaker.

Initial studies on 3-D model-based coding have scaled the wireframe model by fitting the orthographic projection of the wireframe model to a frontal view of the actual speaker by means of affine transformations and a set of manually selected feature points [Aiz 89, Aiz 93, Kan 91]. The four feature points used by Aizawa *et al.* [Aiz 89, Aiz 93], tip of the chin, temples, and a point midway between the left and right eyebrows, are shown in Figure 24.7. The points (x_1', x_2') in frame k that correspond to the selected feature points (x_1, x_2) on the orthographic projection of the wireframe model are marked interactively. The parameters of an affine transform, $x_1' = ax_1 + bx_2 + c$ and $x_2' = dx_1 + ex_2 + f$, are then estimated using a least squares procedure to obtain the best fit at the selected feature points. This transformation is subsequently applied to the coordinates of all vertices for scaling. The depth at each vertex is modified according to the scaling factor $\sqrt{(a^2 + e^2)/2}$. In an attempt to automatic scaling, Huang *et al.* [Hua 91] propose using

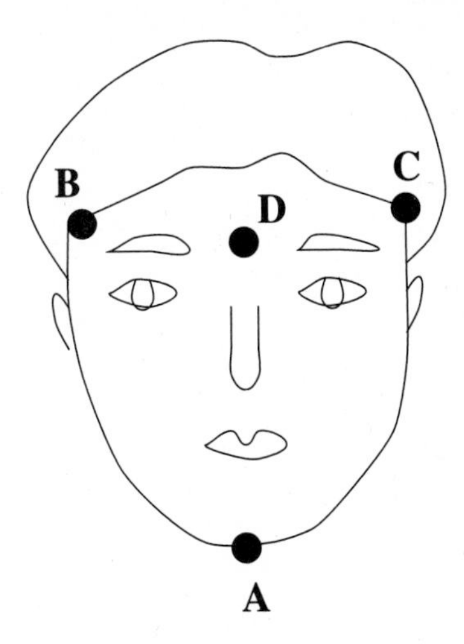

Figure 24.7: Feature points to be used in scaling the wireframe.

spatial and temporal gradients of the image to estimate the maximum height and width of the actual face and scale the wireframe model accordingly.

An alternative approach is to use snakes or ellipses to model the boundary of the face. Recently, Reinders *et al.* [Rei 92] consider automated global and local modification of the 2-D projection of the wireframe model in the x_1 and x_2 directions. They segment the image into background, face, eyes, and mouth, and approximate the contours of the face, eyes, and mouth with ellipses. Then local transformations are performed using elastic matching techniques. Waite and Welsh use snakes to find the boundary of the head, which is claimed to be a robust method [Wel 90]. However, all of the above methods have applied an approximate scaling in the z-direction (depth) since they are based on a single frame.

2. *Motion Analysis*

The facial motion can be analyzed into two components, the global motion of the head and the local motion due to facial expressions, such as motion of the mouth, eyebrows, and eyes. The global motion of the head can be characterized by the six rigid motion parameters. These parameters can be estimated using point correspondence or optical flow based approaches, which were discussed in Chapters 9-12. The MBASIC and the flexible-wireframe-model based approaches that are presented in the next two subsections are representative of the respective approaches. These methods also provide depth estimates at the selected feature points, which may be used for improved scaling of the depth parameters of the generic wireframe model.

A popular approach to characterize local motion is to describe it in terms of the so-called facial action units (AUs) [Aiz 93, Cho 94] which are based on the Facial Action Coding system (FACS) [Ekm 77]. FACS describes facial expressions in terms of AUs that are related to movement of single muscles or clusters of muscles. According to FACS, a human facial expression can be divided into approximately 44 basic AUs, and all facial expressions can be synthesized by an appropriate combination of these AUs. Several algorithms have been proposed to analyze local motion using AUs [For 89, Kan 91, Aiz 93, Li 93, Cho 94]. Among these Forchheimer [For 89] used a least squares estimation procedure to find the best combination of AUs to fit the residual motion vector field (the displacement field after compensating for the global motion). That is, the vector **a** of AU parameters is estimated from

$$\Delta d = Aa$$

where $\mathbf{\Delta d}$ denotes the residual displacement vector field, and $\mathbf{A}$ is the matrix of displacement vectors for each AU. Recently, Li *et al.* [Li 93] proposed a method to recover both the local and global motion parameters simultaneously from the spatio-temporal derivatives of the image using a similar principle.

Figure 24.8: Demonstration of facial expression synthesis using AUs.

Facial action units are demonstrated in Figure 24.8 using a frame of the sequence "Miss America." The picture illustrates the synthesis of the AUs 2, 17, and 46, corresponding to "outer brow raiser," "chin raiser," and "blinking." Deformable contour models have also been used to track the nonrigid motion of facial features [Ter 90]. Huang *et al.* [Hua 91] used splines to track features, such as eyes, eyebrows, nose and the lips.

3. *Synthesis*

   Synthesis of facial images involves transformation of the wireframe model according to a particular set of global and local motion parameters followed by texture mapping. Texture mapping is a widely studied topic in computer graphics to obtain realistic images [Yau 88, Aiz 93]. It refers to mapping a properly warped version of the texture observed in the first frame onto the surface of the deformed wireframe model.

   Texture mapping under orthographic projection can be described as follows:
   i) collapse the initial wireframe model onto the image plane (by means of orthographic projection) to obtain a collection of triangles,
   ii) map the observed texture in the first frame into the respective triangles,
   iii) rotatate and translate the initial wireframe model according to the given global and local motion parameters, and then collapse again to obtain a set of deformed triangles for the next frame, and
   iv) map the texture within each triangle in the first frame into the corresponding triangle by means of appropriate decimation or interpolation.

   Texture mapping is illustrated in Figure 24.9, where Figure 24.9.a shows one of the collapsed triangles in the initial frame. Figure 24.9.b depicts the corresponding triangle with its appropriately warped texture. Texture warping can be accomplished with orthographic or perspective projection techniques.

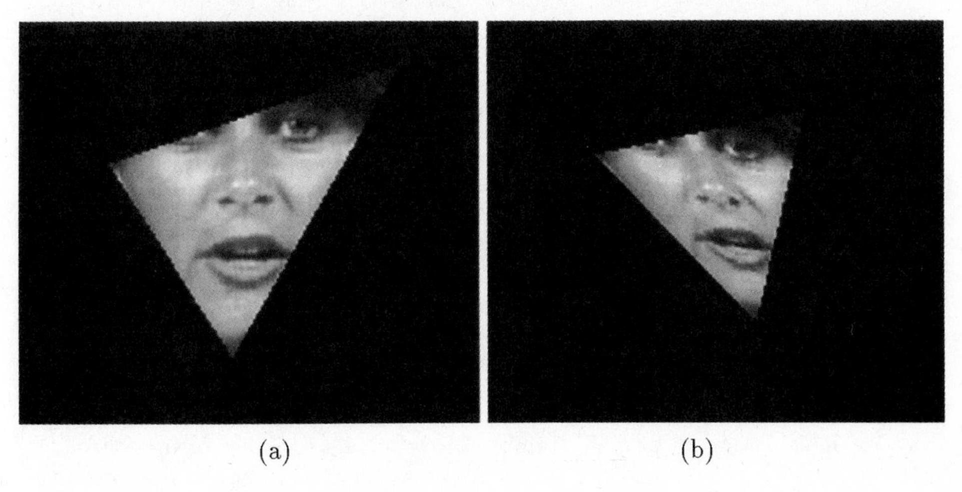

(a) (b)

Figure 24.9: Texture mapping: a) before and b) after processing.

3-D knowledge-based coding schemes have been shown to yield higher compression ratios when applied to typical videophone scenes than do waveform coding and/or 2-D object-based coding, since 3-D models provide a more compact description of the scene for these special class of images. Clearly, knowledge-based schemes offer a compromise between generality and higher compression ratio. Practical implementations of object-based schemes usually feature a fall-back mode, such as DCT frame coding, to cope with frame-to-frame accumulation of image analysis and synthesis errors or sudden scene changes.

24.2.2 MBASIC Algorithm

The MBASIC algorithm is a simple knowledge-based coding scheme that is based on the source model of a known wireframe model (up to a scaling factor) subject to 3-D rigid motion (to model the global motion of the head). Facial action unit analysis has been used to model the local motion deformations. The algorithm can be summarized through the following steps [Aiz 89]:

- *Scale the wireframe model.* First, an edge detection is performed to find the boundaries of the face. Certain extreme points on the edge map, such as the corners of the two ears, the middle of the chin, and the forehead, are then detected to compute the maximum horizontal and vertical distances. The x_1 and x_2 scaling factors are then calculated to match the respective distances on the generic wireframe model to these values. The scaling factor in the X_3 direction is approximated by the mean of the x_1 and x_2 scaling factors

- *Determine N matching point pairs.* Mark seven to ten characteristic points on the rigid parts of the face, such as tip of the nose, around the eye brows, etc., on the initial frame. Find the best matching points in the next frame using block matching or another technique.

- *Estimate the global 3-D motion and structure parameters.* The two-step iteration process described in Chapter 9 is used to estimate the 3-D motion and structure parameters. That is, we initialize the unknown depth parameters using the values obtained from the scaled wireframe model. Next, we estimate the 3-D motion parameters given the depth parameters from (9.10), and estimate the depth parameters from (9.11) using the 3-D motion parameters estimated in the previous step, iteratively.

- *Estimate action units.* After compensating for the global motion of the head, the AUs that best fit the residual motion field can be estimated.

Since scaling of the wireframe model in the X_3 (depth) direction is approximate (because it is based on a single frame), there is an inevitable mismatch of the initial depth (X_3) parameters of the wireframe model and the actual speaker in the image sequence. The two-step motion estimation procedure has been found to be sensitive to these errors if it exceeds 10%. To overcome this problem, an improved iterative algorithm (see Section 9.2.2) can be employed for 3-D motion and structure parameter estimation in the above procedure.

24.2.3 Estimation Using a Flexible Wireframe Model

Many existing methods consider fitting (scaling) a generic wireframe to the actual speaker using only the initial frame of the sequence [Aiz 89, Rei 92]. Thus, the scaling in the z-direction (depth) is necessarily approximate. Furthermore, although the utility of photometric cues in 3-D motion and structure estimation are well known [Ver 89, Pea 90, Pen 91, Dri 92], photometric information has not commonly been used in the context of motion estimation in knowledge-based coding. In this section, we introduce a recent formulation where the 3-D global and local motion estimation and the adaptation of the wireframe model are considered simultaneously within an optical-flow-based framework, including the photometric effects of motion. The adaptation of the wireframe model serves two purposes that cannot be separated: to reduce the misfit of the wireframe model to the speaker in frame $k-1$, and to account for the local motion from frame $k-1$ to frame k without using any *a priori* information about the AUs.

The source model is a flexible wireframe model whose local structure is characterized by the normal vectors of the patches which are related to the coordinates of the nodes. Geometrical constraints that describe the propagation of the movement of the nodes are introduced, which are then efficiently utilized to reduce the number of independent structure parameters. A stochastic relaxation algorithm has been used to determine optimum global motion estimates and the parameters describing the structure of the wireframe model. The simultaneous estimation formulation is motivated by the fact that estimation of the global and local motion and adaptation of the wireframe model, including the depth values, are mutually related; thus, a combined optimization approach is necessary to obtain the best results. Because an optical-flow-based criterion function is used, computation of the synthesis error is not necessary from iteration to iteration, and thus, results in an efficient implementation. The synthesis error at the conclusion of the iterations is used to validate the estimated parameters, and to decide whether a texture update is necessary.

In the following we summarize the estimation of the illuminant direction, the formulation of the simultaneous motion estimation and wireframe adaptation problem including the photometric effects of motion, and the algorithm for the proposed simultaneous estimation, including an efficient method to update the nodes of the wireframe model.

Estimation of the Illuminant Direction

Photometric effects refer to the change in shading due to 3-D motion of the object. For example, in the case of rotational motion, because the surface normals change, the shading of the objects varies even if the external illumination remains constant. Recently, Pentland [Pen 91] showed that the changes in image intensity because of photometric effects can dominate intensity changes due to the geometric effects of motion (changes in projected surface geometry due to 3-D motion). Similar discussions can be found in [Ver 89, Pea 90, Dri 92]. Here we briefly discuss estimation of the illuminant direction with the aim of incorporating photometric effects into the above optical-flow-based formulation.

Recall from Section 2.3 that the image intensity $s_c(x_1, x_2, t)$ can be expressed as

$$s_c(x_1, x_2, t) = \rho \mathbf{N}(t) \cdot \mathbf{L} \tag{24.1}$$

where $\mathbf{L} = (L_1, L_2, L_3)$ is the unit vector in the mean illuminant direction and $\mathbf{N}$ is the unit surface normal of the scene at position $(X_1, X_2, X_3(X_1, X_2))$ given by

$$\mathbf{N}(t) = (-p, -q, 1)/(p^2 + q^2 + 1)^{1/2} \tag{24.2}$$

and $p = \frac{\partial X_3}{\partial x_1}$ and $q = \frac{\partial X_3}{\partial x_2}$ are the partial derivatives of depth $X_3(x_1, x_2)$ with respect to the image coordinates x_1 and x_2, respectively.

Note that the illuminant direction $\mathbf{L}$ can also be expressed in terms of tilt and slant angles as

$$\mathbf{L} = (L_1, L_2, L_3) = (\cos\tau \sin\sigma, \sin\tau \sin\sigma, \cos\sigma) \tag{24.3}$$

where τ, the tilt angle of the illuminant, is the angle between $\mathbf{L}$ and the $X_1 - X_3$ plane, and σ, the slant angle, is the angle between $\mathbf{L}$ and the positive X_3 axis. In order to incorporate the photometric effects of motion into 3-D knowledge-based coding, the illuminant direction $\mathbf{L}$ must be known or estimated from the available frames.

A method to estimate the tilt and slant angles of the illuminant, based on approximating the 3-D surface by spherical patches, was proposed by Zheng *et al.* [Zhe 91]. They estimate the tilt angle as

$$\tau = \arctan\left(\frac{E\{\hat{L}_1/\sqrt{\hat{L}_1^2 + \hat{L}_2^2}\}}{E\{\hat{L}_2/\sqrt{\hat{L}_1^2 + \hat{L}_2^2}\}}\right) \tag{24.4}$$

where $E\{.\}$ denotes expectation over the spatial variables (which is approximated by 4×4 local averaging), and $\hat{L}_1$ and $\hat{L}_2$ are the x_1 and x_2 components of the local estimate of the tilt of the illuminant, respectively, computed as

$$\begin{bmatrix} \hat{L}_1 \\ \hat{L}_2 \end{bmatrix} = (B^t B)^{-1} B^t \begin{bmatrix} \delta I_1 \\ \delta I_2 \\ \vdots \\ \delta I_N \end{bmatrix}, \quad \text{and} \quad B = \begin{bmatrix} \delta x_{11} & \delta x_{21} \\ \delta x_{12} & \delta x_{22} \\ \vdots & \vdots \\ \delta x_{1N} & \delta x_{2N} \end{bmatrix}$$

Here, δI_i, $i = 1, \ldots, N$ is the difference in image intensity along a particular direction $(\delta x_{1i}, \delta x_{2i})$, and N is the number of directions (typically $N = 8$ for each 4×4 window).

The slant angle σ can be uniquely estimated from

$$\frac{E\{I\}}{E\{I^2\}} = f_3(\sigma) \tag{24.5}$$

since $f_3(\sigma)$ (defined in [Zhe 91]) is a monotonically decreasing function of σ, where $E\{I\}$ and $E\{I^2\}$ are the expected values of the image intensities and the square of the image intensities, respectively, estimated from the image area where the wireframe model is fitted.

Finally, the surface albedo can be estimated from

$$\rho = \frac{E\{I\} \cdot f_1(\sigma) + \sqrt{E\{I^2\} \cdot f_2(\sigma)}}{f_1^2(\sigma) + f_2(\sigma)} \tag{24.6}$$

where $f_1(\sigma)$ and $f_2(\sigma)$ are seventh-order polynomials in $\cos\sigma$ as defined in [Zhe 91].

Incorporation of the Photometric Effects into the OFE

Since we represent the 3-D structure of a head-and-shoulders scene by a wireframe model and the surface of the wireframe model is composed of planar patches, the variation in the intensity of a pixel due to photometric effects of motion will be related to a change in the normal vector of the patch to which this pixel belongs.

Assuming that the mean illuminant direction $\mathbf{L} = (L_1, L_2, L_3)$ remains constant, we can represent the change in intensity due to photometric effects of motion, based on the photometric model (2.32), as

$$\frac{ds_c(x_1, x_2, t)}{dt} = \rho \mathbf{L} \cdot \frac{\mathbf{dN}}{\mathbf{dt}} \tag{24.7}$$

Then, substituting Equations (24.7) and (2.33) into the optical flow equation (5.5), and expressing 2-D velocities in terms of the 3-D motion parameters, we include the photometric effects into the optical-flow-based formulation as [Boz 94]

$$\frac{\partial s_c}{\partial x_1}(\Omega_3 x_2 - \Omega_2 X_3 + V_1) + \frac{\partial s_c}{\partial x_2}(-\Omega_3 x_1 + \Omega_1 X_3 + V_2) + \frac{\partial s_c}{\partial t} =$$
$$\rho \mathbf{L} \cdot \left[\frac{(-p', -q', 1)^T}{\sqrt{p'^2 + q'^2 + 1}} - \frac{(-p, -q, 1)^T}{\sqrt{p^2 + q^2 + 1}} \right] \tag{24.8}$$

The term on the right-hand side of Equation (24.8) may be significant, especially if the change in the surface normal has components either toward or away from the illuminant direction [Pen 91].

Structure of the Wireframe Model

Next, we introduce geometrical constraints about the structure of the wireframe model and discuss formulation of the simultaneous estimation problem. The wireframe model is composed of triangular patches which are characterized by the

(X_1, X_2, X_3) coordinates of their respective vertices. Given the (X_1, X_2, X_3) coordinates of the vertices of a patch, we can write the equation of the plane containing this patch. Let $P_1^{(i)} = (X_{11}^{(i)}, X_{21}^{(i)}, X_{31}^{(i)})$, $P_2^{(i)} = (X_{12}^{(i)}, X_{22}^{(i)}, X_{32}^{(i)})$ and $P_3^{(i)} = (X_{13}^{(i)}, X_{23}^{(i)}, X_{33}^{(i)})$ denote the vertices of the ith patch, and $P^{(i)} = (X_1^{(i)}, X_2^{(i)}, X_3^{(i)})$ be any point on this patch. Then,

$$\overrightarrow{P^{(i)}P_1^{(i)}} \cdot (\overrightarrow{P_2^{(i)}P_1^{(i)}} \times \overrightarrow{P_3^{(i)}P_1^{(i)}}) = 0$$

gives the equation of the plane containing $P_1^{(i)}$, $P_2^{(i)}$ and $P_3^{(i)}$, where $\overrightarrow{P^{(i)}P_1^{(i)}}$, $\overrightarrow{P_2^{(i)}P_1^{(i)}}$, and $\overrightarrow{P_3^{(i)}P_1^{(i)}}$ are the vectors from the former point to the latter, respectively. We can express this equation in the form

$$X_3^{(i)} = p_i X_1^{(i)} + q_i X_2^{(i)} + c_i \tag{24.9}$$

where

$$p_i = -\frac{(X_{22}^{(i)} - X_{21}^{(i)})(X_{33}^{(i)} - X_{31}^{(i)}) - (X_{32}^{(i)} - X_{31}^{(i)})(X_{23}^{(i)} - X_{21}^{(i)})}{(X_{12}^{(i)} - X_{11}^{(i)})(X_{23}^{(i)} - X_{21}^{(i)}) - (X_{22}^{(i)} - X_{21}^{(i)})(X_{13}^{(i)} - X_{11}^{(i)})}$$

$$q_i = -\frac{(X_{32}^{(i)} - X_{31}^{(i)})(X_{13}^{(i)} - X_{11}^{(i)}) - (X_{12}^{(i)} - X_{11}^{(i)})(X_{33}^{(i)} - X_{31}^{(i)})}{(X_{12}^{(i)} - X_{11}^{(i)})(X_{23}^{(i)} - X_{21}^{(i)}) - (X_{22}^{(i)} - X_{21}^{(i)})(X_{13}^{(i)} - X_{11}^{(i)})}$$

and

$$\begin{aligned} c_i &= X_{31}^{(i)} + X_{11}^{(i)} \frac{(X_{22}^{(i)} - X_{21}^{(i)})(X_{33}^{(i)} - X_{31}^{(i)}) - (X_{32}^{(i)} - X_{31}^{(i)})(X_{23}^{(i)} - X_{21}^{(i)})}{(X_{12}^{(i)} - X_{11}^{(i)})(X_{23}^{(i)} - X_{21}^{(i)}) - (X_{22}^{(i)} - X_{21}^{(i)})(X_{13}^{(i)} - X_{11}^{(i)})} \\ &+ X_{21}^{(i)} \frac{(X_{32}^{(i)} - X_{31}^{(i)})(X_{13}^{(i)} - X_{11}^{(i)}) - (X_{12}^{(i)} - X_{11}^{(i)})(X_{33}^{(i)} - X_{31}^{(i)})}{(X_{12}^{(i)} - X_{11}^{(i)})(X_{23}^{(i)} - X_{21}^{(i)}) - (X_{22}^{(i)} - X_{21}^{(i)})(X_{13}^{(i)} - X_{11}^{(i)})} \end{aligned}$$

Using Equation (24.9), the X_3 coordinate of any point on the ith patch can be expressed in terms of the parameters p_i, q_i, and c_i and the X_1 and X_2 coordinates of the point. Then, we can eliminate X_3 from Equation (24.8) by substituting (24.9) into (24.8) with $X_1^{(i)} = x_1^{(i)}$, $X_2^{(i)} = x_2^{(i)}$, where the patch index i is determined for each (x_1, x_2) according to the orthographic projection.

It is important to note that the parameters p_i, q_i, and c_i of each planar patch of the wireframe are not completely independent of each other. Each triangular patch is either surrounded by two (if it is on the boundary of the wireframe) or three other triangles. The requirement that the neighboring patches must intersect at a straight line imposes a constraint on the structure parameters of these patches in the form

$$p_i x_1^{(ij)} + q_i x_2^{(ij)} + c_i = p_j x_1^{(ij)} + q_j x_2^{(ij)} + c_j \tag{24.10}$$

where p_j, q_j, and c_j denote the parameters of the jth patch, and $(x_1^{(ij)}, x_2^{(ij)})$ denote the coordinates of a point that lie at the intersection of the ith and jth patches.

Problem Statement

The 3-D global motion parameters Ω_1, Ω_2, Ω_3, V_1, V_2, and the structure parameters p_i, q_i, c_i, can be simultaneously estimated by minimizing the sum square error in the optical flow equation (24.8) over all pixels in a frame, given by

$$E = \sum_i \sum_{(x_1,x_2)\in i^{th} patch} e_i^2(x_1,x_2) \tag{24.11}$$

where

$$\begin{aligned} e_i(x_1,x_2) &= \frac{\partial s_c}{\partial x_1}(\Omega_3 x_2 - \Omega_2(p_i x_1 + q_i x_2 + c_i) + V_1) \\ &+ \frac{\partial s_c}{\partial x_2}(-\Omega_3 x_1 + \Omega_1(p_i x_1 + q_i x_2 + c_i) + V_2) + \frac{\partial s_c}{\partial t} \\ &- \rho(L_1,L_2,L_3) \cdot \left(\frac{(-\frac{-\Omega_2+p_i}{1+\Omega_2 p_i}, -\frac{\Omega_1+q_i}{1-\Omega_1 q_i}, 1)}{((\frac{-\Omega_2+p_i}{1+\Omega_2 p_i})^2 + (\frac{\Omega_1+q_i}{1-\Omega_1 q_i})^2 + 1)^{1/2}} - \frac{(-p_i,-q_i,1)}{(p_i^2+q_i^2+1)^{1/2}} \right) \end{aligned}$$

with respect to Ω_1, Ω_2, Ω_3, V_1, V_2, p_i, q_i, c_i, and $i = 1, \ldots,$ number of patches, subject to the geometrical constraints given by (24.10). It is assumed that the values of ρ and (L_1, L_2, L_3) are estimated *a priori*. The constraints, Equation (24.10), not only enable us to preserve the structure of the wireframe during the adaptation, but also facilitate reducing the number of independent unknowns in the optimization process, as described in the following.

The Algorithm

The criterion function E, defined by Equation (24.11), can be minimized using a stochastic relaxation algorithm (see Chapter 8) to find the global optima of Ω_1, Ω_2, Ω_3, V_1, V_2, p_i, q_i, c_i. Each iteration consists of perturbing the state of the system in some random fashion in a manner consistent with the constraint equations (24.10). The constraints are enforced as follows: At each iteration cycle, we visit all patches of the wireframe model in sequential order. If, at the present iteration cycle, none of the neighboring patches of patch i has yet been visited (e.g., the initial patch), then p_i, q_i, c_i are all independently perturbed. If only one of the neighboring patches, say patch j, has been visited (p_j, q_j, c_j have already been updated), then two of the parameters, say p_i and q_i, are independent and perturbed. The dependent variable c_i is computed from Equation (24.10) as

$$c_i = p_j x_1^{(ij)} + q_j x_2^{(ij)} + c_j - p_i x_1^{(ij)} - q_i x_2^{(ij)} \tag{24.12}$$

where $(x_1^{(ij)}, x_2^{(ij)})$ is one of the nodes common to both patches i and j that is either in the boundary or has already been updated in the present iteration cycle. If two of the neighboring patches, say patches j and k, have already been visited, i.e., the variables p_j, q_j, c_j and p_k, q_k, c_k have been updated, then only one variable, say p_i, is independent and perturbed. In this case, c_i can be found from Equation (24.12), and q_i can be evaluated as

$$q_i = \frac{p_k x_1^{(ik)} + q_j x_2^{(ik)} + c_k - p_i x_1^{(ik)} - c_k}{x_2^{(ik)}} \tag{24.13}$$

where $(x_1^{(ik)}, x_2^{(ik)})$ is one of the nodes common to both patches i and k that is either in the boundary or has already been updated in the present iteration cycle.

The perturbation of the structure parameters p_i, q_i, and c_i for each patch i results in a change in the coordinates of the nodes of the updated wireframe. The new coordinates $(X_1^{(n)}, X_2^{(n)}, X_3^{(n)})$ of the node n can be computed given the updated structure parameters of three patches that intersect at node n. Let the patches i, j, and k intersect at node n. Then the relations

$$\begin{aligned} p_i X_1^{(n)} + q_i X_2^{(n)} + c_i &= p_j X_1^{(n)} + q_j X_2^{(n)} + c_j \\ p_i X_1^{(n)} + q_i X_2^{(n)} + c_i &= p_k X_1^{(n)} + q_k X_2^{(n)} + c_k \end{aligned} \tag{24.14}$$

specify $X_1^{(n)}$ and $X_2^{(n)}$. Therefore,

$$\begin{bmatrix} X_1^{(n)} \\ X_2^{(n)} \end{bmatrix} = \begin{bmatrix} p_i - p_j & p_j - p_k \\ q_i - q_j & q_j - q_k \end{bmatrix}^{-1} \begin{bmatrix} c_j - c_i \\ -c_j + c_k \end{bmatrix} \tag{24.15}$$

The new $X_3^{(n)}$ can be computed from Equation (24.9) given $X_1^{(n)}$, $X_2^{(n)}$, and the p_i, q_i, c_i for any patch passing through that node. It is this updating of the coordinates of the nodes that allows the adaptation of the wireframe model to lower the misfit error and accommodate the presence of local motion, such as the motion of the eyes and the mouth.

The proposed algorithm can be summarized as follows:

1. Estimate the illumination direction using Equations (24.4) and (24.5), and the surface albedo using Equation (24.6).

2. Initialize the coordinates $(X_1^{(n)}, X_2^{(n)}, X_3^{(n)})$, of all nodes n, using an approximately scaled initial wireframe model. Determine the initial values of p_i, q_i, and c_i for all patches. Set the iteration counter $m = 0$.

3. Determine the initial motion estimates using Equation (9.10) based on a set of selected feature correspondences and their depth values obtained from the wireframe model.

4. Compute the value of the cost function E given by (24.11).

5. If $E < \epsilon$, stop.

   Else, set $m = m + 1$, and

Perturb the motion parameters $\mathbf{\Omega} = [\Omega_1\ \Omega_2\ \Omega_3\ V_1\ V_2]^T$ as

$$\mathbf{\Omega}^{(m)} \longleftarrow \mathbf{\Omega}^{(m-1)} + \alpha^m \mathbf{\Delta}, \tag{24.16}$$

where the components of $\mathbf{\Delta}$ are zero-mean Gaussian with the variance $\sigma^{2^{(m)}} = E$; and the structure parameters p_i, q_i, and c_i through the procedure:

```
Define count_i as the number of neighboring patches to patch i
whose structure parameters have been perturbed. Set count_i=0,
for all patches i.
Perturb p_1, q_1, c_1 as
```

$$\begin{aligned} p_1^{(m)} &\longleftarrow p_1^{(m-1)} + \alpha^m \Delta_1 \\ q_1^{(m)} &\longleftarrow q_1^{(m-1)} + \alpha^m \Delta_1 \\ c_1^{(m)} &\longleftarrow c_1^{(m-1)} + \alpha^m \Delta_1 \end{aligned} \tag{24.17}$$

where $\Delta_i = N_i(0, \sigma_i^{2^{(m)}})$, i.e., zero mean Gaussian with variance $\sigma_i^{2^{(m)}}$, where $\sigma_i^{2^{(m)}} = \sum_{(x,y)\in\text{patch } i} e_i^2(x,y)$.

```
increment count_j, for all j denoting neighbors of patch 1.
for( i=2 to number of patches)
 {
 if(count_i==1) {
   Perturb p_i and q_i.
   Increment count_m, for all m denoting neighbors of patch i.
   Compute c_i using Equation 24.12, where x_ij and y_ij are
   the coordinates of a fixed or a precomputed node on
   the line of intersection between patches i and j. }

 if(count_i==2) {
   Perturb p_i.
   Increment count_m, for all m denoting neighbors of patch i.
   Compute c_i using Equation 24.12 and q_i using Equation 24.13,
   where (x_ij,y_ij) and (x_ik,y_ik) denote coordinates of
   a fixed or a precomputed node on the line of intersection
   between patches i,j and i,k respectively. }

 If p_i, q_i, and c_i for at least three patches intersecting
 at a node are updated, then update the coordinates of the node
 by using Equation 24.15.
 }
```

6. Go to (4).

The synthesis error, which may be due to (i) misfit of the wireframe, (ii) error in global motion estimation, (iii) error in local motion estimation, and (iv) the photometric effects of the motion, can also be coded and transmitted to improve upon image quality.

24.3 Examples

We have implemented six algorithms using the first and fourth frames of the Miss America sequence (at the frame rate 10 frames/s): a scaled-down version of the H.261 algorithm, the OBASC algorithm (Section 24.1.2), two variations of the AFMC algorithm (Section 24.1.3), the global motion compensation part of the MBASIC algorithm, and the flexible 3-D wireframe model-based algorithm.

The first two algorithms are executed on QCIF frames (176 pixels × 144 lines). The specific implementation of the H.261 coder that we used is the PVRG-P64 coder with the target bitrate of 16 kbps. The reconstructed fourth frame is shown in Figure 24.10. In the OBASC algorithm, the size of the model failure region, hence the amount of color information that needs to be transmitted, is adjusted to meet the same target bitrate. Figure 24.11 (a) and (b) show the model failure region and the reconstructed fourth frame using the OBASC algorithm. Visual comparison of the images confirm that the PVRG-P64 reconstructed image suffers from annoying blocking artifacts. The PSNR values are shown in Table 24.1.

The latter four algorithms are executed on CIF format images. The reconstructed images are then converted into QCIF format to compare their PSNR with those of the first two algorithms. Note that Table 24.1 reports the PSNR for these images in both QCIF and CIF formats. In all cases, model-failure regions are encoded using a DCT based approach with a target bitrate of 16 kbps. In the 3-D model-based algorithms, we have used the extended CANDIDE wireframe, depicted in Figure 24.5, with 169 nodes. The 2-D triangular mesh model that is used in the AFMC has been generated by computing the orthographic projection of CANDIDE into the image plane. The wireframe has been fitted to the first frame by a least squares scaling procedure using 10 preselected control points. Figure 24.12 (a) depicts the wireframe model overlayed onto the original fourth frame. Two alternative approaches have been tried with the AFMC method: i) motion vectors at the nodes of the 2-D mesh are estimated using the Lucas-Kanade (LK) method, which are then used to compute affine motion parameters (AFMC-LK method), and ii) a hexagonal search procedure starting with the LK motion estimates is employed (AFMC-Hex method). The PSNR for both are shown in Table 24.1. The reconstructed fourth frame using the AFMC-Hex method is depicted in Figure 24.12 (b).

Among the 3-D model-based methods, we have implemented the global-motion compensation module of the MBASIC algorithm (Global-3D) and the method based on the 3-D flexible wireframe model (Flexible-3D). The fourth frame after global motion compensation by the five global motion parameters (3 rotation and 2 translation) is depicted in Figure 24.13 (a). Finally, the result obtained by the Flexible-3D method which provides both global and local motion compensation is shown in Figure 24.13 (b). We conclude by noting that the 2-D model-based implementations are more general and robust [Tek 95]. Methods based on the 3-D object models may provide more compact motion representations, with possibly lower accuracy. Note however that, the method based on the flexible 3-D wireframe model facilitates incorporation of photometric effects, which may sometimes prove significant.

Table 24.1: Comparison of model-based motion-compensation results.

Method	PSNR (dB)	
	QCIF	CIF
Frame Difference	33.01	31.24
PVRG-64	34.84	-
OBASC	38.01	-
AFMC-LK	38.19	36.47
AFMC-Hex	41.08	37.40
Global-3D	36.01	34.24
Flexible-3D	38.67	36.53

Figure 24.10: A scaled-down implementation of the H.261 algorithm.

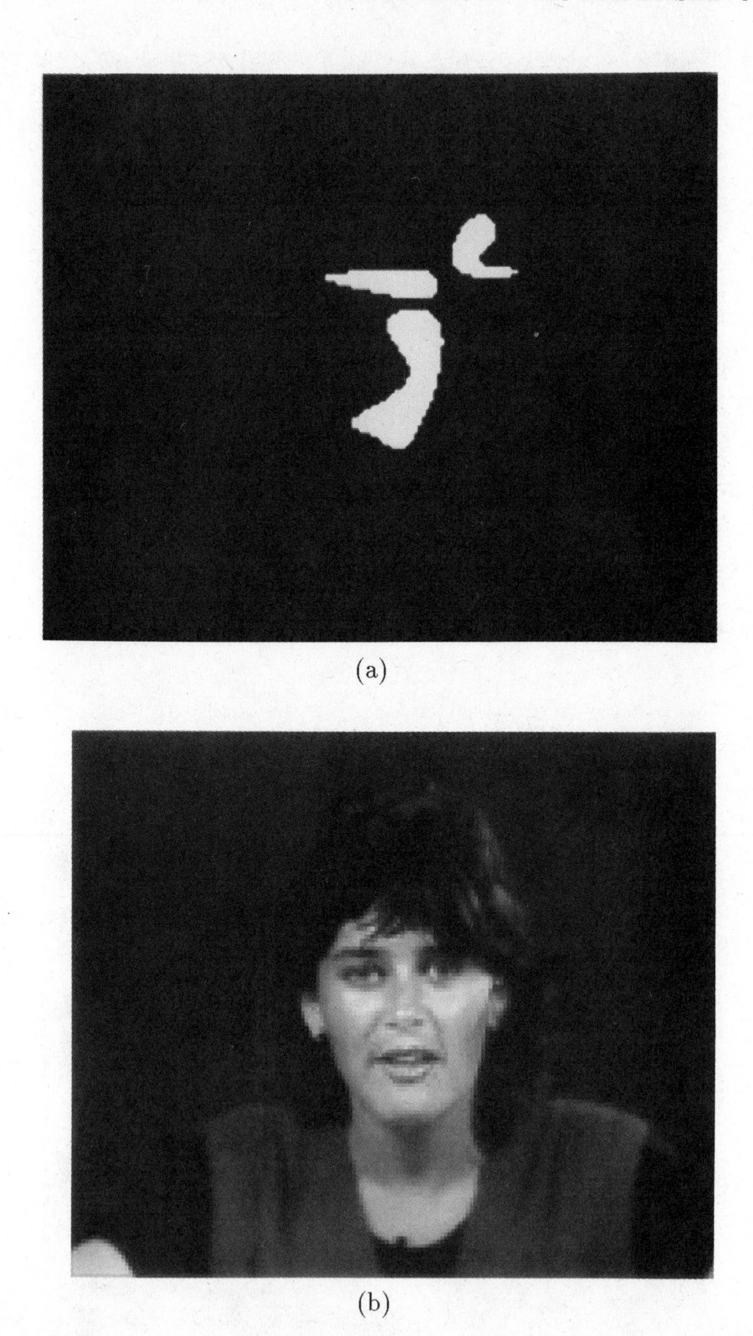

(a)

(b)

Figure 24.11: Flexible 2-D object based method: a) the model-failure region and b) the decoded the third frame of Miss America. (Courtesy Francoise Aurtenechea)

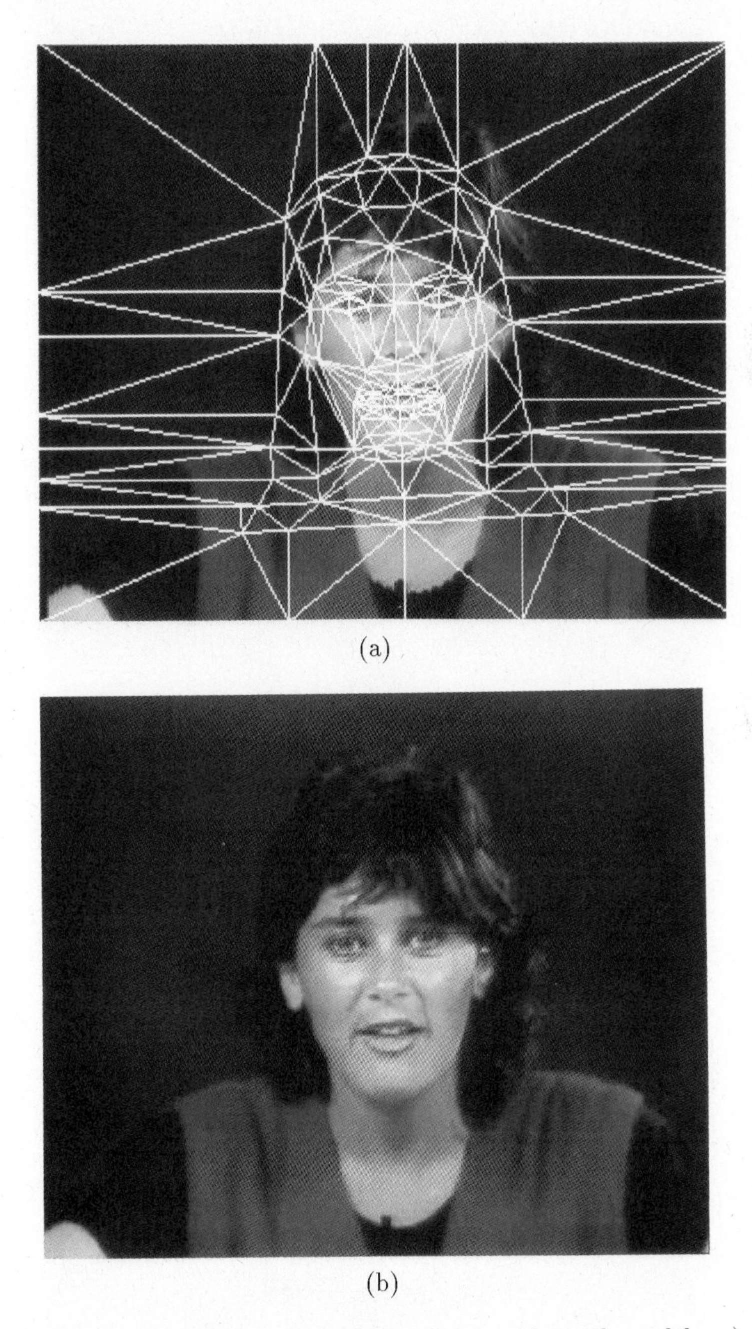

(a)

(b)

Figure 24.12: 2-D object-based coding using deformable mesh models: a) an irregular mesh fitted to the first frame of Miss America and b) the decoded third frame. (Courtesy Yucel Altunbasak)

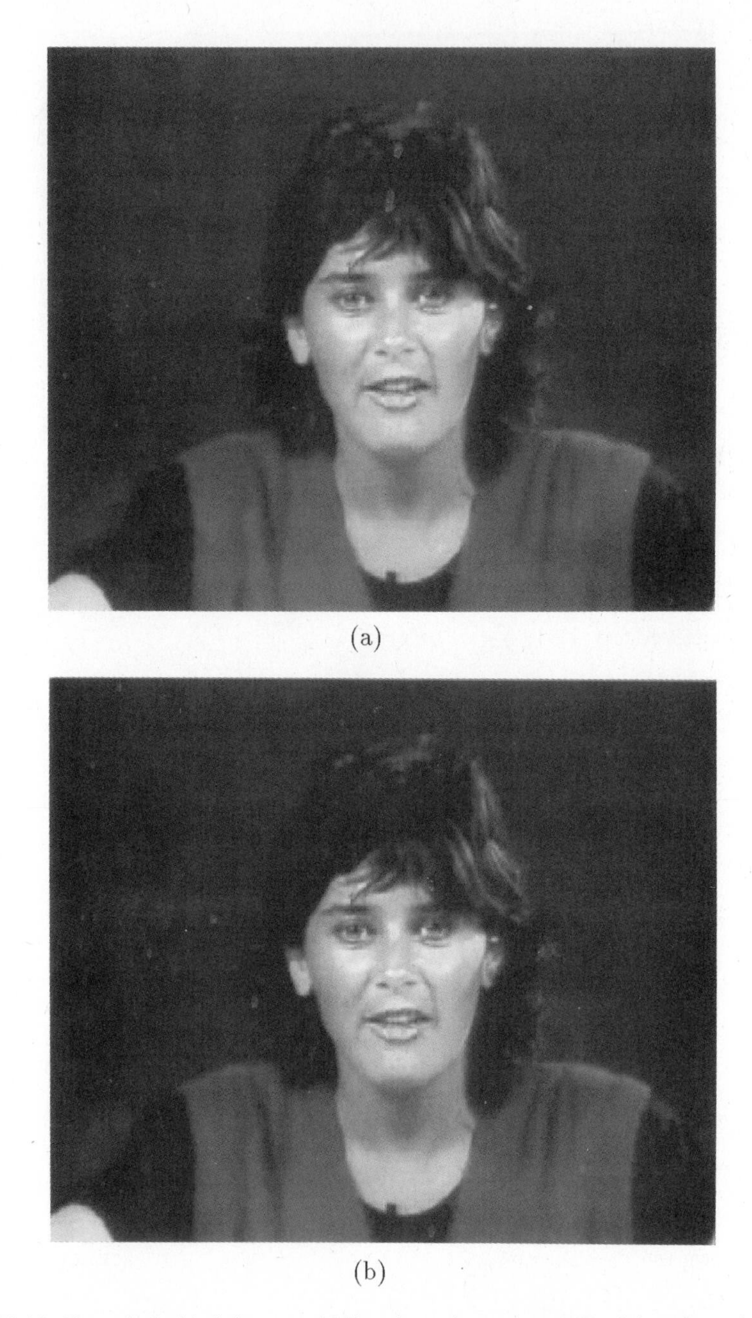

(a)

(b)

Figure 24.13: Decoded third frame of Miss America using 3-D object-based coding: a) global motion compensation using the two-step iteration (part of the MBASIC algorithm) and (b) global and local motion compensation using a flexible wireframe model. (Courtesy Gozde Bozdagi)

Bibliography

[Aiz 89] K. Aizawa, H. Harashima, and T. Saito, "Model-based analysis-synthesis image coding (MBASIC) system for a person's face," *Signal Proc.: Image Comm.*, no. 1, pp. 139–152, Oct. 1989.

[Aiz 93] K. Aizawa, C. S. Choi, H. Harashima, and T. S. Huang, "Human facial motion analysis and synthesis with application to model-based coding," in *Motion Analysis and Image Sequence Processing*, M. I. Sezan and R. L. Lagendijk, eds., Norwell, MA: Kluwer, 1993.

[Aiz 95] K. Aizawa and T. S. Huang, "Model-based image coding: Advanced video coding techniques for very low bit-rate applications," *Proc. IEEE*, vol. 83, no. 2, pp. 259–271, Feb. 1995.

[Boz 94] G. Bozdağı, A. M. Tekalp, and L. Onural, "3-D motion estimation and wireframe adaptation including photometric effects for model-based coding of facial image sequences," *IEEE Trans. Circ. and Syst: Video Tech.*, vol. 4, pp. 246–256, Sept. 1994.

[CCI 90] CCITT Recommendation H.261: "Video Codec for Audiovisual Services at $p \times 64Kbit/s$," COM XV-R 37-E, 1990.

[Cho 94] C. S. Choi, K. Aizawa, H. Harashima, and T. Takebe, "Analysis and synthesis of facial image sequences in model-based image coding," *IEEE Trans. Circ. and Syst: Video Tech.*, vol. 4, pp. 257–275, Sept. 1994.

[Die 93] N. Diehl, "Model-based image sequence coding," in *Motion Analysis and Image Sequence Processing*, M. I. Sezan and R. L. Lagendijk, eds., Norwell, MA: Kluwer, 1993.

[Ekm 77] P. Ekman and W. V. Friesen, *Facial Action Coding System*, Consulting Psychologist Press, 1977.

[For 89] R. Forchheimer and T. Kronander, "Image coding-from waveforms to animation," *IEEE Trans. Acoust, Speech Sign. Proc.*, vol. 37, no. 12, pp. 2008–2023, Dec. 1989.

[Fuk 93] T. Fukuhara and T. Murakami, "3-D motion estimation of human head for model-based image coding," *IEE Proc.-I*, vol. 140, no. 1, pp. 26–35, Feb. 1993.

[Ger 94] P. Gerken, "Object-based analysis-synthesis coding of image sequences at very low bit rates," *IEEE Trans. Circ. and Syst: Video Tech.*, vol. 4, pp. 228–237, Sept. 1994.

[Hot 90] M. Hotter, "Object oriented analysis-synthesis coding based on moving two-dimensional objects," *Signal Proc: Image Comm.*, vol. 2, no. 4, pp. 409–428, Dec 1990.

[Hua 91] T. S. Huang, S. C. Reddy, and K. Aizawa, "Human facial motion modeling, analysis, and synthesis for video compression," *SPIE Vis. Comm. and Image Proc'91*, vol. 1605, pp. 234–241, Nov. 1991.

[Kan 86] K. Kanatani, "Structure and motion from optical flow under orthographic projection," *Comp. Vis. Graph. Image Proc.*, vol. 35, pp. 181–199, 1986.

[Kan 91] M. Kaneko, A. Koike, and Y. Hatori, "Coding of facial image sequence based on a 3D model of the head and motion detection," *J. Visual Comm. and Image Rep.*, vol. 2, no. 1, pp. 39–54, March 1991.

[Kla 86] B. Klaus and P. Horn, *Robot Vision*, Cambridge, MA: MIT Press, 1986.

[Koc 93] R. Koch, "Dynamic 3-D scene analysis through synthesis feedback control," *IEEE Trans. Patt. Anal. Mach. Intel.*, vol. 15, pp. 556–568, June 1993.

[Lav 90] F. Lavagetto and S. Zappatore, "Customized wireframe modeling for facial image coding," *Third Int. Workshop on 64kbits/s Coding of Moving Video*, 1990.

[Lee 89] H. Lee and A. Rosenfeld, "Improved methods of estimating shape from shading using light source coordinate system," in *Shape from Shading*, B. K. P. Horn and M. J. Brooks, eds., pp. 323–569, Cambridge, MA: MIT Press, 1989.

[Li 93] H. Li, P. Roivainen, and R. Forchheimer, "3-D motion estimation in model-based facial image coding," *IEEE Trans. Patt. Anal. Mach. Intel.*, vol. 15, no. 6, pp. 545–555, June 1993.

[Li 94] H. Li, A. Lundmark, and R. Forchheimer, "Image sequence coding at very low bitrates: A Review," *IEEE Trans. Image Proc.*, vol. 3, pp. 589–609, Sept. 1994.

[Mor 91] H. Morikawa and H. Harashima, "3D structure extraction coding of image sequences," *J. Visual Comm. and Image Rep.*, vol. 2, no. 4, pp. 332–344, Dec. 1991.

[Mus 89] H. G. Musmann, M. Hotter, and J. Ostermann, "Object-oriented analysis synthesis coding of moving images," *Signal Proc: Image Comm.*, vol. 1, pp. 117–138, 1989.

[Nak 94] Y. Nakaya and H. Harashima, "Motion compensation based on spatial transformations," *IEEE Trans. Circ. and Syst.: Video Tech.*, vol. 4, pp. 339–356, June 1994.

[Ost 90] J. Ostermann, "Modeling of 3D moving objects for an analysis-synthesis coder," in *Sensing and Reconstruction of Three-Dimensional Objects*, B. Girod, ed., *Proc. SPIE*, vol. 1260, pp. 240–249, 1990.

[Ost 93] J. Ostermann, "An analysis-synthesis coder based on moving flexible 3-D objects," *Proc. Pict. Cod. Symp.*, Lausanne, Switzerland, Mar. 1993.

[Pea 90] D. Pearson, "Texture mapping in model-based image coding," *Signal Processing: Image Comm.*, vol. 2, pp. 377–395, 1990.

[Pen 82] A. Pentland, "Finding the illuminant direction," *J. Opt. Soc. Am.*, vol. 72, pp. 448–455, April 1982.

[Pen 91] A. Pentland, "Photometric motion," *IEEE Trans. Patt. Anal. Mach. Intel.*, vol. PAMI-13, no. 9, pp. 879–890, Sept. 1991.

[Rei 92] M. J. T. Reinders, B. Sankur, and J. C. A. van der Lubbe, "Transformation of a general 3D facial model to an actual scene face," *11th Int. Conf. Patt. Rec.*, pp. 75–79, 1992.

[Ryd 87] M. Rydfalk, "CANDIDE: A parametrised face," Rep. LiTH-ISY-I-0866, Dept. Elec. Eng., Linköping Univ., Sweeden, Oct. 1987.

[Sch 93] H. Schiller and M. Hotter, "Investigations on colour coding in an object-oriented analysis-synthesis coder," *Signal Proc.: Image Comm.*, vol. 5, pp. 319–326, Oct. 1993.

[Ter 90] D. Terzopoulos and K. Waters, "Physically-based facial modeling, analysis and animation," *J. of Visualization and Computer*, vol. 1, pp. 73–80, 1990.

[Tek 95] A. M. Tekalp, Y. Altunbasak, and G. Bozdagi, "Two- versus three-dimensional object-based coding," *Proc. SPIE Visual Comm. and Image Proc.*, Taipei, Taiwan, May 1995.

[Wan 94] Y. Wang and O. Lee, "Active mesh - A feature seeking and tracking image sequence representation scheme," *IEEE Trans. Image Proc.*, vol. 3, pp. 610–624, Sept. 1994.

[Wel 90] W. J. Welsh, S. Searby, and J. B. Waite, "Model based image coding," *J. Br. Telecom. Tech.*, vol. 8, no. 3, pp. 94–106, Jul. 1990.

[Wel 91] W. J. Welsh, "Model-based coding of videophone images," *Electronics and Communication Eng. Jour.*, pp. 29–36, Feb. 1991.

[Yau 88] J. F. S. Yau and N. D. Duffy, "A texture mapping approach to 3-D facial image synthesis," *Computer Graphics Forum*, no. 7, pp. 129–134, 1988.

[Zeg 89] K. Zeger and A. Gersho, "Stochastic relaxation algorithm for improved vector quantizer design," *Electronic Letters*, vol. 25, pp. 96–98, July 1989.

[Zhe 91] Q. Zheng and R. Chellappa, "Estimation of illuminant direction, albedo, and shape from shading," *IEEE Trans. Patt. Anal. Mach. Intel.*, vol. PAMI-13, no. 7, pp. 680–702, July 1991.

Chapter 25

DIGITAL VIDEO SYSTEMS

Developments in broadcasting, computers, communication technologies such as the emergence of better image compression algorithms, optical-fiber networks, faster computers, dedicated hardware, and digital recording promise a variety of digital video and image communication products in the very near future. Driving the research and development in the field of digital video are the consumer and commercial applications (ordered according to the bitrate requirement), such as

- Digital TV, including HDTV
  @ 20 Mbps over 6 Mhz taboo channels

- Multimedia, desktop video
  @ 1.5 Mbps CD-ROM or hard disk storage

- Videoconferencing
  @ 384 kbps using p $\times$ 64 kbps ISDN channels

- Videophone and mobile image communications
  @ 10-25 kbps

Other applications include surveillance imaging for military or law enforcement, intelligent vehicle highway systems, harbor traffic control, cine medical imaging, aviation and flight control simulation, and motion picture production. In the following, we overview some of these applications in the order in which international compression standards have become/are becoming available for them. In particular, we discuss video communication over ISDN (Integrated Services Digital Network) lines, multimedia in PC and workstation platforms, digital video broadcasting, and low-bitrate applications, in Sections 25.1 through 25.4, respectively.

25.1 Videoconferencing

Videoconferencing using digital techniques has been in existence for some time. Generally speaking, it refers to interactive distance conferencing using ISDN services with clear sound and sharp full-motion video. Each videoconferencing site employs a codec that converts analog video signals to digital and compresses them for transmission. At the receiver, digital signals are converted back to analog for display.

Early systems required special videoconference rooms with high-cost equipment and T1 links that operate at 1.544 Mbps. These systems were quite expensive for everyday use; for example, a typical T1 system would cost $120,000, and the cost of a T1 link would be about $750/hour. Advances in video compression and the adoption of the CCITT H.261 standard helped the emergence of several newer and better videoconferencing systems which cost about $30,000 to $40,000. The newer systems operate over ISDN lines at p × 64 kbps, ranging from 64 kbps up to 2 Mbps. A typical number for p is 6, putting the bandwidth at 384 kbps. Lower-bandwidth systems that operate at 64 kbps (56 kbps for video and 8 kbps for audio), known as desktop ISDN videophones [Ald 93], are also available. In comparison, the cost of a 128 kbps line is about $35/hour. Some examples of existing videoconferencing equipment are listed in Table 25.1.

Table 25.1: Available videoconferencing products

Vendor	Name	Codec Speed	Max Frame	Comp. Alg.
BT North America	Videocodec VC2200 Videocodec VC2100	56 and 112 kbps 56 kbps to 2048 kbps	30 per sec	H.261
GPT Video Systems	System 261 Twin chan. System 261 Universal	56 and 112 kbps 56 kbps to 2048 kbps	30 per sec	H.261
Compres. Labs.	Rembrandt II/VP	56 kbps to 2048 kbps	30 per sec	H.261, CTX CTX Plus
NEC America	VisualLink 5000 M20 VisualLink 5000 M15	56 kbps to 384 kbps 56 kbps to 2048 kbps	30 per sec	H.261, NEC proprietary
PictureTel Corp.	System 4000	56 kbps to 768 kbps	10 per sec mono	H.261, SG3 SG2/HVQ
Video Telecom	CS350	56 kbps to 768 kbps	15 per sec	H.261, Blue Chip

Besides videoconferencing and desktop videophone, video communication using an ISDN basic rate interface (BRI), operating at a total of 128 kbps, may be used for such applications as distance learning and access to multimedia information services. In distance learning, an instructor teaches students who are at remote locations. The students can interact with the instructor, by direct talking or sharing a whiteboard [Ald 93]. Multimedia information services include image-based electronic library systems and shopping-catalogs that can be browsed from home.

25.2 Interactive Video and Multimedia

Multimedia can mean different things to different people. Here it refers to the ability to provide full-motion interactive digital video in the personal computer (PC) or desktop workstation environment. Multimedia also involves text, graphics, sound, and still-images, which have long existed in the PC or workstation environment. The main components of a multimedia system are:

- data capture tools, such as video recorders and digitizers,
- data editors and authoring tools, for animation, audio and video editing, user-interface development, etc., and
- database storage and retrieval tools, for searching large databases, archiving, and backup.

The difficulty with full-motion video has been in the large data rates required. Multimedia workstations use CD-ROMs and hard disks for video storage at about 1.5 Mbps, which require approximately 30:1 compression of standard TV resolution images. The latest developments in image compression make this possible in real time. Two of the earliest initiatives in this area have been Intel's digital video interactive (DVI) technology, and compact disc-interactive (CD-I), jointly funded by NV Philips and Sony.

DVI technology is a general-purpose hardware for providing full-motion video in the PC environment, and uses CD-ROM for the storage medium. The design goal is to provide multimedia functionality at a cost suitable for desktop computers. It is based on two custom VLSI chips: the 82750PB pixel processor and the 82750DB display processor. The pixel processor mainly performs compression and decompression of images along with functions like YUV-RGB conversion, bilinear interpolation, and so on. It is able to decompress 640 $\times$ 480 JPEG encoded images in less than one second. The display processor retrieves decompressed images from memory, converts them to the format needed for display, and produces timing and control signals to drive various displays. Popular formats such as NTSC, PAL, VGA, and SVGA are supported. This chip set is compatible with 16 or 32 bit microprocessors operating at 6 MHz or higher clock speed, and utilizes 16 Mbyte video RAM (VRAM). The chip-set is programmable; that is, the compression algorithms can be modified according to the application.

The Compact Disc-Interactive (CD-I) system is a CD-based interactive audio/video special-purpose computer. It is based on Motorola's 680X0 processor, and the CD-RTOS, an operating system developed for the CD-I environment. Its original design was capable of providing full-motion video only on part of the screen. However, full-screen, full-motion video has become available with an add-in module [Mee 92]. This module reproduces video from an encoded bitstream with a frame rate of 24 Hz, 25 Hz, or 30 Hz. The frame rate converter transforms this video to 50 Hz or 60 Hz. Finally, the YUV representation of the encoded bitstream is converted into RGB for display. It includes a number of multimedia capabilities, such as a CD file manager for audio, a user communication manager of the video system, and a motion picture file manager. CD-I players currently exist that can be connected to standard TV sets to play back movies that are recorded on CDs. A full-length feature movie usually requires two CDs.

Recently, almost every computer and workstation manufacturer, including Apple, IBM, SUN, and Silicon Graphics, have added full-motion video capabilities. However, no industry-wide standards have been established yet. Recall that the MPEG-1 video compression standard addresses video compression for multimedia stations, but not full compatibility between various multimedia products. As computers and communication equipment move closer towards "grand unification," the line between multimedia systems and videoconferencing and videophone products is getting more and more blurred; that is, most multimedia systems can now be interfaced with LAN (Local Area Networks), WAN (Wide Area Networks), ISDN, or ATM (Asynchronous Transfer Mode) networks for interactive desktop videoconferencing. At present several companies are introducing plug-in cards and software for videoconferencing over LAN, WAN, and ATM networks. A nice feature of these products is that they provide multiplatform support, so that people using different workstations and personal computers may still share a whiteboard, text tools, and full-motion video.

With the newer generation of higher-speed CD-ROMs and faster processors entering the market every day, multimedia remains an active and exciting field. An emerging technology in interactive multimedia is effective human/machine interaction. Stereo vision, talker verification, speech synthesis, tactile interaction, and integration of multiple sensor modalities are active research areas to expand the present capabilities for human/machine interaction.

25.3 Digital Television

TV is arguably the most commonly used image communication system in the world today. However, present TV transmission standards are based on technology that is more than 40 years old. As a result, there has been widespread interest in the consumer electronics industry to develop more advanced TV systems that benefit from recent technological advances. The major advanced TV research programs in the world are tabulated in Table 25.2.

Table 25.2: Major programs for HDTV research.

Japan	NHK
Europe	EUREKA 95
U.S.A.	Grand Alliance
	(AT&T, General Instrument, Mass. Inst. Tech., Philips N. A.
	David Sarnoff Res. Cen., Thomson Cons. El., Zenith)

Development efforts pioneered by the Japanese in late the 70s and early 80s resulted in hybrid advanced TV systems, with digital processing at the transmitter and receiver but using analog transmission. Later, in the early 90s, studies in the U.S. proved the feasibility of the all-digital TV approach, which was believed to be unrealistic not too long ago. All-digital TV will not only offer better image quality, easy conversion between multiple standards, and more channels within the same bandwidth (thanks to advances in digital data compression technology), but more important, it will unite computers with TV sets and telecommunication services with cable TV services in a revolutionary fashion. In the following, we provide an overview of the developments around the world that led to the advent of all-digital TV.

25.3.1 Digital Studio Standards

Although the present TV transmission standards are analog (see Chapter 1), digital TV signals find routine use in TV studios for image processing and digital storage. Digital techniques are commonly used for such tasks as program editing, generating special effects, and standards conversion. Digital storage is preferred at the production level, because consumer-quality video storage devices, such as VHS recorders, introduce degradations that are objectionable in the production studio environment.

In the following, we first summarize the existing analog TV standards. The corresponding CCIR 601 standards for digitization of the respective signals (also known as digital studio standards) are listed in Table 25.3.

- **NTSC** (National Television Systems Committee) - Accepted for B&W in 1941, and extended to color in 1954
  Used in the USA, Canada, Japan, and Latin America
  2:1 interlace, 4:3 aspect ratio
  525 lines/frame, 29.97 frames/sec (262.5 lines/field, 59.94 fields/sec)
  Perceptually 340 lines/frame, 420 resolvable pels/line (Kell factor)
  Analog transmission over 6 MHz channel
  There are 68 channels assigned in the US: 54-88 MHz (ch 2 to 6), 174-216 MHz (ch 7 to 13), 470-806 MHz (ch14 to 69). However, less than 20 channels are used in a locality to prevent interference.

Table 25.3: CCIR 601 standards.

Format	Sampling Frequency (Y/U,V)	Field rate	Interlace/ Aspect	Active lines/frame	Active pixels/frame
NTSC	13.5/6.75	60	2:1/4:3	488	720
PAL/ SECAM	13.5/6.75	50	2:1/4:3	576	720

- **PAL** (Phase Alternation Line) - Accepted in 1967 for color broadcast
  Used in most of Europe, and Australia
  625 lines/frame, 2:1 interlace, 50 fields/sec, 4:3 aspect ratio
  Analog transmission over 8 MHz channel.

- **SECAM** (Systeme Electronique Color Avec Memoire) - Accepted in 1967 for color broadcast, used in France, Russia, and Eastern Europe
  625 lines/frame, 2:1 interlace, 50 fields/sec, 4:3 aspect ratio
  Analog transmission over 8 MHz channel.

Note that these standards are incompatible with each other, and conversion from one to another requires digital techniques.

25.3.2 Hybrid Advanced TV Systems

There were several proposals, before the advent of all-digital TV, that offered various degrees of improvements on the quality of present TV systems. We refer to them as "hybrid advanced TV (ATV) systems." ATV systems generally feature better spatial and/or temporal resolution, wider screens, improved color rendition, and CD-quality stereo sound. They fall under two broad categories: compatible ATV systems and incompatible ATV systems.

1) Compatible system proposals may be summarized as follows:

- IDTV (improved definition TV) systems which use digital processing at the receiver for picture quality improvement using existing analog TV transmission standards. Picture quality may be improved through: i) spatial and temporal resolution enhancement by nonlinear interpolation to recover super-Nyquist frequencies, including motion-compensated deinterlacing and frame rate conversion, ii) luminance-chrominance separation to eliminate cross-luminance and cross-chrominance artifacts due to imperfect Y/C separation in conventional NTSC demodulators, and iii) ghost elimination to reduce time-delayed, attenuated, and distorted versions of the TV signal.

- EQTV (extended-quality TV) systems which require transmission of an augmentation signal. They include systems that feature 16:9 aspect ratio by means of transmitting an augmentation signal at another frequency band. Conventional receivers that do not receive this augmentation signal can display ordinary-quality TV signals.

Early efforts in Europe have been in the direction of EQTV system development. PAL-plus and D2-MAC are two examples of EQTV systems which require transmission of augmentation signals. They have 16:9 aspect ratio, and are compatible with PAL and MAC (Multiplexed Analog Components), respectively. Note that MAC is a 625/50/2:1 analog satellite transmission standard developed by the British around 1970.

2) Incompatible systems are generally known as HDTV (high-definition TV).

CCIR 801, adopted in 1990, defines HDTV as follows: "A high definition TV system is a system designed to allow viewing at about three times picture height such that the transmission system is virtually or nearly transparent to the level of detail that would have been perceived in the original scene by a viewer with average visual acuity." HDTV systems feature a video signal that has about twice the current resolution in both the horizontal and vertical directions, a wider aspect ratio of 16:9, separate luminance and chrominance signals, and CD-quality sound. Initial HDTV proposals, developed first in Japan and later in Europe, were hybrid (mixed analog/digital) systems, which use digital signal/image processing at the transmitter and receiver, but transmission was by analog means. We elaborate on these systems below.

ATV in Japan

Studies towards an advanced TV system were started in the 1970s in Japan at the NHK Laboratories. NHK applied to CCIR, a world standards organization, in 1974 for the standardization of an analog HDTV format, called the MUSE (MUltiple sub-Nyquist Sampling Encoding). CCIR Study Group 11 worked from 1974 to 1986 to achieve a single world standard for production and international exchange of HDTV signals with no apparent success.

The MUSE system is based on a motion adaptive subsampling strategy to reduce the transmission bandwidth requirement by a factor of 3:1. The HDTV signal is defined by the following parameters:

1125 lines/frame, 60 fields/sec, 2:1 interlace, 16:9 aspect
8.1 MHz DBS transmission
Sampling rates: Y at 48.6 MHz and C at 16.2 MHz

It is primarily intended for broadcasting over 24 MHz direct broadcast satellite (DBS) channels, and is not compatible with other transmission standards. The basic idea of the MUSE system is motion-adaptive subsampling. That is, if motion is detected at a certain region, spatial lowpass filtering is applied before subsampling. On the other hand, in still-image areas, subsampling is applied without any lowpass filtering. There is no motion estimation or motion compensation involved in the MUSE system. Major electronic manufacturers in Japan have participated in the project and builded mixed analog/digital HDTV receivers. First broadcast using the MUSE system was realized on Nov. 25, 1991. At present, there are daily HDTV broadcasts via DBS in Japan.

ATV in Europe

European efforts for HDTV research have been organized under the EUREKA-95 project, which resulted in HD-MAC, an analog HDTV standard for DBS transmission developed around 1988. While the Japanese advocated an incompatible HDTV approach, HD-MAC has emerged as a MAC-compatible standard with 1250 lines/frame, 50 fields/sec, 2:1 interlace, and 16:9 aspect ratio.

HD-MAC achieves reduction of bandwidth from 54 MHz to 10.125 MHz by advanced motion-compensated subsampling. Motion information is transmitted digitally to assist the decoding process. The first HD-MAC broadcast was made from the 1992 Winter Olympics in France to selected test centers in Europe via DBS. The HD-MAC system has already been abondoned in Europe in favor of all-digital systems currently under development in the U.S.A.

25.3.3 All-Digital TV

All-digital TV refers to digital representation and processing of the signal as well as its digital transmission. The nature of the digital broadcast removes the synchronicity and real-time requirements of the analog TV, and offers many different options. A digital TV standard will unify the computer/workstation and TV industries in the future; hence, the introduction of the term "telecomputer." Although a standard for all-digital HDTV has not yet been formally approved, there is steady progress toward the adoption of a standard by the FCC before the end of 1995, and there are already some companies offering digital TV services conforming with the present standards using DBS broadcasting. Digital TV broadcast media include:

- Terrestial broadcast
- Direct broadcast satellite
- Cable and broadband ISDN distribution

In the U.S., the Federal Communications Commission (FCC) has ruled that the existing 6 MHz taboo-channels will be used for terrestrial broadcast. For digital terrestrial transmission, a 6 MHz channel can support about 20 Mbps data rate with sophisticated digital vestigial sideband modulation. Considering the parameters of a typical HDTV system, we still need about 545:20 = 28:1 compression to achieve this bitrate. Direct broadcast satellite (DBS) transmission is widely used in Japan and Europe as an analog transmission medium. In the U.S., some companies have already started providing digital TV transmission (at the NTSC resolution) using DBS. Cable distribution systems are heavily employed in the U.S. at present. Each cable channel is allotted 6 MHz, and typically 30 to 50 channels are available. All adjacent channels can be used. Some cable networks offer limited two-way communication capability where upstream data transmission is allowed. With digital transmission and effective compression, cable companies may offer approximately 150 channels using the presently available bandwidth. The broadband ISDN (B-ISDN) offers a unified network capable of providing voice, data, video, LAN, and MAN services [Spe 91]. The H4 access provides approximately 135 Mbps. Asynchronous transfer mode (ATM) is being considered for standardization as the multiplexing and switching vehicle of various services. DBS, cable, and B-ISDN services offer the possibility of more advanced video services, such as video-on-demand [Spe 95]. In the following, we summarize the all-digital HDTV and TV efforts in the U.S.A.

U.S. Grand Alliance

In the U.S.A. efforts to establish a terrestrial HDTV broadcast standard were initiated by the FCC in 1987. At the time, it was generally believed that more than 6 MHz was required to broadcast analog HDTV, and nobody thought that a digital HDTV broadcast would fit within a 6 MHz channel until 1990. In 1993, after only 6 years, it was decided that the HDTV standard in the U.S. will be an all-digital simulcast system. Simulcasting requires every existing TV broadcaster to have a second 6 Mz channel for digital HDTV broadcast. The NTSC source with 4.2 MHz bandwidth will continue to be transmitted through the usual existing 6 MHz channels until the year 2008.

When the tests began at the Advanced TV Test Center (ATTC) in Alexandria, Virginia in 1991, there were six proposals: one NTSC-compatible system, one analog simulcast system, and four digital systems. Early in 1993, all but the four digital system proposals were eliminated. The remaining proposals were:

1) *American TV Alliance* - General Instruments and MIT
   System: DigiCipher
   1050 lines/frame, 29.97 frames/sec, 2:1 interlace
   Horizontal scan rate: 31.469 kHz
   Sampling frequency: 53.65 MHz
   Active pixels: 1408 × 960 luma, 352 × 480 chroma
   Video Comp.: MC 32 × 16, Field/Frame DCT
   RF Modulation: 32/16 QAM

2) *Zenith and AT&T*
   System: Digital Spectrum Compatible (DSC-HDTV)
   787.5 lines/frame, 59.94 frames/sec, progressive (1:1)
   Horizontal scan rate: 47.203 kHz
   Sampling frequency: 75.3 MHz
   Active pixels: 1280 × 720 luma, 640 × 360 (chroma)
   Video Comp.: MC 32 × 16, 8 × 8 leaky pred., DCT, VQ
   RF Modulation: 2/4 level VSB

3) *Advanced TV Research Consortium* - Thomson Consumer Electronics, Philips Consumer Electronics, NBC, David Sarnoff Res. Center, and Compression Labs.

System: AD-HDTV
1050 lines/frame, 29.97 frames/sec, 2:1 interlace
Horizontal scan rate: 31.469 kHz
Sampling frequency: 54 MHz
Active pixels: 1440 $\times$ 960 luma, 720 $\times$ 480 chroma
Video comp.: MPEG++ (MC-DCT)
RF modulation: 32/16 SS-QAM

4) *American TV Alliance* - MIT and General Instruments

System: CC-Digicipher
787.5 lines/frame, 59.94 frames/sec, progressive (1:1)
Horizontal scan rate: 47.203 kHz
Sampling frequency: 75.3 MHz
Active pixels: 1280 $\times$ 720 luma, 640 $\times$ 360 chroma
Video comp.: MC 16 $\times$ 16, 8 $\times$ 8 DCT
RF modulation: 32/16 QAM

The following excerpt from Newslog, *IEEE Spectrum*, April 1993, summarizes the recommendation of the FCC special panel. "On February 11, 1993, a special FCC panel said there were flaws in all five of the systems. It is recommended that the FCC's *Advisory Committee on Advanced Television* hold a new round of testing after the four groups with all-digital systems fixed their problems. Officials from the four groups said they had begun talking about merging their systems into one - an idea that the FCC has been promoting." In May 1993, the four groups agreed to form a "Grand Alliance" to merge their systems into a single system incorporating the best features of each.

In October 1993, the Grand Alliance announced i) that the video compression algorithm will be MPEG-2, main profile, high level, ii) that the MPEG-2 transport mechanism will be used, iii) that the Dolby AC-3 audio system will be used, and iv) that three modulation techniques, 4-level VSB, 6-level VSB, and 32 QAM (quadrature amplitude modulation), will be tested to complete the specification. In order to facilitate interoperability of broadcasting, computer multimedia, computer graphics, industrial imaging, and the National Information Infrastructure multiple scanning formats have been adopted which resembles the open-architecture TV concept [Bov 91]. In an open-architecture video representation, the number of lines in the display depends on the display hardware, and is not coupled to the number of lines employed by the production equipment. The scanning formats supported by the Grand Alliance proposal include progressive and interlaced scanning, at two spatial resolutions 720 lines $\times$ 1280 pixels and 1080 lines $\times$ 1920 pixels. In addition to 60 Hz and 30 Hz frame rates, a 24 Hz film mode is included with both progressive and interlaced scanning for motion-picture source material. The Grand Alliance scanning formats are summarized in Table 25.4. All formats support 16:9 aspect ratio with square pixels.

Table 25.4: Grand Alliance HDTV formats.

Spatial resolution	Line structure	Frame rate
720 × 1280	Progressive	60, 30, 24
1080 × 1920	Interlaced	30
1080 × 1920	Progressive	30, 24

Because there exist multiple scanning formats, transconversion may be required at the transmitter and/or at the receiver. A transconverter at the decoder output performs the necessary format conversion if the display scanning parameters differ from those of the received signal. The decoder accepts both 1080-line interlaced and 720-line progressive formats, and supports bidirectional prediction with motion estimation parameters up to ± 127 horizontal and ± 31 vertical, fully compliant with MPEG-2 requirements. Field tests of the proposed all-digital HDTV standard is currently underway.

Digital DBS TV

Digital transmission of TV signals at today's TV resolution is already becoming available through DBS services using 18 in antennas and digital decoders. These services use MPEG-2 compression starting with RGB source material to provide images sharper than NTSC pictures. Hughes Communications and the United States Satellite Broadcasting Company have just announced digital DBS services, called DIRECTV^{TM} and USSB^{TM}, respectively. At the moment, these services are available only in selected test markets. However, they are expected to become available nationally soon.

Interactive Networks and Video-on-Demand

The available cable TV networks today offer one-way traffic with a fixed set of channels. New high-bandwidth network architectures and protocols, such as fiber optic networks with ATM switching, are needed to provide users with a variety of interactive services. Video-server computers will be integral components of these interactive services to offer customized programming and video-on-demand [Lin 95] [Spe 95]. With the adoption of digital signal formats and transmission standards, video storage is also expected to be dominated by digital technologies, such as CD-ROMs and VTRS [Eto 92].

25.4 Low-Bitrate Video and Videophone

Low-bitrate video (LBV) generally refers to applications that require less than 64 kbps. There are many applications of low-bitrate coding, including:

- *Videophone:* Videophone service on PSTN (Public Switched Telephone Network), mobile, and LANs. Real-time encoder and decoder with easy implementation.
- *Mobile multimedia communication* such as cellular videophones and other personal communication systems.
- *Remote sensing:* One way communication of audio-visual information from a remote location, for example, surveillance, security, intelligent vehicle highway systems (IVHS), harbor traffic management.
- *Electronic newspaper:* Multimedia news service on PSTN, radio channels, and Smart cards.
- *Multimedia videotex:* Videotex is currently a multimedia database environment but lacks capability for full-motion video.
- *Multimedia electronic mail*

These applications require huge compression factors which generally cannot be met satisfactorily with the existing compression standards. Typical compression ratios to reach 10 kbps starting from several video formats are shown in Table 25.5.

Table 25.5: Compression requirements to reach 10 kbps

Frames/s	CCIR 601 (720x576)	CIF (352x288)	QCIF (176x144)
7.5	4979:1	915:1	229:1
10	6637:1	1216:1	304:1
15	9952:1	1824:1	456:1
30	19904:1	3648:1	912:1

Videophone is at the low end of the videoconferencing market, where low cost and operation over the existing subscriber network become extremely important. A basic setup requires at least one desktop monitor to view the video signals, a camera for capturing the video signal, and an audio connection. Due to high cost and long timeframe associated with wide range deployment of a fiber/coaxial-based local subscriber network, the target bitrate for videophone products has been set below 28.8 kbps (using the V.34 modems over the existing PSTN). Although, there are no established video compression standards at such low bitrates, a few representative products have already appeared in the market [Ear 93]. These systems use variations of the MC-DCT coding scheme, which are similar to the H.261 standard;

Table 25.6: Available videophone products

Product	Data Rate	Compression Algorithm
AT&T Videophone 2500	16.8/ 19.2 kbps	MC DCT 10 frames/s (max)
British Telecom/Marconi Relate 2000 Videophone	9.6/ 14.4 kbps	H.261 like 7.5 (3.75) frames/s
COMTECH Labs. STU-3 Secure Videophone	9.6 kbps	MC DCT QCIF resolution
Sharevision	14.4 kbps	MC DCT

that is, using the macroblock concept, DCT, motion estimation/compensation, and run-length and VLC coding. Some examples of early videophones are listed in Table 25.6.

Recognizing the increasing demand for low bitrate applications, especially mobile video communications, efforts for standardization in low bitrate coding have been initiated in 1993 by the ISO/MPEG-4 Ad-Hoc Group and ITU-T/LBC (Expert Group on Low Bitrate Coding). Because of the urgent need to provide a common platform of communication between various products by different vendors (using the available technology) and the need for fundamentally different technologies to provide improved performance and embedded functionality, the work has been divided into two phases: a near-term solution and a far-term solution. The near-term solution has very recently resulted in the ITU Draft Recommendation H.263. The far-term solution refers to a fundamentally new standard expected to be completed by November 1988. This task will be handled by ISO/MPEG-4, with liaison to the ITU.

25.4.1 The ITU Recommendation H.263

The ITU/LBC group has drafted the near-term H.324 specifications, that include audio/video coding, multiplexing, error control, and overall system integration targeted at videophone applications on PSTN and mobile channels. The ITU Draft Recommendation H.263 (frozen in March 1995) specifies the video coding algorithm, which is an "H.261" like (MC-DCT) algorithm, at about 22 kbps (out of 28.8 kbps overall).

The major differences between the H.263 and H.261 standards are:

- Motion estimation with one-half-pixel accuracy, which eliminates the need for loop filtering.

- Overlapped motion compensation to obtain a denser motion field at the expense of more computation.

- Adaptive switching between motion estimation at macroblock (16 × 16) and block (8 × 8) levels.
- Support for sub-QCIF bitstreams.

It has been claimed that the Test Model 5 (TMN5) provides 3-4 dB higher PSNR than the H.261 algorithm at below 64 kbps. It can be used as a milestone to assess the performance of future low-bitrate coding algorithms and standards.

25.4.2 The ISO MPEG-4 Requirements

The MPEG-4 Ad-hoc group was organized in September 1993 with the mission of developing a fundamentally new generic video coding standard at rates below 64 kbps. In November 1994, this mission has been modified as "to provide an audio-visual coding standard allowing for interactivity, high compression, and/or universal accessibility, with high degree of flexibility and extensibility" [Zha 95].

The new MPEG4 vision includes eight functionalities that are not supported by the existing standards. They are:

- Content-based manipulation and bitstream editing
- Content-based multimedia data access tools
- Content-based scalability
- Coding of multiple concurrent data streams
- Hybrid natural and synthetic data coding
- Improved coding efficiency
- Improved temporal access at very low bitrates
- Robustness in error-prone environments

MPEG-4 intends to cover a wide range of applications, including "virtual" conference and classroom; interactive mobile videophone; content-based multimedia database query, searching, indexing, and retrieval; interactive home shopping; wireless monitoring; and so on. At present MPEG-4 structure consists of four elements: syntax, tools, algorithms, and profiles. The syntax is an extensible language that allows selection, description, and downloading of tools, algorithms, and profiles. A tool is a specific method. An algorithm is a collection of tools that implement one or more functionalities. A profile is one or more algorithms to cover a specific class of applications. Interested parties can submit proposals for potential tools, algorithms, and profiles. Deadline for submissions of proposals is October 1995.

Bibliography

[Ald 93] H. Aldermeshian, W. H. Ninke, and R. J. Pilc, "The video communications decade," *AT&T Tech. Jour.*, vol. 72, pp. 15–21, Jan./Feb. 1993.

[Ana 94] D. Anastassiou, "Digital Television," *Proc. IEEE*, vol. 82, pp. 510–519, Apr. 1994.

[Apo 93] J. G. Apostolopoulos and J. S. Lim, "Video compression for digital advanced television systems," in *Motion Analysis and Image Sequence Processing*, M. I. Sezan and R. L. Lagendijk, eds., Kluwer, 1993.

[Bar 84] M. Barton, "Encoding parameters for digital television studios," *West Indian J. Engineering*, vol. 9, no. 1, pp. 29–36, Jan. 1984.

[Bey 92] B. W. Beyers, "Digital television: Opportunities for change," *IEEE Trans. Cons. Electronics*, vol. 38, no. 1, pp. xiii–xiv, Feb. 1992.

[Bov 91] V. M. Bove and A. Lippman, "Open architecture television," *Proc. SMPTE 25th Conf.*, Feb. 1991.

[Byt 92] "HDTV is coming to desktop," *Byte*, July 1992.

[Cha 95] K. Challapali, X. Lebegue, J. S. Lim, W. H. Paik, R. S. Girons, E. Petajan, V. Sathe, P. A. Snopko, and J. Zdepski, "The Grand Alliance system for US HDTV," *Proc. IEEE*, vol. 83, no. 2, pp. 158–174, Feb. 1995.

[Cor 90] I. Corbett, "Moving pictures," *IEE Review*, pp. 257–261, Jul./Aug. 1990.

[Ear 93] S. H. Early, A. Kuzma, and E. Dorsey, "The VideoPhone 2500 - Video technology on the Public Switched Telephone Network," *AT&T Tech. Jour.*, vol. 72, pp. 22–32, Jan./Feb. 1993.

[Eto 92] Y. Eto, "Signal processing for future home-use digital VTR's," *IEEE J. Selected Areas in Comm.*, vol. 10, no. 1, pp. 73–79, Jan. 1992.

[Fla 94] J. Flanagan, "Technologies for multimedia communications," *Proc. IEEE*, vol. 82, pp. 590–603, Apr. 1994.

[Foc 92] P. Fockens and A. Netravali, "The digital spectrum-compatible HDTV system," *Signal Proc.: Image Comm.*, vol. 4, pp. 293–305, 1992.

[Fuj 85] T. Fujio, "High definition television systems," *Proc. IEEE*, vol. 73, no. 4, pp. 646–655, Apr. 1985.

[Hop 94] R. Hopkins, "Choosing and American digital HDTV terrestrial broadcasting system," *Proc. IEEE*, vol. 82, pp. 554–563, Apr. 1994.

[Jos 92] K. Joseph, S. Ng, D. Raychaudhuri, R. Siracusa, J. Zdepski, R. S. Girons, and T. Savatier, "MPEG++: A robust compression and transport system for digital HDTV," *Signal Proc.: Image Comm.*, vol. 4, pp. 307–323, 1992.

[Kun 95] M. Kunt, ed. "Digital Television," Special Issue *Proc. IEEE*, July 1995.

[Lin 95] D. W. Lin, C.-T. Chen, and T. R. Hsing, "Video on phone lines: Technology and applications," *Proc. IEEE*, vol. 83, no. 2, pp. 175–193, Feb. 1995.

[Luc 87] K. Lucas and B. van Rassell, "HDB-MAC a new proposal for high-definition TV transmission," *IEEE Trans. Broadcasting*, vol. BC-33, pp. 170–183, Dec. 1987.

[Mee 92] J. van der Meer, "The full motion system for CD-I," *IEEE Trans. Cons. Electronics*, vol. 38, no. 4, pp. 910–920, Nov. 1992.

[Mil 94] M. D. Miller, "A scenario for the deployment of interactive multimedia cable television systems in the United States in the 1990's," *Proc. IEEE*, vol. 82, pp. 585–589, Apr. 1994.

[Nin 87] Y. Ninomiya, Y. Ohtsuka, Y. Izumi, S. Gohshi and Y. Iwadate, "An HDTV broadcasting system utilizing a bandwidth compression technique-MUSE," *IEEE Trans. Broadcasting*, vol. BC-33, pp. 130–160, Dec. 1987.

[Pan 94] P. Pancha and M. E. Zarki, "MPEG coding for variable bitrate video transmission," *IEEE Comm. Magazine*, pp. 54–66, May 1994.

[Pol 85] L. J. van de Polder, D. W. Parker, and J. Roos, "Evolution of television receivers from analog to digital," *Proc. IEEE*, vol. 73, no. 4, pp. 599–612, Apr. 1985.

[Ros 92] J. Rosenberg, R. E. Kraut, L. Gomez, and C. A. Buzzard, "Multimedia communications for users," *IEEE Comm. Mag.*, pp. 23–36, May 1992.

[San 85] C. P. Sandbank and I. Childs, "The evolution towards high-definition television," *Proc. IEEE*, vol. 73, no. 4, pp. 638–645, April 1985.

[San 90] C. P. Sandbank, *Digital Television*, John Wiley & Sons, 1990.

[Spe 91] *B-ISDN and How It Works*, IEEE Spectrum, pp. 39–44, Aug. 1991.

[Spe 95] "Digital Television," *IEEE Spectrum*, pp. 34–80, Apr. 1995.

[Zha 95] Y.-Q. Zhang, "Very low bitrate video coding standards," *Proc. Vis. Comm. Image Proc.*, SPIE vol. 2501, pp. 1016-1023, May 1995.

Appendix A

MARKOV AND GIBBS RANDOM FIELDS

Markov random fields specified in terms of Gibbs distributions have become a popular tool as *a priori* signal models in Bayesian estimation problems that arise in complex image processing applications such texture modeling and generation [Cro 83, Che 93], image segmentation and restoration [Gem 84, Der 87, Pap 92], and motion estimation [Dub 93]. In this appendix, we provide the definitions of a Markov random field and the Gibbs distribution, and then describe the relationship between them by means of the Hammersley-Clifford theorem. The specification of MRFs in terms of Gibbs distributions led to the name "Gibbs random field" (GRF). We also discuss how to obtain the local (Markov) conditional pdfs from the Gibbs distribution, which is a joint pdf.

A.1 Definitions

A scalar *random field* $\mathbf{z} = \{z(\mathbf{x}),\ \mathbf{x} \in \Lambda\}$ is a stochastic process defined over a lattice Λ. We let ω denote a realization of the random field $\mathbf{z}$. Recall that the random field $z(\mathbf{x})$ evaluated at a fixed location $\mathbf{x}$ is a random variable. It follows that a scalar random field is a collection of scalar random variables, where a random variable is associated with each site of the lattice Λ. A vector random field, such as the velocity or the displacement field, is likewise a collection of random vectors. In the following, we limit our discussion to scalar random fields.

A random field $\mathbf{z}$ can be discrete-valued or real-valued. For a discrete-valued random field, $z(\mathbf{x})$ assumes a set of discrete values, i.e., $z(\mathbf{x}) \in \Gamma = \{0, 1, \ldots, L-1\}$, and for a real-valued random field $z(\mathbf{x}) \in \mathbf{R}$, where $\mathbf{R}$ denotes the set of real numbers.

The first step in defining Markov random fields and Gibbs distributions is to develop a *neighborhood system* on Λ. We let $\mathcal{N}_{\mathbf{x}}$ to denote a neighborhood of a site $\mathbf{x}$ in Λ, which has the properties:

(i) $\mathbf{x} \notin \mathcal{N}_{\mathbf{x}}$, and

(ii) $\mathbf{x}_i \in \mathcal{N}_{\mathbf{x}_j} \leftrightarrow \mathbf{x}_j \in \mathcal{N}_{\mathbf{x}_i}$, for all $\mathbf{x}_i, \mathbf{x}_j \in \Lambda$.

In words, a site $\mathbf{x}$ does not belong to its own set of neighbors, and if $\mathbf{x}_i$ is a neighbor of $\mathbf{x}_j$, then $\mathbf{x}_j$ must be a neighbor of $\mathbf{x}_i$, and vice versa. Two examples of neighborhood $\mathcal{N}_{\mathbf{x}}$ of a site are depicted in Figure A.1. A neighborhood system $\mathcal{N}$ over Λ is then defined as

$$\mathcal{N} = \{\mathcal{N}_{\mathbf{x}}, \mathbf{x} \in \Lambda\}, \tag{A.1}$$

the collection of neighborhoods of all sites.

A.1.1 Markov Random Fields

Markov random fields (MRF) are extensions of 1-D causal Markov chains to 2-D, and have been found useful in image modeling and processing. MRFs have been traditionally specified in terms of local conditional probability density functions (pdf) which limit their utility.

Definition: The random field $\mathbf{z} = \{z(\mathbf{x}), \mathbf{x} \in \Lambda\}$ is called a *Markov random field* (MRF) with respect to $\mathcal{N}$ if

$$p(\mathbf{z}) > 0, \quad \text{for all } \mathbf{z}$$

and

$$p(z(\mathbf{x}_i) \mid z(\mathbf{x}_j), \forall \mathbf{x}_j \neq \mathbf{x}_i) = p(z(\mathbf{x}_i) \mid z(\mathbf{x}_j), \mathbf{x}_j \in \mathcal{N}_{\mathbf{x}_i})$$

where the pdf of a discrete-valued random variable/field is defined in terms of a Dirac delta function.

The first condition states that all possible realizations should have nonzero probability, while the second requires that the local conditional pdf at a particular site $\mathbf{x}_i$ depends only on the values of the random field within the neighborhood $\mathcal{N}_{\mathbf{x}_i}$ of that site.

The specification of an MRF in terms of local conditional pdfs is cumbersome because:

i) the conditional pdfs must satisfy some consistency conditions [Gem 84] which cannot be easily verified,

ii) the computation of the joint pdf $p(\mathbf{z})$ from the local conditional pdfs is not straightforward, and

iii) the relationship between the local spatial characteristics of a realization and the form of the local conditional pdf is not obvious.

Fortunately enough, every MRF can be described by a Gibbs distribution (hence, the name Gibbs random field - GRF), which apparently overcomes all of the above problems.

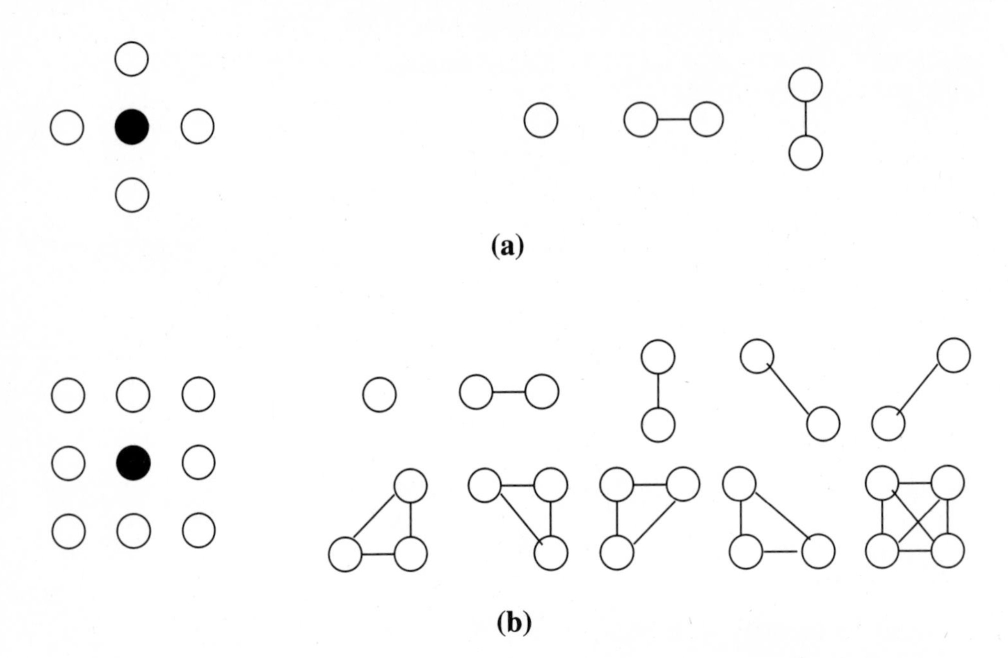

Figure A.1: Examples of neighborhoods: a) 4-pixel neighborhood and associated cliques; b) 8-pixel neighborhood and associated cliques.

Before we can define a Gibbs distribution, we need to define a clique. A clique C, defined over the lattice Λ with respect to the neighborhood system $\mathcal{N}$, is a subset of Λ ($C \subseteq \Lambda$) such that either C consists of a single site or all pairs of sites in C are neighbors. The set of cliques for the 4-pixel and 8-pixel neighborhoods are shown in Figure A.1. Notice that the number of different cliques grows quickly as the number of sites in the neighborhood increases. The set of all cliques is denoted by $\mathcal{C}$.

A.1.2 Gibbs Random Fields

The Gibbs distribution, with a neighborhood system $\mathcal{N}$ and the associated set of cliques $\mathcal{C}$, is defined as:

- for *discrete-valued* random fields,

$$p(\mathbf{z}) = \frac{1}{Q} \sum_{\omega} e^{-U(\mathbf{z}=\omega)/T} \delta(\mathbf{z} - \omega) \tag{A.2}$$

where $\delta(\cdot)$ denotes a Dirac delta function, and the normalizing constant Q, called the *partition function*, is given by

$$Q = \sum_{\omega} e^{-U(\mathbf{Z}=\omega)/T}$$

- or, for *continuous-valued* random fields,

$$p(\mathbf{z}) = \frac{1}{Q} e^{-U(\mathbf{z})/T} \tag{A.3}$$

where the normalizing constant Q is given by

$$Q = \int_{\omega} e^{-U(\mathbf{z})/T} \, d\mathbf{z}$$

and $U(\mathbf{z})$, the Gibbs potential (Gibbs energy), is defined as

$$U(\mathbf{z}) = \sum_{C \in \mathcal{C}} V_C(z(\mathbf{x}) \mid \mathbf{x} \in C)$$

for both the discrete and continuous-valued random fields. Each $V_C(z(\mathbf{x}) \mid \mathbf{x} \in C)$, called the clique potential,depends only on the values $z(\mathbf{x})$ for which $\mathbf{x} \in C$. The parameter T, known as the temperature, is used to control the peaking of the distribution. Note that Gibbs distribution is an exponential distribution, which includes Gaussian as a special case.

Gibbs distribution is a joint pdf of all random variables composing the random field, as opposed to a local conditional pdf. It can be specified in terms of certain desired structural properties of the field which are modeled through the clique potentials. The use of the Gibbs distribution to impose a spatial smoothness constraint is demonstrated in Chapter 8.

A.2 Equivalence of MRF and GRF

The equivalence of an MRF and a GRF is stated by the Hammersley-Clifford theorem, which provides us with a simple and practical way to specify MRFs through Gibbs potentials.

Hammersley-Clifford (H-C) Theorem: Let $\mathcal{N}$ be a neighborhood system. Then $z(\mathbf{x})$ is an MRF with respect to $\mathcal{N}$ if and only if $p(\mathbf{z})$ is a Gibbs distribution with respect to $\mathcal{N}$.

The H-C theorem was highlighted by Besag [Bes 74], based on Hammersley and Clifford's unpublished paper. Spitzer [Spi 71] also provided an alternate proof of the H-C theorem. The work of Geman and Geman [Gem 84] pioneered the popular use of Gibbs distributions to specify MRF models.

The Gibbs distribution gives the joint pdf of $z(\mathbf{x})$, which can easily be expressed in terms of the clique potentials. The clique potentials effectively express the local interaction between pixels and can be assigned arbitrarily, unlike the local pdfs in MRFs, which must satisfy certain consistency conditions. It turns out that the local conditional pdfs induced by a Gibbs distribution can easily be computed in the case of discrete-valued GRFs. This is shown next.

A.3 Local Conditional Probabilities

In certain applications, such as in the Gibbs sampler method of optimization, it may be desirable to obtain the local conditional pdfs from the joint pdf given in terms of Gibbs potential functions. Derivation of the local conditional pdfs of a discrete-valued GRF is shown in the following. Starting with the Bayes rule,

$$\begin{aligned} p(z(\mathbf{x}_i) \mid z(\mathbf{x}_j), \forall \mathbf{x}_j \neq \mathbf{x}_i) &= \frac{p(\mathbf{z})}{p(z(\mathbf{x}_j), \forall \mathbf{x}_j \neq \mathbf{x}_i)} \\ &= \frac{p(z(\mathbf{x}))}{\sum_{z(\mathbf{x}_i)\in\Gamma} p(\mathbf{z})}, \quad \mathbf{x} \in \Lambda \end{aligned} \tag{A.4}$$

where the second line follows from the total probability rule. In particular, let A_γ denote the event $z(\mathbf{x}_i) = \gamma$, for all $\gamma \in \Gamma$, and B stand for a fixed realization of the remaining sites, $\mathbf{x}_j \neq \mathbf{x}_i$; then (A.4) is simply a statement of

$$P(A_\gamma|B) = \frac{P(A_\gamma \cap B)}{\sum_{\gamma\in\Gamma} P(B|A_\gamma)p(A_\gamma)}$$

where $P(\cdot)$ denotes probability (obtained by integrating the pdf $p(\cdot)$ over a suitable range). Substituting the Gibbsian distribution for $p(\cdot)$ in (A.4), the local conditional pdf can be expressed in terms of the Gibbs clique potentials (after some algebra) as

$$p(z(\mathbf{x}_i) \mid z(\mathbf{x}_j), \forall \mathbf{x}_j \neq \mathbf{x}_i) = Q_{\mathbf{x}_i}^{-1} \exp\{-\frac{1}{T} \sum_{C|\mathbf{x}_i\in C} V_C(z(\mathbf{x}) \mid \mathbf{x} \in C)\} \tag{A.5}$$

where

$$Q_{\mathbf{x}_i} = \sum_{z(\mathbf{x}_i)\in\Gamma} \exp\{-\frac{1}{T} \sum_{C|\mathbf{x}_i\in C} V_C(z(\mathbf{x}) \mid \mathbf{x} \in C)\}$$

The use of the expression (A.5) is illustrated by an example in Chapter 8.

For a more detailed treatment of MRFs and GRFs the reader is referred to [Gem 84]. Vigorous treatment of the statistical formulations can also be found in [Bes 74] and [Spi 71].

Bibliography

[Bes 74] J. Besag, "Spatial interaction and the statistical analysis of lattice systems," *J. Royal Stat. Soc. B*, vol. 36, no. 2, pp. 192–236, 1974.

[Che 93] R. Chellappa and A. Jain, *Markov Random Fields: Theory and Application*, Academic Press, 1993.

[Cro 83] G. R. Cross and A. K. Jain, "Markov random field texture models," *IEEE Trans. Patt. Anal. Mach. Intel.*, vol. PAMI-5, pp. 25–39, Jan. 1983.

[Der 87] H. Derin and H. Elliott, "Modeling and segmentation of noisy and textured images using Gibbs random field," *IEEE Trans. Patt. Anal. Mach. Intel.*, vol. PAMI-9, pp. 39–55, Jan. 1987.

[Dub 93] E. Dubois and J. Konrad, "Estimation of 2-D motion fields from image sequences with application to motion-compensated processing," in *Motion Analysis and Image Sequence Processing*, M. I. Sezan and R. L. Lagendijk, eds., Norwell, MA: Kluwer, 1993.

[Gem 84] S. Geman and D. Geman, "Stochastic relaxation, Gibbs distribution, and Bayesian restoration of images," *IEEE Trans. Patt. Anal. Mach. Intel.*, vol. 6, pp. 721–741, Nov. 1984.

[Pap 92] T. N. Pappas, "An adaptive clustering algorithm for image segmentation," *IEEE Trans. Signal Proc.*, vol. SP-40, pp. 901–914, April 1992.

[Spi 71] F. Spitzer, "Markov random fields and Gibbs ensembles," *Amer. Math. Mon.*, vol. 78, pp. 142–154, Feb. 1971.

Appendix B

BASICS OF SEGMENTATION

The goal of this appendix is to introduce those segmentation tools which are employed in Chapter 11 for motion segmentation. In particular, we overview thresholding in Section B.1, clustering in Section B.2, and Bayesian methods in Section B.3, as they apply to image segmentation. Section B.4 generalizes Bayesian formulation to multichannel data.

B.1 Thresholding

Thresholding is a popular tool for image segmentation. Consider an image $s(x_1, x_2)$, composed of a light object on a dark background. Such an image has a bimodal histogram $h(s)$, as depicted in Figure B.1. An intuitive approach to segment the object from the background, based on gray-scale information, is to select a threshold T that separates these two dominant modes (peaks). The segmentation of an image into two regions is also known as binarization.

Then the segmentation mask or the binarized image

$$z(x_1, x_2) = \begin{cases} 1 & \text{if } s(x_1, x_2) > T \\ 0 & \text{otherwise} \end{cases} \tag{B.1}$$

identifies the object and background pixels.

Thresholding techniques can be divided into two broad classes: global and local. In general, the threshold T is a function of

$$T = T(x_1, x_2, s(x_1, x_2), p(x_1, x_2)) \tag{B.2}$$

where (x_1, x_2) are the coordinates of a pixel, $s(x_1, x_2)$ is the intensity of the pixel, and $p(x_1, x_2)$ is some local property of the pixel, such as the average intensity of

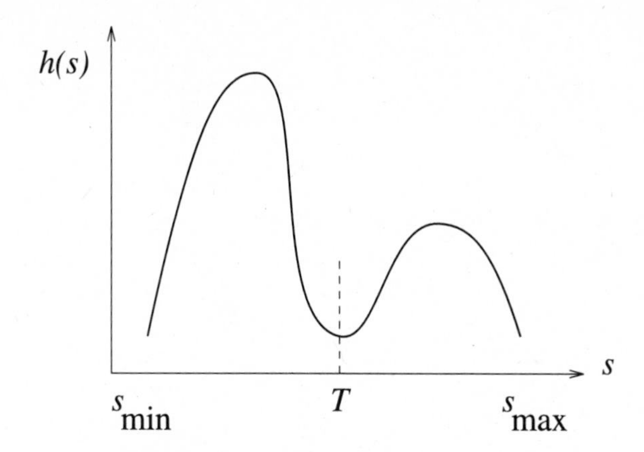

Figure B.1: A bimodal histogram.

a local neighborhood. If T is selected based only on the statistics of individual pixel values $s(x_1, x_2)$ over the entire image, it is called a pixel-dependent global threshold. Alternatively, T can be a global threshold selected based on the statistics of both $s(x_1, x_2)$ and $p(x_1, x_2)$ to reflect certain spatial properties of the image. If, in addition, it depends on (x_1, x_2), it is called a dynamic or adaptive threshold. Adaptive thresholds are usually computed based on a local sliding window about (x_1, x_2). Methods for determining the threshold(s) are discussed in the book by Gonzalez and Woods [Gon 92] as well as in the review papers [Wes 78, Sah 88, Lee 90].

In some applications the histogram has K significant modes (peaks), where $K > 2$. Then we need $K - 1$ thresholds to separate the image into K segments. Of course, reliable determination of the thresholds becomes more difficult as the number of modes increases. In the following, we introduce a method for optimum threshold determination.

B.1.1 Finding the Optimum Threshold(s)

Given a multimodal histogram, $h(s)$, of an image $s(x_1, x_2)$, one can address the problem of finding the optimal thresholds for separating the dominant modes by fitting the histogram with the sum of K probability density functions. Assuming these pdfs are Gaussian, the mixture density model takes the form

$$h(s) = \sum_{\ell=1}^{K} \frac{P_\ell}{\sqrt{2\pi}\sigma_\ell} \exp\left\{-\frac{(s-\mu_\ell)^2}{\sigma_\ell^2}\right\} \tag{B.3}$$

where P_ℓ denotes the *a priori* probability of the particular mode such that $\sum_{\ell=1}^{K} P_\ell = 1$, and μ_ℓ and σ_ℓ denote the mean and the standard deviation of that mode, respectively. The parameters of this mixture density are unknown and have to be identified by minimizing the mean square error between the actual histogram and the model density. Since an analytical solution to this parameter identification problem is usually difficult, numerical methods are employed for the minimization [Sny 90, Gon 92].

Once the parameters of the mixture density model are identified, the optimal threshold for the two-class problem ($K = 2$) can be found by minimizing the overall probability of error

$$E(T) = \frac{P_2}{\sqrt{2\pi}\sigma_2} \int_{-\infty}^{T} \exp\left\{-\frac{(s-\mu_2)^2}{\sigma_2^2}\right\} ds + \frac{P_1}{\sqrt{2\pi}\sigma_1} \int_{T}^{\infty} \exp\left\{-\frac{(s-\mu_1)^2}{\sigma_1^2}\right\} ds \quad \text{(B.4)}$$

with respect to T. If $\sigma_1 = \sigma_2 = \sigma$, then the optimum threshold is given by [Gon 92]

$$T = \frac{\mu_1 + \mu_2}{2} + \frac{\sigma^2}{\mu_1 - \mu_2} \ln \frac{P_2}{P_1} \quad \text{(B.5)}$$

Observe that if $P_1 = P_2$, then the optimal threshold is simply the average of the two class means.

B.2 Clustering

Clustering refers to classification of objects into groups according to certain properties of these objects. In image segmentation, it is expected that feature vectors from similar-appearing regions would form groups, known as clusters, in the feature space. If we consider segmentation of an image into K classes, then the segmentation label field, $z(x_1, x_2)$, assumes one of the K values at each pixel, i.e., $z(x_1, x_2) = \ell$, $\ell = 1, \ldots, K$. In the case of scalar features, such as pixel intensities, clustering can be considered as a method of determining the $K - 1$ thresholds that define the decision boundaries in the 1-D feature space. With M-dimensional vector features, the segmentation corresponds to partitioning the M-dimensional feature space into K regions. A standard procedure for clustering is to assign each sample to the class of the nearest cluster mean [Col 79]. In the unsupervised mode, this can be achieved by an iterative procedure, known as the K-means algorithm, because the cluster means are also initially unknown. In the following, we describe the K-means algorithm assuming that we wish to segment an image into K regions based on the gray values of the pixels.

Let $\mathbf{x} = (x_1, x_2)$ denote the coordinates of a pixel, and $s(\mathbf{x})$ its gray level. The K-means method aims to minimize the performance index

$$J = \sum_{\ell=1}^{K} \sum_{\mathbf{x} \in \Lambda_\ell^{(i)}} ||s(\mathbf{x}) - \mu_\ell^{(i+1)}||^2 \quad \text{(B.6)}$$

where $\Lambda_\ell^{(i)}$ denotes the set of samples assigned to cluster ℓ after the ith iteration, and μ_ℓ denotes the mean of the ℓth cluster. This index measures the sum of the distances of each sample from their respective cluster means. The K-means algorithm, given by the following steps, usually converges to a local minimum of the performance index J.

K-Means Algorithm:

1. Choose K initial cluster means, $\mu_1^{(1)}, \mu_2^{(1)}, \ldots, \mu_K^{(1)}$, arbitrarily.

2. At the ith iteration assign each pixel, $\mathbf{x}$, to one of the K clusters according to the relation

$$\mathbf{x} \in \Lambda_j(i) \text{ if } ||s(\mathbf{x}) - \mu_j^{(i)}|| < ||s(\mathbf{x}) - \mu_\ell^{(i)}||$$

for all $\ell = 1, 2, \ldots, K$, $\ell \neq j$, where $\Lambda_\ell^{(i)}$ denotes the set of samples whose cluster center is $\mu_\ell^{(i)}$. That is, assign each sample to the class of the nearest cluster mean.

3. Update the cluster means $\mu_\ell^{(i+1)}$, $\ell = 1, 2, \ldots, K$ as the sample mean of all samples in $\Lambda_\ell^{(i)}$

$$\mu_\ell^{(i+1)} = \frac{1}{N_\ell} \sum_{\mathbf{x} \in \Lambda_\ell^{(i)}} s(\mathbf{x}), \quad \ell = 1, 2, \ldots, K$$

where N_ℓ is the number of samples in $\Lambda_\ell^{(i)}$.

4. If $\mu_\ell^{(i+1)} = \mu_\ell^{(i)}$ for all $\ell = 1, 2, \ldots, K$, the algorithm has converged, and the procedure is terminated. Otherwise, go to step 2.

Example:

The case of $K = 2$ with scalar ($M = 1$) features is demonstrated in Figure B.2. Here, "x" denotes the scalar feature points. The procedure begins with specifying two cluster means μ_1 and μ_2, denoted by the filled dots, arbitrarily. Then all feature points that are closer to μ_1 are labeled as "$z = 1$" and those closer to μ_2 as "$z = 2$." Then the average of all feature points that are labeled as "1" gives the new value of μ_1, and the average of those labeled as "2" gives the new value of μ_2. The procedure is repeated until convergence.

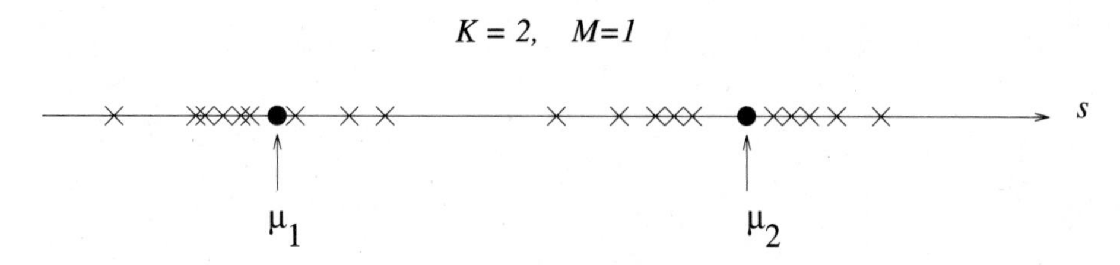

Figure B.2: K-means method with scalar data and $K = 2$.

The biggest obstacle to overcome in clustering procedures is the determination of the correct number of classes, which is assumed to be known. In practice, the value of K is determined by trial and error. To this effect, measures of clustering quality have been developed, such as the within-cluster and between-cluster scatter measures. These measures are derived from the within-cluster and between-cluster scatter matrices [Fuk 72]. Different values of K can be tried with the K-means algorithm until a desired clustering quality is achieved.

Although we have presented the K-means algorithm here for the case of a scalar first-order (pixel-dependent) image feature, it can be straightforwardly extended to the case of vector features and/or higher-order (region-based) image features. The procedure provides cluster means, which can be used for other purposes, as a by-product of the segmentation. However, it does not incorporate any spatial smoothness constraint on the estimated segmentation labels to ensure spatial connectivity of the image segments. The Bayesian segmentation, which is described next, can be considered as a clustering method employing a statistical spatial smoothness constraint.

B.3 Bayesian Methods

In this section, we discuss the *maximum a posteriori probability* (MAP) approach, which belongs to the class of Bayesian methods. The MAP approach is motivated by the desire to obtain a segmentation that is spatially connected and robust in the presence of noise in the image data. The MAP criterion functional consists of two parts, the *class conditional probability* distribution, which is characterized by a model that relates the segmentation to the data, and the *a priori probability* distribution, which expresses the prior expectations about the resulting segmentation. The *a priori* model is usually selected to encourage spatially connected regions.

Derin and Elliott [Der 87] have proposed a MAP approach for the segmentation of monochromatic images, and have successfully used Gibbs random fields (GRF) as *a priori* probability models for the segmentation labels. The GRF prior model expresses our expectation about the spatial properties of the segmentation. In order to eliminate isolated regions in the segmentation that arise in the presence of noise, the GRF model can be designed to assign a higher probability for segmentations that have contiguous, connected regions. Thus, estimation of the segmentation is not only dependent on the image intensity, but also constrained by the expected spatial properties imposed by the GRF model. The solution to the MAP segmentation problem is obtained via Monte-Carlo type methods. In particular, we use the sub-optimal method of *iterated conditional mode* (ICM), trading optimality for reduced computational complexity.

B.3.1 The MAP Method

In the MAP formulation, we take into account the presence of observation noise in the image explicitly by modeling observed monochrome image data as $g(\mathbf{x}) = s(\mathbf{x}) + v(\mathbf{x})$, where $v(\mathbf{x})$ denotes the observation noise. In vector notation $\mathbf{g}$ denotes an N-dimensional vector obtained by lexicographical ordering of the image data. We wish to estimate a segmentation label field, denoted by the N-dimensional vector $\mathbf{z}$. A label $z(\mathbf{x}) = \ell$, $\ell = 1, \ldots, K$, implies that the pixel (site) $\mathbf{x}$ belongs to the ℓ-th class among the K classes.

The desired estimate of the segmentation label field, $\hat{\mathbf{z}}$, is defined as the one that maximizes the *a posteriori* pdf $p(\mathbf{z} \mid \mathbf{g})$ of the segmentation label field, given the observed image $\mathbf{g}$. Using the Bayes rule,

$$p(\mathbf{z} \mid \mathbf{g}) \propto p(\mathbf{g} \mid \mathbf{z})p(\mathbf{z}) \tag{B.7}$$

where $p(\mathbf{g} \mid \mathbf{z})$ represents the conditional pdf of the data given the segmentation labels, i.e., the class-conditional pdf. The term $p(\mathbf{z})$ is the *a priori* probability distribution that can be modeled to impose a spatial connectivity constraint on the segmentation. In the following, we first provide a model for the *a priori* pdf, then proceed to examine the assumptions used in characterizing the conditional pdf.

A Priori Probability Model

A spatial connectivity constraint on the segmentation field can be imposed by modeling it as a discrete-valued Gibbs random field (see Appendix A and Chapter 8). Then the *a priori* pdf $p(\mathbf{z})$ of the label field can be expressed by an impulse train with Gibbsian weighting

$$p(\mathbf{z}) = \frac{1}{Q} \sum_{\omega \in \Omega} \exp\{-U(\mathbf{z})/T\} \delta(\mathbf{z} - \omega) \tag{B.8}$$

where Ω is the finite sample space of the random vector $\mathbf{z}$, $\delta(\cdot)$ denotes a Dirac delta function, Q is called the partition function (normalizing constant), T is the temperature parameter, and $U(\mathbf{z})$ is the Gibbs potential defined by

$$U(\mathbf{z}) = \sum_{C \in \mathcal{C}} V_C(\mathbf{z}) \tag{B.9}$$

Here $\mathcal{C}$ is the set of all cliques, and V_C is the individual clique potential function.

The single pixel clique potentials can be defined as

$$V_C(z(\mathbf{x})) = \alpha^\ell \;\text{ if }\; z(\mathbf{x}) = \ell \;\text{ and }\; \mathbf{x} \in C, \quad \text{for } \ell = 1, \ldots, K$$

They reflect our *a priori* knowledge of the probabilities of different region types. The smaller α^ℓ the higher the likelihood of region ℓ, $\ell = 1, \ldots, K$.

Spatial connectivity of the segmentation can be imposed by assigning the following two-pixel clique potential,

$$V_C(z(\mathbf{x}_i), z(\mathbf{x}_j)) = \begin{cases} -\beta & \text{if } z(\mathbf{x}_i) = z(\mathbf{x}_j) \text{ and } \mathbf{x}_i,\ \mathbf{x}_j \in C \\ \beta & \text{if } z(\mathbf{x}_i) \neq z(\mathbf{x}_j) \text{ and } \mathbf{x}_i,\ \mathbf{x}_j \in C \end{cases} \tag{B.10}$$

where β is a positive parameter. The larger the value of β, the stronger the smoothness constraint.

Conditional Probability Model

The original image, $\mathbf{s}$, will be modeled by a mean intensity function, denoted by the vector μ, plus a zero-mean, white Gaussian residual process, $\mathbf{r}$, with variance σ_r^2, that is,

$$\mathbf{s} = \mu + \mathbf{r} \tag{B.11}$$

where all vectors are respective N-dimensional lexicographic orderings of the respective arrays. Then,

$$\mathbf{g} = \mu + \eta \tag{B.12}$$

where $\eta \doteq \mathbf{r}+\mathbf{v}$ is a zero-mean Gaussian process with variance $\sigma_\eta^2 \doteq \sigma_r^2+\sigma_v^2$, and σ_v^2 denotes the variance of the observation noise which is also taken to be a zero-mean, white Gaussian process. In what follows, we refer to this combined term, η, simply as the additive noise term. In the context of segmenting an image into K regions, the elements of μ attain the values of the class means associated with each one of the regions.

The original formulation by Derin and Elliott [Der 87] for nontextured images models the mean intensity of each image region as a constant, denoted by the scalar μ_ℓ, $\ell = 1, 2, \ldots, K$. That is the elements of μ attain K distinct values, μ_ℓ, $\ell = 1, 2, \ldots, K$. Based on the model of the observed image in (B.12), the conditional probability distribution is expressed as

$$p(\mathbf{g}|\mathbf{z}) \propto \exp\left\{-\sum_{\mathbf{x}} \frac{[g(\mathbf{x}) - \mu_{z(\mathbf{x})}]^2}{2\sigma_\eta^2}\right\} \tag{B.13}$$

where $z(\mathbf{x}) = \ell$ designates the assignment of site $\mathbf{x}$ to region ℓ. Notice that (B.13) is the probability distribution that would be used in the *maximum likelihood* (ML) segmentation of the image.

Given (B.8) and (B.13), the *a posteriori* density has the form

$$p(\mathbf{z} \mid \mathbf{g}) \propto \exp\left\{-\sum_{\mathbf{x}} \frac{1}{2\sigma_\eta^2}\left[g(\mathbf{x}) - \mu_{z(\mathbf{x})}\right]^2 - \sum_C V_C(\mathbf{z})\right\} \tag{B.14}$$

We maximize the posterior probability (B.14) to find an estimate of μ_ℓ, $\ell = 1, 2, \ldots, K$, and the desired segmentation labels $\mathbf{z}$. This maximization can be performed by means of simulated annealing. Observe that if we turn off the spatial smoothness constraints, that is, neglect the second term, the result would be identical to that of the K-means algorithm. Thus, the MAP estimation follows a procedure that is similar to that of the K-means algorithm, i.e., we start with an initial estimate of the class means and assign each pixel to one of the K classes by maximizing (B.14), then we update the class means using these estimated labels, and iterate between these two steps until convergence.

B.3.2 The Adaptive MAP Method

The MAP method can be made adaptive by letting the cluster means μ_i slowly vary with the pixel location $\mathbf{x}$. An adaptive clustering approach for monochromatic images that is based on a more realistic image model has been proposed by Pappas [Pap 92]. In earlier works on MAP segmentation, the image has been modeled as consisting of a finite number of regions, each with a constant mean intensity [Der 87]. The adaptive method allows for slowly space-varying mean intensities for each region.

Because the uniform mean intensity $\mu_{z(\mathbf{x})}$ may not be adequate in modeling actual image intensities within each region, a space variant mean intensity $\mu_{z(\mathbf{x})}(\mathbf{x})$ will be used in the class-conditional pdfs. Then the modified conditional pdf model becomes

$$p(\mathbf{g}|\mathbf{z}) \propto \exp\left\{-\sum_{\mathbf{x}} \frac{[g(\mathbf{x}) - \mu_{z(\mathbf{x})}(\mathbf{x})]^2}{2\sigma_\eta^2}\right\} \tag{B.15}$$

There are several advantages in introducing space-varying class means. First, a more realistic image model yields better segmentation. Second, in parts of the image where the local intensity contrast is low between two perceptually distinct regions, using a global mean usually yields a single label to both regions. Third, in the case of spatially varying class means, fewer regions are required to segment an image into perceptually meaningful regions.

The adaptive algorithm follows a procedure that is similar to the nonadaptive algorithm, except that the cluster means $\mu_\ell(\mathbf{x})$ at the site $\mathbf{x}$ for each region ℓ are estimated as the sample mean of those pixels with label ℓ within a local window about the pixel $\mathbf{x}$ (as opposed over the entire image). The local window is depicted in Figure B.3.

To reduce the computational burden of the algorithm to a reasonable level, the following simplifications are usually performed: i) The space-varying mean estimates are computed on a sparse grid, and then interpolated. ii) The optimization is performed via the ICM method. Note that the ICM is equivalent to maximizing the local *a posteriori* pdf

$$\begin{aligned} p(z(\mathbf{x}_i) \quad | \quad & g(\mathbf{x}_i), z(\mathbf{x}_j), \text{ all } \mathbf{x}_j \in \mathcal{N}_{\mathbf{x}_i}) \\ & \propto \exp\left\{-\frac{1}{2\sigma_\eta^2}[g(\mathbf{x}_i) - \mu_{z(\mathbf{x}_i)}(\mathbf{x}_i)]^2 - \sum_C V_C(\mathbf{z})\right\} \end{aligned}$$

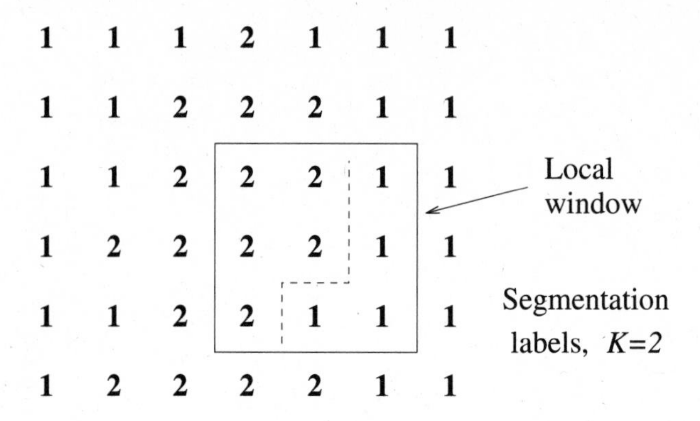

Figure B.3: Estimation of space-varying means.

B.3.3 Vector Field Segmentation

In most applications we deal with the segmentation of multichannel data such as color images or motion vector fields. The Bayesian segmentation algorithms (MAP and adaptive MAP) can be generalized to segment multichannel data. This extension involves modeling the multichannel data with a vector random field, where the components of the vector field (each individual channel) are assumed to be conditionally independent given the segmentation labels of the pixels. Note that we have a scalar segmentation label field, which means each vector is assigned a single label, as opposed to segmenting the channels individually. The class-conditional probability model for the vector image field is taken as a multivariate Gaussian distribution with a space-varying mean function.

We assume that M channels of multispectral image data are available and denote them by a P-dimensional ($P = N \cdot M$) vector $[\mathbf{g}_1, \mathbf{g}_2, \ldots, \mathbf{g}_M]^t$, where $\mathbf{g}_j$ corresponds to the jth channel. A single segmentation field, $\mathbf{z}$, which is consistent with all M channels of data and is in agreement with the prior knowledge, is desired. By assuming the conditional independence of the channels given the segmentation field, the conditional probability in (B.13) becomes

$$p(\mathbf{g}_1, \mathbf{g}_2, \ldots, \mathbf{g}_M \mid \mathbf{z}) = p(\mathbf{g}_1 \mid \mathbf{z}) p(\mathbf{g}_2 \mid \mathbf{z}) \cdots p(\mathbf{g}_M \mid \mathbf{z}) \tag{B.16}$$

The extension of the Bayesian methods for multichannel data following the model (B.16) is straightforward.

Bibliography

[Col 79] G. B. Coleman and H. C. Andrews, "Image segmentation by clustering," *Proc. IEEE*, vol. 67, pp. 773–785, May 1979.

[Der 87] H. Derin and H. Elliott, "Modeling and segmentation of noisy and textured images using Gibbs random field," *IEEE Trans. Patt. Anal. Mach. Intel.*, vol. PAMI-9, pp. 39–55, Jan. 1987.

[Fuk 72] K. Fukunaga, *Introduction to Statistical Pattern Recognition*, Academic Press, New York, 1972.

[Gon 92] R. C. Gonzalez and R. E. Woods, *Digital Image Processing*, Addison-Wesley, 1992.

[Har 85] R. M. Haralick and L. G. Shapiro, "Survey: Image segmentation techniques," *Comp. Vis. Graph Image Proc.*, vol. 29, pp. 100–132, 1985.

[Lee 90] S. U. Lee, S. Y. Chung, and R. H. Park, "A comparative performance study of several global thresholding techniques for segmentation," *Comp. Vis. Graph Image Proc.*, vol. 52, pp. 171–190, 1990.

[Pap 92] T. N. Pappas, "An adaptive clustering algorithm for image segmentation," *IEEE Trans. on Signal Proc.*, vol. SP-40, pp. 901–914, Apr. 1992.

[Sah 88] P. K. Sahoo, S. Soltani, A. K. C. Wong, and Y. C. Chan, "A survey of thresholding techniques," *Comp. Vis. Graph Image Proc.*, vol. 4, pp. 233–260, 1988.

[Sny 90] W. Snyder, G. Bilbro, A. Logenthiran, and S. Rajala, "Optimal thresholding - A new approach," *Patt. Recog. Lett.*, vol. 11, pp. 803–810, 1990.

[Wes 78] J. S. Weszka, "A survey of threshold selection techniques," *Comp. Graph Image Proc.*, vol. 7, pp. 259–265, 1978.

Appendix C

KALMAN FILTERING

A Kalman filter estimates the state of a dynamic system recursively at each time, in the linear minimum mean square error (LMMSE) sense, given a time series of vector or scalar observations that are linearly related to these state variables. If the state variables and the noise are modeled as uncorrelated, Gaussian random processes, then the Kalman filter is the minimum mean square error estimator. This appendix briefly reviews the basics of Kalman filtering and extended Kalman filtering (EKF). The reader is referred to [Bro 83, Fau 93, Men 87] for a more detailed treatment of the subject and the derivation of the equations.

C.1 Linear State-Space Model

Let $\mathbf{z}(k)$ denote an M-dimensional state vector of a dynamic system at time k. We assume that the propogation of the state in time can be expressed in the form of a linear state-transition equation given by

$$\mathbf{z}(k) = \mathbf{\Phi}(k, k-1)\mathbf{z}(k-1) + \mathbf{w}(k), \quad k = 1, \ldots, N \tag{C.1}$$

where $\mathbf{\Phi}(k, k-1)$ is the state-transition matrix and $\mathbf{w}(k)$ is a zero-mean, white random sequence, with the covariance matrix $\mathbf{Q}(k)$, representing the state model error.

Suppose that a time-series of observations or measurements, $\mathbf{y}(k)$, are available, which are linearly related to the state variables as

$$\mathbf{y}(k) = \mathbf{H}(k)\mathbf{z}(k) + \mathbf{v}(k), \quad k = 1, \ldots, N \tag{C.2}$$

where $\mathbf{H}(k)$ is called the observation matrix, and $\mathbf{v}(k)$ denotes a zero-mean, white observation noise sequence, with the covariance matrix $\mathbf{R}(k)$, which is uncorrelated with $\mathbf{w}(k)$.

The Kalman filter minimizes, at each time k, the trace of the error covariance matrix conditioned on all observations up to time k, defined as

$$\mathbf{P}_a(k) = \begin{bmatrix} E\left\{e_1(k)^2|\mathbf{y}(k),\ldots,\mathbf{y}(1)\right\} & \cdots & E\left\{e_1(k)\, e_M(k)|\mathbf{y}(k),\ldots,\mathbf{y}(1)\right\} \\ \vdots & \vdots & \vdots \\ E\left\{e_M(k)\, e_1(k)|\mathbf{y}(k),\ldots,\mathbf{y}(1)\right\} & \cdots & E\left\{e_M(k)^2|\mathbf{y}(k),\ldots,\mathbf{y}(1)\right\} \end{bmatrix} \tag{C.3}$$

where

$$e_i(k) = z_i(k) - \hat{z}_i(k), \quad i = 1, \ldots, M \tag{C.4}$$

$\hat{z}_i(k)$ denotes an estimate of the state $z_i(k)$, and $E\{\cdot\}$ denotes the expectation operator. It consists of two steps: i) the prediction step, comprising the state and error covariance prediction, and ii) an update step, comprising Kalman gain computation followed by state and error covariance update. They are described below, where the subscripts "b" and "a" denote before and after the update, respectively.

State-prediction equation

$$\hat{\mathbf{z}}_b(k) = \mathbf{\Phi}(k, k-1)\hat{\mathbf{z}}_a(k-1) \tag{C.5}$$

Covariance-prediction equation

$$\mathbf{P}_b(k) = \mathbf{\Phi}(k, k-1)\mathbf{P}_a(k-1)\mathbf{\Phi}(k, k-1)^T + \mathbf{Q}(k) \tag{C.6}$$

Kalman-gain equation

$$\mathbf{K}(k) = \mathbf{P}_b(k)\mathbf{H}^T(k)(\mathbf{H}(k)\mathbf{P}_b(k)\mathbf{H}^T(k) + \mathbf{R}(k))^{-1} \tag{C.7}$$

State-update equation

$$\hat{\mathbf{z}}_a(k) = \hat{\mathbf{z}}_b(k) + \mathbf{K}(k)[\mathbf{y}(k) - \mathbf{H}(k)\hat{\mathbf{z}}_b(k)] \tag{C.8}$$

Covariance-update equation

$$\mathbf{P}_a(k) = \mathbf{P}_b(k) - \mathbf{K}(k)\mathbf{H}(k)\mathbf{P}_b(k) \tag{C.9}$$

Observe that the covariance prediction, Kalman gain, and covariance update equations are independent of the observations, and can be processed off-line provided that state propagation and the observation models are known. Two special cases are of interest:

1. Steady-state filter: If the state propagation and the observation matrices are time-invariant, that is, $\mathbf{\Phi}(k, k-1) = \mathbf{\Phi}$, $\mathbf{Q}(k) = \mathbf{Q}$, $\mathbf{H}(k) = \mathbf{H}$, and $\mathbf{R}(k) = \mathbf{R}$, then the Kalman gain matrix converges to a steady-state value $\mathbf{K}$ after some recursions. In this case, the covariance and the gain equations can be turned off after convergence, resulting in computational savings. Note that the steady-state filter is a linear time-invariant (LTI) filter.

2. Recursive least squares filter: If there is no temporal dynamics, that is, $\mathbf{\Phi}(k, k-1) = \mathbf{I}$ and $\mathbf{Q}(k) = 0$ for all k, then the state consists of constant (but unknown) parameters, and the Kalman filter reduces to the recursive least squares parameter identification filter.

C.2 Extended Kalman Filtering

In some applications, the state-transition equation, the observation equation, or both may be nonlinear in the state variables, given by

$$\mathbf{z}(k) = \mathbf{f}(\mathbf{z}(k-1)) + \mathbf{w}(k), \quad k = 1, \ldots, N \tag{C.10}$$

and

$$\mathbf{y}(k) = \mathbf{h}(\mathbf{z}(k)) + \mathbf{v}(k), \quad k = 1, \ldots, N \tag{C.11}$$

where $\mathbf{f}(\cdot)$ and $\mathbf{h}(\cdot)$ are nonlinear, vector-valued functions of the state.

In order to apply the Kalman filtering equations presented in Section C.1, Equations (C.10) and/or (C.11) must be linearized about the best available estimate of the state vector at that instant. To this effect, we expand $\mathbf{f}(\mathbf{z}(k-1))$ and $\mathbf{h}(\mathbf{z}(k))$ into first-order Taylor series about $\hat{\mathbf{z}}_a(k-1)$ and $\hat{\mathbf{z}}_b(k)$, respectively, as follows:

$$\mathbf{f}(\mathbf{z}(k-1)) \approx \mathbf{f}(\hat{\mathbf{z}}_a(k-1)) + \frac{\partial \mathbf{f}(\mathbf{z}(k-1))}{\partial \mathbf{z}(k-1)}\Big|_{\mathbf{z}(k-1)=\hat{\mathbf{z}}_a(k-1)}(\mathbf{z}(k-1) - \hat{\mathbf{z}}_a(k-1)) \tag{C.12}$$

$$\mathbf{h}(\mathbf{z}(k)) \approx \mathbf{h}(\hat{\mathbf{z}}_b(k)) + \frac{\partial \mathbf{h}(\mathbf{z}(k))}{\partial \mathbf{z}(k)}\Big|_{\mathbf{z}(k)=\hat{\mathbf{z}}_b(k)}(\mathbf{z}(k) - \hat{\mathbf{z}}_b(k)) \tag{C.13}$$

Substituting (C.12) into (C.10), we obtain the linearized state-transition equation

$$\mathbf{z}(k) \approx \mathbf{\Phi}(k, k-1)\ \mathbf{z}(k-1) + \mathbf{u}(k) + \mathbf{w}(k), \quad k = 1, \ldots, N \tag{C.14}$$

where $\mathbf{u}(k)$ is a deterministic input, given by

$$\mathbf{u}(k) = \mathbf{f}(\hat{\mathbf{z}}_a(k-1)) - \mathbf{\Phi}(k, k-1)\hat{\mathbf{z}}_a(k-1)$$

and the state-transition matrix of the linearized model is

$$\mathbf{\Phi}(k, k-1) = \frac{\partial \mathbf{f}(\mathbf{z}(k-1))}{\partial \mathbf{z}(k-1)}\Big|_{\mathbf{z}(k-1)=\hat{\mathbf{z}}_a(k-1)}$$

Likewise, substituting (C.13) into (C.11) results in the modified observation equation

$$\tilde{\mathbf{y}}(k) \approx \mathbf{H}(k)\ \mathbf{z}(k) + \mathbf{v}(k), \quad k = 1, \ldots, N \tag{C.15}$$

where $\tilde{\mathbf{y}}(k)$ denotes the modified observation given by

$$\tilde{\mathbf{y}}(k) = \mathbf{y}(k) - \mathbf{h}(\hat{\mathbf{z}}_b(k)) + \mathbf{H}(k)\hat{\mathbf{z}}_b(k)$$

and the observation matrix of the linearized system is

$$\mathbf{H}(k) = \frac{\partial \mathbf{h}(\mathbf{z}(k))}{\partial \mathbf{z}(k)}\Big|_{\mathbf{z}(k)=\hat{\mathbf{z}}_b(k)}$$

The Kalman filtering equations (C.5)-(C.9) apply to the case of the linearized models (C.14) and (C.15) straightforwardly. However, neither optimality nor convergence of this extended Kalman filter can be guaranteed. Indeed, the success of the filter strongly depends on the goodness of the predicted estimate about which a linearization has been performed.

In some cases, additional performance improvement can be achieved using an iterated extended Kalman filter (IEKF), which carries out the linearization in an iterative fashion about the best available estimate from the previous iteration at that instant.

Bibliography

[Bro 83] R. G. Brown, *Introduction to Random Signal Analysis and Kalman Filtering*, New York, NY: Wiley, 1983.

[Fau 93] O. Faugeras, *Three-Dimensional Computer Vision*, Cambridge, MA: MIT Press, 1993.

[Men 87] J. Mendel, *Lessons in Digital Estimation Theory*, Englewood Cliffs, NJ: Prentice Hall, 1987.

Index